Modeling Semi-arid Water–Soil–Vegetation Systems in a Changing Climate

Modeling Semi-arid Water–Soil–Vegetation Systems in a Changing Climate

Xixi Wang
Old Dominion University, Norfolk, VA, USA

IOP Publishing, Bristol, UK

ISBN 978-0-7503-4020-5 (ebook)
ISBN 978-0-7503-4018-2 (print)
ISBN 978-0-7503-4021-2 (myPrint)
ISBN 978-0-7503-4019-9 (mobi)

DOI 10.1088/978-0-7503-4020-5

Version: 20220801

IOP ebooks

British Library Cataloguing-in-Publication Data: A catalogue record for this book is available from the British Library.

Published by IOP Publishing, wholly owned by The Institute of Physics, London

IOP Publishing, Temple Circus, Temple Way, Bristol, BS1 6HG, UK

US Office: IOP Publishing, Inc., 190 North Independence Mall West, Suite 601, Philadelphia, PA 19106, USA

Contents

Preface

Modeling Semi-arid Water–Soil–Vegetation Systems in a Changing Climate is a reference text for researchers, graduates, professionals, and managers in dryland management, ecoenvironment, ecohydrology, semi-arid ecosystem resilience, and water resources. It can be used for graduate-level courses, thesis and/or dissertation works, and practical projects.

Water–soil–vegetation dynamic nexuses affect and are affected by human activity and climate change. For a given semi-arid ecosystem, inappropriate land management practices can result in soil and vegetation degradation, which in turn will likely alter natural hydrological processes, leading to more frequent and severe flooding and drought. In response, an altered hydrological condition tends to promote soil erosion by wind and water, which in turn can cause further vegetation degradation or even desertification. Such nexuses will likely become more interwoven in a changing climate because a non-stationary climate can further deteriorate the already-altered hydrological conditions. So far our scientific understanding is incomplete regarding how such nexuses can sustain or break the equilibria between water, soil, and/or vegetation in terms of ecosystem resilience.

This reference book conceptualizes the interrelated physical processes of semi-arid water–soil–vegetation systems, introduces mathematical models for simulating these processes, and presents example modeling studies of selected semi-arid ecosystems. Currently, there is no book focusing on such a topic, although some incomplete information is scattered in various peer-reviewed journals and project reports. This reference book provides a systematic elucidation of this important topic and serves as a one-stop information source. By reading this reference book, the reader will gain in-depth mechanistic knowledge of water–soil–vegetation dynamics while obtaining hands-on experience of modeling approaches and tools. Upon having read this reference book, the reader will be able to apply the materials to conduct advanced research on this topic and develop practical measures for protecting fragile semi-arid vegetation ecosystems and confronting climate change. The broader application is to prevent land degradation and desertification as induced by climate change and inappropriate human activities (e.g. over-grazing and over-exploitation).

This reference book includes both principles and applications, emphasizes nexuses rather than individual processes, deciphers the dynamics and interrelations of the physical processes of semi-arid water–soil–vegetation ecosystems, addresses the impacts of climate change from practical management perspectives, and maximizes readability using illustrations, graphs, and annotated equations.

Acknowledgements

I would like to thank Ms Emily Tapp, Commissioning Editor at IOP Publishing, for her solicitation and assistance in publishing this reference book and the highly appreciate proofreading comments of Ms Claire Roy, Editorial Assistant at IOP Publishing. In addition, I want to express my heartfelt gratitude to my parents, Fuying Wang and Erheiyan Zhou, for my upbringing and their selfless support, and to my siblings, Limei Wang, Lixia Wang, Lifang Wang, and Jun Wang, for their encouragement, sacrifice, and support in my process of growth. Further, I thank my wife, Peilian Cui, and my daughter, Yueying Wang, for their unconditional love and support, without which this reference book would not have been possible.

Xixi Wang
Norfolk, VA, USA
March 3, 2022

Author biography

Xixi Wang

Xixi Wang, PhD, PE, is currently Professor and Graduate Program Director in the Department of Civil and Environmental Engineering at Old Dominion University (ODU), Norfolk, VA, USA. Prior to his current position, Xixi worked as an Associate Professor at ODU, an Assistant Professor at Tarleton State University in Stephenville, TX, a Research Scientist at the Energy and Environmental Research Center (EERC) at the University of North Dakota in Grand Forks, ND, a Civil Engineer at Michael Baker Jr Inc in Alexandria, VA, a Graduate Research and Teaching Assistant at Iowa State University in Ames, IA, and an Assistant Professor at Tsinghua University in Beijing, China. Xixi has extensive teaching and research experience in hydrology, hydraulics, and water resources. His current research interests include the effects of climate change versus human activity on water resources, water–soil–vegetation nexus and equilibrium in changing climate, and watershed hydrology and stormwater management. Xixi has authored or co-authored a variety of peer-reviewed journal papers, conference presentations, technical reports, books, and book chapters. He earned his Bachelor's (1989) and Master's (1993) degrees in Hydrology and Water Resources Engineering from Tsinghua University, and holds a doctoral (2001) degree in Agricultural Engineering (with a minor in Statistics) from Iowa State University. He has been a registered professional engineer in North Dakota (#5145) since 2003 and Texas (#99798) between 2007 and 2011. Xixi is a 2022–2023 Fulbright Scholar at University of Canberra, Australia.

IOP Publishing

Modeling Semi-arid Water–Soil–Vegetation Systems in a Changing Climate

Xixi Wang

Chapter 1

Overview

Following an overview of the knowledge gaps, science questions, and practice needs of concern in arid and semi-arid ecosystems, this chapter presents a modeling framework for addressing these concerns, and describes the structure and use of this book.

1.1 Problems and challenges

A dryland area receives precipitation that is much less than evapotranspiration (UNEP 1991, Davies *et al* 2016). Its aridity index (AI), the ratio of the average annual precipitation to the average annual potential evapotranspiration, is less than 0.65. Herein, the dryland can be hyper-arid (AI < 0.05), arid (AI = 0.05–0.2), semi-arid (AI = 0.2–0.5), or dry-subhumid (AI = 0.5–0.65) (Wikipedia 2022). Drylands cover around 41.3% (55 288 000 km^2) of the Earth's land surface, with 1.3% (1 740 000 km^2) hyper-arid, 12% (15 910 000 km^2) arid, 18% (23 739 000 km^2) semi-arid, and 10% (13 899 000 km^2) dry-subhumid (see figures 1.1(a) and (b) for a breakdown by region). For a given type of dryland, the total area in a geographic region can be larger or smaller than that in another geographic region, whereas for a given geographic region, the percent of one type of dryland can be higher or lower than that of another type of dryland. Globally, regardless of the type of dryland, Asia and Russia have the largest while Africa has the second largest percent coverage of drylands (figure 1.1(c)). For a given type of dryland, regardless of the geographic region, semi-arid drylands account for the highest while arid drylands account for the second highest percent of the total area of this type of drylands (figure 1.1(c)). This book focuses on how to model the soil–water–vegetation processes of semi-arid drylands, because semi-arid drylands account for 27.9% of land area in Europe and 58.3% in North America, with a global average of 44.3%, and have relatively high populations among the dryland types (figure 1.2).

doi:10.1088/978-0-7503-4020-5ch1

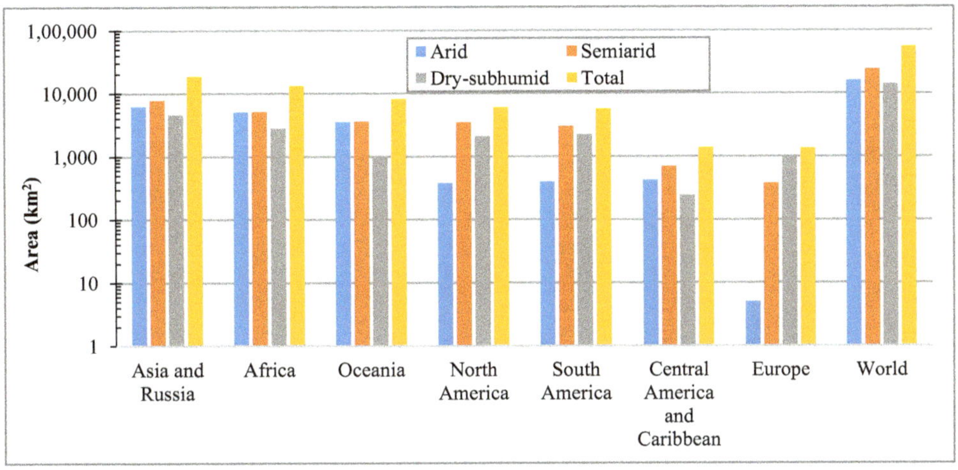

(a)

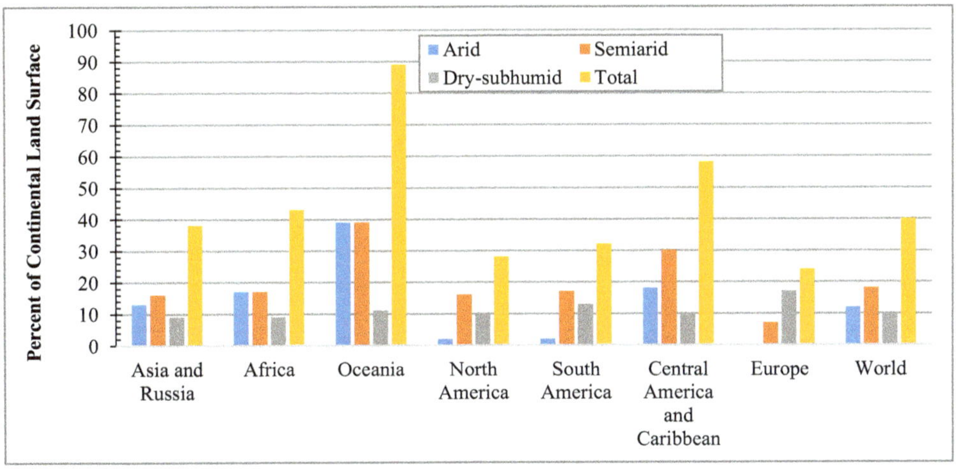

(b)

Figure 1.1. Drylands by continents and types. Plots showing (a) the areas, (b) percent coverages of the continental land surface areas, (c) percents of continental lands to global areas by types of drylands, and (d) ratios of the areas of type-specific drylands to the total areas of all three types of drylands by continent.

Drylands are primarily covered by grasses and forests, and secondarily by farming crops and urban settlements. They have cool–dry, hot–dry, and moderate–rainy seasons each year; temperature and wind speed fluctuate largely season by season, day by day, and diurnally. Drylands have geomorphological features of mountains, plains, pediments, and valleys and floodplains with braided patterns. They have loose sandy and loamy soils with low organic contents and water–nutrient holding capacities. Drylands regulate global heat–water–carbon cycles,

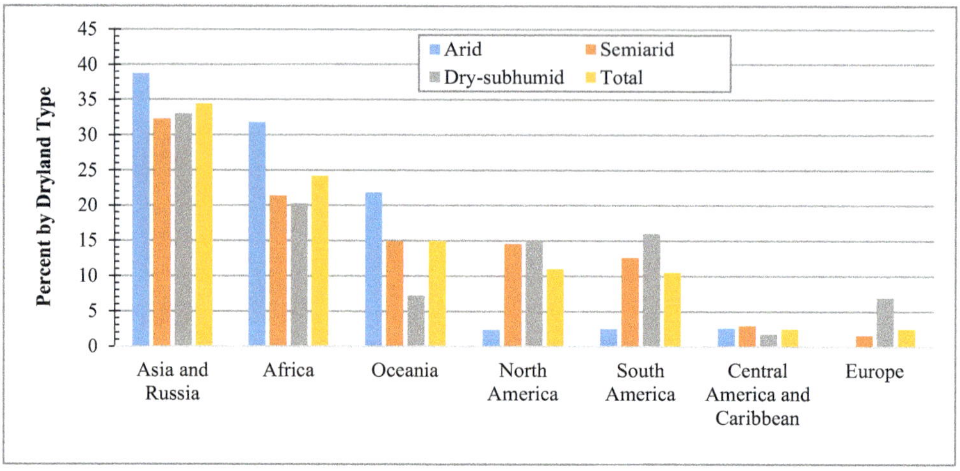

(c)

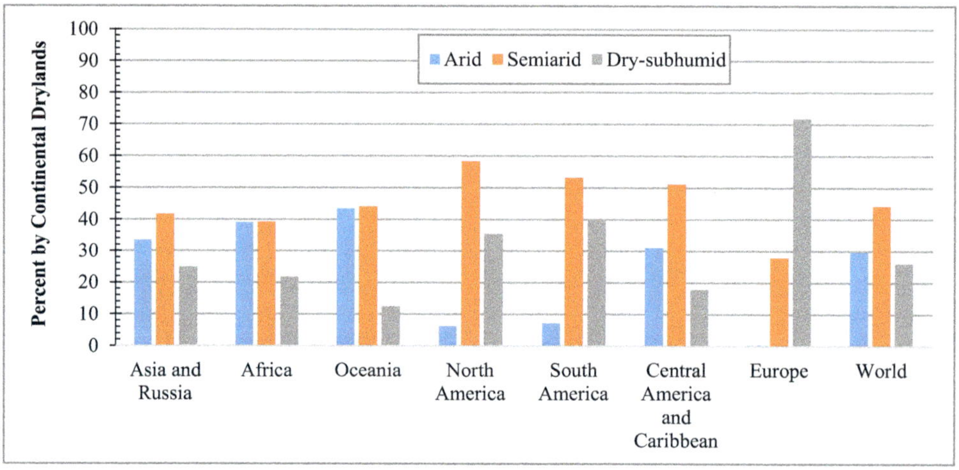

(d)

Figure 1.1. (Continued.)

provide habitats for animals, fowls, micro-organisms, and plants, and sequester green-house gases (e.g. CO_2 and CH_4), and also supply about 2×10^9 people (figure 1.2) with commodities, fruits, meats, and dairy products (UNEMG 2011, Wale and Dejenie 2013).

However, about 10% (i.e. 6×10^6 km^2) of the global drylands have been somewhat degraded and more are susceptible to degradation, risking the aforementioned ecological and economic functions (UNEMG 2011). As shown in figure 1.3, the primary reasons for degradation are inappropriate land use practices (e.g. cultivation, deforestation, over-exploitation, over-grazing, and urbanization) and climate change (Radstake *et al* 2010). The land practices remove grasses and

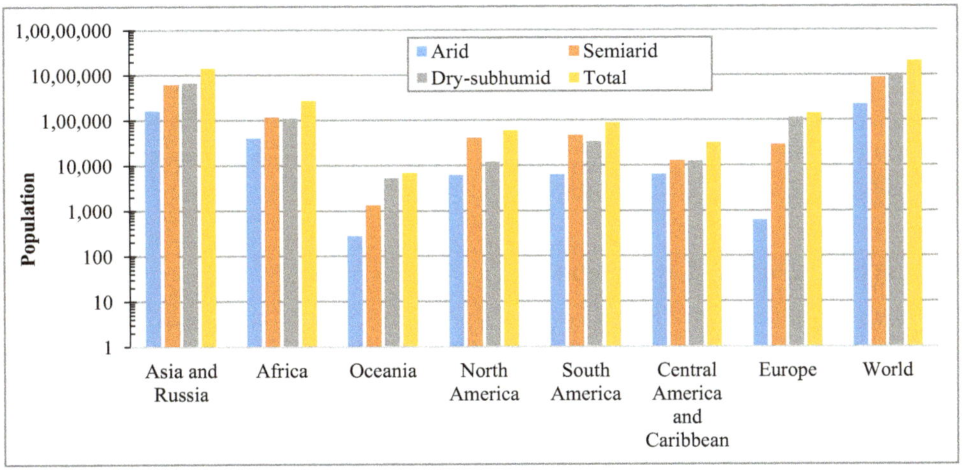

Figure 1.2. Populations living in drylands by continent. About 2×10^9 people depend on drylands.

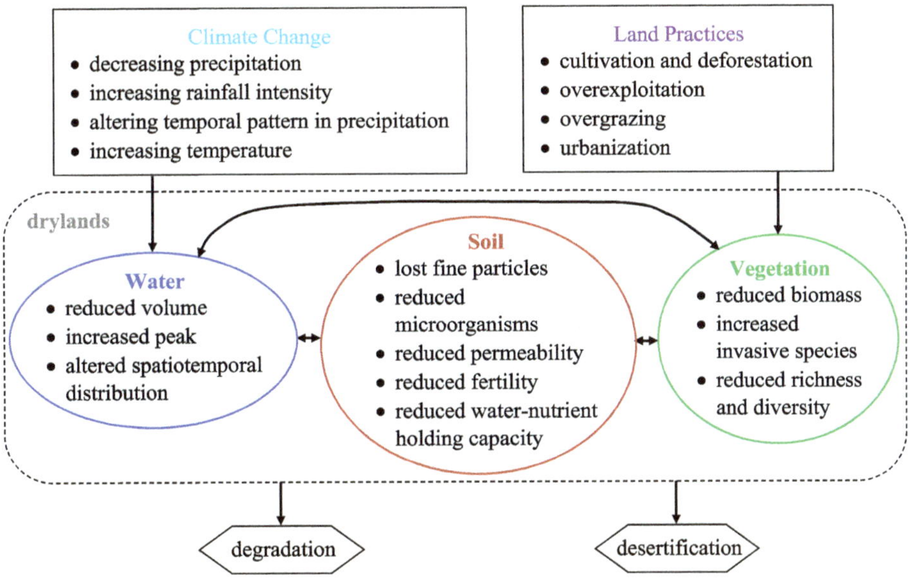

Figure 1.3. Problems of drylands. Triggering factors for dryland degradation and desertification.

plants, reduce above-ground biomass, and change species richness and composition. As a result, soils will be directly exposed to erosive forces from wind and water and subject to erosion, reducing the contents of fine particles and organic matter, water permeability, and water and nutrient holding capacities. This in turn exacerbates the damage to vegetation growth and community structure. On the other hand, due to climate change, the available water for vegetation transpiration tends to decrease but the runoff peak for soil erosion will increase, deteriorating soil stability and vegetation productivity. For a given dryland ecosystem, it will likely start to degrade if its water–soil–vegetation interactions are restricted to a threshold low level because of water shortages, soil erosion, and vegetation loss. If such a restriction becomes more severe and/or persists for a long period of time, the ecosystem may become a desert. Herein, the challenges are to tackle the water–soil–vegetation interactions and determine the threshold.

1.2 Hypotheses and questions

This book aims to verify the following hypotheses on semi-arid ecosystems:

- Actual evapotranspiration is largely controlled by soil water content and vegetation canopy architecture.
- Infiltration into soils is dependent not only on soil and biocrust properties, but also vegetation density and maturity. Soils with a dense vegetation coverage tend to be better wetted and have relatively high moisture. Plant roots enhance soil water holding capacity and facilitate transpiration, leading to an abnormal rainfall–runoff relation: runoff does not always increase with an increase of rainfall.
- Topsoil detachment by water depends on not only the presence or absence of biocrust and organic matter but also vegetation density and maturity.
- Vegetation degradation can be measured using the vegetation degradation index (table 1.1), which is computed as

$$I_{vd} = \sum_{i=1}^{5} \frac{D_i A_i}{A} \tag{1.1}$$

grazing score for i^{th} degradation level [−] (Table 1.1)
area with D_i [ha]
vegetation degradation index [−]
total area [ha]
degradation level [−]

Table 1.1. Grazing scores and vegetation statuses.

Degradation level	1	2	3	4	5
Grazing intensity	1.0–0.8	0.8–0.6	0.6–0.4	0.4–0.2	<0.2
Grazing score[a]	1	2	3	4	5
Degradation index	0–1	1–2	2–3	3–4	>4
Vegetation status	undegraded	lightly degraded	moderately degraded	severely degraded	extremely degraded

[a]Ratio of biomass with grazing to that without grazing.

- Climate change will exacerbate vegetation degradation because of alterations of natural hydrology that include: (i) a reduced infiltration-to-precipitation ratio and soil water content; (ii) an increased ratio of evaporation from water bodies and soil surfaces to transpiration through plants; (iii) an increased Bowen ratio of sensible heat flux to latent heat flux; and (iv) an increased ratio of soil heat flux to net solar radiation.

- Vegetation degradation tends to induce topsoil erosion, while loss of the top 10–20 cm organic-matter-rich soils will further increase vegetation degradation. Plants depend on water and nutrients supplied by the topsoil. On the other hand, plant roots can increase the erosive resistance of topsoil, lowering erosion risk. Thus, the spatial pattern of vegetation degradation is expected to be consistent with that of topsoil erosion.
- The topsoil layer with a plentiful supply of organic matter is thin (<40 cm). It will become thinner and less fertile with increasing erosion. The formation of the organic-matter-rich topsoil took thousands of years through a chain of physical and chemical processes, namely calciumization, humus accumulation, dissolving, and deposition. Because such processes cannot occur below a threshold depth below the ground surface, the layer of calcic castanozem soil is very thin. Once erosion occurs, it will take a long time for the exposed soils to have a measurable organic matter content.
- Water erosion is predominant during the growing season because water-drops from high-intensity rainfall can splash and detach soil particles, whereas wind erosion may dominate during the nongrowing season because of prevalent dry soils and gusty winds (figure 1.4).

This book intends to answer the following science questions on semi-arid ecosystems:

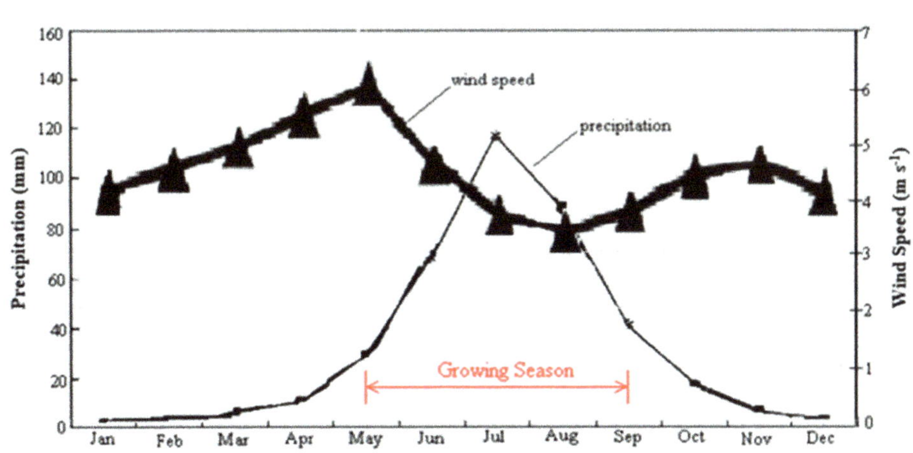

Figure 1.4. Typical semi-arid climate. Annual average monthly precipitation and wind speeds.

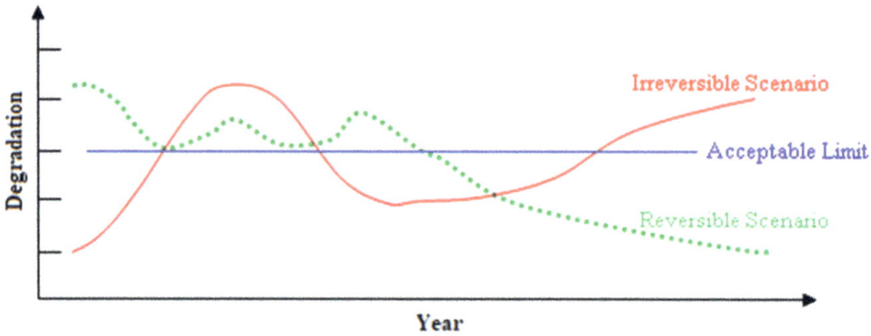

Figure 1.5. Degradation. Illustration of reversible (resilient) versus irreversible (non-resilient) shifts.

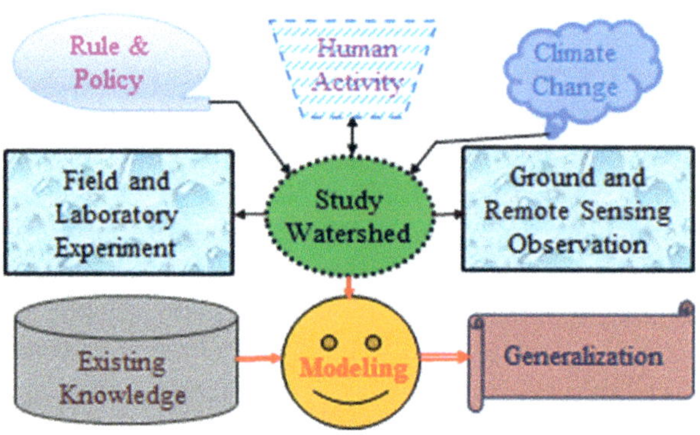

Figure 1.6. The adopted holistic approach. The framework for modeling water–soil–vegetation interactions.

- What are the functional relationships among vegetation degradation, soil water, organic matter content, and topsoil erosion?
- What are the thresholds of abiotic and biotic factors that trigger reversible and irreversible vegetation degradation (figure 1.5)?
- How do water–soil–vegetation interactions affect the plant growth, biomass production, and resilience of a semi-arid ecosystem of interest?

1.3 Modeling framework

To verify the hypotheses and answer the questions, this book adopts the holistic approach shown in figure 1.6. For a given experimental watershed (i.e. testbed), data on human activities and land management practices are collected from government agencies, stakeholders, and existing documents. In addition, the watershed is instrumented for observations of climate, water, soils, and vegetation. In this regard, *in situ* monitoring, laboratory tests and experiments, and remote sensing analyses are carried out to collect more data for developing and validating algorithms. Further, the collected data and prior knowledge are infused together to model water–soil–vegetation interactions as affected by climate change and land practices (e.g.

grazing). Moreover, the modeling results are extrapolated to examine the resilience and the possible catastrophic shifts of semi-arid ecosystems that would trigger an irreversible degradation.

1.4 The structure of this book

This book consists of six chapters and three appendices as follows:

- Chapter 1 (Overview) introduces the problems, challenges, hypotheses, and science questions as well as the general modeling framework for addressing them.
- Chapter 2 (Physical processes) presents the governing ecohydrological characteristics and processes of semi-arid ecosystems.
- Chapter 3 (Conceptualizations) presents conceptualizations of the semi-arid ecosystem processes discussed in chapter 2.
- Chapter 4 (Mathematical models) introduces existing models and their application conditions as well as elaborates what new algorithms are needed to improve existing models and how the algorithms can be developed.
- Chapter 5 (Case studies) presents eight case studies to elucidate the ecohydrological processes and concepts presented in chapters 2 and 3 as well as to demonstrate how to apply the models presented in chapter 4 to examine water–soil–vegetation interactions in semi-arid ecosystems.
- Chapter 6 (Overall conclusions and discussion) summarizes and discusses the findings from the previous five chapters as well as highlights future research needs related to semi-arid hydro-ecosystems.

The appendices present derivations of some equations and formulas.

References

Davies J, Barchiesi S, Ogali C J, Welling R, Dalton J and Laban P 2016 *Water in Drylands: Adapting to Scarcity Through Integrated Management* (Gland: International Union for Conservation of Nature (IUCN))

Radstake F, Ariens M, Luna H and Quitazol-Gonzalez J 2010 *Dryland Ecosystems: Introducing an Integrated Management Approach in the People's Republic of China* (Manila: Asian Development Bank) pp 1–44

UNEMG 2011 *Global Drylands: A UN System-Wide Response* (New York: The United Nations Environment Management Group (UNEMG)) pp 1–132

UNEP 1991 *Status of Desertification and Implementation of the United Nations Plan of Action to Combat Desertification* (Nairobi: United Nations Environment Programme (UNEP))

Wale H A and Dejenie T 2013 Dryland ecosystems: their features, constraints, potentials and managements *Res. J. Agri. Environ. Manag.* **2** 277–88

Wikipedia 2022 Drylands

Chapter 2

Physical processes

Given that semi-arid ecosystems receive precipitation below potential evapotranspiration (i.e. climatic demand), the hydrological cycle is dominated by vertical fluxes rather than horizontal movements. Usually, such an ecosystem has a thin layer of topsoil that can hold water and nutrients for vegetation growth and the soils beneath this layer have no organic matter and are infertile, so are susceptible to climatic variability and anthropogenic activity. This chapter presents the governing ecohydrological characteristics and processes of semi-arid ecosystems.

2.1 Characteristics of semi-arid ecosystems

Based on the Köppen climate classification (Wikipedia 2021), the climates of semi-arid ecosystems fall between those of deserts and humid regions. For a given semi-arid ecosystem, its climate can be either hot or cold, depending on the latitude and geographic location. The hot semi-arid ecosystems are mainly found in the zone of 20°–30° S, where Africa, Australia, and South Asia are located, whereas the cold semi-arid ecosystems are mostly in the zone above 30° N, where North America and Eurasia are located. In terms of the topology of the landscapes, the hot semi-arid ecosystems tend to be around the fringes of subtropical deserts, while the cold semi-arid ecosystems typically border a humid continental or Mediterranean climate.

Semi-arid ecosystems receive a mean annual precipitation of 150–700 mm (Gallart *et al* 2002). For a given year, the precipitation has a very large seasonal variability, with clustered short-duration heavy storms separated by consecutive sunny days or months. In contrast, such ecosystems have a mean annual potential evapotranspiration (PET) of more than 1000 mm (Wang *et al* 2016). PET, which is also termed climatic water demand, depends on climate only and is independent of landscape condition. For the hot semi-arid ecosystems, large values of PET can be attributed to hot temperature, low humidity, and large solar radiation, whereas for

doi:10.1088/978-0-7503-4020-5ch2

the cold semi-arid ecosystems, large values of PET are induced by high-speed wind, low humidity, and high altitude. The values for the modified aridity index (i.e. the ratio of annual PET to annual precipitation) are between 2.0 and 20.0, indicating that the annual precipitation can satisfy only 5% of the climatic water demand in one year and up to 50% in another year (Deblauwe *et al* 2008, Garcia-Franco *et al* 2018). The modified aridity index is more favored than the conventional aridity index (the humidity index) because the former indicates directly how many times the PET is larger than the annual precipitation. The conventional aridity index is the reciprocal of the modified aridity index. Hereafter, the modified aridity index and aridity index are used interchangeably unless specified otherwise.

For a given semi-arid ecosystem, its land surface can be bare and/or vegetated (figure 2.1). The bare surface may expand while the vegetated surface can recede in a dry year, and vice versa in a wet year. The expanding–receding dynamics can not only reshape the overall border between these two types of surfaces, but also may result in irregular vegetation patches, each of which is sandwiched by bare soils. A bare land surface may be covered by loose topsoil, physical crust, or biocrust. The loose topsoil, usually dominated by silt (0.004–0.062 mm in diameter), very fine sand (0.062–0.125 mm in diameter), and fine sand (0.125–0.25 mm in diameter) particles with a very small percentage of clay (<0.004 mm in diameter) particles, is classified as sandy loam or loamy sand (USDA-NRCS 2004). It is susceptible to wind and/or water erosion of the fine particles that can hold water and nutrients for vegetation rejuvenation, leading to a gap-graded or uniform grain size distribution. For this reason, the organic matter content in the loose topsoil is generally very low and can have a large spatiotemporal variability, depending on the erosion–deposition history of fine particles. In contrast, the physical crust results from mechanical forces of raindrop splashing, livestock trampling, and human disturbances. These external forces harden the previously loose soils due to accumulation of salts and silica (Cantón *et al* 2014). In comparison, the biocrust (i.e. biological crust) is formed by intimate associations of soil particles with cyanobacteria, algae, fungi, lichens and bryophytes, uncountable bacteria, and mosses. It is omnipresent as living cover in plant inter-spaces (i.e. the open spaces surrounding vascular plants). Although both types of crusts can increase the resistance to erosion of the soil beneath, the biocrust has wider ecohydrological effects than the physical crust (Mager and Thomas 2011, Bu *et al* 2014, Bu *et al* 2015, Chamizo *et al* 2018). On the other hand, a vegetated land surface is mostly covered by fertile topsoil with a relatively large holding capacity of water and nutrients. The fertile topsoil has a well-graded grain size distribution, which has a coefficient of uniformity $C_u > 6$ and a coefficient of curvature $C_c = 1–3$ (Todd and Mays 2005, Charbeneau 2006). C_u is the ratio of the grain diameter of 60% finer by weight (d_{60}) to that of 10% finer (d_{10}), while C_c is the ratio of the grain diameter of 30% finer (d_{30}) to the multiplication of d_{10} and d_{60}. Moreover, regardless of the surfaces, the vertical soil profile is almost homogeneous and isotropic in terms of texture, structure, and soil–water properties (Wang *et al* 2021). The soils in the rootzone (i.e. top 30 cm depth) have slightly larger saturated soil moisture and field capacity than those in the deeper layer, but the relative permeability of these two horizons depends on the geographic location: the upper

Figure 2.1. Typical land surface of semi-arid ecosystems. Vegetation inter-patches in a (a) dry year and (b) wet year.

horizon is more permeable at one site but less permeable at another site than the lower horizon. Generally, the lower-horizon soils have a much smaller organic matter content and holding capacity than the upper-horizon soils.

The vegetation of semi-arid ecosystems has a patchy pattern, low population density (i.e. number of individual stems per unit ground surface area), short height, small leaf area index (LAI) (i.e. one-sided green leaf area per unit ground surface area), and shallow root depth (Deblauwe *et al* 2008). For a given ecosystem, these physiological properties are related closely to precipitation and management practices (e.g. livestock grazing). For instance, in a drier year, the vegetation inter-patches tend become more prevalent (e.g. figure 2.1), the height and LAI will be smaller, and the roots can be restricted. The vegetation patches can vary from small clumps of grasses (e.g. 0.5–2.0 m^2) to large forces of mulga (*Acacia aneura*) trees (e.g. 100–1000 m^2) (Dunkerley 2002, Ludwig *et al* 2005). The inter-patching pattern is governed by the self-organization of vegetation, an adaptive process of vegetation to limited water and nutrients (Rietkerk *et al* 2002).

2.2 Water–heat–carbon cycles

This section presents cycles of water, heat, and carbon in the four typical land surfaces of semi-arid ecosystems, namely bare land surface with loose topsoil, bare land surface with a physical crust, bare land surface with a biocrust, and vegetated land surface. The cycles are profound for understanding and modeling semi-arid ecosystems.

2.2.1 Bare land surface with loose topsoil

For the bare land surface covered by loose topsoil, the soils above 5 cm depth are dry most of the time but can be spontaneously saturated when heavy storms occur, whereas the soils beneath are usually moist (Wang 2015). The dry soil layer (DSL) can retard underlying soils from further drying. During a non-rainy time period, the soil water is vaporized at the top of the moist soils (figure 2.2(a)). Subsequently, the water vapor moves up through the DSL into the ambient atmosphere if an upward soil temperature gradient exists, is entrapped within the DSL if a downward soil temperature gradient exists, and condenses into liquid if the DSL soil temperature falls below a threshold (e.g. 10 °C) (Sakai *et al* 2009). The condensed liquid would vaporize again once the soil temperature rises above the threshold. Herein, the soil temperature gradient is the ratio of the difference of temperatures between a deeper and a shallower point to the vertical distance between these two points. The threshold depends on atmospheric pressure and pore air humidity (i.e. ratio of volumetric water vapor to pore air). In contrast, during a rainy time period, if the rainfall is large enough to saturate the DSL, three hydrological processes will occur (figure 2.2(b)): the liquid water in the DSL is raised up to the ground surface where it vaporizes and evaporates, moves down to replenish the moisture of the moist soils, and drains as intermediate flow into streams. Otherwise, the soil water is vaporized at both the wetted ground surface and top of the moist soils. The water vapor can be directly released (i.e. evaporated) into ambient atmosphere, transported up through the DSL, entrapped in the DSL, and/or condensed into liquid (figure 2.2(c)). The evaporation will dry up the surface soils, forming another DSL (figure 2.2(d)). The soil water is vaporized at the tops of the two moist soil layers and the water vapor moves up through or entraps in the DSLs and/or condenses into liquid. Eventually, the wet layer will disappear as the vaporization progresses and the water cycle goes back to the one shown in figure 2.2(a). Regardless, the seepage from waterbodies

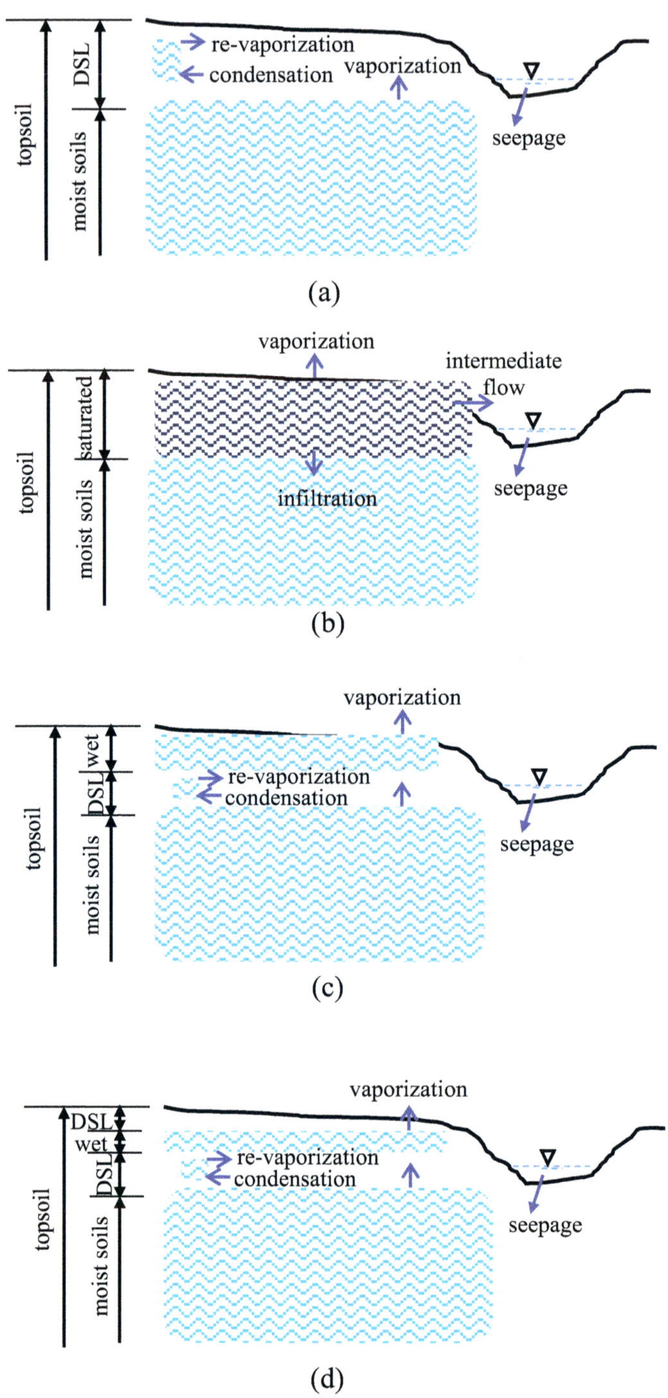

Figure 2.2. The water cycle for bare land surface covered by loose topsoil. Illustrations for (a) a non-rainy time period, (b) a large storm that saturates the dry soil layer (DSL), (c) a storm that only wets the DSL, and (d) the progressively drying stage.

(e.g. streams and ponds) may replenish the moisture of the moist soils. Moreover, heat is either absorbed by the topsoil to raise soil temperature and vaporize soil water or emitted from water vapor condensation and evaporation. The loose topsoil does not sequester, but may release, carbon dioxide (CO_2). The releasing occurs when the topsoil is wetted and thus its respiration becomes strong.

2.2.2 Bare land surface with physical crust

For the bare land surface covered by a physical crust, the soils beneath the crust are usually moist. Depending on the mechanical force, the crust can have a thickness of millimeters to centimeters. During a non-rainy time period, the soil water is vaporized at the top of the moist soils (figure 2.3(a)) and then the water vapor moves up through the crust into the ambient atmosphere if an upward soil temperature gradient exists, or is entrapped at the bottom of the crust if a downward soil temperature gradient exists (de la Piedra 2012). During a rainy time period, four hydrological processes may occur (figure 2.3(b)): vaporization, runoff, penetration, and infiltration. If the rainfall is insufficient to penetrate through the crust, the subsequent infiltration will not occur. Once the water on the ground surface is dried out, the water cycle will go back to the one shown in figure 2.3(a). Again, the seepage from waterbodies (e.g. streams and ponds) may replenish the moisture of the moist soils. The seepage from waterbodies may replenish the moisture of the moist soils. Moreover, the heat is either absorbed by the crust to raise its temperature and

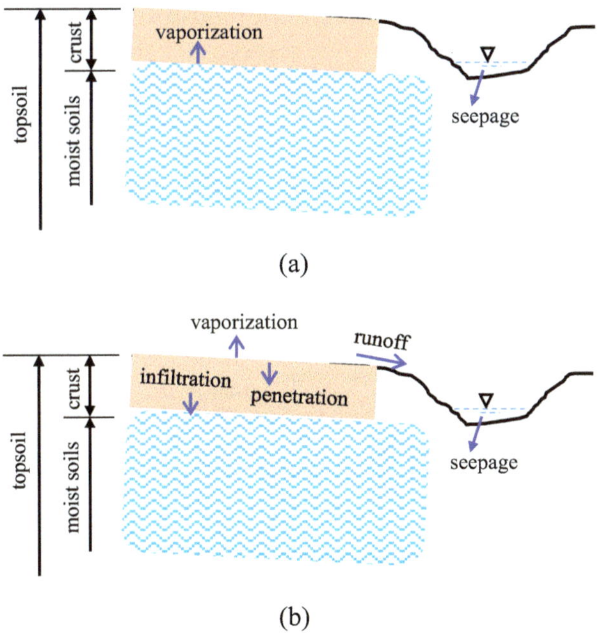

Figure 2.3. The water cycle for bare land surface covered by either a physical or biological crust. Illustrations for a time period of (a) non-raining and (b) raining.

vaporize soil water from the moist soils. As with the loose topsoil, the physical crust does not sequester CO_2, but may release the greenhouse gas when the hardened crust is wetted and thus its respiration becomes strong.

2.2.3 Bare land surface with biocrust

For a bare land surface covered by a biocrust, the soils beneath the crust are moist most of the time. A biocrust has a thickness of 1–2 cm, depending on its composition and successional stage, discussed in the following context. The water cycles are similar to those shown in figure 2.3. However, as shown in figure 2.4, the biocrust can play more roles than the physical crust because the former is living and evolving. The micro-organisms in the biocrust consume water, fix CO_2 and nitrogen (N_2) gases, and release oxygen (O_2) gas. For a biocrust of interest, its structure can be characterized by micro-organism composition, abundance of dominant phototropic species, chlorophyll *a* content, color, successional stage, community mottled and patchy pattern, and spatial cover and distribution. The biocrust can have a dominant phototropic species of cyanobacteria, lichens, or mosses, which corresponds to one of the three successional stages (Belnap 2006). As the crust matures (i.e. grows through the successional stages), the biomass of cyanobacteria, lichens, and mosses increases, which in turn increases the aggregate stability, shear strength, and roughness of the ground surface. In the first stage, filamentous cyanobacteria that live just below the soil surface aggregate soil particles through releasing extracellular polymeric substances (EPS), forming sheaths, and excreting polysaccharides (Adessi *et al* 2018), while lichens and mosses live on top and colonize the soil surface in the second and third stage, respectively (Belnap 2006). Cyanobacteria are less likely to clog soil pores than lichens and mosses. Cyanobacteria biocrusts may be further classified into one of the six levels of development (LODs) in terms of the visual classification system presented by Belnap *et al* (2008). A lower LOD biocrust tends to have a lighter color, fewer lichens and/or mosses, and a smoother surface than a higher LOD biocrust, and vice versa. Li *et al* (2016a) and Becerra-Absalón *et al* (2019) used DNA extraction techniques to examine the microbial species. They found that the cyanobacteria biocrusts have more bacteria and

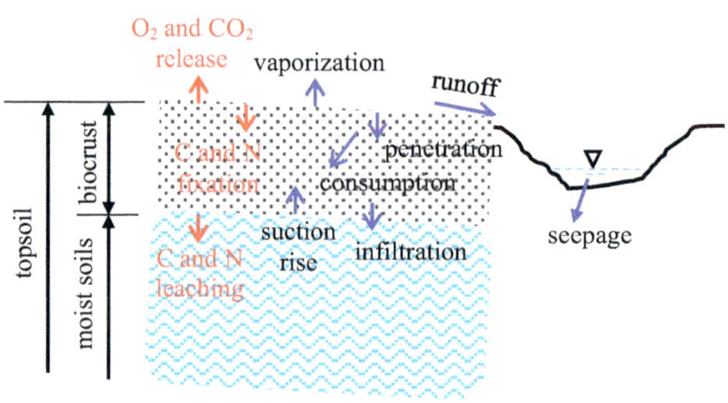

Figure 2.4. Illustration of the water–carbon cycle for bare land surface covered by a biological crust.

phototrophs but fewer fungi than the lichen and moss biocrusts, while the species compositions of the latter two biocrusts are statistically the same. The species composition of cyanobacteria biocrusts seem to be sensitive to site-specific environments. Román *et al* (2019) used a nondestructive methodology to quantify chlorophyll *a* content, which is a well-proven proxy of biocrust development and photosynthetic species status. As expected, those authors found that the moss biocrust has a much higher chlorophyll *a* content than the lichen biocrust, which in turn has a marginally higher chlorophyll *a* content than the cynanobacteria biocrust. Li *et al* (2016a) found that at the community level, the micro-organisms within a cyanobacteria biocrust tend to be homogeneously distributed, but those within a lichen and moss biocrusts tend to have a mottled and patchy distribution pattern. For biocrusts at the second and/or third stage, the well-developed (i.e. moss and lichen) patches are usually separated by less-developed (i.e. cyanobacteria) patches, with a well-developed patch surrounded by less-developed patches. Navas Romero *et al* (2019) analyzed the variations of biocrust coverage in the central west of Argentina. They found that the arid system has a higher biocrust coverage than the hyper-arid system, which in turn has a higher biocrust coverage than the semi-arid system, and that the hyper-arid system has the highest coverage of cyanobacteria biocrusts, while the arid system has the highest coverage of lichen and moss biocrusts. The biocrust coverage of an ecosystem likely depends on its characteristics and environments regardless of spatial scales. Gao *et al* (2017a) found that for a given ecosystem, the biocrust and vegetation coverages are complementary, that is, the biocrust coverage tends to increase with a decrease of the vegetation coverage, and vice versa.

Biocrusts are found to increase the beneficial micro-organisms and water–carbon–nutrient holding capacity of rootzone soils, enhancing the sustainability and resilience of fragile ecosystems in arid and semi-arid regions. Steven *et al* (2014) employed an rRNA gene and shotgun metagenomics sequencing technique (Janda and Abbott 2007) to compare the microbial communities inhabiting the root zones of the dominant shrub and the interspace biocrusts in a Mojave desert shrubland in Nevada. They found that most of the numerically abundant bacteria and fungi are present in both habitats (i.e. the biocrusts and root zones) although the proportional abundance of those members differ significantly between the habitats. Adessi *et al* (2018) measured the soil retention curves (matric potential versus relative water content) for intact biocrusts versus EPS-extracted biocrusts (i.e. the same biocrust soils without living micro-organisms), revealing that the intact biocrusts have a much larger water holding capacity. The presence of EPS allows the matric potential to remain unchanged until 80%, whereas without EPS, the matric potential starts decreasing at 20% of the saturated water content. This indicates that biocrusts can delay the onset of restrictive matric potentials due to a larger water holding capacity. In addition, Belnap *et al* (2013) performed *in situ* artificial rainfall simulations at 83 experiment plots (covered by cyanobacteria biocrusts) within a cool desert region of the Colorado Plateau. They found that the eroded soil per unit plot area consistently decreases with increasing LOD (i.e. a plot covered by the biocrusts at an earlier LOD is more susceptible to water erosion than a plot covered by the biocrusts at a later LOD). Gao *et al* (2017a) performed artificial laboratory rainfall simulations using undisturbed biocrusts box-sampled at 59 sites across the Loess Plateau of China. Those sites were covered by biocrusts ranging from cyanobacteria to moss. They found that the soil erodibility is almost halved with the presence of a biocrust (regardless of its successional stage) relative to without the biocrust. Further, Gao *et al* (2017a) found that the organic matter contents (OMCs) are twice or more higher in the biocrusts than in bare soils and that the OMCs are higher in the moss than the cyanobacteria biocrusts. The soils underneath the cyanobacteria biocrusts have a one and half times higher OMC, while the top 10 cm soils underneath the moss biocrusts have a four times higher OMC than the bare soils. This can be attributed to the biocrusts' stage-dependent roles in carbon sequestration. Cantón *et al* (2014) performed *in situ* artificial rainfall simulations on four 1 m square plots in the Tabernas Desert of Spain, which were covered by a physical crust, incipient cyanobacteria crust, well-developed cyanobacteria crust, and lichen crust, respectively. They found that the OMC is higher in a more mature than a less mature (e.g. lichen versus well-developed cyanobacteria) biocrust and that the OMCs in the biocrusts are higher than in the physical crust. The OMC in the top 1 cm soils underneath the well-developed cyanobacteria crust is highest, while the OMCs in the soils underneath the other three crusts are barely different. The carbon loss is highest from the physical crust, while it is higher from a more-matured than a less-matured biocrust; the removal of the lichen biocrust induces a highest carbon loss, followed by the removals of the physical and cyanobacteria crusts. Regardless, most of the carbon losses are associated with soil erosion and sedimentation. Chamizo *et al* (2017) observed organic carbon losses under natural rainfall in the Tabernas Desert of Spain. Their findings are consistent with those of Gao *et al* (2017a) and Cantón *et al* (2014). A few studies such as Chamizo *et al* (2018) showed that as with the aforementioned effects on organic carbon, biocrusts can also increase nutrients (e.g. nitrogen and phosphorus) in the underlying soils and for nearby plants. Moreover, Belnap (2006) reviewed the pre-2004 studies on the hydrological effects of biocrusts, leading to the two major conclusions that (i) the presence of biocrusts can increase infiltration in comparison with the absence of biocrusts or the presence of physical crusts and (ii) a biocrust at a later successional stage (or higher LOD) tends to have a rougher surface, is more favorable to infiltration, and has a larger soil water retention than a biocrust at an earlier stage (or lower LOD). The first conclusion is consistently supported by studies published since Belnap's review (Bu *et al* 2014, Bu *et al* 2015, Chamizo *et al* 2018), but the second conclusion is not always supported (Li *et al* 2016a). Based on *in situ* artificial rainfall simulations in the Loess Plateau of China, Zhao and Xu (2013) found that infiltration is decreasing with increasing successional stages, whereas based on observations of natural rainfalls in the Tabernas desert of Spain, Chamizo *et al* (2017) found an opposite relationship between infiltration and biocrust stage. The biocrusts' presence was found to increase (Zhao and Xu 2013, Cantón *et al* 2014, Fick *et al* 2019) or decrease (Bu *et al* 2015, Xiao *et al* 2015, Chamizo *et al* 2017) runoff relative to their absence; this inconsistency has yet to be explained (Belnap 2006, Kuske *et al* 2012, Faist *et al* 2017).

2.2.4 Vegetated land surface

For the land surface occupied by vegetation, the vertical soil profile can be subdivided into two horizons, namely the rootzone and deep layer. The rootzone extends from the ground surface to the end of the deepest root. Its depth depends on the plant species, development stage, and growing conditions as well as soil–water properties and water availability (Fan *et al* 2017). In semi-arid ecosystems such as the Eurasian steppe grasslands (Singh *et al* 1998, Kurc and Small 2007), the rootzone depth varies from centimeters to meters, with a typical value of 30 cm (Wang *et al* 2021). On the other hand, the deep layer refers to the soils below the rootzone. Its soil water has a relatively high and stable content; it may either be replenished when the rootzone has a water content above the field capacity or be suctioned up to help plants survive when the water content of the rootzone is very low (e.g. near the wilting point). During a non-rainy time period, the plant roots uptake water from the rootzone soils, which in turn is transported through the plant stem to the leaves, where the liquid water is transpired (i.e. vaporized and released) into the ambient atmosphere through the leaf stomata (figure 2.5(a)). In contrast, during a rainy time period, in addition to transpiration, five more hydrological processes, including interception, evaporation, runoff, infiltration, and percolation, may also occur (figure 2.5(b)). The interception is the rainfall that is retained by plant leaves, branches, and stems. It does not reach the ground and is mainly lost to evaporation. The interception is a function of LAI and storm characteristics (e.g. intensity, duration, and amount) (Wang *et al* 2012). If the throughfall, which is equal to the total rainfall minus interception, exceeds the storage capacity of ground depressions and/or the soil infiltration capacity, runoff will be generated. The infiltration replenishes the moisture of rootzone soils and/or percolates into the deep layer. Regardless, the seepage from waterbodies (e.g. streams and ponds) may replenish the moisture of both soil horizons. Moreover, plant leaves absorb CO_2 and light energy for photosynthesis, a process of creating chemical energy (i.e. sugars) in carbohydrate molecules while releasing oxygen (O_2) gas (figure 2.5). The roots uptake nutrients (e.g. N and P) from the rootzone soils for the plant to stay healthy, whereas they also release CO_2 from respiration. The C and N that cannot be taken up by living roots or are formed from dead roots may leach into the deep layer when percolation occurs. Heat energy is used in photosynthesis, warming soils, and evaporation, while it can be released from the respiration of plant leaves and roots.

In a semi-arid ecosystem of interest, the aforementioned four types of land surfaces usually co-exist as patches (Ludwig *et al* 2005). For most natural situations, vegetation and biocrust patches are mutually sandwiched; the trampling by vehicles and animals, however, may result in a physical biocrust patch between two vegetated patches. Loose topsoil patches can be formed due to vegetation degradation or crust destruction. Although biocrust inter-patches are an important water source for vegetation patches, these two types of patches may compete for water, nutrients, and light, leading to a smaller interpatch when it is wet but a larger interpatch when it is dry. A physical crust or loose topsoil patch rarely evolves into a vegetation patch unless some cultivation measures (e.g. seedlings) are taken.

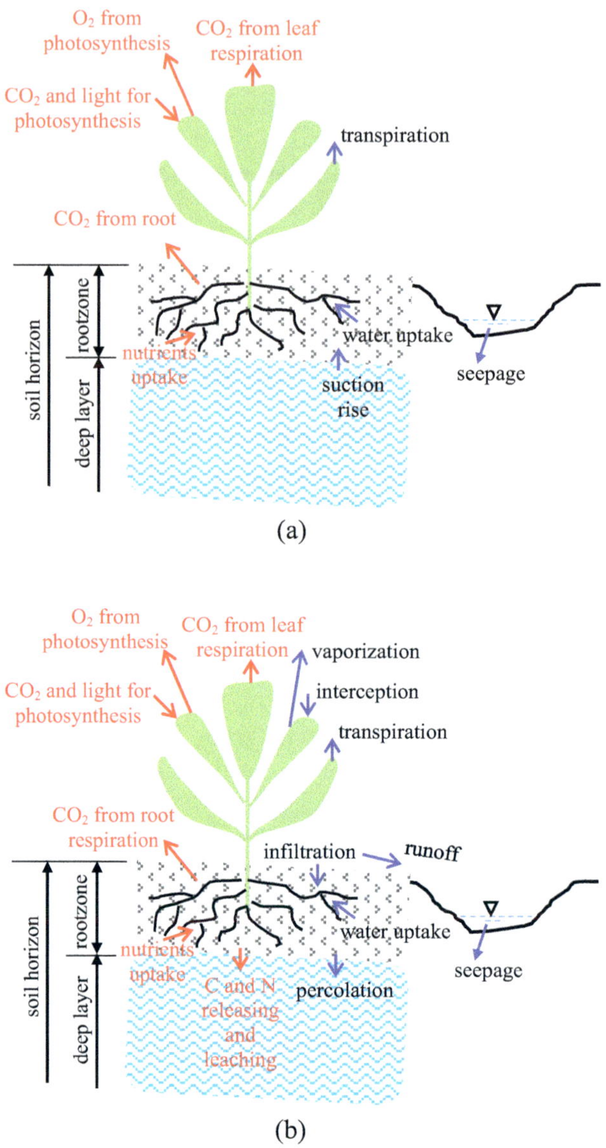

Figure 2.5. The water–carbon cycle for vegetated land surface. Illustrations for a time period of (a) non-raining and (b) raining.

2.3 The abiotic and biotic properties of semi-arid soils

Semi-arid soils have a moisture regime between ustic and xeric (Monger and Khresat 2005). As detailed by USDA-NRCS (2014), the ustic moisture regime has overall higher soil moisture and soil temperature than the xeric moisture regime. For the ustic moisture regime, the soil moisture in some or most parts is dry for at least 90 or more cumulative days in a normal year and moist for more than 180 cumulative or

90 consecutive days per year. The soil temperature at the reference depth (i.e. 50 cm from the soil surface) is above 5 °C for 45 or more cumulative days per year, with an annual mean of above 22 °C and a variation of above 6 °C between summer and winter in a year. In contrast, for the xeric moisture regime the soil moisture in all parts is dry in summer and moist in winter for 45 or more consecutive days in four months. The soil temperature is above 6 °C for more than half of the cumulative days per year or higher than 8 °C for 90 or more consecutive days, with an annual mean of below 22 °C and a variation of around 6 °C between summer and winter in a year. Semi-arid soils have a large variety of surface epipedons (e.g. anthropic formed in human-altered or human-transported material) and subsurface horizons (e.g. calcic with secondary calcium carbonate or other carbonate accumulated to a significant extent). Each epipedon or horizon is characterized by a specific formation history and materials. Semi-arid soils can be classified into alfsols, entisols, inceptisols, ultisols, oxisols, vertisols, and andisols. This section presents the abiotic (i.e. physical and geochemical) and biotic (i.e. biological) properties of semi-arid soils that are closely related to the water–heat–carbon cycles described in section 2.2.

2.3.1 The abiotic properties

For a given soil, its texture is determined by the contents of particles with various sizes (figure 2.6) and is the most profound physical property. The soil classification can be identified from the US Department of Agriculture (USDA) soil classification triangle (figure 2.7) in terms of the combination of percent sand particles, percent silt particles, and percent clay particles. The proportions of these three types of particles are determined by excluding particles larger than 2.0 mm. That is, the summation of the proportions should be equal to one.

In addition, the soil structure refers to how the particles are arranged spatially and can be measured using five physical properties, namely porosity (n), void ratio (e), dry density (ρ_d), particle density (ρ_s), and pore-size distribution index (λ). The former four properties are defined as

$$\underset{\text{porosity [–]}}{n} = \frac{\overset{\text{volume of pores [L}^3\text{]}}{V_v}}{\underset{\text{volume of soil [L}^3\text{]}}{V_t}} \tag{2.1}$$

$$\underset{\text{void ratio [–]}}{e} = \frac{\overset{\text{volume of pores [L}^3\text{]}}{V_v}}{\underset{\text{volume of particles [L}^3\text{]}}{V_s}} = \frac{\overset{\text{porosity [–]}}{n}}{1-n} \tag{2.2}$$

$$\underset{\text{dry density [M L}^3\text{]}}{\rho_d} = \frac{\overset{\text{mass of particles [M]}}{m_s}}{\underset{\text{volume of soil [L}^3\text{]}}{V_t}} \tag{2.3}$$

inches	U.S. Standard Sieve No.	mm	USDA
		4026 –	
		2048 –	boulders
		1024 –	
		512 –	
12		300 –	
10		256 –	
6		128 –	cobbles
3		75 –	
		64 –	
		32 –	
1		25.4 –	
0.75		19 –	
		16 –	
0.5		12.7 –	gravel
0.375		9.5 –	
		8 –	
0.25		6.35 –	
	4	4.76 –	
		4 –	
	10	2 –	
		1 –	coarse sand
		0.5 –	
	40	0.425 –	medium sand
		0.25 –	
			fine sand
		0.125 –	
	200	0.074 –	very fine sand
		0.0625 –	
		0.05 –	
		0.031 –	
		0.0156 –	
		0.0078 –	silt
		0.005 –	
		0.0039 –	
		0.001 –	clay

Figure 2.6. Soil particle sizes. (Reproduced from (USDA 1951), public domain.)

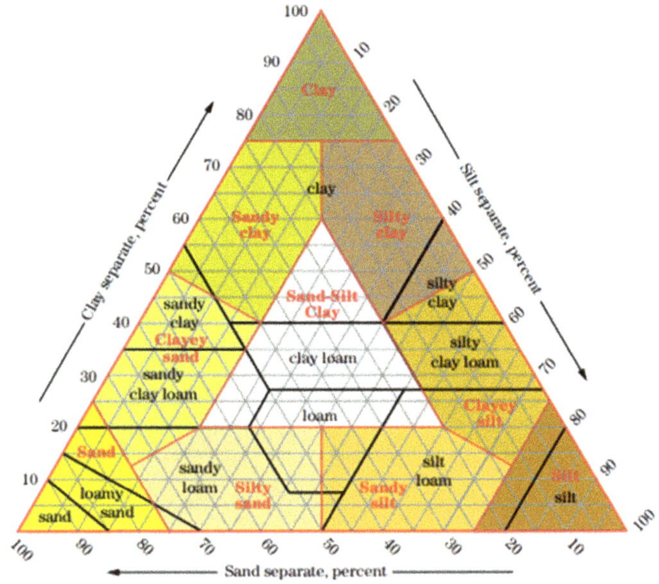

Figure 2.7. The United States Department of Agriculture (USDA) soil classification triangle. The percentages are determined by excluding particles larger than 2.0 mm. (Reproduced from (USDA 1951), public domain.)

$$\rho_s = \frac{m_s}{V_s} \tag{2.4}$$

particle density [M L³] → ρ_s; mass of particles [M] → m_s; volume of particles [L³] → V_s

The capillary rise theory (Charbeneau 2006) indicates that the suction head (ψ) is inversely related to pore radius (r) and soil water holding capacity (θ_c) (figure 2.8(a)). As presented by Nimmo (2004), the pore-size distribution defines the functional relationship between $\frac{\partial \theta_c}{\partial r}$ (i.e. the slope of the θ_c–r curve) and r (figure 2.8(b)). This function is integrated from the smallest radius r_{min} to the maximum radius r_{max} to derive the pore-size distribution index as (Cary and Hayden 1973)

$$\lambda = \int_{r_{min}}^{r_{max}} \frac{\partial \theta_c}{\partial r} dr \tag{2.5}$$

pore-size distribution index [–] → λ; maximum pore radius [L] → r_{max}; minimum pore radius [L] → r_{min}; soil-water holding capacity [–] → θ_c; pore radius [L] → r

ρ_s ranges from 2.60–2.75 g cm^{-3} and is almost independent of soil classifications. It is related to the parent material of a soil: smaller for soils with a higher organic matter content and larger for soils with a higher iron-oxide content. A value of 2.65 g cm^{-3} is good enough for most applications. In contrast, the other four physical

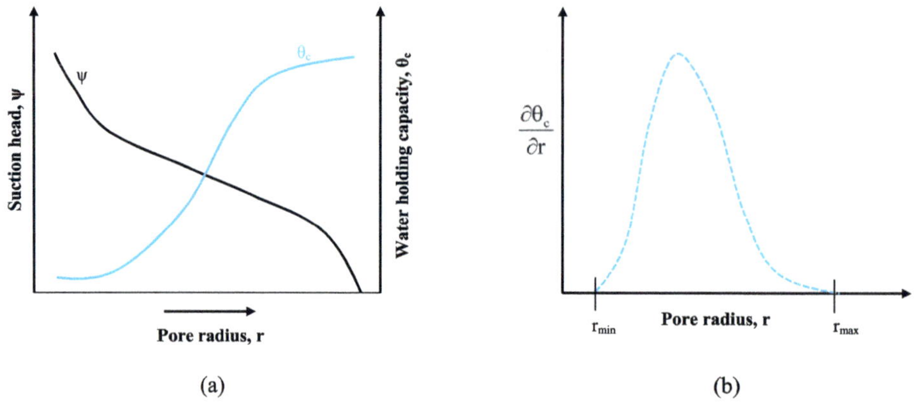

Figure 2.8. Derivation of the pore-size distribution for a hypothetical soil. Illustrations of the (a) functional relationships of suction head and water holding capacity versus pore radius and (b) corresponding pore-size distribution function.

Table 2.1. Typical physical properties for sandy soils.[a]

Classification	Porosity n [–]	Void ratio e [–]	Dry density ρ_d [g cm^{-3}]	Pore-size distribution index λ [–]
Loamy sand	0.456 ± 0.015	0.840 ± 0.025	1.44 ± 0.04	1.16 ± 0.49
Sand	0.467 ± 0.011	0.877 ± 0.019	1.41 ± 0.03	3.28 ± 1.20
Sandy clay	0.443 ± 0.012	0.796 ± 0.019	1.48 ± 0.03	0.24 ± 0.06
Sandy clay loam	0.438 ± 0.014	0.780 ± 0.022	1.49 ± 0.04	0.36 ± 0.14
Sandy loam	0.450 ± 0.016	0.820 ± 0.026	1.46 ± 0.04	1.41 ± 0.65

[a]The values for n, ρ_d, and λ are from table 3.12 of Wang (2021), while the values for e are calculated using equation (2.2).

properties (i.e. n, e, ρ_d, and λ) depend on soil classifications. Table 2.1 shows the typical values for sandy soils that are common in semi-arid ecosystems.

Further, soil texture and structure conjunctively determine the five soil–water properties, namely residual soil moisture (θ_r), wilting point (θ_ψ), field capacity (θ_{fc}), saturated soil moisture (θ_{sat}), and saturated hydraulic conductivity (K_s). The typical values of θ_ψ, θ_{fc}, θ_{sat}, and K_s for sandy soils are shown in table 2.2. θ_r is the volumetric water content that cannot be reduced by a further increase in negative suction head and has quite a small value of < 0.001–0.12. θ_ψ is the minimum amount of water in the soil at which the plant can survive, whereas θ_{sat} is the maximum amount of water that can be stored in the soil. The suction head is around 1500 kPa (15 bars) at θ_ψ and zero at θ_{sat}. When the soil is fully saturated, its pores are fully occupied by water and thus $\theta_{sat} = n$. However, because not all trapped air can be removed by water, the naturally saturated soil moisture is roughly equal to $0.95\theta_{sat}$. θ_{fc} is the maximum amount of water that can be retained by the soil with no drainage by water gravity and is the optimal moisture level for vegetation because of

Table 2.2. Typical soil–water properties for sandy soils.[a]

Classification	Wilting point θ_ψ [–]	Field capacity θ_{fc} [–]	Saturated soil moisture θ_{sat} [–]	Saturated hydraulic conductivity K_s [m d^{-1}]
Loamy sand	0.059 ± 0.027	0.123 ± 0.022	0.456 ± 0.015	2.237 ± 0.789
Sand	0.044 ± 0.017	0.093 ± 0.019	0.467 ± 0.011	3.004 ± 0.741
Sandy clay	0.252 ± 0.034	0.363 ± 0.037	0.443 ± 0.012	0.040 ± 0.033
Sandy clay loam	0.180 ± 0.032	0.286 ± 0.043	0.438 ± 0.014	0.240 ± 0.162
Sandy loam	0.083 ± 0.040	0.192 ± 0.038	0.450 ± 0.016	1.228 ± 0.704

[a] Source: table 3.12 of Wang (2021).

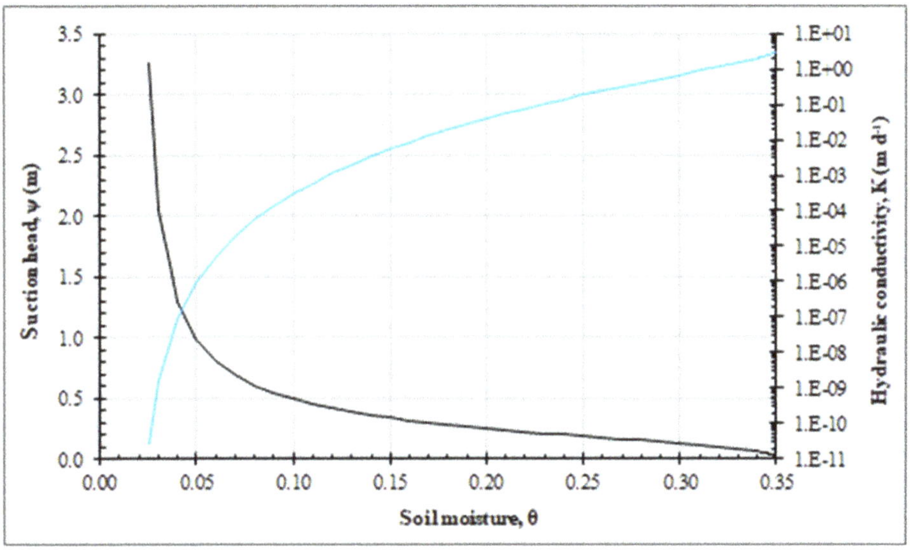

Figure 2.9. The retention curve and hydraulic conductivity function for typical semi-arid soil.

the zero stresses from both water and oxygen (Hack-ten Broeke *et al* 2016). K_s is the Darcy velocity of water flow in the soil for a unit hydraulic head gradient (Charbeneau 2006). It depends on the intrinsic permeability of the soil as well as the specific weight and viscosity of water. K_s, in m d^{-1}, can be computed as

$$K_s = 86400 K_{soil} \frac{\gamma_w}{\mu_w} \tag{2.6}$$

saturated hydraulic conductivity [m d^{-1}] → K_s; specific weight of water [N m^{-3}] → γ_w; viscosity of water [N s m^{-2}] → μ_w; soil-specific permeability [m^2] → K_{soil}

Moreover, because semi-arid soils are unsaturated most of the time, water movements in such a soil are mainly driven by the suction head rather than the force of gravity. The relations between the suction head (ψ) and soil moisture (θ) as well as between unsaturated hydraulic conductivity (K) and θ are also important physical properties of the soil (figure 2.9). Hereafter, given that ψ is negative, for description purposes, an increase of ψ means that its absolute value increases and a decrease implies that its absolute value decreases. A larger absolute value of ψ indicates that the soil is drier with a higher suction head. Water moves from a location with a lower (i.e. larger negative) suction head to a location with a higher (i.e. smaller negative) suction head. The ψ–θ relation is called the retention curve, which can also be influenced by chemical properties of the soil. Semi-arid soils have low contents of organic matter, cation exchange capacity, and iron and aluminum oxide, while they are rich in various trace elements, including cadmium (Cd), copper (Cu), cobalt (Co), chromium (Cr), nickel (Ni), lead (Pb), mercury (Hg), manganese (Mn), arsenic (As), selenium (Se), molybdenum (Mo), and boron (B) (Rezapour and Samadi 2011). For such a soil, soluble salts are present in most parts and calcium carbonate exists in most sublayers.

2.3.2 The biotic properties

Semi-arid soils have low biological (i.e. microbial and faunal) activity because of their prevalent dryness and low organic matter content. The soils with a biocrust may have relatively higher microbial activity, whereas the soils occupied by vegetation can have relatively higher faunal activity. Loose topsoil and soils with physical crusts are visually free of biological activity. Faunal activity is carried out by a variety of earthworms, nematodes, and arthropods. Soil organisms break down organic matter into inorganic nutrients for plants and maintain soil structure, whereas earthworms can till and make soils more favorable for vegetation growth. In addition, bacteria affect the nitrogen cycle in soil. For instance, bacteria decompose dead plants and animal waste to produce nitrates that are usable by growing plants. Nitrifying bacteria on the roots of peas and beans fix nitrogen from the atmosphere into organic substances, whereas denitrifying bacteria convert nitrates and nitrites in soil into gases, through which nitrogen is released into the atmosphere. The nitrification process fertilizes soils and prompts vegetation growth. In contrast, the denitrification process reduces soil fertility and tends to degrade the semi-arid ecosystem of interest. Further, microbes consume organic materials in soil to form a temporary carbon reservoir with byproducts of CO_2, water, and humus. The complete decomposition of humus may take years, depending on the respiration of microbes, which is faster at higher carbon concentrations, warmer soil temperatures, and more soil water.

2.4 Land–atmosphere interactions

Semi-arid land and atmosphere exchange heat and mass (i.e. water, carbon, and nutrients) (figure 2.10). These exchanges are interrelated and driven by heat and water (Goutorbe *et al* 1993, Gentine *et al* 2019). The solar radiation is reflected by land surface and transferred into soils. The reflection depends on the land surface type and cover, and can be measured by albedo (table 2.3), whereas the absorption depends on the heat capacity and thermal conductivity of soil. The absorbed radiation raises soil temperature to drive the physical, chemical, and biological processes in soil, which are discussed in sections 2.2 and 2.3. Once soil temperature is

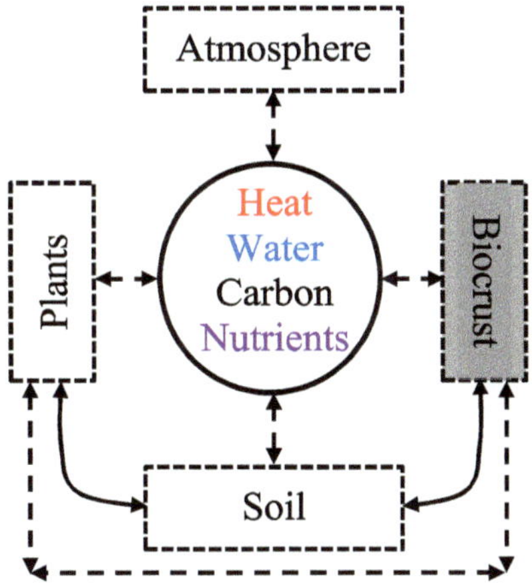

Figure 2.10. Diagram showing land–atmosphere interactions.

Table 2.3. The albedos for typical semi-arid lands.[a]

Land surface	Albedo (%)
Crop	15–25
Desert	27–29
Forest	5–15
Grass	15–25
Meadow	10–20
Sandy soil, dry	25–45
Sandy soil, wet	20–30
Snow	40–85
Water	5–10

[a] Compiled from numerous sources.

higher than the ambient air temperature, some heat can be emitted from the soil into the atmosphere. Table 2.4 shows the heat capacities and thermal conductivities for the soil components, namely sand particles, silt particles, clay particles, water, and air. For a given soil, its heat capacity and thermal conductivity are computed as

$$C_{sh} = \zeta_{sand} C_{sand} + \zeta_{silt} C_{silt} + \zeta_{clay} C_{clay} + \zeta_{water} C_{water} + \zeta_{air} C_{air} \tag{2.7}$$

C_{sh}: heat capacity of soil [J cm^{-3} °C^{-1}]
C_{sand}: heat capacity of sand particles [J cm^{-3} °C^{-1}]
C_{silt}: heat capacity of silt particles [J cm^{-3} °C^{-1}]
C_{clay}: heat capacity of clay particles [J cm^{-3} °C^{-1}]
C_{water}: heat capacity of water [J cm^{-3} °C^{-1}]
C_{air}: heat capacity of air [J cm^{-3} °C^{-1}]
ζ_{sand}: volumetric fraction of sand particles [–]
ζ_{silt}: volumetric fraction of silt particles [–]
ζ_{clay}: volumetric fraction of clay particles [–]
ζ_{water}: volumetric fraction of water [–]
ζ_{air}: volumetric fraction of air [–]

Table 2.4. The thermal properties of soil components.[a]

Property	Sand (0.05–2.0 mm)	Silt (0.002–0.05 mm)	Clay (<0.002 mm)	Water (Ice)	Air[b]
Heat capacity (J cm^{-3} $°C^{-1}$)	2.128	2.496	2.385	4.18 (1.95)	0.001 212
Thermal conductivity (J cm^{-1} $°C^{-1}$ d^{-1})	7603	216	2523	520 (2100)	7–80

[a] Sources: Farouki (1981), Kroes *et al* (2017).
[b] The actual values vary with atmospheric pressure and temperature. These are values at standard atmospheric pressure and 20 °C.

$$\lambda_{sh} = \zeta_{sand}\lambda_{sand} + \zeta_{silt}\lambda_{silt} + \zeta_{clay}\lambda_{clay} + \zeta_{water}\lambda_{water} + \zeta_{air}\lambda_{air} \tag{2.8}$$

where (annotations printed around the equation):
λ_{sh}: thermal conductivity of soil [J cm^{-1} $°C^{-1}$ d^{-1}]
λ_{sand}: thermal conductivity of sand particles [J cm^{-1} $°C^{-1}$ d^{-1}]
λ_{silt}: thermal conductivity of silt particles [J cm^{-1} $°C^{-1}$ d^{-1}]
λ_{clay}: thermal conductivity of clay particles [J cm^{-1} $°C^{-1}$ d^{-1}]
λ_{water}: thermal conductivity of water [J cm^{-1} $°C^{-1}$ d^{-1}]
λ_{air}: thermal conductivity of air [J cm^{-1} $°C^{-1}$ d^{-1}]
ζ_{sand}: volumetric fraction of sand particles [–]
ζ_{silt}: volumetric fraction of silt particles [–]
ζ_{clay}: volumetric fraction of clay particles [–]
ζ_{water}: volumetric fraction of water [–]
ζ_{air}: volumetric fraction of air [–]

The land receives precipitation from the atmosphere in the form of liquid (i.e. rain), solids (i.e. snow, graupel, sleet, hail), and/or a mixture (i.e. freezing rain) (Means 2019). A portion of the precipitation can be intercepted by vegetation and the remaining reaches the ground as throughfall, which may be converted into runoff, evaporate into the atmosphere, penetrate through crusts, and infiltrate in beneath soils. The intercepted precipitation is mainly lost to evaporation. The soil water can be transpired by plants and evaporated from bare soils. In semi-arid ecosystems, runoff is barely generated except by sporadically heavy storms. It is ultimately lost to evaporation, infiltration, and consumption. Streams have a poorly incised channel and are intermittent with a zero flow most of the time. Evaporation and transpiration consume the heat received by the land (Lyons *et al* 1993), leading to spatially varying temperature gradients for heat redistribution.

Plants fix CO_2 from the atmosphere by photosynthesis to assimilate carbon into leaves, stems, and roots. The carbon in dead plants is stored in the soil. On the other hand, the respiration of soil micro-organisms and plant roots releases CO_2 into the atmosphere. Land surfaces with loose topsoil and physical crusts are unfavorable for organisms and have negligible respiration, whereas land surfaces covered by biocrusts and vegetation can be strong in respiration. The difference between the released and fixed CO_2 is the net ecosystem exchange (NEE), which is a determinant of plant biomass (Wang *et al* 2021). Removal of vegetation will increase the release while decrease the fixing of CO_2 and mobilize the carbon stored in soil, releasing more CO_2 into the atmosphere and increasing climate change. This is also true for

Table 2.5. The annual global carbon budget.[a]

Budget term	Amount (GT)
Storage	122 618 139
Atmosphere	839
Continental crusts	12 576 000
Ocean	38 000
Permafrost	1700
Soils and vegetation	1600
Dry	500
Moist	900
Wet	200
Emission	693
Industrial	385
Land use change	185
Soil, vegetation, and organism respiration	123
Fixation	119
Vegetation photosynthesis	119

[a] Compiled using data presented by Janowiak *et al* (2017).

Table 2.6. The carbon budget in a semi-arid shortgrass steppe ecosystem.[a]

Budget term	Storage in roots	Storage in stems and leaves	Storage in soils	Photosynthesis	Grass respiration	Soil respiration
Amount (g m^{-2} d^{-1})	901	68	7045	206	63	115

[a] Compiled using data presented by Janowiak *et al* (2017).

the destruction of biocrusts. Thus, it is vital to protect vegetation and biocrusts to sustain semi-arid ecosystems. Table 2.5 shows the global carbon budget. The global carbon storage in drylands is about 500 GT. The carbon stocks and fluxes in a semi-arid shortgrass steppe ecosystem in the state of Colorado are shown in table 2.6.

2.5 Soil water dynamics

Soil water is vital for sustaining semi-arid ecosystems (Wang *et al* 2021). It is replenished by precipitation and lost to evaporation and transpiration. For such an ecosystem, soil moisture, the volumetric percentage of soil pores occupied by water, is far below the field capacity most of the time (figures 2.11 and 2.12). Sporadically, heavy storms may wet the topsoil for a short time period but usually do not have much influence on soils below the rootzone (e.g. 30 cm). Soil moisture pulses in late spring (i.e. March and April) because of melting snow. The pulse tends to be larger in a year with a larger snowpack than in a year with a smaller snowpack, and its magnitude and duration depend on the melting rate as controlled by snowpack

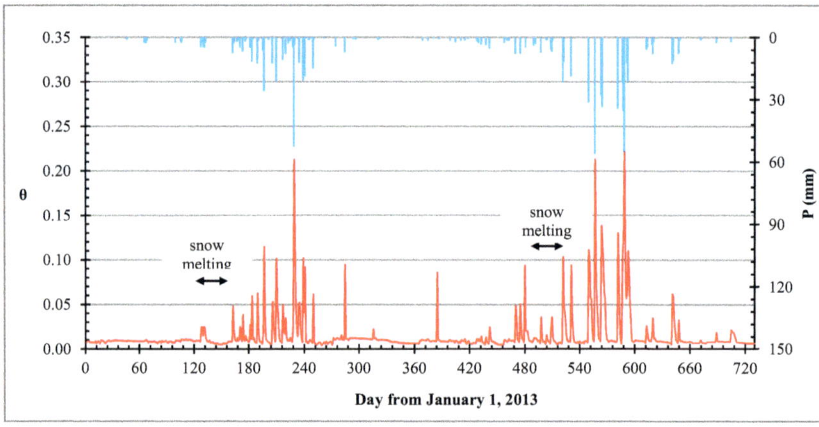

(a)

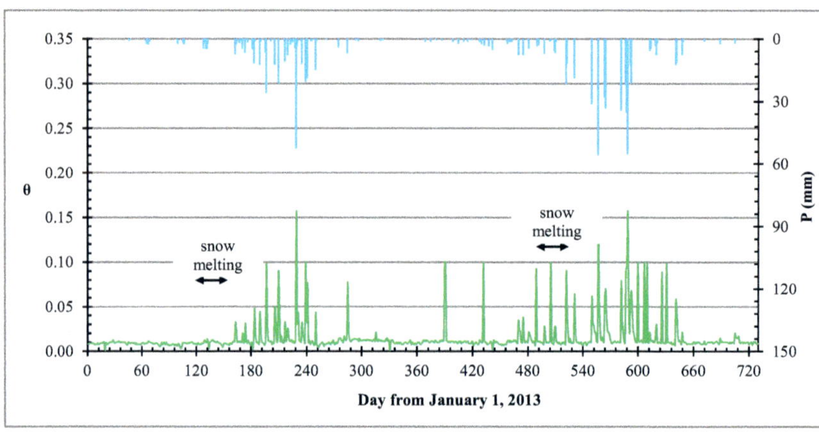

(b)

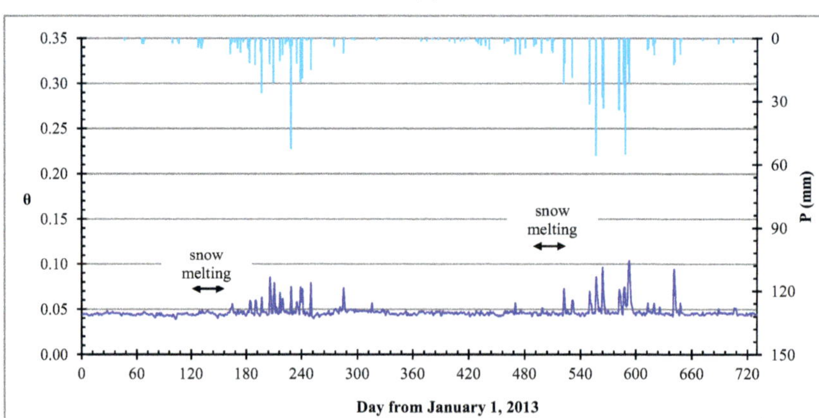

(c)

Figure 2.11. Instantaneous soil moisture, θ, at a semi-arid bare site with loose topsoil (122° 37′ 30″ E, 43° 20′ 56″ N). Fluctuations at the depth of (a) 5 cm, (b) 10 cm. and (c) 20 cm. The soil–water properties are: $\theta_r = 0.02$, $\theta_\psi = 0.05$, $\theta_{fc} = 0.15$, and $\theta_{sat} = 0.38$. P is precipitation.

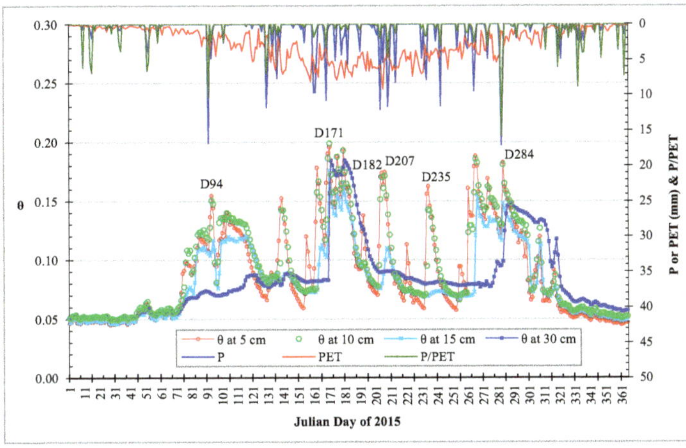

(a)

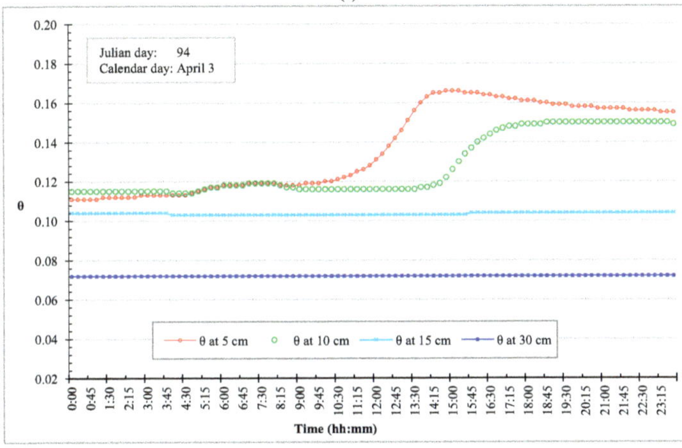

(b)

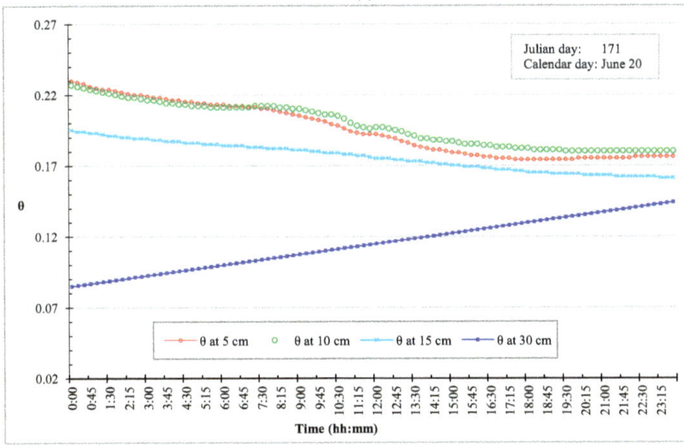

(c)

Figure 2.12. Soil moisture, θ, at a semi-arid site covered by steppe grass and biocrust patches (117° 52′ 00″ E, 44° 20′ 00″ N). Fluctuations of (a) daily moisture and (b)–(g) instantaneous moisture within the selected days. The soil–water properties are: $\theta_r = 0.02$, $\theta_\psi = 0.05$, $\theta_{fc} = 0.15$, and $\theta_{sat} = 0.38$. P is precipitation and PET is potential evapotranspiration.

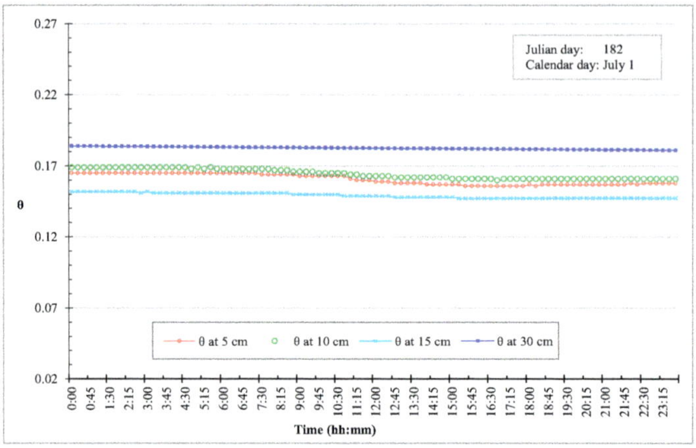

(d)

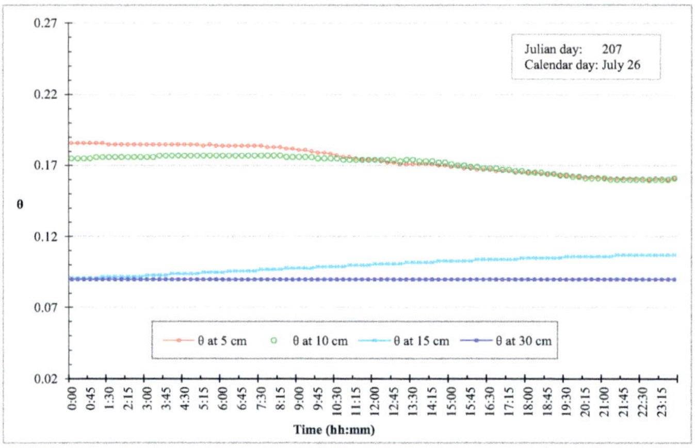

(e)

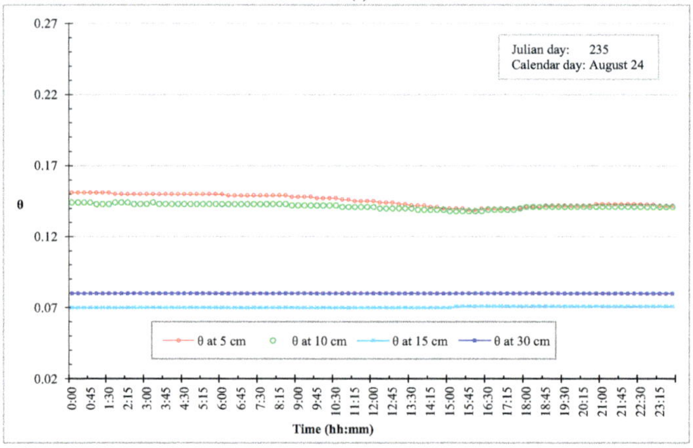

(f)

Figure 2.12. (Continued.)

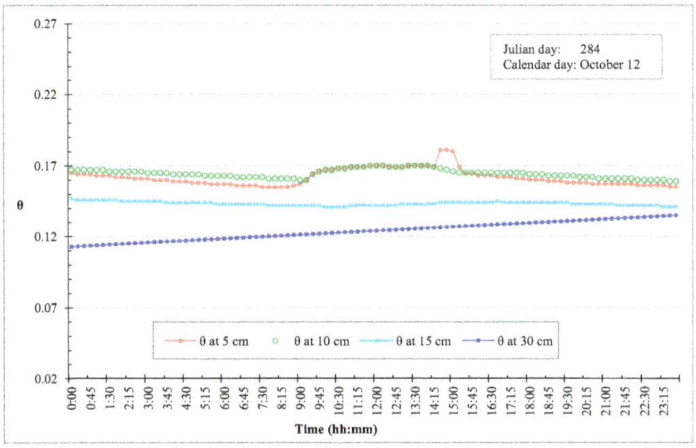

(g)

Figure 2.12. (Continued.)

temperature fluctuation (figure 2.11). Such soil moisture pulse plays crucial roles in vegetation rejuvenation and protecting the topsoil from aeolian erosion (Wang *et al* 2014a). Given that PET is usually much larger than precipitation (P), soil moisture depends on P/PET rather than PET: the larger the ratio, the higher the soil moisture, and vice versa (figure 2.12(a)).

For a land surface with loose topsoil, the moisture of the top 5 cm soils instantaneously responds to rainfall, whereas the moisture of soils below 20 cm can be influenced slightly at a lag time of hours to days, depending on the antecedent wetness condition (figure 2.11). The moisture of soils between 5 and 20 cm has a moderate response to rainfall. Once the rain stops, the top 10 cm soils will quickly be dried up to a moisture level close to the residual soil moisture (θ_r), forming a DSL that can retard moisture loss from beneath soils. In figure 2.11(c), the moisture of soils below 20 cm reaches a minimum of around 0.05, which is the wilting point of the soils, and remains almost constant between two successive storm events. In contrast, soil moisture is higher for a land surface with than without a biocrust (Chamizo *et al* 2016, Nevins *et al* 2020). The biocrust has a rougher surface, providing more micro-storage of water for penetration to replenish the moisture of the soils underneath, while the biocrust can intercept partial solar radiation, reducing soil water evaporation. Thus, after a rain event stops, it takes a long time period (e.g. months) for soil moisture to decrease to θ_r (Chamizo *et al* 2017). A biocrust with a higher LOD can better maintain the water underneath soil than that with a lower LOD. Similarly, the physical crust can also maintain the moisture of the soils underneath, but it has a smaller penetration rate and is less effective in retarding soil water evaporation than a biocrust.

For a vegetated land surface, the moisture of rootzone soils has a quick response to storms with a rainfall of larger than the canopy interception capacity (figure 2.12(a)). Once rain stops, the moisture starts to decrease by transpiration. The below-rootzone (e.g. 30 cm depth) soil moisture increases gradually following storms with a large rainfall and recess steadily to a level above the wilting point. The steady recession may be attributed to transpiration being much slower than evaporation, shading effects can reduce water loss from soils surrounding plant roots, and a vegetation patch can receive runoff from its adjacent biocrust patches to replenish moisture of underneath soils. On a rainy day, the moisture has a larger fluctuation in the top 15 cm than the soils beneath (figure 2.12(b), (c), (g)), whereas on a non-rainy or drizzly day, the moisture decreases slightly with a lapse of time in top 15 cm soils mainly due to transpiration and stays almost constant in the soils beneath probably due to few roots (figure 2.12(d), (e), (f)). At the earlier growing stage, because vegetation roots have a much larger density within the top 15 cm than the soils beneath, the moisture of the top 15 cm soils may be consumed by vegetation to a level below that of the soils beneath (figure 2.12(d)).

2.6 Vegetation community composition

For a semi-arid ecosystem, its productivity can be quantitatively measured as net primary production (NPP) (Luo *et al* 2014). NPP is defined as the rate of accumulation of organic matter by vegetation (i.e. biomass increase) and equals the difference between the carbon assimilated by plant photosynthesis and the carbon consumed by plant autotrophic respiration. That is, NPP can be expressed as the total net organic matter or carbon of vegetation stems, leaves, and roots at harvest. Given the socioeconomic importance of ecosystem productivity, extensive studies (Wang *et al* 2014b, 2014c) have been conducted to (i) measure or estimate NPP, (ii) develop NPP prediction models, and (iii) examine NPP responses to climate change and land management (e.g. livestock grazing). NPP fluctuates with precipitation, with an overall positive interrelationship, but a year with a larger annual precipitation might not have a higher productivity and vice versa. For instance, as shown in figure 2.13(a), the annual precipitation in the year 2000 was about 290 mm but the NPP was only 160 g C m^{-2}; the NPP in the year 2007 was even lower, although the annual precipitation was almost 190 mm. For a given year, the precipitation in winter and early spring seems to play more a important role in increasing biomass than that in the later seasons (Luo *et al* 2014). Other factors influencing NPP include livestock grazing, soil texture and salinity, topographic gradient, and depth to water table (Tang *et al* 2015, Wang *et al* 2016); so the NPP may exhibit a clear spatial pattern. In figure 2.13(b) the NPP tends to decrease from upstream to downstream because the downstream part of the watershed has a larger grazing intensity, higher soil salinity, and shallower water table than the upstream part (Luo *et al* 2014). For the same reasons, the NPP has a larger inter-annual variability for a lower than an upper sub-watershed.

In principle, NPP is closely related to heat–water–carbon cycling processes, which in turn depend on the vegetation community composition (VCC). For a semi-arid ecosystem, its VCC can be characterized by species physique, species richness, and species diversity. The species physique reflects the overall feature of the heights, densities, and NPPs of all vegetation species presented in the community, whereas the species richness measures the number of species. The species diversity takes into

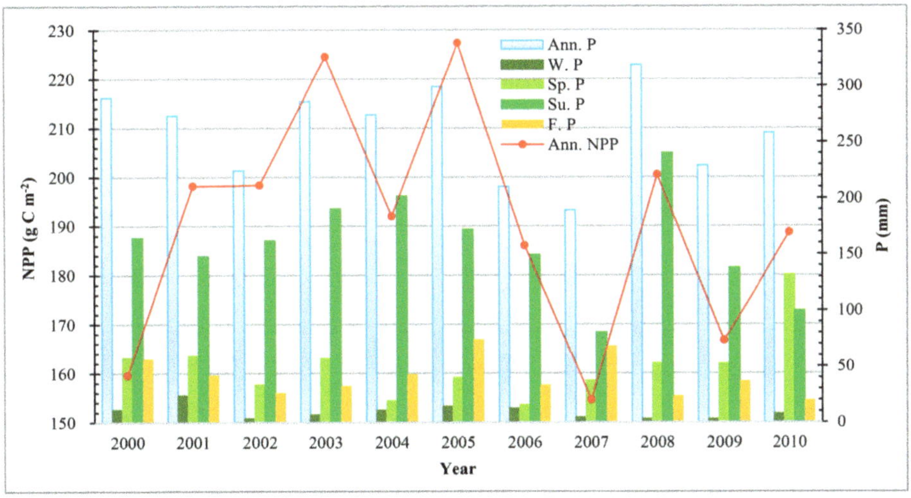

(a)

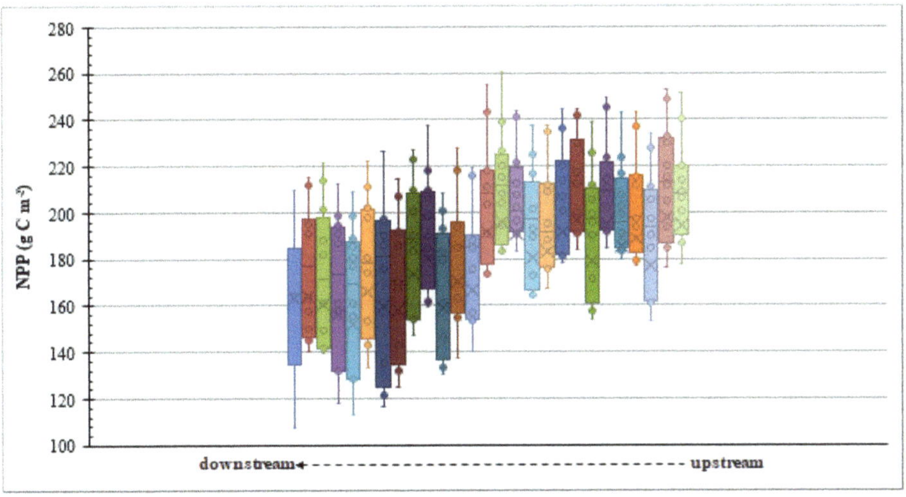

(b)

Figure 2.13. Net primary production, NPP, of steppe grasses in the semi-arid 5365 km^2 Balagaer River watershed (44° 00′ to 44° 15′ N, 117° 40′ to 117° 48′ E), located in the northeast Inner Mongolian Autonomous Region of China. Plots showing the (a) watershed-averaged NPP versus precipitation and (b) sub-watershed NPP inter-annual fluctuations between 2000 and 2010.

account both the number of species and the evenness of species in relation to one another. These VCC characteristics are commonly represented by dimensionless indices. Currently, the data on semi-arid VCC are very scarce in the existing literature. Table 2.7 shows the indices for three typical semi-arid ecosystems. Although these sample ecosystems are compatible in terms of species types, the

Table 2.7. The compositions of three typical semi-arid ecosystems.

Ecosystem	Physique index, SP_I	Richness index, SR_I	Diversity index, SD_I	Data source
Forests in Sierra Nevada, CA, USA (119° 34′ 27″ W, 38° 16′ 09″ N)		0.57	2.02	Richter *et al* (2019)
Meadows in northeast Inner Mongolia, China (122° 38′ 30″ E, 43° 20′ 06″ N)	0.43	0.61	1.48	Duan *et al* (2011)
Sand dunes in northeast Inner Mongolia, China (122° 38′ 30″ E, 43° 20′ 06″ N)	0.49	0.57	1.51	Duan *et al* (2011)

ecosystem in California has a more even distribution of vegetation species than the ecosystems in Inner Mongolia. The ecosystem of sand dunes have fewer species types but a more even distribution of species than the ecosystem of meadows. Future research needs to assess the compositions of semi-arid ecosystems across the world.

The species physique index, SP_I, can be determined as

$$SP_I = \frac{1}{3}\sum_{j=1}^{N_s}\left(h_{r,j} + \rho_{r,j} + m_{r,j}\right) \tag{2.9}$$

(SP_I: species physique index [–]; N_s: number of species [–]; j: counter of species [–]; $h_{r,j}$: relative height [–]; $\rho_{r,j}$: relative population density [–]; $m_{r,j}$: relative biomass [–])

$$h_{r,i} = \frac{\bar{h}_i}{\sum_{j=1}^{N_s}\bar{h}_j} \tag{2.10}$$

($h_{r,i}$: relative height [–]; i: identifier of species [–]; $\bar{h}_i$, $\bar{h}_j$: average heights [L]; N_s: number of species [–]; j: counter of species [–])

$$\rho_{r,i} = \frac{\rho_i}{\sum_{j=1}^{N_s}\rho_j} \tag{2.11}$$

($\rho_{r,i}$: relative population density [–]; i: identifier of species [–]; ρ_i, ρ_j: population density [L^{-2}], number of individuals per unit land area; N_s: number of species [–]; j: counter of species [–])

$$m_{r,i} = \frac{\bar{m}_i}{\sum_{j=1}^{N_s} \bar{m}_j} \tag{2.12}$$

identifier of species [–]; relative biomass [–]; number of species [–]; average masses [M]; counter of species [–]

The species richness index, SR_I, can be determined as (Hartnett and Wilson 1999)

$$SR_I = \frac{N_s}{\sqrt{n}} \tag{2.13}$$

species richness index [–]; number of species [–]; sample size [–]

The species diversity index, SD_I, can be determined as (Shannon 1948)

$$SD_I = \sum_{j=1}^{N_s} \left[\left| \frac{n_j}{n} \right| \ln\left(\frac{n_j}{n} \right) \right] \tag{2.14}$$

number of individuals of species j [–]; number of species [–]; species diversity index [–]; counter of species [–]; sample size [–]

2.7 Equilibrium and shift

Semi-arid ecosystems can have various states from fully vegetated (i.e. healthy) to bare (i.e. frail) (Karssenberg *et al* 2017). Each of the states signifies an equilibrium among the abiotic and biotic factors. The equilibrium state, at which the factors are invariant with time, is the intersection of the nullclines (i.e. zero-gradient lines) for all factors. The fully vegetated state is achieved when water and nutrient stresses are minimal, whereas the bare state occurs when the stresses such as those caused by drought, flood, and overgrazing are large and long-lasting enough to suppress vegetation growth. A semi-arid ecosystem adapts to gradual changes in one or more driving factors (i.e. drivers) while it stays in an equilibrium state as long as the drivers do not exceed their thresholds. Otherwise, the ecosystem can be shifted from the current to a new equilibrium state. Such a transition between equilibrium states is usually slow, offering a valuable time window to reverse land degradation to an unexpected state (e.g. desert). Nevertheless, rapid transitions may occur if some drivers incur an abrupt change and/or the vegetation species are very sensitive to a small disturbances in the surrounding environment. The state shifts of a semi-arid ecosystem can be reversible or irreversible. A reversible shift temporarily changes the ecosystem from one equilibrium state to another; once the stresses are removed, the ecosystem tends to 'bounce' back to its inception state. In contrast, an irreversible shift permanently changes the ecosystem to a new equilibrium state without being restored back to its inception state although stresses disappear.

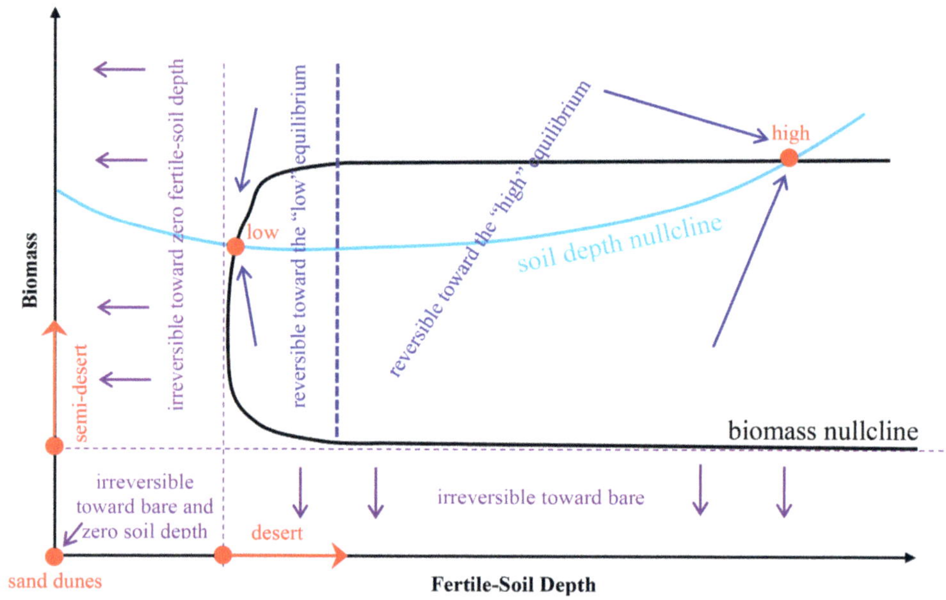

Figure 2.14. Shifts of semi-arid ecosystems. Illustration of a hypothetical ecosystem. The red solid dots signify the stable equilibrium states. On a nullcline, the state variable does not change with time.

If biomass is used to represent the overall characteristics of biotic factors (e.g. plant respiration and assimilation) and fertile-soil depth is used to represent the combined effect of abiotic factors (e.g. climate and soil erosion), the possible shifts can be depicted as in figure 2.14. For a given semi-arid ecosystem, it may have five stable equilibrium states, namely sand dunes, desert, semi-desert, low, and high. If both biomass and fertile-soil depth go below the corresponding thresholds, the ecosystem will keep on degrading toward sand dunes; when biomass or fertile-soil depth becomes smaller than its threshold, the ecosystem will likely continue to degrade toward semi-desert or desert. These three shifts are irreversible and will keep on deteriorating the ecosystem. On the other hand, if both biomass and fertile-soil depth are above the corresponding thresholds, once stresses disappear, the ecosystem will likely go back to either low or high stable equilibrium state. The low stable equilibrium state has a smaller biomass and a shallower fertile-soil depth than the high stable equilibrium state. Usually, the nullcline of fertile-soil depth has an inflection point, at which the slope changes. For description purpose, the paired values at the inflection point are referred to as 'inflection biomass' and 'inflection depth', respectively. If the fertile-soil depth is smaller than the 'inflection depth', the ecosystem tends to 'bounce' back to the low stable equilibrium state; otherwise, it would probably be reinstated to the high stable equilibrium state.

Figure 2.15 shows the possible state shifts of a semi-arid steppe grassland located in northeast China. For a grazing intensity of 1.0 kg m^{-2} $year^{-1}$, the grassland would incur an irreversible degradation once it has a biomass of below 0.1 kg m^{-2} and/or a fertile-soil depth of less than 5.0 cm. This is the optimal grazing intensity found by

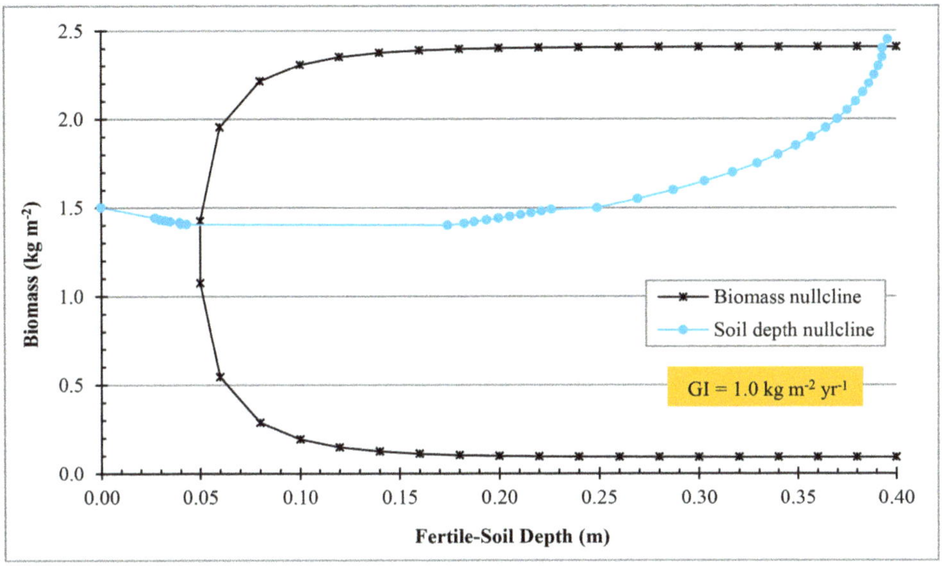

(a)

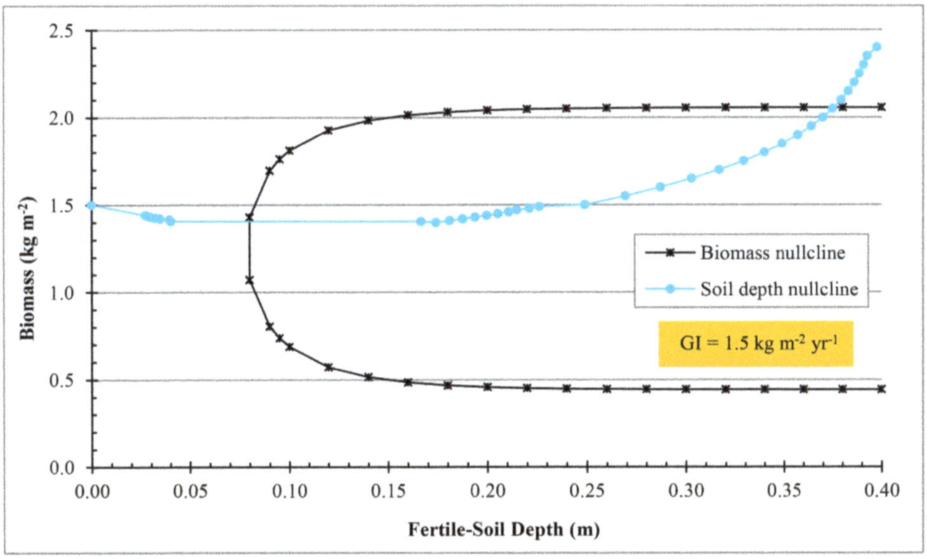

(b)

Figure 2.15. Predicted shifts of the semi-arid 5365 km^2 Balagaer River watershed (44° 00′ to 44° 15′ N, 117° 40′ to 117° 48′ E), located in the northeast Inner Mongolian Autonomous Region of China. Shifts for a livestock grazing intensity, in kg m^{-2} $year^{-1}$, of (a) GI = 1.0 and (b) GI = 1.5.

Wang *et al* (2021). As expected, for a higher grazing intensity of 1.5 kg m^{-2} $year^{-1}$, the degradation thresholds will be increased to 0.5 kg m^{-2} and 8.0 cm for biomass and fertile-soil depth, respectively. On average, these thresholds can be marginally met in winter and spring, and certainly satisfied in summer and fall (biomass in kg m^{-2} = 0.00339NPP—0.1 in figure 2.13) (Luo *et al* 2014, Wang *et al* 2014b). However, some parts of the grassland are incurring a vegetation removal rate of up to 75% by husbandry animals (Wang *et al* 2016) and losing topsoil at a rate of above 0.07 mm $year^{-1}$ due to wind and water erosion (Wang *et al* 2014a), likely triggering irreversible degradations if the ongoing land management practices are not adjusted appropriately because climate change will decrease soil moisture, increase rainfall intensity, and thus intensify topsoil loss (Wang *et al* 2013). It is needed to maintain a fertile-soil depth of 20 cm or thicker for the grassland to 'bounce' back to its high stable equilibrium state after an enforced sway.

2.8 Summary and discussion

A semi-arid ecosystem can have one of the five stable equilibrium states, namely sand dunes, desert, semi-desert, low, and high (figure 2.14). However, the ecosystem usually exists at an intermediate equilibrium state that falls between two of the five stable equilibrium states because of changes in the biotic and abiotic drivers. Such changes affect water–heat–carbon cycles, which in turn influence soil respiration as well as plant assimilation and respiration. Interacting with atmosphere dynamics, these phenological processes shape the land surface and vegetation community, leading to intermediate states. Once the drivers improve, e.g. precipitation increases and grazing intensity reduces, the ecosystem will start evolving to the better (i.e. upper-bound) stable equilibrium state; otherwise, it would be deteriorating toward the worse (i.e. lower-bound) stable equilibrium state. For instance, an intermediate equilibrium state between the low and high stable equilibrium states can be pushed toward either of these two stable equilibrium states. The evolution and deterioration processes may be gradual or rapid, depending on how and to what extent the drivers are changed. Nonetheless, the ecosystem can consist of patched land surfaces that are covered by loose topsoil, biocrusts, and vegetation, which have distinctly different ecohydrological characteristics. The borders between patches are dynamic and can be relocated as a result of changing drivers. The land surface patches can be mutually beneficial or advantageous, so the occupation area of a patch may be expanding and shrinking interchangeably. For instance, a biocrust patch supplies water and nutrients to its adjacent vegetation patches, prompting vegetation growth to conquer parts of the biocrust patch. Conversely, the expanding vegetation patches tend to deplete the available water and nutrients, releasing parts of vegetation patches back to the biocrust patch. In practice, semi-arid ecosystems need to be well conceptualized for determining the thresholds of the drivers that trigger an unwanted shift, tracing intermediate states, and developing sustainable and resilient measures.

References

Adessi A, de Carvalho C, de Philippis R, Branquinho C and da Silva J M 2018 Microbial extracellular polymeric substances improve water retention in dryland biological soil crusts *Soil Biol. Biochem.* **116** 67–9

Becerra-Absalón I, Muñoz-Martín M Á, Montejano G and Mateo P 2019 Differences in the cyanobacterial community composition of biocrusts from the drylands of central Mexico: are there endemic species? *Front. Microbiol.* **10** 1–21

Belnap J 2006 The potential roles of biological soil crusts in dryland hydrologic cycles *Hydrol. Process.* **20** 3159–78

Belnap J, Phillips S L, Witwicki D L and Miller M E 2008 Visually assessing the level of development and soil surface stability of cyanobacterially dominated biological soil crusts *J. Arid. Environ.* **72** 1257–64

Belnap J, Wilcox B P, van Scoyoc V and Philips S L 2013 Successional stage of biological soil crusts: an accurate indicator of ecohydrological condition *Ecohydrology* **6** 474–82

Bu C, Wu S, Han F, Yang Y and Meng J 2015 The combined effects of moss-dominated biocrusts and vegetation on erosion and soil moisture and implications for disturbance on the Loess Plateau, China *PLosS ONE* **10** e0127394

Bu C, Wu S and Yang K 2014 Effects of physical soil crusts on infiltration and splash erosion in three typical Chinese soils *Int. J. Sediment Res.* **29** 491–501

Cantón Y, Román J R, Chamizo S, Rodríguez-Caballero E and Moro M J 2014 Dynamics of organic carbon losses by water erosion after biocrust removal *J. Hydrol. Hydromech.* **62** 258–68

Cary J W and Hayden C W 1973 An index for soil pore size distribution *Geoderma* **9** 249–56

Chamizo S, Cantón Y, Rodríguez-Caballero E and Domingo F 2016 Biocrusts positively affect the soil water balance in semiarid ecosystems *Ecohydrology* **9** 1208–21

Chamizo S, Mugnai G, Rossi F, Certini G and de Philippis R 2018 Cyanobacteria inoculation improves soil stability and fertility on different textured soils: gaining insights for applicability in soil restoration *Front. Environ. Sci.* **6** 49

Chamizo S, Rodríguez-Caballero E, Román J R and Cantón Y 2017 Effects of biocrust on soil erosion and organic carbon losses under natural rainfall *Catena* **148** 117–25

Charbeneau R J 2006 *Groundwater Hydraulics and Pollutant Transport* (Long Grove, IL: Waveland)

de la Piedra S C 2012 The role of physical and biological soil crusts on the water balance in semiarid ecosystems *Doctoral thesis* University of Almería

Deblauwe V, Barbier N, Couteron P, Lejeune O and Bogaert J 2008 The global biogeography of semi-arid periodic vegetation patterns *Global Ecol. Biogeogr.* **17** 715–23

Duan L, Liu T, Wang X, Wang G, Ma L and Luo Y 2011 Spatio-temporal variations in soil moisture and physicochemical properties of a typical semiarid sand-meadow-desert landscape as influenced by land use *Hydrol. Earth Syst. Sci.* **15** 1865–77

Dunkerley D L 2002 Infiltration rates and soil moisture in a groved mulga community near Alice Springs, arid central Australia: evidence for complex internal rainwater redistribution in a runoff–runon landscape *J. Arid. Environ.* **51** 199–219

Faist A M, Herrick J E, Belnap J, van Zee J W and Barger N N 2017 Biological soil crust and disturbance controls on surface hydrology in a semi-arid ecosystem *Ecosphere* **8** 1–24

Fan Y, Miguez-Macho G, Jobbágy E G, Jackson R B and Otero-Casal C 2017 Hydrologic regulation of plant rooting depth *Proc. Natl Acad. Sci. USA* **114** 10572–7

Farouki O T 1981 *CRREL Monograph 81-1: Thermal Properties of Soils* (Hanover, NH: United States Army Corps of Engineers)

Fick S E, Barger N N and Duniway M C 2019 Hydrological function of rapidly induced biocrusts *Ecohydrology* **12** 1–38

Gallart F, Solé A, Puigdefàbregas J and Lázaro R 2002 Badland systems in the Mediterranean *Dryland Rivers: Hydrology and Geomorphology of Semi-Arid Channels* ed L J Bull and M J Kirkby (Hoboken, NJ: Wiley) pp 299–326

Gao L, Bowker M A, Xu M, Sun H, Tuo D and Zhao Y 2017a Biological soil crusts decrease erodibility by modifying inherent soil properties on the Loess Plateau, China *Soil Biol. Biochem.* **105** 49–58

Garcia-Franco N H, Hübner E R and Wiesmeier M 2018 Climate-smart soil management in semiarid regions *Soil Management and Climate Change: Effects on Organic Carbon, Nitrogen Dynamics, and Greenhouse Gas Emissions* ed M Ángeles-Muñoz and R Zornoza (Waltham, MA: Academic) pp 349–68

Gentine P, Massmann A, Lintner B R, Alemohammad S H, Fu R, Green J K, Kennedy D and de Arellano J V 2019 Land–atmosphere interactions in the tropics—a review *Hydrol. Earth Syst. Sci.* **23** 4171–97

Goutorbe J P *et al* 1993 A large scale study of land–atmosphere interactions in the semiarid tropics (HAPEX-Sahel) *Exchange Processes at the Land Surface for a Range of Space and Time Scales* (Yokohama: IAHS)

Hack-ten Broeke M J D, Kroes J G, Bartholomeus R P, van Dam J C, de Wilt A J W, Supit I, Walvoort D J J, van Bakel P J T and Ruijtenberg R 2016 Quantification of the impact of hydrology on agricultural production as a result of too dry, too wet or too saline conditions *Soil* **2** 391–402

Hartnett D C and Wilson W T 1999 Mycorrhizae influence plant community structure and diversity in tallgrass prairie *Ecology* **80** 1187–95

Janda J M and Abbott S L 2007 16 rRNA gene sequencing for bacterial identification in the diagnostic laboratory: pluses, perils, and pitfalls *J. Clin. Microbiol.* **45** 2761–4

Janowiak M *et al* 2017 *Considering Forest and Grassland Carbon in Land Management* (Washington DC: United States Department of Agriculture (USDA) Forest Service (FS))

Karssenberg D, Bierkens M F P and Rietkerk M 2017 Catastrophic shifts in semiarid vegetation–soil systems may unfold rapidly or slowly *Am. Nat.* **190** E146–55

Kroes J G, van Dam J C, Bartholomeus R P, Groenendijk P, Heinen M, Hendriks R F A, Mulder H M, Supit I and van Walsum P E V 2017 *SWAP Version 4: Theory Description and User Manual* (Wageningen: Wageningen Environmental Research)

Kurc S A and Small E E 2007 Soil moisture variations and ecosystem-scale fluxes of water and carbon in semiarid grassland and shrubland *Water Resour. Res.* **43** W06416

Kuske C R, Yeager C M, Johnson S and Ticknor L O 2012 Response and resillence of soil biocrust bacterial communities to chronic physical disturbance in arid shrublands *ISME J.* **6** 886–97

Li H, Li R, Rossi F, Li D, de Philippis R, Hu C and Liu Y 2016a Differentiation of microbial activity and functional diversity between various biocrust elements in a heterogeneous community *Catena* **147** 138–45

Ludwig J A, Wilcox B P, Breshears D D, Tongway D J and Imeson A C 2005 Vegetation patches and runoff-erosion as interacting ecohydrological processes in semiarid landscapes *Ecology* **86** 288–97

Luo Y, Wang X, Li F, Gao R, Duan L and Liu T 2014 Responses of grass production to precipitation in a mid-latitude typical steppe watershed *Trans. ASABE* **57** 1595–610

Lyons T J, Schwerdtfeger P, Hacker J M, Foster I J, Smith R C G and Huang X 1993 Land–atmosphere interaction in a semiarid region: the bunny fence experiment *Bull. Am. Meterol. Soc.* **74** 1327–34

Mager D M and Thomas A D 2011 Extracellular polysaccharides from cyanobacterial soil crusts: a review of their role in dryland processes *J. Arid. Environ.* **75** 91–7

Means T 2019 Rain, snow, sleet, and other types of precipitation https://thoughtco.com/precipitation-types-3444529?print

Monger H C and Khresat S A 2005 Tropical soils: arid and semiarid *Encyclopedia of Soils in the Environment* ed D Hillel, J H Hatfield and D S Powlson *et al* (Waltham, MA: Elsevier/Academic)

Navas Romero A L, Herrera Moratta M A, Vento B, Rodriguez R A and Martinez Carretero E E 2019 Variations in the coverage of biological soil crusts along a gradient of aridity in the center-west of Argentina, bioRxiv: https://biorxiv.org/content/10.1101/725986v1

Nevins C J, Strauss S L and Inglett P W 2020 Biological soil crusts enhance moisture and nutrients in the upper rooting zone of sandy soil agroecosystems *J. Soil Sci. Plant Nutr.* **183** 615–26

Nimmo J R 2004 Porosity and pore size distribution *Encyclopedia of Soils in the Environment* ed D Hillel vol 3 (London: Elsevier) pp 295–303

Rezapour S and Samadi A 2011 Soil quality response to long-term wastewater irrigation in Inceptisols from a semi-arid environment *Nutr. Cycling Agroecosyst.* **91** 269–80

Richter C, Rejmánek M, Miller J E D, Welch K R, Weeks J M and Safford H 2019 The species diversity × fire severity relationship is hump-shaped in semiarid yellow pine and mixed conifer forests *Ecosphere* **10** e02882

Rietkerk M, Boerlijst M C, van Langevelde F, HilleRisLambers R, va de Koppel J, Kumar L, Prins H H and Roos A M 2002 Self-organization of vegetation in arid ecosystems *Am. Nat.* **160** 524–30

Román J R, Rodríguez-Caballero E, Rodríguez-Lozano B, Roncero-Ramos B, Chamizo S, Águila-Carricondo P and Cantón Y 2019 Spectral response analysis: an indirect and non-destructive methodology for the chlorophyll quantification of biocrusts *Remote Sens.* **11** 1350

Sakai M, Toride N and Šimůnek J 2009 Water and vapor movement with condensation and evaporation in a sandy column *Soil Sci. Soc. Am. J.* **73** 707–17

Shannon C E 1948 A mathematical theory of communication *Bell Syst. Tech. J.* **27** 479–523

Singh J S, Milchunas D G and Lauenroth W K 1998 Soil water dynamics and vegetation patterns in a semiarid grassland *Plant Ecol.* **134** 77–89

Steven B, Gallegos-Graves L V, Yeager C, Belnap J and Kuske C R 2014 Common and distinguishing features of the bacterial and fungal communities in biological soil crusts and shrub root zone soils *Soil Biol. Biochem.* **69** 302–12

Tang G, Arnone J A III, Verburg P S J, Jason R L and Sun L 2015 Trends and climatic sensitivities of vegetation phenology in semiarid and arid ecosystems in the US Great Basin during 1982–2011 *Biogeosciences* **12** 6985–97

Todd D K and Mays L W 2005 *Groundwater Hydrology* (Hoboken, NJ: Wiley)

USDA-NRCS 2004 *National Engineering Handbook* (Washington, DC: US Department of Agriculture Natural Resources Conservation Service)

USDA-NRCS 2014 *Keys to Soil Taxonomy* (Washington, DC: United States Department of Agriculture (USDA) Natural Resources Conservation Service (NRCS))

USDA 1951 *Soil Survey Manual: USDA Agricultural Handbook No. 18* (Washington, DC: US Government Printing Office)

Wang X 2015 Vapor flow resistance of dry soil layer to soil water evaporation in arid environment: an overview *Water* **7** 4552–74

Wang X 2021 *Water Resources and Hydraulics* (New York: Cambridge University Press)

Wang X, Gao R and Yang X 2021 Responses of soil moisture to climate variability and livestock grazing in a semiarid Eurasian steppe *Sci. Total Environ.* **781** 146705

Wang X, Li F, Gao R, Luo Y and Liu T 2014b Predicted NPP spatiotemporal variations in a semiarid steppe watershed for historical and trending climates *J. Arid. Environ.* **104** 67–79

Wang X, Liu T, Li F, Gao R, Yang X, Duan L, Luo Y and Li R 2014a Simulated soil erosion from a semiarid typical steppe watershed using an integrated aeolian and fluvial prediction model *Hydrol. Process.* **28** 325–40

Wang X, Liu T and Yang W 2012 Development of a robust runoff-prediction model by fusing the Rational equation and a modified SCS-CN method *Hydrol. Sci. J.* **57** 1118–40

Wang X, Luo Y, Liu T, Gao R, Li F and Duan L 2014c Estimated grazing removal rate of grasses in a semiarid Eurasian steppe watershed as influenced by climate *18th CIGR World Congress, Session I Land and Water Engineering, 2nd Inter-Regional Conference on Land and Water Challenges Ecohydrological Process and Pasture Management Issues (Hohhot, China)*

Wang X, Pedram S, Liu T, Gao R, Li F and Luo Y 2016 Estimated grass grazing removal rate in a semiarid Eurasian steppe watershed as influenced by climate *Water* **8** 339

Wang X, Yang X, Liu T, Li F, Gao R, Duan L and Luo Y 2013 Trend and extreme occurrence of precipitation in a mid-latitude Eurasian steppe watershed at various time scales *Hydrol. Process.* **28** 5547–60

Wikipedia 2021 Köppen climate classification https://en.wikipedia.org/wiki/K%C3%B6ppen_climate_classification

Xiao B, Zhao Y, Wang Q and Li C 2015 Development of artificial moss-dominated biological soil crusts and their effects on runoff and soil water content in a semi-arid environment *J. Arid. Environ.* **117** 75–83

Zhao Y and Xu M 2013 Runoff and soil loss from revegetated grasslands in the hilly loess plateau region, China: influence of biocrust patches and plant canopies *J. Hydrol. Eng.* **18** 387–93

Chapter 3

Conceptualizations

This chapter presents conceptualizations of the semi-arid ecosystem processes discussed in chapter 2. For each of the processes, it elucidates reasonable assumptions, algorithms, and governing equations and solutions. Moreover, the variables that connect two or more processes are analyzed to scrutinize the water–soil–vegetation interactions.

3.1 Approximation of semi-arid ecosystems

A semi-arid ecosystem can be approximated using lumped or process-based approaches. Lumped approaches treat the ecosystem as a 'black box' with biomass as the output (i.e. dependent) variable and fertile-soil depth as the input (i.e. independent) variable. The effects of climatic factors are reflected by soil formation and topsoil erosion (Wakatsuki and Rasyidin 1992), while the impacts of livestock production are represented by grazing pressure (i.e. biomass removal per unit area per year). Herein, the fertile-soil depth refers to the layer of soils that have sufficient holding capacities of water and nutrients and are favorable for plant growth. Although lumped approaches may have different formulizations and complexities, they share three common basic assumptions that (i) biomass is positively related to fertile-soil depth and plant growth rate, (ii) biomass is negatively related to grazing pressure, and (iii) fertile-soil depth is positively related to soil formation rate but negatively related to biomass and soil erosion rate. In comparison with others, the differential equations presented by Karssenberg *et al* (2017) are well structured for ecosystem states and executed easily as a lumped approach. The equations have typical parameters listed in table 3.1 and can be expressed as

Table 3.1. Typical values of the parameters in equations (3.1) to (3.6).[1]

Variable	Parameter	Definition	Value
Biomass	i [–]	Reduction factor (negative) at zero soil depth	–0.7
	d [m]	Soil depth range parameter	0.04
	r [$year^{-1}$]	Plant growth rate	2.1
	c [kg m^{-2}]	Carrying capacity (i.e. biomass for an infinitely thick soil)	2.9
Fertile-soil depth	W_0 [m $year^{-1}$]	Soil formation rate of bedrock with no soil cover	10^{-5}–5.0×10^{-4}
	a [m^{-1}]	Soil formation exponent	4.0
	b [kg m^{-2}]	Reduction effect of erosion by biomass	0.28
	E_∞ [m $year^{-1}$]	Erosion rate at zero biomass and an infinite soil depth	0.021
	E_0 [m $year^{-1}$]	Erosion rate at zero soil depth	0.084
	κ [m]	Dependence of erosion on soil depth	0.05
	C [m $year^{-1}$]	Soil loss rate by creep	$\log_{10}(C) = 0.017S_0 - 1.88$; $C = 0.762\sin^2\theta$

[1]Compiled from a variety of sources (Kirkby 1964, Schumm 1964, Owens 1969, Wakatsuki and Rasyidin 1992, Karssenberg *et al* 2017).

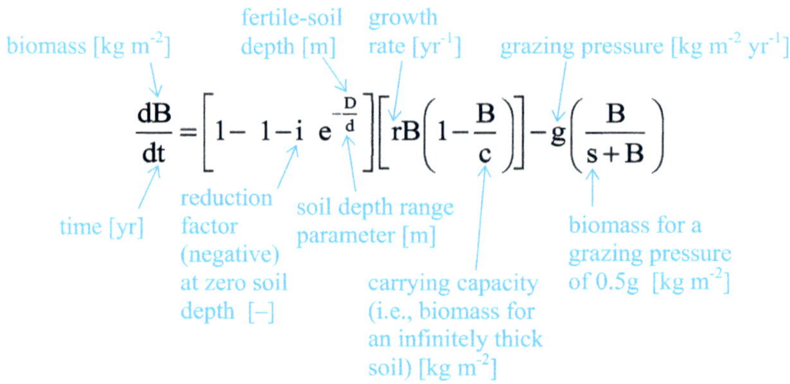

$$\frac{dB}{dt} = \left[1 - (1-i)\,e^{-\frac{D}{d}}\right]\left[rB\left(1-\frac{B}{c}\right)\right] - g\left(\frac{B}{s+B}\right) \tag{3.1}$$

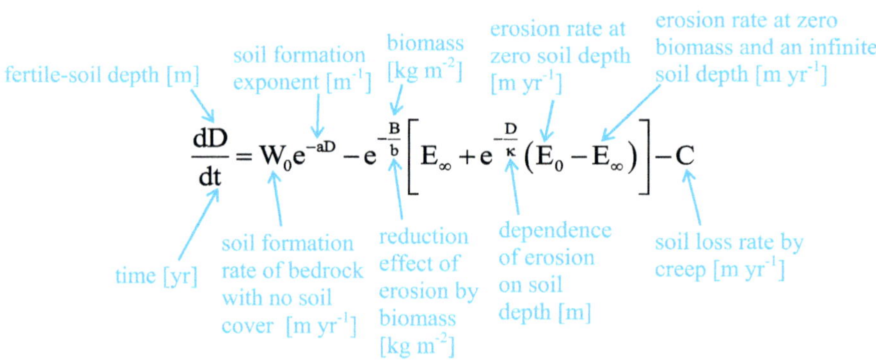

$$\frac{dD}{dt} = W_0 e^{-aD} - e^{-\frac{B}{b}}\left[E_\infty + e^{-\frac{D}{\kappa}}\left(E_0 - E_\infty\right)\right] - C \tag{3.2}$$

When the ecosystem reaches a steady or equilibrium state, $\frac{dB}{dt} = 0$, leading to the biomass nullcline, and $\frac{dD}{dt} = 0$, leading to the fertile-soil depth nullcline. As illustrated in figure 2.14, the stable equilibrium states of the ecosystem occur at the intersections between these two nullclines. The ecosystem will become catastrophic and/or incur an irreversible degradation if these two nullclines do not intersect.

The biomass nullcline is defined as

$$B' = \frac{(c-s) \pm \sqrt{(c-s)^2 + \frac{4c}{r}\left[rs - \frac{g}{1-(1-i)e^{-\frac{D'}{d}}}\right]}}{2} \tag{3.3}$$

steady-state biomass [kg m^{-2}] → B'
carrying capacity (i.e., biomass for an infinitely thick soil) [kg m^{-2}] → c
biomass for a grazing pressure of 0.5g [kg m^{-2}] → s
grazing pressure [kg m^{-2} yr^{-1}] → g
steady-state fertile-soil depth [m] → D'
soil depth range parameter [m] → d
growth rate [yr^{-1}] → r
reduction factor (negative) at zero soil depth [–] → i

The fertile-soil depth nullcline is defined as

$$W_0 e^{-aD'} - e^{-\frac{B'}{b}}\left[E_\infty + e^{-\frac{D'}{\kappa}}(E_0 - E_\infty)\right] - C = 0 \tag{3.4}$$

soil formation rate of bedrock with no soil cover [m yr^{-1}] → W_0
steady-state fertile-soil depth [m] → D'
soil formation exponent [m^{-1}] → a
steady-state biomass [kg m^{-2}] → B'
reduction effect of erosion by biomass [kg m^{-2}] → b
erosion rate at zero biomass and an infinite soil depth [m yr^{-1}] → E_∞
erosion rate at zero soil depth [m yr^{-1}] → E_0
dependence of erosion on soil depth [m] → κ
soil loss rate by creep [m yr^{-1}] → C

Equations (3.1) and (3.2) can be solved simultaneously using the backward finite-difference scheme expressed as

$$B_t = B_{t-1} + \left[1-(1-i)e^{-\frac{D_t}{d}}\right]\left[rB_{t-1}\left(1-\frac{B_{t-1}}{c}\right)\right] - g_t\left(\frac{B_{t-1}}{s_t + B_{t-1}}\right) \tag{3.5}$$

biomass in current year [kg m^{-2}] → B_t
biomass in previous year [kg m^{-2}] → B_{t-1}
fertile-soil depth in current year [m] → D_t
growth rate [yr^{-1}] → r
grazing pressure in current year [kg m^{-2} yr^{-1}] → g_t
reduction factor (negative) at zero soil depth [–] → i
soil depth range parameter [m] → d
carrying capacity (i.e., biomass for an infinitely thick soil) [kg m^{-2}] → c
biomass for a grazing pressure of 0.5g_t [kg m^{-2}] → s_t

$$D_t = D_{t-1} + W_0 e^{-aD_{t-1}} - e^{-\frac{B_{t-1}}{b}} \left[E_\infty + e^{-\frac{D_{t-1}}{\kappa}} \left(E_0 - E_\infty \right) \right] - C \tag{3.6}$$

fertile-soil depth in current year[m]; soil formation exponent [m^{-1}]; biomass in previous year [kg m^{-2}]; erosion rate at zero soil depth [m yr^{-1}]; soil loss rate by creep [m yr^{-1}]

fertile-soil depth in previous year[m]; soil formation rate of bedrock with no soil cover [m yr^{-1}]; reduction effect of erosion by biomass [kg m^{-2}]; dependence of erosion on soil depth [m]; erosion rate at zero biomass and an infinite soil depth [m yr^{-1}]

In contrast, process-based approaches partition a semi-arid ecosystem into three subsystems (figure 3.1), namely ground surface, rootzone, and below rootzone. The ground-surface subsystem provides water ponding storage and interfaces with the atmosphere, whereas the rootzone subsystem shelters vegetation roots and supplies water and nutrients for vegetation. The below-rootzone subsystem is the unsaturated soils between the rootzone bottom and shallow water table. The atmosphere, three subsystems, and shallow aquifer are interrelated by heat–water–carbon fluxes. If the land surface is bared with loose topsoil, physical crusts, or biocrusts, the ecosystem interacts with atmosphere through the ground-surface subsystem; however, if the land surface is occupied by vegetation, the rootzone subsystem interacts directly

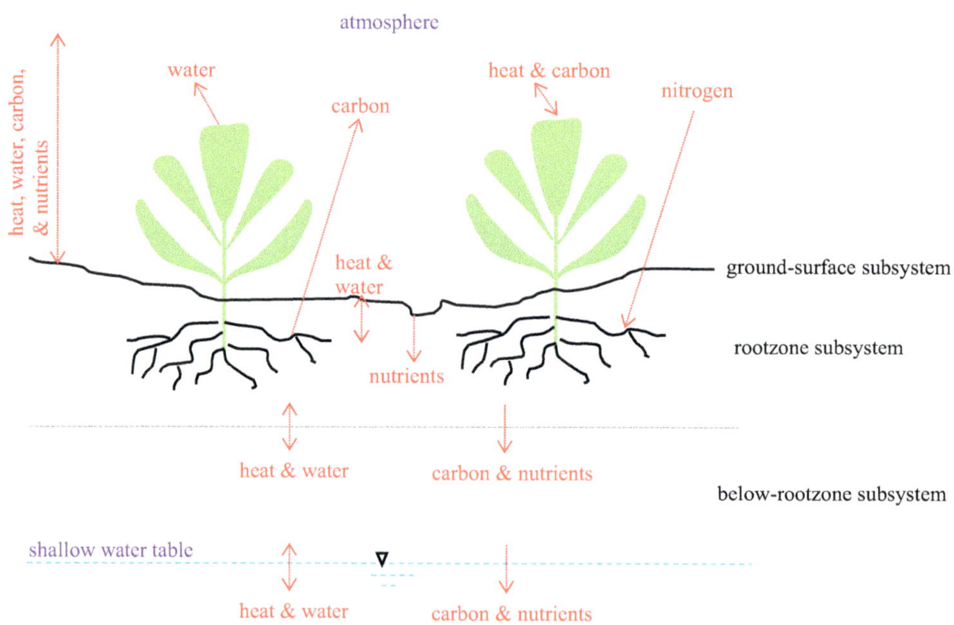

Figure 3.1. The three subsystems of a semi-arid ecosystem. Heat and mass fluxes linking the subsystems.

with the atmosphere, transpiring soil water and heat through leaf stomata, assimilating carbon dioxide (CO_2) and heat by photosynthesis, fixing nitrogen gases by roots, and releasing CO_2 by root respiration. The following sections present formulations for as well as linkages among the abiotic and biotic processes by subsystem.

3.2 Ground-surface processes

This section presents how to describe the processes of water, heat, and carbon in the ground-surface subsystem. The governing equations for these processes can be expressed as

$$\frac{dy_{gs}}{dt} = i_P - i_L - i_I - e_{gs} - r_o + r_I \tag{3.7}$$

water storage [m]; time [s]; rainfall intensity [m s^{-1}]; interception rate [m s^{-1}]; penetration rate [m s^{-1}]; evaporation rate [m s^{-1}]; efflux [m s^{-1}]; influx [m s^{-1}]

$$\frac{dH_{gs}}{dt} = (1-\alpha) r_{sl} - g_{gs\leftrightarrow r} - \lambda (i_L + e_{gs}) \tag{3.8}$$

heat storage [J m^{-2}]; time [s]; albedo [–] (Table 2.3); solar radiation [W m^{-2}]; heat exchange with rootzone [W m^{-2}]; positive to rootzone; latent heat of water evaporation [= 2.162 × 10^{9} J m^{-3}]; interception rate [m s^{-1}]; evaporation rate [m s^{-1}]

$$\frac{dC_{gs}}{dt} = c_p i_p + c_I r_I - \frac{C_{gs}}{y_{gs}} (r_o + i_I) \tag{3.9}$$

carbon storage [kg m^{-2}]; time [s]; rainfall carbon concentration [kg m^{-3}]; rainfall intensity [m s^{-1}]; runoff carbon concentration [kg m^{-3}]; influx [m s^{-1}]; water storage [m]; efflux [m s^{-1}]; penetration rate [m s^{-1}]

For a time step of Δt, equations (3.7)–(3.9) can be integrated to obtain the computational formulas expressed as

$$y_{gs,t} = y_{gs,t-\Delta t} + P - L - I - E_{gs} - R_o + R_I \tag{3.10}$$

water storage at current time [m]
water storage at previous time [m]
rainfall [m]
interception [m]
penetration [m]
evaporation [m]
effluent [m]
influent [m]

$$H_{gs,t} = H_{gs,t-\Delta t} + (1-\alpha)R_{sl} - G_{gs\leftrightarrow r} - \lambda(L + E_{gs}) \tag{3.11}$$

heat storage at current time [J m^{-2}]
heat storage at previous time [J m^{-2}]
albedo [–] (Table 2.3)
solar radiation [J m^{-2}]
heat exchange with rootzone [J m^{-2}]; positive to rootzone
latent heat of water evaporation [= 2.162 × 10^9 J m^{-3}]
interception [m]
evaporation [m]

$$C_{gs,t} = C_{gs,t-\Delta t} + c_p P + c_I R_I - \frac{C_{gs,t}}{y_{gs,t}}(R_o + I) \tag{3.12}$$

carbon storage at current time [kg m^{-2}]
carbon storage at previous time [kg m^{-2}]
rainfall carbon concentration [kg m^{-3}]
rainfall [m]
influx carbon concentration [kg m^{-3}]
influent [m]
carbon storage at previous time [kg m^{-2}]
water storage at previous time [m]
effluent [m]
penetration [m]

3.2.1 Interception

Interception refers to the process that vegetation canopy temporally stores a portion of rainfall. It depends on both rainfall characteristics (e.g. intensity and duration) and vegetation features (e.g. leaf area index or LAI). For a semi-arid ecosystem of interest, the interception of bare surfaces is zero, whereas the interception of vegetated surfaces can be estimated as the multiplication of an empirical equation developed by Corbett and Crouse (1968) and the relative leaf area index (i.e. ratio of LAI to LAI_{opt}). LAI_{opt} is the optimum (i.e. rational maximum) of LAI and can be determined as suggested by Anten *et al* (1995). The resulting formula is expressed as rainfall [mm] leaf area index [–]

interception [mm]

$$L = (0.026P + 0.9398)\frac{LAI}{LAI_{opt}} \quad (R^2 = 0.997) \tag{3.13}$$

optimal leaf area index [–]

3.2.2 Penetration

Penetration refers to the process that water seeps through the ground surface downward into the soils underneath. It can be determined using three families of methods (Charbeneau 2006, Wang 2021). The first family of methods, such as the US Department of Agriculture (USDA) Soil Conservation Service (SCS) curve number (CN) method and its variants, do not consider how water moves in the soils underneath, whereas the second family of methods, such as the Green–Ampt (G–A) method, assumes the soils underneath to be one single vertically homogeneous profile and computes penetration based on the thickness of the top saturated soils. The third family of methods, such as the Richards equation, determine penetration as a function of water movement in the soils underneath without a homogeneous assumption. The first and second families of methods are inappropriate for land surfaces with physical and biological crusts because these methods cannot explicitly consider the effects of crusts on penetration. In contrast, the third family of methods is good for all types of land surfaces because these methods treat each soil layer separately. The loose topsoil and crusts, which have distinctly different properties from the soils underneath, can be characterized as a separate layer. This subsection presents the methods that are applicable for semi-arid ecosystems.

3.2.2.1 The SCS-CN method and its improved variants

For this family of methods, the penetration is computed as

rainfall [mm] direct runoff [mm]

penetration [mm]

$$I = P - L - Q \tag{3.14}$$

interception [mm]

The SCS-CN method (USDA-NRCS 2004) computes direct runoff as

rainfall [mm] initial abstraction [mm]

direct runoff [mm]

$$Q = \begin{cases} \dfrac{(P - I_a)^2}{(P - I_a) + S} & P > I_a \\ 0 & P \le I_a \end{cases} \tag{3.15}$$

potential maximum retention after runoff begins [mm]

$$S = 25.4\left(\frac{1000}{CN} - 10\right) \tag{3.16}$$

potential maximum retention after runoff begins [mm] → S

curve number [–] → CN

$$I_a = 0.2S \tag{3.17}$$

initial abstraction [mm] → I_a

potential maximum retention after runoff begins [mm] → S

The initial abstraction, I_a, includes canopy storage or interception, soil penetration during early parts of the storm, and surface depression storage. For a rainfall event, the total maximum possible loss is the summation of I_a and S. The value of CN depends on land use and land cover (LULC), hydrological soil groups (HSGs), and antecedent moisture conditions (AMCs) or antecedent runoff conditions (ARCs) (Wang *et al* 2012, Wang 2021). In terms of water transmission rates, USDA-NRCS (2004) classifies soils into four HSGs, signified as A, B, C, and D. HSG A soils are highly permeable with a saturated hydraulic conductivity of $K_s > 3.5$ m d^{-1}; HSG B soils are moderately permeable with $K_s = 0.8$–3.5 m d^{-1}; HSG C soils are fairly permeable with $K_s = 0.04$–0.8 m d^{-1}; and HSG D soils are poorly permeable with $K_s < 0.04$ m d^{-1}. In addition, the SCS defines three ARCs: ARC I is for dry conditions that correspond to the wilting point, whereas ARC III is for wet conditions that correspond to field capacity. ARC II is for average conditions that correspond to a soil moisture value between the wilting point and field capacity. In principle, ARC II can be interpreted as the soil moisture condition associated with the median curve number CN_{II}, whereas ARC I and ARC III are defined as the soil moisture conditions associated with the curve numbers of 10% and 90% exceedance frequencies, respectively (Hjelmfelt 1991). The curve numbers for ARC I (CN_I) and for ARC III (CN_{III}) can be estimated based on CN_{II} (USDA-NRCS 2004). The potential maximum retention for ARC I, S_I, can be computed by substituting CN_I for CN in equation (3.16), while the potential maximum retention for ARC III, S_{III}, can be computed by substituting CN_{III} for CN. The potential maximum retention for ARC II, S_{II}, can be computed by substituting CN_{II} for CN. Table 3.2 gives CN_{II} for rural environments.

The curve number for ARC I can be computed as (USDA-NRCS 2004)

$$CN_I = CN_{II} - \frac{20(100 - CN_{II})}{100 - CN_{II} + e^{[2.533 - 0.0636(100 - CN_{II})]}} \tag{3.18}$$

curve number for antecedent runoff condition I [–] → CN_I

curve number for antecedent runoff condition II [–] → CN_{II}

Table 3.2. Values of CN_{II}, curve number for antecedent runoff condition II.[1]

	Hydrological soil group (HSG)			
Land cover and hydrological condition	A	B	C	D
Pasture, grassland, or range—continuous forage for grazing				
Poor (<50% ground cover or heavily grazed with no mulch)	68	79	86	89
Fair (50%–75% ground cover and not heavily grazed)	49	69	79	84
Good (>75% ground cover and lightly or occasionally grazed)	39	61	74	80
Meadow—continuous grass, protected from grazing and generally mowed for hay	30	58	71	78
Brush–brush weed–grass mixture with brush the major element				
Poor (<50% ground cover)	48	67	77	83
Fair (50%–75% ground cover)	35	56	70	77
Good (>75% ground cover)	30	48	65	73
Woods–grass combination (orchard or tree farm) (50% woods and 50% grass or pasture)				
Poor (<50% ground cover)	57	73	82	86
Fair (50%–75% ground cover)	43	65	76	82
Good (>75% ground cover)	32	58	72	79
Woods				
Poor (forest litter, small trees, and brush are destroyed by heavy grazing or regular burning)	45	66	77	83
Fair (woods are grazed but not burned, and some forest litter covers the soil)	36	60	73	79
Good (woods are protected from grazing, and litter and brush adequately cover the soil)	30	55	70	77
Farmsteads—buildings, lanes, driveways, and surrounding lots	59	74	82	86

[1]Source: USDA-NRCS (2004), public domain.

The curve number for ARC III can be computed as (USDA-NRCS 2004)

$$CN_{III} = CN_{II} e^{[0.00673(100 - CN_{II})]} \tag{3.19}$$

curve number for antecedent runoff condition III [–]

curve number for antecedent runoff condition II [–]

To overcome two intrinsic issues of the SCS-CN method, namely unreasonable sudden jump and baseless assumption of initial abstraction (equation (3.17)), the author (Wang *et al* 2008, Wang *et al* 2012) proposed two improved variants, namely MoCN and MoRE, and showed that the MoRE method performs better than the MoCN method and both variants have a better performance than the SCS-CN method.

The MoCN method computes direct runoff using equation (3.15), with S and I_a computed as

potential maximum retention after runoff begins [mm]; potential maximum retention for ARC I [mm]; soil moisture [–]; wilting point [–]; exponent constant [–]; saturated soil moisture [–]

$$S = S_I \left[1 - \left(\frac{\theta - \theta_\psi}{\theta_{sat} - \theta_\psi} \right)^b \right] \tag{3.20}$$

exponent constant [–]; potential maximum retention for ARC III [mm]; potential maximum retention for ARC I [mm]; field capacity [–]; saturated soil moisture [–]

$$b = \frac{\ln\left(1 - \frac{S_{III}}{S_I}\right)}{\ln\left(\frac{\theta_{fc}}{\theta_{sat}}\right)} \tag{3.21}$$

soil moisture [–]; wilting point [–]; field capacity [–]; antecedent 5-d rainfall [mm]

$$\theta = \begin{cases} \theta_\psi & \text{for } P_5 < 12.5 \text{ mm (dormant season) or } P_5 < 35.5 \text{ mm (growing season)} \\ \theta_\psi + \frac{\theta_{fc} - \theta_\psi}{15.5} \cdot (P_5 - 12.5) & \text{for } 12.5 \text{ mm} \le P_5 < 28.0 \text{ mm (dormant season)} \\ \theta_\psi + \frac{\theta_{fc} - \theta_\psi}{18.0} \cdot (P_5 - 35.5) & \text{for } 35.5 \text{ mm} \le P_5 < 53.5 \text{ mm (growing season)} \\ \theta_{fc} & \text{for } P_5 \ge 28.0 \text{ mm (dormant season) or } P_5 \ge 53.5 \text{ mm (growing season)} \end{cases} \tag{3.22}$$

initial abstraction [mm]; calibration constant [0.09 ~ 11.36]; potential maximum retention after runoff begins [mm]; calibration constant [0 ~ 2.82]; rainfall [mm]

$$I_a = \lambda_1 \cdot S \cdot \left(\frac{P}{P + S} \right)^\alpha \tag{3.23}$$

The MoRE method computes direct runoff as

$$
Q=\begin{cases}
P\left(10^{\left[-0.143\left[\log_{10}\left(\frac{S}{P}\right)\right]^{2}-0.598\log_{10}\left(\frac{S}{P}\right)-0.296\right]}\right) & \text{for } \frac{S}{P}\geq 1.0\\
P\left(10^{\left[-0.296\exp^{\left[1.837\cdot\log_{10}\left(\frac{S}{P}\right)\right]}\right]}\right) & \text{for } 0<\frac{S}{P}<1.0\\
P & \text{for } \frac{S}{P}=0
\end{cases}
\tag{3.24}
$$

(Annotations: Q — direct runoff [mm]; S — potential maximum retention after runoff begins [mm] (Eqs. 3.20 and 3.21); P — rainfall [mm])

3.2.2.2 The G–A method

As illustrated in figure 3.2, the G–A method assumes that the soil profile is homogeneous, the initial soil moisture is uniformly distributed throughout the

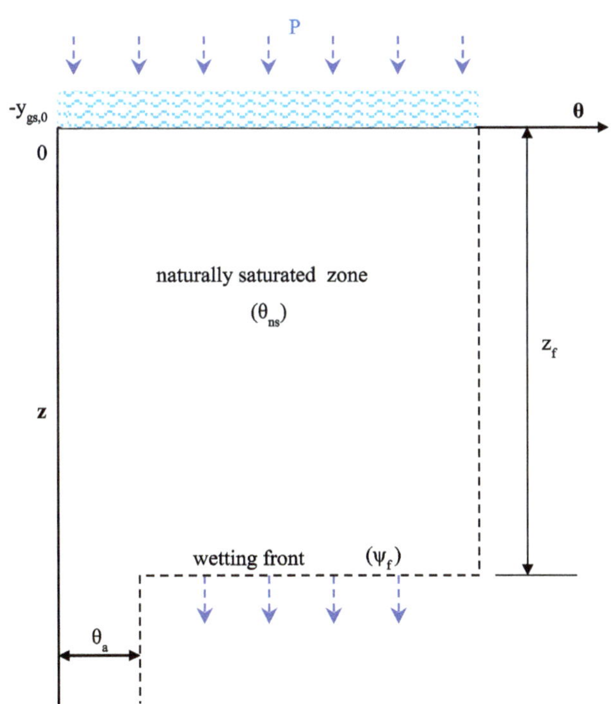

Figure 3.2. Soil moisture profile assumed by the G–A method. z [L]: depth from ground surface; θ [–]: soil moisture; θ_{ns} [–]: naturally saturated soil moisture; θ_a [–]: antecedent soil moisture; ψ_f [L]: suction head at wetting front; P [L T^{-1}]: rainfall intensity; and $y_{gs,0}$ [L]: water storage at beginning.

profile, the soils above the wetting front are naturally saturated, and the soils below the wetting front have the initial soil moisture. For semi-arid ecosystems, these assumptions may not be valid most of the time except for during heavy storms.

For a time period Δt, the penetration is computed as

interception [mm]; rainfall [mm]; penetration capacity during Δt [mm]; penetration [mm]; water storage at beginning of Δt [mm]

$$I = \min\left\{P - L + y_{gs,0}, F_{\Delta t}\right\} \tag{3.25}$$

As shown in appendix A, the penetration capacity can be estimated as

saturated hydraulic conductivity [mm s^{-1}]; saturated soil moisture [–]; suction head at wetting front [mm]; initial soil moisture of soil profile [–]; penetration capacity during Δt [mm]; time period [s]; water storage at beginning of Δt [m]; total penetration prior to Δt [mm]

$$F_{\Delta t} = \frac{K_s}{2}(\Delta t) + \left(|\psi_f| + y_{gs,0}\right)\left(0.95\theta_{sat} - \theta_a\right)\ln\left[1 + \frac{F_{\Delta t}}{F_0 + \left(|\psi_f| + y_{gs,0}\right)\left(0.95\theta_{sat} - \theta_a\right)}\right] \tag{3.26}$$

soil pore-size distribution index [–]; air-entry pressure head [mm]; suction head at wetting front [mm]

$$\psi_f = \frac{3\lambda + 2}{3\lambda + 1}\left(\frac{\psi_b}{2}\right) \tag{3.27}$$

3.2.2.3 Richards equation

The well-known Richards equation (Charbeneau 2006) is expressed as

unsaturated hydraulic conductivity [L T^{-1}]; soil moisture [–]; suction head [L]; time [T]; source (< 0) and sink (> 0) [T^{-1}]; soil depth (positive downward) [L]

$$\frac{\partial \theta}{\partial t} = S' - \frac{\partial K}{\partial z}\left(1 + \frac{\partial |\psi|}{\partial z}\right) - K\frac{\partial^2 |\psi|}{\partial z^2} \tag{3.28}$$

For a soil column of interest, evapotranspiration and penetration are the typical source and sink, respectively. Equation (3.28) has three variables, θ, ψ, and K, and thus additional two functions that relate these three variables are needed to have a unique solution. In practice, the retention curve (θ–ψ) can be defined using a modified van Genuchen (Schaap and van Genuchten 2006), which in turn is submitted into the permeability model developed by Mualem (1976) to derive the Mualem–van Genuchten formula for K.

The modified van Genuchen retention curve (Schaap and van Genuchten 2006) is defined as

$$\theta = \begin{cases} \theta_r + (\theta_{sat} - \theta_r)\left(1 + |\alpha\psi|^n\right)^{-m} & \text{for } |\psi| \geq |\psi_b| \\ \theta_{sat} & \text{for } |\psi| < |\psi_b| \end{cases} \tag{3.29}$$

θ: soil moisture [–]; θ_r: residual soil moisture [–]; θ_{sat}: saturated soil moisture [–]; α: soil hysteresis factor [L^{-1}]; n, m: exponents [–]; ψ: suction head [L]; ψ_b: air-entry pressure head [L]

$$\alpha = \frac{\xi}{|\psi_b|} \tag{3.30}$$

α: soil hysteresis factor [L^{-1}]; ξ: drying or wetting curve factor [–]; ψ_b: air-entry pressure head [L]

$$n = \lambda + 1 \tag{3.31}$$

n: exponent in Eq. (3.29) [–]; λ: soil pore-size distribution index [–]

$$m = 1 - \frac{1}{n} \tag{3.32}$$

m: exponent in Eq. (3.29) [–]; n: exponent in Eq. (3.29) [–]

The Mualem–van Genuchten formula (Wang *et al* 2021) is expressed as

$$K = \begin{cases} K_s \left[\dfrac{\left(1 + |\alpha\psi|^n\right)^m - |\alpha\psi|^{mn}}{\left(1 + |\alpha\psi_b|^n\right)^m - |\alpha\psi_b|^{mn}} \right]^2 \left(1 + |\alpha\psi|^n\right)^{-m(\beta+2)} \left(1 + |\alpha\psi_b|^n\right)^{2m} & \text{for } |\psi| \geq |\psi_b| \\ K_s & \text{for } |\psi| < |\psi_b| \end{cases} \tag{3.33}$$

K: unsaturated hydraulic conductivity [L T^{-1}]; K_s: saturated hydraulic conductivity [L T^{-1}]; n: exponent (Eq. 3.31) [–]; m: exponent (Eq. 3.32) [–]; β: pore connectivity parameter (0 ~ 1) [–]; α: soil hysteresis factor (Eq. 3.30) [L^{-1}]; ψ_b: air-entry pressure head [mm]

The reduction impact of frost soil temperature on K can be assumed to follow a linear function between two thresholds: one is the lowest soil temperature, T_1, below which soil water starts freezing, whereas another is the highest soil temperature, T_2, above which soil ice starts melting and below which all soil water except for residual water is frozen. That is, the hydraulic conductivity is equal to K at T_1 and zero at T_2. At a soil temperature T, the hydraulic conductivity can be estimated as

unsaturated hydraulic conductivity above soil temperature T_1 [$L\ T^{-1}$]

unsaturated hydraulic conductivity at soil temperature T [$L\ T^{-1}$]

$$K' = \begin{cases} K & \text{for } T \geq T_1 \\ K\dfrac{T - T_2}{T_1 - T_2} & \text{for } T_2 < T < T_1 \\ 0 & \text{for } T \leq T_2 \end{cases} \tag{3.34}$$

As shown in appendix B, substituting equation (3.29) into equation (3.28) and referencing the discretized soil profile in figure 3.3, yields the backward finite-difference scheme expressed as

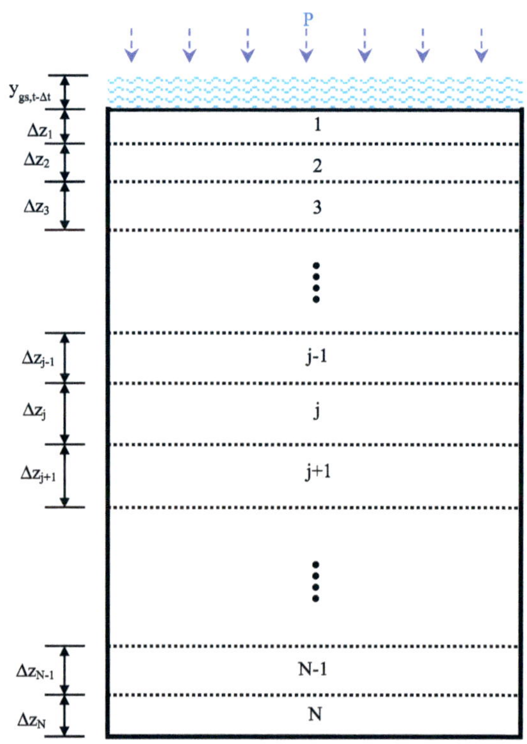

Figure 3.3. Discretization of a soil profile into N blocks. Δz_j [L]: thickness of block j; and $y_{gs,t-\Delta t}$ [L]: water storage at previous time.

$$
\begin{aligned}
|\psi|_{t,j} = |\psi|_{t-\Delta t,j} + \frac{\Delta t}{(\alpha mn)(\theta_r - \theta_{sat})} \frac{\left(1+|\alpha\psi|^{n}_{t-\Delta t,j}\right)^{m+1}}{|\alpha\psi|^{n-1}_{t-\Delta t,j}}
\Bigg[& S'_{t-\Delta t,j} - \frac{2\left(K_{t-\Delta t,j} - K_{t-\Delta t,j-1}\right)}{(\Delta z)_j + (\Delta z)_{j-1}} \left(1 + \frac{2\left(|\psi|_{t-\Delta t,j} - |\psi|_{t-\Delta t,j-1}\right)}{(\Delta z)_j + (\Delta z)_{j-1}}\right) \\
& - K_{t-\Delta t,j} \frac{\dfrac{2\left(|\psi|_{t-\Delta t,j+1} - |\psi|_{t-\Delta t,j}\right)}{(\Delta z)_{j+1} + (\Delta z)_j} - \dfrac{2\left(|\psi|_{t-\Delta t,j} - |\psi|_{t-\Delta t,j-1}\right)}{(\Delta z)_j + (\Delta z)_{j-1}}}{\dfrac{(\Delta z)_{j+1}}{2} + (\Delta z)_j + \dfrac{(\Delta z)_{j-1}}{2}} \Bigg]
\end{aligned}
\tag{3.35}
$$

- $|\psi|_{t,j}$: suction head at current time and block j [L]
- $|\psi|_{t-\Delta t,j}$: suction head at previous time and block j [L]
- Δt: computational time step [T]
- α: soil hysteresis factor (Eq. 3.30) [L^{-1}]
- n: exponent (Eq. 3.31) [–]
- m: exponent (Eq. 3.32) [–]
- θ_{sat}: saturated soil moisture [–]
- θ_r: residual soil moisture [–]
- $K_{t-\Delta t,j-1}$: unsaturated hydraulic conductivity at previous time and block j-1 [L T^{-1}]
- $S'_{t-\Delta t,j}$: source (< 0) and sink (> 0) at previous time and block j [T^{-1}]
- $|\psi|_{t-\Delta t,j+1}$: suction head at previous time and block j+1 [L]
- $|\psi|_{t-\Delta t,j-1}$: suction head at previous time and block j-1 [L]
- $K_{t-\Delta t,j}$: unsaturated hydraulic conductivity at previous time and block j [L T^{-1}]
- $(\Delta z)_{j+1}$: thickness of block j+1 [L]
- $(\Delta z)_j$: thickness of block j [L]
- $(\Delta z)_{j-1}$: thickness of block j-1 [L]

The upper boundary condition is defined as

$$
|\psi|_{t-\Delta t,0} = -\left(P - L + y_{gs,t-\Delta t}\right) \tag{3.36}
$$

- $|\psi|_{t-\Delta t,0}$: pressure head on ground surface [L]
- P: rainfall [L]
- L: interception [L]
- $y_{gs,t-\Delta t}$: water storage at beginning of Δt [L]

The lower boundary condition can be zero flux or constant pressure head, which can be specified as

$$
|\psi|_{t-\Delta t,N} = |\psi|_{t-\Delta t,N-1} \tag{3.37}
$$

- $|\psi|_{t-\Delta t,N}$: suction head at bottom block [L]
- $|\psi|_{t-\Delta t,N-1}$: suction head at second bottom block [L]

In equation (3.35) $K_{t\text{-}\Delta t,j\text{-}1}$ is computed by substituting $|\psi|_{t\text{-}\Delta t,j\text{-}1}$ for $|\psi|$ in equation (3.33), while $K_{t\text{-}\Delta t,j}$ is computed by substituting $|\psi|_{t\text{-}\Delta t,j}$ for $|\psi|$. In addition, $(\Delta z)_0$ and $K_{t-\Delta t,0}$ can be specified to consider effects of physical and biological crusts if such a crust is present.

3.2.3 Evaporation

The evaporation from time (t – Δt) to t is computed as

$$E_{gs,t} = \min\left\{P + y_{gs,t-\Delta t}, PET\right\} \tag{3.38}$$

evaporation [mm] → $E_{gs,t}$; rainfall [mm] → P; water storage at beginning of Δt [mm] → $y_{gs,t-\Delta t}$; potential evapotranspiration [mm] → PET

The potential evapotranspiration can be computed using the Penman–Monteith equation (Allen *et al* 1998), which is expressed as

$$PET = \frac{\Delta\left(R_{sl} - g_{gs\leftrightarrow r}\right) + \rho_a c_a \left(\frac{e_s - e_a}{r_a}\right)}{\lambda[\Delta + \gamma]} \tag{3.39}$$

potential evapotranspiration [mm] → PET; slope of vapor pressure curve [kPa °C^{-1}] → Δ; solar radiation [MW m^{-2}] → R_{sl}; heat exchange with rootzone [MW m^{-2}]; positive to rootzone → $g_{gs\leftrightarrow r}$; air density [kg m^{-3}] → ρ_a; air specific heat [= 10^{-3} MJ kg^{-1} °C^{-1}] → c_a; saturation vapor pressure [kPa] → e_s; actual vapor pressure [kPa] → e_a; aerodynamic resistance to vapor transport [s m^{-1}] → r_a; latent heat of water evaporation [= 2.257 MJ kg^{-1}] → λ; psychrometric constant [kPa °C^{-1}] → γ

$$\Delta = \frac{4098\left(0.6108 e^{\frac{17.27 T_a}{T_a + 237.3}}\right)}{\left(T_a + 237.3\right)^2} \tag{3.40}$$

slope of vapor pressure curve [kPa °C^{-1}] → Δ; mean air temperature [°C] → T_a

$$T_a = \frac{T_{min} + T_{max}}{2} \tag{3.41}$$

mean air temperature [°C] → T_a; minimum air temperature [°C] → T_{min}; maximum air temperature [°C] → T_{max}

$$g_{gs\leftrightarrow r} = c_s D_{rz} \frac{T_{a,t} - T_{a,t-\Delta t}}{\Delta t} \tag{3.42}$$

heat exchange with rootzone [MW m^{-2}]; positive to rootzone → $g_{gs\leftrightarrow r}$; soil heat capacity [MJ m^{-3} °C^{-1}] → c_s; rootzone depth [m] → D_{rz}; mean air temperature at current time [°C] → $T_{a,t}$; mean air temperature at previous time [°C] → $T_{a,t-\Delta t}$; time period [s] → Δt

atmosphere pressure [kPa]

air density [kg m^{-3}]

$$\rho_a = \frac{1000\,p_a}{R_a\left(T_a + 273.15\right)} \tag{3.43}$$

dry air gas constant [= 287.058 J kg^{-1} K^{-1}]

mean air temperature [°C]

mean air temperature [°C]

saturation vapor pressure [kPa]

$$e_s = 0.611\left(10^{\frac{7.5T_a}{T_a+237.3}}\right) \tag{3.44}$$

saturation vapor pressure [kPa]

air relative humidity [%]

actual vapor pressure [kPa]

$$e_a = e_s\left(\frac{RH}{100}\right) \tag{3.45}$$

atmosphere pressure [kPa]

psychrometric constant [kPa °C^{-1}]

$$\gamma = \frac{p_a}{1405.72} \tag{3.46}$$

zero-plane displacement height [m]

above-ground height of wind measurement [m]

above-ground height of humidity measurement [m]

aerodynamic resistance to vapor transport [s m^{-1}]

$$r_a = \frac{\ln\left(\frac{z_m - d_0}{z_{0m}}\right)\ln\left(\frac{z_h - d_0}{z_{0h}}\right)}{\kappa^2 u_{zm}} \tag{3.47}$$

roughness length governing momentum transfer [m]

roughness length governing heat and vapor transfer [m]

von Karman constant [=0.41]

wind speed at above-ground height z_m [m s^{-1}]

$$d_0 = \begin{cases} 0 & \text{bare surface} \\ 0.67 h_{plant} & \text{vegetated surface} \end{cases} \tag{3.48}$$

zero-plane displacement height [m] (d_0); plant height [m] (h_{plant})

$$z_{0m} = \begin{cases} h_{soil} & \text{bare surface} \\ 0.123 h_{plant} & \text{vegetated surface} \end{cases} \tag{3.49}$$

roughness length governing momentum transfer [m] (z_{0m}); soil surface roughness height [m] (h_{soil}); plant height [m] (h_{plant})

$$z_{0h} = \begin{cases} h_{soil} & \text{bare surface} \\ 0.0123 h_{plant} & \text{vegetated surface} \end{cases} \tag{3.50}$$

roughness length governing heat and vapor transfer [m] (z_{0h}); soil surface roughness height [m] (h_{soil}); plant height [m] (h_{plant})

3.2.4 Effluent

The ground-surface subsystem can be hypothesized as a ponding reservoir controlled by a weir (Kroes *et al* 2017). The effluent out of the subsystem is the overflow through the weir and thus can be computed as

$$R_0 = C_{weir}\left(y_{gs,t-\Delta t} - y_0\right)^{\varepsilon_{weir}} \tag{3.51}$$

effluent during Δt [m] (R_0); water storage at beginning of Δt [m] ($y_{gs,t-\Delta t}$); discharge exponent [–] (ε_{weir}); discharge coefficient [$m^{-\varepsilon weir}$] (C_{weir}); weir crest height [m] (y_0)

3.3 Rootzone processes

This section presents how to describe processes of water, heat, and carbon in the rootzone subsystem. The water movement is governed by the Richards equation (equation (3.28)).

The heat and carbon transport processes can be approximated by the one-dimensional convection equations (Philip 1957, Charbeneau 2006, Vanderborght *et al* 2017) expressed as

$$C_{sh}\frac{\partial T}{\partial t}=\frac{\partial}{\partial z}\left(\lambda_{sh}\frac{\partial T}{\partial z}\right) \tag{3.52}$$

temperature of soil [°C]
thermal conductivity of soil [J cm^{-1} °C^{-1} d^{-1}]
heat capacity of soil [J cm^{-3} °C^{-1}]
time [s]
soil depth (positive downward) [cm]

$$\frac{\partial C_{rz}}{\partial t}=S_c^{'}-\left(K\frac{\partial C_{rz}}{\partial z}+C_{rz}\frac{\partial K}{\partial z}\right)\left(1+\frac{\partial|\psi|}{\partial z}\right)-KC_{rz}\frac{\partial^2|\psi|}{\partial z^2} \tag{3.53}$$

carbon concentration in soil water [kg m^{-3}]
unsaturated hydraulic conductivity (Eqs. 3.33 and 3.34) [m s^{-1}]
suction head [m]
carbon source (< 0) and sink (> 0) per unit volume of soil water [kg m^{-3} s^{-1}]
soil depth (positive downward) [m]

For a time step of Δt and in reference to figure 3.2, equations (3.52) and (3.53) can be solved using the finite-difference schemes expressed as

$$T_{t,j}=T_{t-\Delta t,j}+\frac{\Delta t}{C_{sh,t-\Delta t,j}}\left[\frac{\lambda_{sh,t-\Delta t,j+1/2}\dfrac{2\left(T_{t-\Delta t,j+1}-T_{t-\Delta t,j}\right)}{(\Delta z)_{j+1}+(\Delta z)_j}-\lambda_{sh,t-\Delta t,j-1/2}\dfrac{2\left(T_{t-\Delta t,j}-T_{t-\Delta t,j-1}\right)}{(\Delta z)_j+(\Delta z)_{j-1}}}{\dfrac{(\Delta z)_{j+1}}{2}+(\Delta z)_j+\dfrac{(\Delta z)_{j-1}}{2}}\right] \tag{3.54}$$

current temperature of soil block j [°C]
previous temperature of soil block j [°C]
time period [s]
average of thermal conductivities of soil blocks j and j-1 [J cm^{-1} °C^{-1} d^{-1}]
previous temperature of soil block j+1 [°C]
previous temperature of soil block j-1 [°C]
average of thermal conductivities of soil blocks j+1 and j [J cm^{-1} °C^{-1} d^{-1}]
thickness of block j+1 [cm]
thickness of block j [cm]
thickness of block j-1 [cm]

$$C_{rz,t,j} = C_{rz,t-\Delta t,j} + (\Delta t)\left[S'_{c,t-\Delta t,j} - \left(\begin{array}{l} K_{t-\Delta t,j-1/2} \dfrac{2\left(C_{rz,t-\Delta t,j} - C_{rz,t-\Delta t,j-1}\right)}{(\Delta z)_j + (\Delta z)_{j-1}} + \\ C_{rz,t-\Delta t,j-1/2} \dfrac{2\left(K_{t-\Delta t,j} - K_{t-\Delta t,j-1}\right)}{(\Delta z)_j + (\Delta z)_{j-1}} \end{array} \right) \left(1 + \dfrac{2\left(|\psi|_{t-\Delta t,j} - |\psi|_{t-\Delta t,j-1}\right)}{(\Delta z)_j + (\Delta z)_{j-1}}\right) - K_{t-\Delta t,j} C_{rz,t-\Delta t,j} \dfrac{\dfrac{2\left(|\psi|_{t-\Delta t,j+1} - |\psi|_{t-\Delta t,j}\right)}{(\Delta z)_{j+1} + (\Delta z)_j} - \dfrac{2\left(|\psi|_{t-\Delta t,j} - |\psi|_{t-\Delta t,j-1}\right)}{(\Delta z)_j + (\Delta z)_{j-1}}}{\dfrac{(\Delta z)_{j+1}}{2} + (\Delta z)_j + \dfrac{(\Delta z)_{j-1}}{2}} \right] \quad (3.55)$$

current carbon concentration in soil water of block j [kg m^{-3}]
previous carbon concentration in soil water of block j [kg m^{-3}]
average of hydraulic conductivities of soil blocks j and j-1 [m s^{-1}]
previous carbon concentration in soil water of block j-1 [kg m^{-3}]
previous suction head in block j [m]
previous suction head in block j-1 [m]
Previous carbon source (< 0) and sink (> 0) per unit volume of soil water for block j [kg m^{-3} s^{-1}]
hydraulic conductivity of soil block j-1 [m s^{-1}]
time period [s]
average carbon concentrations in soil water of blocks j and j-1 [kg m^{-3}]
hydraulic conductivity of soil block j [m s^{-1}]
thickness of block j+1 [cm]
thickness of block j [cm]
thickness of block j-1 [cm]

3.3.1 Water source and sink

In equation (3.28) S′ includes soil water evaporation through bare surface and extraction by plant roots for transpiration through vegetated surface.

The soil water evaporation from a soil block (figure 3.2) can be estimated by introducing a soil resistance factor into the Penman–Monteith equation (3.39), leading to a new equation expressed as

$$E_{soil} = \frac{\Delta\left(R_{sl} - g_{gs\leftrightarrow r}\right) + \rho_a c_a \left(\dfrac{e_s - e_a}{r_a}\right)}{\lambda\left[\Delta + \gamma\left(1 + \dfrac{r_s}{r_a}\right)\right]} \quad (3.56)$$

potential evapotranspiration [mm]
slope of vapor pressure curve [kPa °C^{-1}]
solar radiation [MW m^{-2}]
heat exchange with rootzone [MW m^{-2}]; positive to rootzone
air density [kg m^{-3}]
air specific heat [= 10^{-3} MJ kg^{-1} °C^{-1}]
saturation vapor pressure [kPa]
actual vapor pressure [kPa]
aerodynamic resistance to vapor transport [s m^{-1}]
latent heat of water evaporation [= 2.257 MJ kg^{-1}]
soil resistance factor to vapor transport [s m^{-1}]
psychrometric constant [kPa °C^{-1}]

Equations (3.40)–(3.50) are used to determine the variables in equation (3.56) except for the soil resistance factor, which can be estimated using the formulas developed by Wang (2015) as thickness of dry soil layer (DSL) above the block of interest [m]

soil resistance factor to vapor transport [s m^{-1}]

$$r_s = \frac{\delta}{D_v \tau} \tag{3.57}$$

molecular diffusivity of water vapor in air [m^2 s^{-1}]

tortuosity factor of vapor flow through the soil matrix [–]

theoretical maximum DSL thickness [~ 0.05 m]

threshold upper bound of soil moisture for DSL [–]

surface soil moisture [–]

thickness of dry soil layer (DSL) above the depth of interest [m]

$$\delta = \begin{cases} \delta_{max} \dfrac{\theta_{DSL} - \theta_0}{\theta_{DSL} - \theta_r} & \text{for } \theta_0 < \theta_{DSL} \\ 0 & \text{for } \theta_0 \geq \theta_{DSL} \end{cases} \tag{3.58}$$

residual soil moisture [–]

surface soil temperature [K]

molecular diffusivity of water vapor in air [m^2 s^{-1}]

$$D_v = 2.12\left(\frac{T_{ss}}{273.15}\right)^2 \tag{3.59}$$

saturated soil moisture [–]

tortuosity factor of vapor flow through the soil matrix [–]

$$\tau = (\theta_{sat} - \theta_r)^{\frac{2\lambda+3}{\lambda}} (\theta_{sat})^{-\frac{3}{\lambda}} \tag{3.60}$$

residual soil moisture [–]

soil pore-size distribution index [–]

The soil water transpiration from a soil block (figure 3.2) is a positive function of root density (figure 3.4) and can be estimated as (van Ittersum *et al* 2003, Boogaard *et al* 2014)

$$T_{soil,j} = \frac{T_p}{\int_0^{z_{rz}} \xi(z)\,dz} \int_{z_{j,U}}^{z_{j,L}} \left[\alpha_d \alpha_o \alpha_s \xi(z)\right] dz \tag{3.61}$$

$T_{soil,j}$: transpiration of soil water from block j [mm]
T_p: potential transpiration of soil water [mm]
z_{rz}: gross thickness of rootzone [mm]
$z_{j,L}$: depth from ground surface to bottom of block j [mm]
$z_{j,U}$: depth from ground surface to top of block j [mm]
α_d: reduction factor due to drought [–]
α_o: reduction factor due to wetness [–]
α_s: reduction factor due to salinity [–]
$\xi(z)$: root density (i.e., root length divided by inclusive soil volume) at depth z [mm^{-2}] (e.g., Figure 3.4)

By substituting a plant resistance factor, which can be found in Allen *et al* (1998), for r_s in equation (3.62), the potential transpiration can be determined as

$$T_p = \frac{\Delta\left(R_{sl} - g_{gs\leftrightarrow r}\right) + \rho_a c_a \left(\frac{e_s - e_a}{r_a}\right)}{\lambda\left[\Delta + \gamma\left(1 + \frac{r_c}{r_a}\right)\right]} \tag{3.62}$$

T_p: potential transpiration [mm]
Δ: slope of vapor pressure curve [kPa °C^{-1}]
R_{sl}: solar radiation [MW m^{-2}]
$g_{gs\leftrightarrow r}$: heat exchange with rootzone [MW m^{-2}]; positive to rootzone
ρ_a: air density [kg m^{-3}]
c_a: air specific heat [= 10^{-3} MJ kg^{-1} °C^{-1}]
e_s: saturation vapor pressure [kPa]
e_a: actual vapor pressure [kPa]
r_a: aerodynamic resistance to vapor transport [s m^{-1}]
r_c: plant resistance factor to vapor transport [s m^{-1}]
λ: latent heat of water evaporation [= 2.257 MJ kg^{-1}]
γ: psychrometric constant [kPa °C^{-1}]

Depending on the plant species and potential transpiration, the reduction factors due to drought and wetness (i.e. oxygen) stresses can be determined from a plant-specific curve (e.g. figures 3.4 and 3.5) (Feddes *et al* 1978) as

$$\alpha_d = \begin{cases} 1 & \text{for } \theta \geq \theta_L \text{ or } \theta_H \\ \dfrac{\theta - \theta_\psi}{\theta_L - \theta_\psi} \text{ or } \dfrac{\theta - \theta_\psi}{\theta_H - \theta_\psi} & \text{for } \theta_\psi < \theta < \theta_L \text{ or } \theta_H \\ 0 & \text{for } \theta \leq \theta_\psi \end{cases} \tag{3.63}$$

α_d: reduction factor due to drought [–]
θ_L: threshold soil moisture below which potential transpiration (T_p) starts to be restricted when T_p is low [–]
θ_H: threshold soil moisture below which potential transpiration (T_p) starts to be restricted when T_p is high [–]
θ_ψ: field capacity [–]

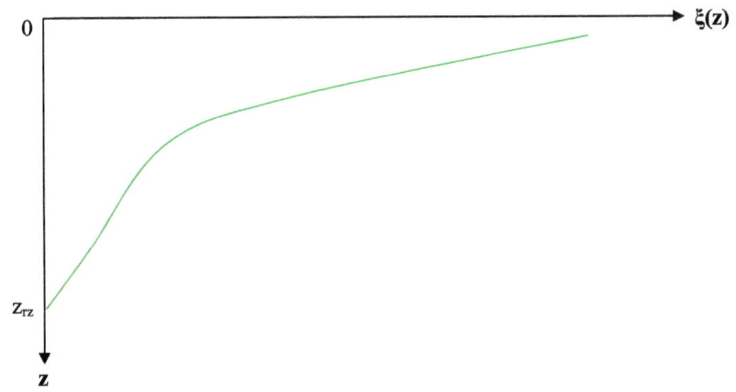

Figure 3.4. Root density, $\xi(z)$. Illustration of $\xi(z)$ versus z, depth from ground surface. z_{rz} [L]: gross thickness of rootzone.

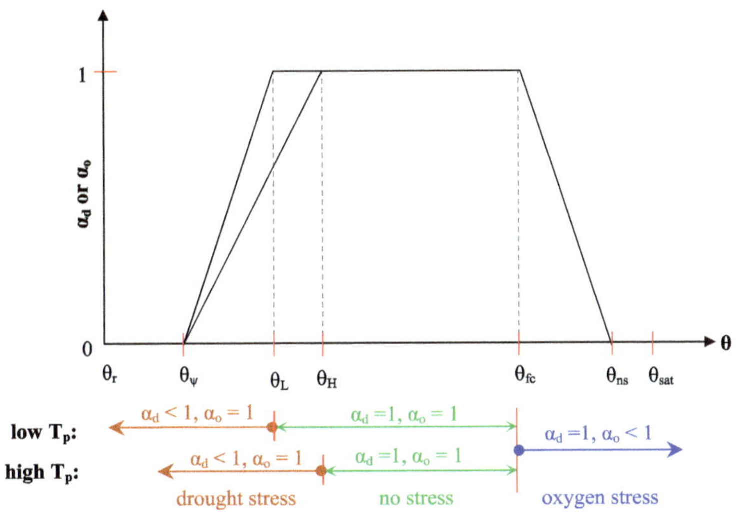

Figure 3.5. Transpiration reduction factors due to drought (α_d) and oxygen (α_o) stresses. Illustration of α_d and α_o versus soil moisture (θ). θ_r [−]: residual soil moisture; θ_ψ [−]: wilting point; θ_{fc} [−]: field capacity; θ_{ns} [−]: naturally saturated soil moisture; θ_{sat} [−]: saturated soil moisture; and T_p [L]: potential transpiration.

field capacity [−]

reduction factor due to wetness [−]

$$\alpha_o = \begin{cases} 1 & \text{for } \theta \le \theta_{fc} \\ \dfrac{\theta - \theta_{ns}}{\theta_{fc} - \theta_{ns}} & \text{for } \theta_{fc} < \theta < \theta_{ns} \\ 0 & \text{for } \theta \ge \theta_{ns} \end{cases} \tag{3.64}$$

naturally saturated soil moisture [−]

3.3.2 Carbon source and sink

In equation (3.53) S_c' is the summation of (i) carbon sinks (i.e. gains) from decayed plant roots, stems, and leaves as well as decomposed micro-organisms, and (ii) carbon sources (i.e. losses) to soil and plant respiration.

The carbon addition to the rootzone soils is a function of the relative dry mass of dead plants and micro-organisms, soil moisture, and soil temperature. The relative dry mass is the dry mass of dead materials per unit dry mass of the inclusive soils. If the soils are too dry, too cold, or too hot, the decomposition processes will cease (Thongjoo *et al* 2005). The function may be physically based when the biochemical reactions of decomposition are quantitatively described, which is ideal in principle but may be a big challenge in practice. Thus, for simplicity, an empirical form of the function (e.g. figure 3.6) can be expressed as

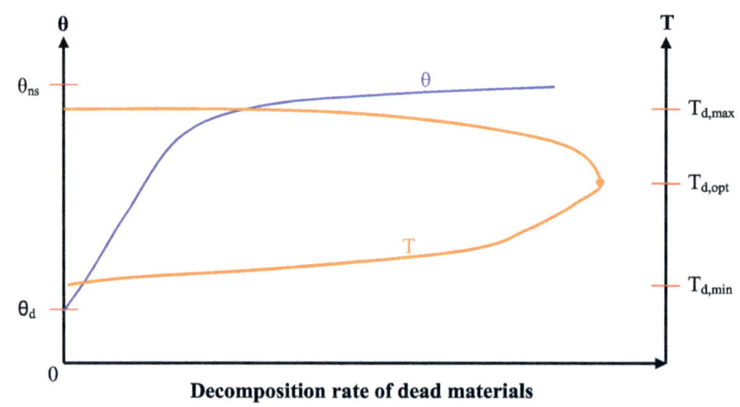

Figure 3.6. Decomposition of dead plants and micro-organisms. Illustration of decomposition rate versus soil moisture (θ) and soil temperature (T). θ_d [–]: minimum soil moisture for decomposition to incept; θ_{sat} [–]: saturated soil moisture; $T_{d,min}$ [°C]: minimum soil temperature for decomposition to incept; $T_{d,max}$ [°C]: maximum soil temperature for decomposition to cease; and $T_{d,opt}$ [°C]: optimal soil temperature for decomposition.

$$S'_{c,d} = \begin{cases} 0 & \text{if } \theta \le \theta_d \text{ or } T \le T_{d,min} \text{ or } T \ge T_{d,max} \\ a_d \left(\dfrac{W_d}{W_{d,s}} \right)^{b_{d,1}} \left(\dfrac{\theta - \theta_d}{\theta_{sat} - \theta_d} \right)^{b_{d,2}} \left(\dfrac{T - T_{d,min}}{T_{d,max} - T_{d,min}} \right)^{b_{d,3}} & \text{otherwise} \end{cases} \tag{3.65}$$

carbon gain rate [kg C ha^{-1} d^{-1}]
mass of dead materials [kg DM]
minimum soil moisture for decomposition to incept [–]
soil moisture [–]
minimum soil temperature for decomposition to incept [°C]
maximum soil temperature for decomposition to cease [°C]
soil temperature [°C]
scale coefficient [kg C ha^{-1} d^{-1}]
mass of inclusive soils [kg DM]
mass shape factor [–]
saturated soil moisture [–]
moisture shape factor [–]
temperature shape factor [–]

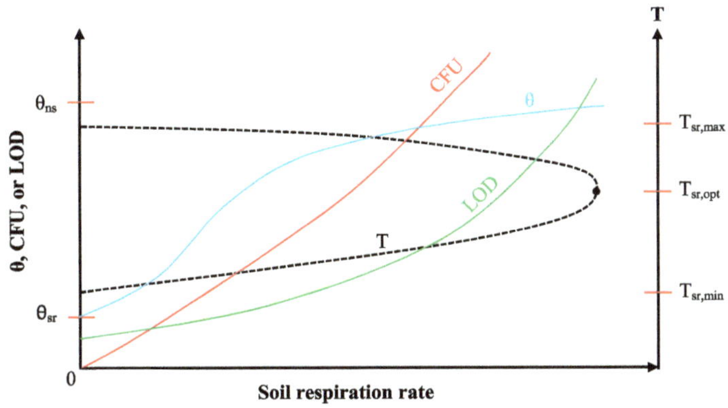

Figure 3.7. Soil respiration. Illustration of soil respiration rate versus soil moisture (θ), soil temperature (T), number of colony-forming units (CFUs), and level of development (LOD). θ_{sr} [−]: minimum soil moisture for soil respiration to incept; θ_{sat} [−]: saturated soil moisture; $T_{sr,min}$ [°C]: minimum soil temperature for soil respiration to incept; $T_{sr,max}$ [°C]: maximum soil temperature for soil respiration to cease; and $T_{sr,opt}$ [°C]: optimal soil temperature for soil respiration.

The carbon loss to soil respiration is a function of number of total bacterial and fungal colony-forming units (CFUs), level of development (LOD) of the organism community, soil moisture, and soil temperature. CFUs are determined by adding a soil sample to solution of sodium pyrophosphate to incubate at 25 °C or 30 °C for up to 10 d (Vieira and Nahas 2005), while LOD is discussed in section 2.2.3. Similarly, if the soils are too dry, too cold, or too hot, the respiration process will cease (Thongjoo *et al* 2005). An empirical form of the function (e.g. figure 3.7) can be expressed as

$$S'_{c,sr} = \begin{cases} 0 & \text{if } \theta \le \theta_{sr} \text{ or } T \le T_{sr,min} \text{ or } T \ge T_{sr,max} \\ a_{sr}\,(CFU)^{b_{sr,1}}\,(LOD)^{b_{sr,2}} \left(\dfrac{\theta-\theta_{sr}}{\theta_{sat}-\theta_{sr}}\right)^{b_{sr,3}} \left(\dfrac{T-T_{sr,min}}{T_{sr,max}-T_{sr,min}}\right)^{b_{sr,4}} & \text{otherwise} \end{cases} \tag{3.66}$$

carbon loss to soil respiration [kg C ha^{-1} d^{-1}]; colony-forming units [−]; level of development [−]; soil moisture [−]; minimum soil moisture for soil respiration to incept [−]; minimum soil temperature for soil respiration to incept [°C]; maximum soil temperature for soil respiration to cease [°C]; soil temperature [°C]; scale coefficient [kg C ha^{-1} d^{-1}]; colony shape factor [−]; level shape factor [−]; saturated soil moisture [−]; moisture shape factor [−]; temperature shape factor [−]

The carbon loss to plant respiration is a function of maintenance coefficients and dry masses of organs, reduction factor for senescence (f_{senes}), and air temperature (T_a in equation (3.41)). f_{senes} is crop-specific and depends on development stage (Kroes *et al* 2017). An empirical form of the function can be expressed as (Penning de Vries and van Laar 1982)

$$S'_{c,pr} = f_{senes}\left(c_{m,leaf}W_{leaf} + c_{m,stem}W_{stem} + c_{m,stor}W_{stor} + c_{m,root}W_{root}\right)\left(Q_{10}\right)^{\frac{T_a - 25^{\circ}C}{10^{\circ}C}} \tag{3.67}$$

- $S'_{c,pr}$: carbon loss to plant respiration [kg C ha^{-1} d^{-1}]
- f_{senes}: reduction factor for senescence [–]
- $c_{m,leaf}$: maintenance coefficient of leaf [kg C kg^{-1} d^{-1}]
- W_{leaf}: dry mass of leaf [kg ha^{-1}]
- $c_{m,stem}$: maintenance coefficient of stem [kg C kg^{-1} d^{-1}]
- W_{stem}: dry mass of stem [kg ha^{-1}]
- $c_{m,stor}$: maintenance coefficient of storage organ [kg C kg^{-1} d^{-1}]
- W_{stor}: dry mass of storage organ [kg ha^{-1}]
- $c_{m,root}$: maintenance coefficient of root [kg C kg^{-1} d^{-1}]
- W_{root}: dry mass of root [kg ha^{-1}]
- Q_{10}: rate increase per 10 °C increase in T_a (~ 2) [–]
- T_a: daily average air temperature [°C]

3.4 Below-rootzone processes

For the below-rootzone subsystem, the water movement is governed by equation (3.28), while the heat and carbon transports are governed by equations (3.52) and (3.53), respectively. The source and sink terms are zero, that is, $S' = 0$ and $S_c' = 0$.

3.5 Plant growth and senescence

Plants can be annual or perennial. Annual plants (e.g. maize and soybean) complete their life cycle in one growing season or two and usually go through ten growth stages (Zadoks *et al* 1974), namely: (0) sowing and/or germination; (1) main shoot leaf production; (2) tiller production; (3) stem elongation; (4) booting; (5) heading; (6) anthesis; (7) grain milk; (8) grain dough; and (9) ripening. In contrast, perennial plants (e.g. grasses and trees) live longer than two years, stay in a vegetative stage during growing seasons, and may or may not become dormant in nongrowing seasons depending on the plant species. Unlike annual plants, perennial plants do not experience the above growth stages; instead, a perennial plant regerminates each year with its roots alive regardless of the seasons. Both types of plants incur senescence or biological aging (i.e. the gradual deterioration of functional characteristics) (Wikipedia 2021) and can be harvested, mowed, or grazed before they are mature.

3.5.1 Growth

Photosynthesis utilizes light radiation to assimilate CO_2 to grow roots and shoots (i.e. leaves, stems, and storage organs). Herein, the assimilation rate is dependent on CO_2 concentration and photosynthetically active radiation (PAR), which has a wavelength band of 400–700 nm (Kroes *et al* 2017) and refers to the radiation absorbed by plants to carry out the photosynthesis process. It reaches a maximum at light saturation and when free of stresses, and is plant species-specific. Oberbauer and Oechel (1989) report the maximum CO_2 assimilation rates of 19 vascular plant species. If a plant is subject to water, CO_2, temperature, and/or nitrogen stresses, its daily assimilation rate can be computed as (Yamori *et al* 2011, Kroes *et al* 2017) daily gross assimilation rate [kg C ha^{-1} d^{-1}]; CO_2 correction factor [–]; actual transpiration (Eq. 3.61) [mm]; maximum instantaneous assimilation rate [kg C ha^{-1} d^{-1}]

$$A_{ssim} = \frac{30}{64} f_{Tssim} f_{CO2ssim} \min\left(\frac{T_{soil,rz}}{T_p}, f_{Nssim}\right) A_{ssim,max} \tag{3.68}$$

air-temperature stress factor [–]; potential transpiration (Eq. 3.62) [mm]; nitrogen stress factor [–]

The air-temperature stress factor can be determined as (Kroes *et al* 2017)

weighted average daily air temperature [°C]; air temperature stress factor [–]

$$f_{Tssim} = \begin{cases} 0 & \text{for} \quad T_a \le 0°C \\ 0.12T_a & \text{for } 0°C < T_a \le 5°C \\ 0.03T_a + 0.55 & \text{for } 5°C < T_a \le 15°C \\ 1.0 & \text{for } 15°C < T_a \le 25°C \\ \frac{8}{3} - \frac{T_a}{15} & \text{for } 25°C < T_a \le 40°C \\ 0 & \text{for} \quad T_a > 40°C \end{cases} \tag{3.69}$$

minimum air temperature [°C]; maximum air temperature [°C]; weighted average daily air temperature [°C]

$$T_a = 0.25T_{min} + 0.75T_{max} \tag{3.70}$$

Rising CO_2 has a negligible influence on C_4 plant species but inhibits shoot nitrate assimilation in a wide variety of C_3 plants (Bloom *et al* 2012). Empirically, some stress would occur once the CO_2 concentration in the ambient atmosphere was higher than 310 ppm. The stress can be considered using a CO_2 correction factor of $f_{CO_2assim} = 0.9$ to 1.0. In addition, water is vital to assimilation and its stress can be reflected by the ratio of actual to potential transpiration: the smaller the ratio, the larger the stress, and vice versa. Further, nitrogen assimilation, the formation of organic nitrogen compounds (e.g. amino acids) from inorganic nitrogen compounds (e.g. nitrite and nitrate) in the environment, is an important aspect of the overall assimilation. The nitrogen stress is very sensitive to plant species: the species (e.g. legumes) that can fix nitrogen gas (N_2) are less likely to be limited by nitrogen availability than the species (e.g. maize) that cannot fix N_2. Regressing the data presented by van der Meer *et al* (1986), yields the equation for nitrogen stress factor of intensive grasslands expressed as nitrogen supply [kg]

nitrogen stress factor [–]

$$f_{Nssim} = -0.74\left(\frac{N_{sup}}{800}\right)^2 + 1.32\left(\frac{N_{sup}}{800}\right) + 0.41 \tag{3.71}$$

As discussed in section 2.4, the net ecosystem exchange (i.e. assimilation rate) and biomass gain are computed as

daily gross assimilation rate (Eq. 3.68) [kg C ha^{-1} d^{-1}]

net ecosystem exchange [kg C ha^{-1} d^{-1}]

$$NEE = A_{ssim} - S'_{c,pr} \tag{3.72}$$

carbon loss to plant respiration (Eq. 3.67) [kg C ha^{-1} d^{-1}]

overall conversion factor [kg DM kg^{-1} C]

gross biomass gain [kg DM ha^{-1} d^{-1}]

$$w_{gross} = C_e\left(NEE\right) \tag{3.73}$$

net ecosystem exchange [kg C ha^{-1} d^{-1}]

partitioning factor for root [–]

root biomass gain [kg DM ha^{-1} d^{-1}]

$$w_{root} = \omega_{root} w_{gross} \tag{3.74}$$

gross biomass gain [kg DM ha^{-1} d^{-1}]

partitioning factor for leaf [–]

partitioning factor for root [–]

leaf biomass gain [kg DM ha^{-1} d^{-1}]

$$w_{leaf} = \omega_{leaf}\left(1 - \omega_{root}\right) w_{gross} \tag{3.75}$$

gross biomass gain [kg DM ha^{-1} d^{-1}]

partitioning factor for stem [–]

partitioning factor for root [–]

stem biomass gain [kg DM ha^{-1} d^{-1}]

$$w_{stem} = \omega_{stem}\left(1 - \omega_{root}\right) w_{gross} \tag{3.76}$$

gross biomass gain [kg DM ha^{-1} d^{-1}]

$$w_{stor} = \omega_{stor}\left(1-\omega_{root}\right)w_{gross} \tag{3.77}$$

partitioning factor for storage organ [–]
partitioning factor for root [–]
storage-organ biomass gain [kg DM ha⁻¹ d⁻¹]
gross biomass gain [kg DM ha⁻¹ d⁻¹]

The partitioning factors for leaf, stem, and storage organ are subject to

$$\omega_{leaf} + \omega_{stem} + \omega_{stor} = 1 \tag{3.78}$$

partitioning factor for stem [–]
partitioning factor for leaf [–]
partitioning factor for storage organ [–]

The overall conversion factor in equation (3.73) can be computed as (Kroes *et al* 2017)

$$C_e = \left[\left(\frac{\omega_{leaf}}{C_{e,leaf}} + \frac{\omega_{stem}}{C_{e,stem}} + \frac{\omega_{stor}}{C_{e,stor}}\right)\left(1-\omega_{root}\right) + \frac{\omega_{root}}{C_{e,root}}\right]^{-1} \tag{3.79}$$

partitioning factor for leaf [–]
overall conversion factor [kg DM kg⁻¹ C]
partitioning factor for stem [–]
partitioning factor for storage organ [–]
partitioning factor for root [–]
leaf conversion factor [kg DM kg⁻¹ C]
stem conversion factor [kg DM kg⁻¹ C]
storage-organ conversion factor [kg DM kg⁻¹ C]
root conversion factor [kg DM kg⁻¹ C]

3.5.2 Senescence and net growth

Plant organs die off due to natural aging and stresses of temperature, water, and nitrogen. At a growing time of interest, for a given organ, its death rate, in kg DM d^{-1}, can be estimated as a multiplication of the organ's biomass, in kg DM, and a relative death rate, in kg DM kg^{-1} DM d^{-1}. The net growth equals to the biomass gain minus the death rate.

The net growth of root is computed as

$$w'_{root} = w_{root} - \varsigma_{root} W_{root} \tag{3.80}$$

net root biomass gain [kg DM ha^{-1} d^{-1}]; root biomass gain (Eq. 3.74) [kg DM ha^{-1} d^{-1}]; root biomass [kg DM ha^{-1} d^{-1}]; relative death rate of root [kg DM kg^{-1} DM d^{-1}]

$$d'_{root} = \frac{w'_{root}}{w'_{root,p}} z_{rz} \tag{3.81}$$

net increase of root depth [mm]; net root biomass gain [kg DM ha^{-1} d^{-1}]; gross thickness of rootzone [mm]; potential root biomass gain [kg DM ha^{-1} d^{-1}]

The net growth of leaf is computed as

$$w'_{leaf} = w_{leaf} - \varsigma_{leaf} W_{leaf} \tag{3.82}$$

net leaf biomass gain [kg DM ha^{-1} d^{-1}]; leaf biomass gain (Eq. 3.75) [kg DM ha^{-1} d^{-1}]; leaf biomass [kg DM ha^{-1} d^{-1}]; relative death rate of leaf [kg DM kg^{-1} DM d^{-1}]

$$w'_{LAI} = w'_{leaf} S_{leaf} \tag{3.83}$$

net increase of leaf area index or LAI [d^{-1}]; net leaf biomass gain [kg DM ha^{-1} d^{-1}]; specific leaf area [ha kg^{-1} DM d^{-1}]

The net growth of stem is computed as

$$w'_{stem} = w_{stem} - \varsigma_{stem} W_{stem} \tag{3.84}$$

net stem biomass gain [kg DM ha^{-1} d^{-1}]; stem biomass gain (Eq. 3.76) [kg DM ha^{-1} d^{-1}]; stem biomass [kg DM ha^{-1} d^{-1}]; relative death rate of stem [kg DM kg^{-1} DM d^{-1}]

$$w'_{GAI,stem} = w'_{stem} S_{stem} \tag{3.85}$$

net increase of green area index or GAI for stem [d^{-1}]; net stem biomass gain [kg DM ha^{-1} d^{-1}]; specific green area of stem [ha kg^{-1} DM d^{-1}]

The net growth of storage organ is computed as

storage-organ biomass gain (Eq. 3.77) [kg DM ha^{-1} d^{-1}]
storage-organ biomass [kg DM ha^{-1} d^{-1}]
net storage-organ biomass gain [kg DM ha^{-1} d^{-1}]

$$w'_{stem} = w_{stem} - \varsigma_{stem} W_{stem} \tag{3.86}$$

relative death rate of storage organ [kg DM kg^{-1} DM d^{-1}]

net storage-organ biomass gain [kg DM ha^{-1} d^{-1}]
net increase of green area index or GAI for storage organ [d^{-1}]

$$w'_{GAI,stor} = w'_{stor} S_{stor} \tag{3.87}$$

specific green area of storage organ [ha kg^{-1} DM d^{-1}]

3.6 Topsoil erosion

For semi-arid ecosystems, the total topsoil erosion is a result of both aeolian and fluvial processes and interactions thereof (Wang *et al* 2014). The partition between these two types of erosions is a function of soil moisture level, land surface condition, and climate. Regardless of whether the land surface of interest is bare or vegetated, the long-term average erosion rates can be illustrated as figure 3.8. The aeolian (i.e. wind) erosion rate tends to decrease with increase of soil moisture and

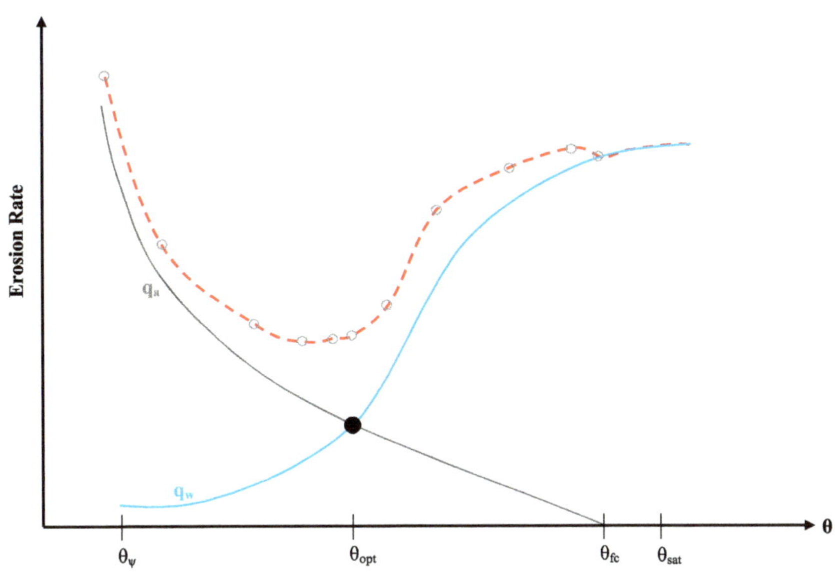

Figure 3.8. Topsoil erosion rates. Illustration of long-term average rates versus soil moisture (θ). θ_ψ [−]: wilting point; θ_{opt} [−]: optimal soil moisture; θ_{fc} [−]: field capacity; θ_{sat} [−]: saturated soil moisture; q_a [g m^{-2} s^{-1}]: aeolian erosion rate; q_w [g m^{-2} s^{-1}]: fluvial erosion rate; and q_{aw} [g m^{-2} s^{-1}]: total erosion rate.

approaches zero at field capacity, whereas the fluvial (i.e. water) erosion rate tends to increase with an increase of soil moisture and plateaus at saturated soil moisture. At a soil moisture level of around wilting point, the wind erosion can increase exponentially, while the water erosion will become very small because of a low runoff potential. At the soil moisture at which there is a maximum interaction between aeolian and fluvial processes, the wind and water erosion rates are the same while the overall erosion rate attains a minimum; hereinafter this value is designated 'optimal' soil moisture, θ_{opt}, for description purposes. The overall erosion rate is the sum of aeolian and fluvial rates.

The total soil erosion rate is computed as

aeolian erosion rate [g m^{-2} s^{-1}]
fluvial erosion rate [g m^{-2} s^{-1}]
total erosion rate [g m^{-2} s^{-1}]

$$q_{aq} = q_a + q_w \tag{3.88}$$

The erosion depth rate can be determined by dividing the apparent dry density of topsoil by the two sides of equation (3.88).

3.6.1 Aeolian erosion

Aeolian (i.e. wind) erosion rate can be computed as (Chepil 1946, 1956, Marshall 1971)

effective wind shear stress [N m^{-2}]
threshold wind shear stress [N m^{-2}]
aeolian erosion rate [g m^{-2} s^{-1}]
wind-erosion coefficient when soil is bare and dry soil [g m^{-5} s^{2}]
air density (Eq. 3.43) [kg m^{-3}]

$$q_a = C_a \left(\frac{\tau_* - \tau_{*c}}{\rho_a} \right)^{\frac{3}{2}} \tag{3.89}$$

The threshold wind shear stress depends on soil properties (e.g. texture, structure, and moisture) (Gillette 1978, Fullen 1985, Chen *et al* 1996, Fécan *et al* 1999, Shao and Lu 2000, Lu and Shao 2001, Dong *et al* 2002, Ishizuka *et al* 2005) and vegetation physiological characters (e.g. individual height, population density, and coverage) (Fullen 1985, Musick and Gillette 1990, Stockton and Gillette 1990, Wolfe and Nickling 1993, Lancaster and Baas 1998, Okin 2008, Yan *et al* 2011). It can be determined as (Bagnold 1941)

air density (Eq. 3.43) [kg m^{-3}]
threshold wind shear stress [N m^{-2}]
threshold surface friction shear velocity at surface soil moisture θ and vegetation roughness density λ [m s^{-1}]

$$\tau_{*c} = \rho_a \left(u_{*\theta\lambda} \right)^2 \tag{3.90}$$

Regressing the data presented by Lancaster and Baas (1998) yields

$$u_{*\theta\lambda} = (4.19\lambda + 1)\,u_{*\theta} \tag{3.91}$$

$u_{*\theta\lambda}$: threshold surface friction shear velocity at surface soil moisture θ and vegetation roughness density λ [m s^{-1}]

λ: vegetation roughness density [–]

$u_{*\theta}$: threshold surface friction shear velocity at surface soil moisture θ and bare surface [m s^{-1}]

The vegetation roughness density is defined as (Raupach *et al* 1993)

$$\lambda = \frac{nA_v}{A} = D_v A_v \tag{3.92}$$

λ: vegetation roughness density [–]

n: number of vegetation individuals [–]

A_v: upwind projected (i.e., mean frontal-silhouette) area per vegetation individual [m^2]

A: total soil surface area [m^2]

D_v: vegetation population density (i.e., number of vegetation individuals per unit soil surface area) [m^{-2}]

$u_{*\theta}$ can be estimated as (Fécan *et al* 1999)

$$u_{*\theta} = f_w u_{*\psi} \tag{3.93}$$

$u_{*\theta}$: threshold surface friction shear velocity at surface soil moisture θ and bare surface [m s^{-1}]

f_w: dimensionless coefficient that reflects effects of surface soil moisture [–]

$u_{*\psi}$: threshold surface friction shear velocity at wilting point and bare surface [m s^{-1}]

$$f_w = \begin{cases} 1 & \text{for } \omega \le \omega' \\ \sqrt{1 + 1.21(\omega - \omega')^{0.68}} & \text{for } \omega > \omega' \end{cases} \tag{3.94}$$

f_w: dimensionless coefficient that reflects effects of surface soil moisture [–]

ω: water content [kg kg^{-1}]

ω′: threshold water content [kg kg^{-1}]

The water content can be computed from soil moisture as (Charbeneau 2006)

$$\omega = \frac{\rho_w}{\rho_s}(1 + e)\theta \tag{3.95}$$

ω: water content [kg kg^{-1}]

ρ_w: water density [kg m^{-3}]

e: void ratio [–]

ρ_s: soil particle density [kg m^{-3}]

θ: soil moisture [–]

ω' reflects the effects of clay content and can be computed as (Fécan *et al* 1999)

$$\omega' = 0.0014\left(P_{clay}\right)^2 + 0.17P_{clay} \tag{3.96}$$

threshold water content [kg kg^{-1}] (ω'); clay content [%] (P_{clay})

$u_{*\psi}$ can be computed as (Shao and Lu 2000)

$$u_{*\psi} = \sqrt{\frac{a_1}{\rho_a}\left(\rho_s g d_s + \frac{a_2}{d_s}\right)} \tag{3.97}$$

threshold surface friction shear velocity at wilting point and bare surface [m s^{-1}] ($u_{*\psi}$); dimensionless constant [= 0.0123] (a_1); constant [= 0.0003 kg s^{-2}] (a_2); diameter of surface soil particle of interest [m] (d_s); air density (Eq. 3.43) [kg m^{-3}] (ρ_a); soil particle density [kg m^{-3}] (ρ_s); gravitational acceleration [= 9.81 m s^{-2}] (g)

On the other hand, τ_* is a function of climatic factors (e.g. wind speed) (Chepil *et al* 1962, Skidmore 1986), soil surface roughness (Wolfe and Nickling 1993, Chen *et al* 1996), and vegetation structural characteristics (Musick and Gillette 1990, Stockton and Gillette 1990). τ_* is defined as (Marshall 1971)

$$\tau_* = \rho_a\left(u_*\right)^2 \tag{3.98}$$

effective wind shear stress [N m^{-2}] (τ_*); air density (Eq. 3.43) [kg m^{-3}] (ρ_a); surface friction shear velocity [m s^{-1}] (u_*)

Assuming that wind speed follows a power-function distribution from soil surface to zero-plane displacement height, yields (Dixon and Hall 2010)

$$u_* = \begin{cases} u'_*\left[\dfrac{z_0}{\max\left(d, z_0\right)}\right]^{\frac{1}{7}} & \text{for } d > 0 \\ u'_* & \text{for } d = 0 \end{cases} \tag{3.99}$$

surface friction shear velocity [m s^{-1}] (u_*); surface roughness length [m] (z_0); zero-plane displacement height [m] (d); maximum function (max); friction shear velocity at a height of max(d, z_0) [m s^{-1}] (u'_*)

Based on Lyles and Allison (1975) and Wolfe and Nickling (1993), the friction shear velocity at the zero-plane displacement height (i.e. the height at which wind speed is zero) can be computed as

$$u_*' = \frac{\kappa u}{\ln\left[\dfrac{z}{\max(d, z_0)}\right]} \tag{3.100}$$

von Karman constant [= 0.41]
friction shear velocity at a height or max(d, z_0) [m s^{-1}]
wind speed at height z [m s^{-1}]
height above soil surface [m]
maximum function
surface roughness length [m]
zero-plane displacement height [m]

The zero-plane displacement height is a function of average vegetation height (Dong *et al* 2002) and LAI. It can be estimated as

$$d = \begin{cases} 0 & \text{for bare surface} \\ \left(0.5 + \dfrac{LAI - LAI_{min}}{LAI_{max} - LAI_{min}}\right) h & \text{for vegetated surface} \end{cases} \tag{3.101}$$

leaf area index [–]
average vegetation height [m]
zero-plane displacement height [m]
theoretically maximum leaf area index [–]
theoretically minimum leaf area index [–]

The surface roughness length can be computed as (Wolfe and Nickling 1993, Chen *et al* 1996, Dong *et al* 2002)

$$z_0 = \begin{cases} 0.5 d_s & \text{for bare surface and } \omega \le 0.045 \\ 6144.1 \omega^{-6.1} d_s & \text{for bare surface and } \omega > 0.045 \\ 0.5 \lambda h & \text{for vegetated surface} \end{cases} \tag{3.102}$$

diameter of surface soil particle of interest [m]
water content (Eq. 3.95) [kg kg^{-1}]
surface roughness length [m]
average vegetation height [m]
vegetation roughness density [–]

3.6.2 Fluvial erosion

Fluvial (i.e. water) erosion depends on precipitation, runoff, topographic gradient, and vegetation structure and coverage (Fullen and Reed 1986, Kirkbride and Reeves 1993, Yoder and Lown 1995, Roo *et al* 1996, Gao *et al* 2002). Its rate can be determined as

$$q_w = C_f \left(\frac{Q}{P}\right)^{b_1} \left[1 + (S_N)^{b_2}\right] e^{-b_3 \frac{LAI - LAI_{min}}{LAI_{max} - LAI_{min}}} \tag{3.103}$$

fluvial erosion rate [g m^{-2} s^{-1}]
direct runoff [mm]
calibration coefficients [–]
leaf area index [–]
water erosion modulus for flat, saturated, and bare soil surface [g m^{-2} s^{-1}]
rainfall [mm]
surface slope factor [–]
theoretically maximum leaf area index [–]
theoretically minimum leaf area index [–]

The surface slope factor is defined by Nearing (1997) as

$$S_N = -1.55 + \frac{17}{1 + e^{2.3 - 6.1\sin(\alpha)}} \tag{3.104}$$

surface slope factor [–]
surface slope angle [°]

Fullen and Reed (1986) measured precipitation, runoff, and water erosion from eight 25 m^2 bare plots located in England for two consecutive years. The plots had slope angles of 7.8° to 13.1° and predominant soils of fine sand (particle diameter from 0.063 to 0.25 mm) and medium sand (particle diameter from 0.25 to 0.5 mm). Fitting equation (3.103) to those measured data, yields $b_1 = 0.7$ and $b_2 = 0.8$, which are also supported by the data presented by Dai and Tan (1996). The coefficient b_3 depends on vegetation structural characteristics and population density (Fullen 1991, Simanton *et al* 1991, Fullen 1998) and can be computed as

$$b_3 = \frac{nA_h}{A} = D_v A_h \tag{3.105}$$

calibration coefficient in Eq. (3.103) [–]
number of vegetation individuals [–]
stem cross-sectional area per vegetation individual [m^2]
total soil surface area [m^2]
vegetation population density (i.e., number of vegetation individuals per unit soil surface area) [m^{-2}]

3.6.3 Effects of physical and biological crusts on erosion

As discussed in section 2.1 and subsections 2.2.2 and 2.2.3, crusts can increase the soil resistance to erosion (Chamizo *et al* 2017, Gao *et al* 2017, 2020). For aeolian erosion the threshold wind shear stress (i.e. τ_{*c} in equation (3.90)) will be increased, while for fluvial erosion the water erosion modulus (i.e. C_f in equation (3.103)) is decreased.

To consider such effects of crusts on topsoil erosion, $u_{*\psi}$ in equation (3.97) and C_f in equation (3.103) can be adjusted as

$$u_{*\psi,\mathrm{crust}} = f_{\mathrm{crust,a}} u_{*\psi} \qquad (3.106)$$

where $u_{*\psi,\mathrm{crust}}$: threshold surface friction shear velocity at wilting point and bare surface with a crust [m s^{-1}]; $f_{\mathrm{crust,a}}$: crust resistance factor to wind erosion [> 1.0]; $u_{*\psi}$: threshold surface friction shear velocity at wilting point and bare surface [m s^{-1}]

$$C_{f,\mathrm{crust}} = f_{\mathrm{crust,w}} C_f \qquad (3.107)$$

where $C_{f,\mathrm{crust}}$: water erosion modulus for flat, saturated, and bare soil surface with a crust [g m^{-2} s^{-1}]; $f_{\mathrm{crust,w}}$: crust resistance factor to water erosion [< 1.0]; C_f: water erosion modulus for flat, saturated, and bare soil surface [g m^{-2} s^{-1}]

In principle, for a given physical crust the resistance factors in equations (3.106) and (3.107) are a function of the total content of salts and silica in the crust, whereas for a given biocrust, the resistance factors are a function of the level of development (LOD) of micro-organisms in the crust. Although these functions may not be expressed in a generic form due to a large data gap in existing literature, they can be conceptually illustrated as figure 3.9. The LOD can be one of the six stages defined by Belnap *et al* (2008) and the total content of salts and silica can be determined as

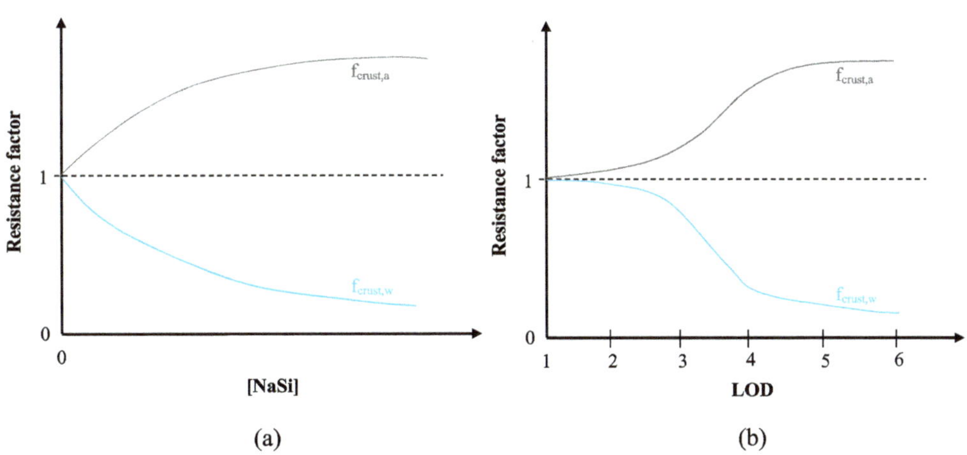

Figure 3.9. Crust resistance factors to wind ($f_{\mathrm{crust,a}}$ in equation (3.106)) and water ($f_{\mathrm{crust,w}}$ in equation (3.107)) erosion. Illustration for bare surface with a (a) physical crust and (b) biocrust. [NaSi] [mEq g^{-1} DM crust]: total content of salts and silica in the physical crust; and LOD [–]: level of development of micro-organisms in the biocrust, which can be one of the six stages proposed by Belnap *et al* (2008).

$$[\mathrm{NaSi}] = \frac{[\mathrm{NaCl}]}{28} + \frac{[\mathrm{SiO_2}]}{7.25} \tag{3.108}$$

where $[\mathrm{NaSi}]$ is the total content of salts and silica in physical crust [mEq g^{-1} DM crust]; $[\mathrm{NaCl}]$ is the content of salts in physical crust [mg NaCl g^{-1} DM crust]; $[\mathrm{SiO_2}]$ is the content of silica in physical crust [mg SiO_2 g^{-1} DM crust]; 28 is the equivalent weight of salts [mg NaCl mEq^{-1}]; 7.25 is the equivalent weight of silica [mg SiO_2 mEq^{-1}].

3.7 Summary and discussion

As detailed above and illustrated in figure 3.10, the components of a semi-arid ecosystem are interconnected by a variety of dynamic heat, water, and carbon fluxes.

First, the atmosphere and ground surface patches, which can be bare surfaces with loose topsoil, bare surfaces with physical crusts, bare surfaces with biocrusts, and/or vegetated surfaces, directly exchange heat, water, and carbon. The precipitation and solar radiation from atmosphere drive the abiotic and biotic processes of the ecosystem. Vertically, bare surfaces lose water to the atmosphere by evaporation of open water and topsoil water, whereas vegetated surfaces transpire rootzone soil water through leaf stomates. Laterally, the water and nutrients in a bare surface patch tend to move to its adjacent vegetated surface patches because these latter

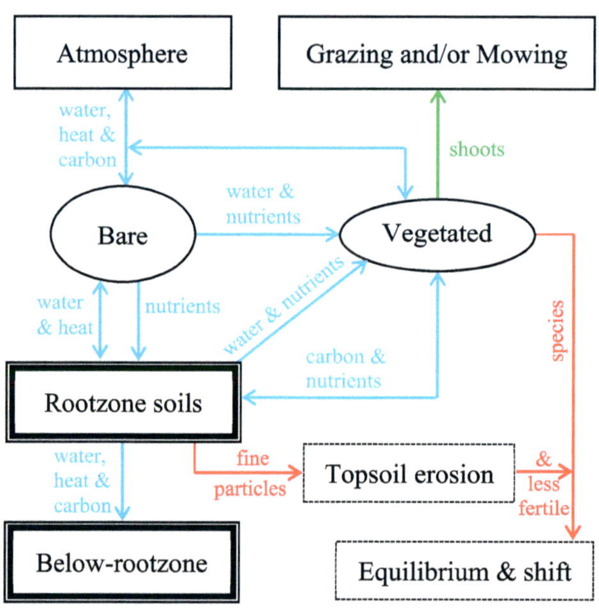

Figure 3.10. Interconnections among the components of a semi-arid ecosystem. Illustration of the heat, water, and carbon fluxes.

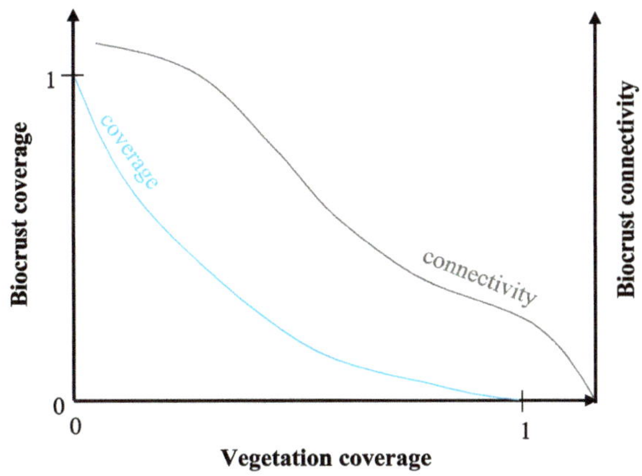

Figure 3.11. Expansion and recession of biocrusts versus vegetated area. Illustration of the dynamic coverage and connectivity of biocrust versus vegetated patches.

patches have a relatively lower elevation than the former patch. For a given year, the biocrust coverage versus vegetation coverage increases in a drier year but decreases in a wetter year, while the connectivity of biocrust patches (i.e. ratio of the number of patches to the total surface area of interest) tends to decrease with an increase of vegetation coverage (Zhao and Xu 2013) (figure 3.11).

Second, the penetration water from ground surfaces replenishes the moisture of, and transports nutrients downward into, rootzone soils. The soil water in the rootzone can be lost to evaporation and/or taken up by vegetation roots for transpiration. Plants also take up nutrients from rootzone soils. Living plants assimilate CO_2 and/or N_2 gases from the atmosphere into their organs and rootzone soils, while the respiration of plants and soil micro-organisms releases CO_2 and/or N_2 gases into atmosphere. The net difference of fixed and released CO_2 is the gain of plant biomass. Deceased micro-organisms and dead plants can be decomposed into carbon and nutrients to be stored in rootzone soils.

Third, the extra heat, water, and carbon in the rootzone would be transported downward into below-rootzone soils. Although this zone has minimal influence on the ecosystem under normal conditions, it does provide 'buffering' resources to protect the ecosystem from extreme and long-lasting droughts.

Moreover, the topsoil can be eroded by wind and/or water, reducing the fertile-soil thickness. In this regard, physical crusts, biocrusts, and vegetation canopies play an important role in topsoil stabilization by forming a surface 'armor', increasing surface roughness and resistance, and decreasing runoff. The stabilization and hydrological effects of biocrusts have been an ongoing research topic (Chamizo *et al* 2012, Gao *et al* 2017) and can be illustrated in figure 3.12. For an ecosystem of interest, the topsoil erosion and livestock overgrazing can deteriorate the vegetation community composition (VCC), leading to a worse equilibrium condition or even triggering an irreversible shift toward desertification (figure 2.14). Thus, the conceptualizations in this chapter are profound for understanding and modeling the abiotic and biotic processes of semi-arid ecosystems.

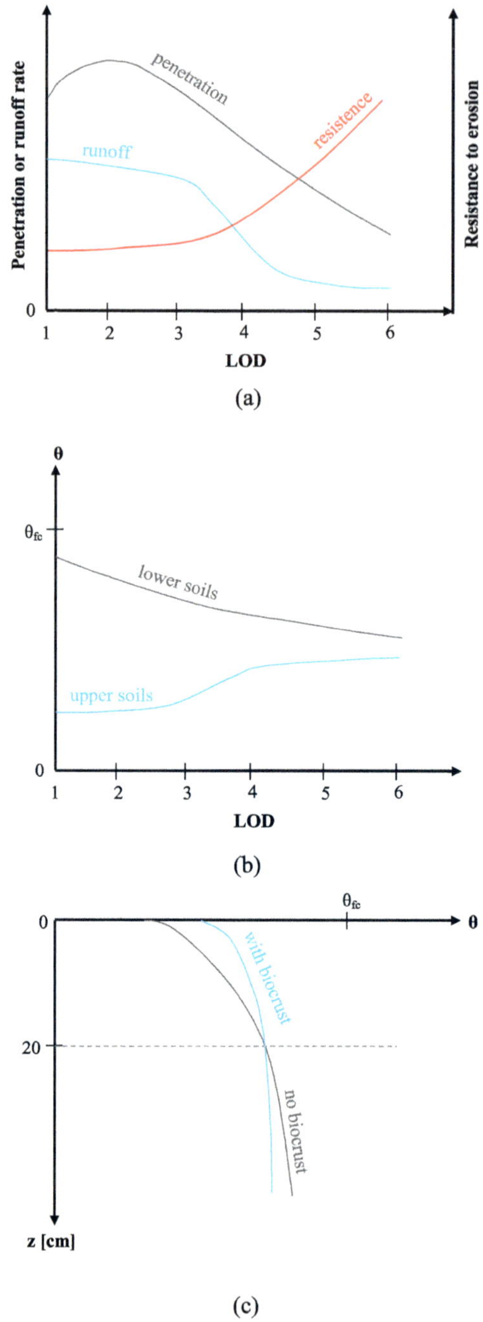

Figure 3.12. The effects of biocrusts on topsoil stabilization and hydrology. Illustration of changes of (a) penetration, runoff, and resistance with level of development or LOD of micro-organisms in the biocrust, (b) soil moisture, θ, with LOD, and (c) θ with soil depth, z. θ_{fc} [–]: field capacity; and LOD [–]:one of the six stages proposed by Belnap *et al* (2008).

References

Allen R G, Pereira L S, Raes D and Smith M 1998 *Crop Evapotranspiration Guidelines for Computing Crop Water Requirements* (Washington, DC: United Nations (UN) Food and Agriculture Organization (FAO))

Anten N, Schieving F, Werger M and Schuffelen P 1995 Optimal leaf area indices in C_3 and C_4 mono- and dicotyledonous species at low and high nitrogen availability *Physiol. Plant.* **95** 541–50

Bagnold R A 1941 *The Physics of Blown Sand and Desert Dunes* (London: Methuen)

Belnap J, Phillips S L, Witwicki D L and Miller M E 2008 Visually assessing the level of development and soil surface stability of cyanobacterially dominated biological soil crusts *J. Arid. Environ.* **72** 1257–64

Bloom A J, Salvador J, Asensio R, Randall L, Rachmilevitch S, Cousins A B and Carlisle E A 2012 CO_2 enrichment inhibits shoot nitrate assimilation in C_3 but not C_4 plants and slows growth under nitrate in C_3 plants *Ecology* **93** 355–67

Boogaard H L, de Wit A J W, de Roller J A, van Diepen C A, Rötter R P, Cabrera J M C A and van Laar H H 2014 *WOFOST Control Centre 2.1 and WOFOST 7.1.7* (Wageningen: Wageningen University and Research Centre)

Chamizo S, Cantón Y, Lázaro R, Solé-Benet A and Domingo F 2012 Crust composition and disturbance drive infiltration through biological soil crusts in semiarid ecosystems *Ecosystems* **15** 148–61

Chamizo S, Rodríguez-Caballero E, Román J R and Cantón Y 2017 Effects of biocrust on soil erosion and organic carbon losses under natural rainfall *Catena* **148** 117–25

Charbeneau R J 2006 *Groundwater Hydraulics and Pollutant Transport* (Long Grove, IL: Waveland Press)

Chen W, Dong Z, Li Z and Yang Z 1996 Wind tunnel test of the influence of moisture on the erodibility of loessial sandy loam soils by wind *J. Arid. Environ.* **34** 391–402

Chepil W S 1946 Dynamics of wind erosion: V. cumulative intensity of soil drifting across eroding fields *Soil Sci.* **61** 257

Chepil W S 1956 Influence of moisture on erodibility of soil by wind *Soil Sci. Soc. Am. Proc.* **20** 288–92

Chepil W S, Siddoway F H and Armbrust D V 1962 Climatic factor for estimating wind erodibility of farm fields *J. Soil Water Conserv.* **17** 162–5

Corbett E S and Crouse R P 1968 *Rainfall Interception by Annual Grass and Chaparral* U.S. Department of Agriculture Forest Service Research Paper PSW-48 (Berkeley, CA: Pacific Southwest Forest and Range Experiment Station)

Dai D and Tan Y 1996 Soil erosion and sediment yield in the Upper Yangtze River basin *The Exeter Symp. Erosion and Sediment Yield Global and Regional Perspectives* (Rennes: International Association of Hydrological Sciences (IAHS)) p 236

Dixon S L and Hall C A 2010 *Fluid Mechanics and Thermodynamics of Turbomachinery* (Oxford: Butterworth–Heinemann)

Dong Z, Liu X and Wang X 2002 Wind initiation thresholds of the moistened sands *Geophys. Res. Lett.* **29** 1585

Fécan F, Marticorena B and Bergametti G 1999 Parameterization of the increase of the aeolian erosion threshold wind friction velocity due to soil moisture for arid and semi-arid areas *Ann. Geophys.* **17** 149–57

Feddes R A, Kowalik P J and Zaradny H 1978 *Simulation of Field Wate Use and Crop Yield. Simulation Monographs* (Wageningen: Wageningen Centre for Agricultural Publishing and Documentation) p 189

Fullen M A 1985 Wind erosion of arable soils in east Shropshire (England) during Spring 1983 *Catena* **12** 111–20

Fullen M A 1991 A comparison of runoff and erosion rates on bare and grassed loamy sand soils *Soil Use Manag.* **7** 136–9

Fullen M A 1998 Effects of grass ley set-aside on runoff, erosion and organic matter levels in sandy soils in east Shropshire, UK *Soil Tillage Res.* **46** 41–9

Fullen M A and Reed A H 1986 Rainfall, runoff and erosion on bare arable soils in east Shropshire, England *Earth Surf. Processes Landforms* **11** 413–25

Gao L, Bowker M A, Xu M, Sun H, Tuo D and Zhao Y 2017 Biological soil crusts decrease erodibility by modifying inherent soil properties on the Loess Plateau, China *Soil Biol. Biochem.* **105** 49–58

Gao L, Sun H, Xu M and Zhao Y 2020 Biocrusts resist runoff erosion through direct physical protection and indirect modification of soil properties *J. Soils Sediments* **20** 133–42

Gao Q, Ci L and Yu M 2002 Modeling wind and water erosion in northern China under climate and land use changes *J. Soil Water Conserv.* **57** 46–55

Gillette D 1978 Tests with a portable wind tunnel for determining wind erosion threshold velocities *Atmos. Environ.* **12** 2309–13

Hjelmfelt A T 1991 Investigation of curve number procedure *J. Hydraul. Eng.* **117** 725–37

Ishizuka M, Mikami M, Yamada Y, Zeng F and Gao W 2005 An observational study of soil moisture effects on wind erosion at a Gobi site in the Taklimakan Desert *J. Geophys. Res.* **110** D18S03

Karssenberg D, Bierkens M F P and Rietkerk M 2017 Catastrophic shifts in semiarid vegetation–soil systems may unfold rapidly or slowly *Am. Nat.* **190** E146–55

Kirkbride M P and Reeves A D 1993 Soil erosion caused by low-intensity rainfall in Angus, Scotland *Appl. Geogr.* **13** 299–311

Kirkby M J 1964 Discussion of a paper by Schumm (1964) *Z. Geomorph. Supp.* **5** 237

Kroes J G, van Dam J C, Bartholomeus R P, Groenendijk P, Heinen M, Hendriks R F A, Mulder H M, Supit I and van Walsum P E V 2017 *SWAP Version 4: Theory Description and User Manual* (Wageningen: Wageningen Environmental Research)

Lancaster N and Baas A 1998 Influence of vegetation cover on sand transport by wind: field studies at Owens Lake, California *Earth Surf. Processes Landforms* **23** 69–82

Lu H and Shao Y 2001 Toward quantitative prediction of dust storms: an integrated wind erosion modeling system and its applications *Environ. Model. Softw.* **16** 233–49

Lyles L and Allison B 1975 Wind erosion: uniformly spacing nonerodible elements eliminates effects of wind direction variability *J. Soil. Water Conserv.* **30** 225–6

Marshall J K 1971 Drag measurements in roughness arrays of varying density and distribution *Agric. Meteorol.* **8** 269–92

Mualem Y 1976 A new model for predicting the hydraulic conductivity of unsaturated porous media *Water Resour. Res.* **12** 513–22

Musick H B and Gillette D A 1990 Field evaluation of relationships between a vegetation structural parameter and sheltering against wind erosion *Land Degrad. Rehab.* **2** 87–94

Nearing M A 1997 A single, continuous function for slope steepness influence on soil loss *Soil Sci. Am. J.* **61** 917–9

Oberbauer S F and Oechel W C 1989 Maximum CO_2-assimilation rates of vascular plants on an Alaskan arctic tundra slope *Ecography* **12** 312–6

Okin G S 2008 A new model of wind erosion in the presence of vegetation *J. Geophys. Res.* **113** F02S10

Owens I F 1969 Causes and rates of soil creep in the Chilton Valley, Cass, New Zealand *Arct. Alp. Res.* **1** 213–20

Penning de Vries F W T and van Laar H H 1982 *Simulation of Plant Growth and Crop Production* (Wageningen: Centre for Agricultural Publishing and Documentation)

Philip J R 1957 The theory of infiltration: the infiltration equation and its solution *Soil Sci.* **83** 345–57

Raupach M R, Gillette D A and Leys J F 1993 The effect of roughness elements on wind erosion threshold *J. Geophys. Res.* **98** 3023–9

Roo A P D, Wesseling C G and Ritsema C J 1996 LISEM: a single-event physically based hydrological and soil erosion model for drainage basins (part I: theory, input and output) *Hydrol. Processes* **10** 1107–17

Schaap M G and van Genuchten M T 2006 Modified Mualem–van Genuchten formulation for improved description of the hydraulic conductivity near saturation *Vadose Zone J.* **5** 27–34

Schumm S A 1964 Seasonal variations in erosion rates and processes on hillslopes in western Colorado *Z. Geomorph. Supp.* **5** 215–37

Shao Y and Lu H 2000 A simple expression for wind erosion threshold friction velocity *J. Geophys. Res.* **105** 22437–43

Simanton J R, Weltz M A and Larsen H D 1991 Rangeland experiments to parameterize the water erosion prediction project model: vegetation canopy cover effects *J. Range Manag.* **44** 276–82

Skidmore E L 1986 Wind erosion climatic erosivity *Clim. Change* **9** 195–208

Stockton P H and Gillette D A 1990 Field measurement of the sheltering effect of vegetation on erodible land surfaces *Land Degrad. Dev.* **2** 77–85

Thongjoo C, Miyagawa S and Kawakubo N 2005 Effects of soil moisture and temperature on decomposition rates of some waste materials from agriculture and agro-industry *Plant Prod. Sci.* **8** 475–81

USDA-NRCS 2004 *National Engineering Handbook* (Washington, DC: US Department of Agriculture Natural Resources Conservation Service)

van der Meer H G, Ryden J C and Ennik G C (ed) 1986 *Nitrogen Fluxes in Intensive Grassland Systems. Developments in Plant and Soil Sciences* (New York: Springer)

van Ittersum M K, Leffelaar P A, van Keulen H, Kropff M J, Bastiaans L and Goudriaan J 2003 On approaches and applications of the Wageningen crop models *Eur. J. Agron.* **18** 201–34

Vanderborght J, Fetzer T, Mosthaf K, Smits K M and Helmig R 2017 Heat and water transport in soils and across the soil–atmosphere interface: 1. Theory and different model concepts *Water Resour. Res.* **53** 1057–79

Vieira F C S and Nahas E 2005 Comparison of microbial numbers in soils by using various culture media and temperatures *Microbiol. Res.* **160** 197–202

Wakatsuki T and Rasyidin A 1992 Rates of weathering and soil formation *Geoderma* **52** 251–63

Wang X 2015 Vapor flow resistance of dry soil layer to soil water evaporation in arid environment: an overview *Water* **7** 4552–74

Wang X 2021 *Water Resources and Hydraulics* (New York: Cambridge University Press)

Wang X, Gao R and Yang X 2021 Responses of soil moisture to climate variability and livestock grazing in a semiarid Eurasian steppe *Sci. Total Environ.* **781** 146705

Wang X, Liu T, Li F, Gao R, Yang X, Duan L, Luo Y and Li R 2014 Simulated soil erosion from a semiarid typical steppe watershed using an integrated aeolian and fluvial prediction model *Hydrol. Processes* **28** 325–40

Wang X, Liu T and Yang W 2012 Development of a robust runoff-prediction model by fusing the Rational equation and a modified SCS-CN method *Hydrol. Sci. J.* **57** 1118–40

Wang X, Shang S, Yang W and Melesse A M 2008 Simulation of an agricultural watershed using an improved curve number method in SWAT *Trans. ASABE* **51** 1323–39

Wikipedia 2021 Senescence https://en.wikipedia.org/wiki/Senescence

Wolfe S A and Nickling W G 1993 The protective role of sparse vegetation in wind erosion *Prog. Phys. Geog.* **17** 50–68

Yamori W, Nagai T and Makino A 2011 The rate-limiting step for CO_2 assimilation at different temperatures is influenced by the leaf nitrogen content in several C_3 crop species *Plant, Cell Environ.* **34** 764–77

Yan Y, Xu X, Xin X, Yang G, Wang X, Yan R and Chen B 2011 Effect of vegetation coverage on aeolian dust accumulation in a semiarid steppe of northern China *Catena* **87** 351–6

Yoder D and Lown J 1995 The future of RUSLE: inside the new revised universal soil loss equation (special issue: water research and management in semiarid environments) *J. Soil Water Conserv.* **50** 484

Zadoks J C, Chang T T and Konzak C F 1974 A decimal code for the growth stages of cereals *Weed Res.* **14** 415–21

Zhao Y and Xu M 2013 Runoff and soil loss from revegetated grasslands in the hilly loess plateau region, China: influence of biocrust patches and plant canopies. *J. Hydrol. Eng.* **18** 387–93

Chapter 4

Mathematical models

This chapter introduces existing models and their application conditions. In this regard, the strengths and weaknesses of the models are compared to develop a set of criteria for model selection. Also, this chapter elaborates what new algorithms are needed to improve existing models and how the algorithms can be developed.

4.1 Overview

Since the 1980s a variety of mathematical models have been developed for various modeling applications. In terms of the primary dependent variables (i.e. outputs), these models can be grouped into five categories, namely climatic, meteorological, hydrologic, phenological, and soil erosion. The dependent variables of the models in one category may be treated as independent variables (i.e. drivers) of the models in another category, and vice versa. For instance, the climatic models simulate the transfer of energy and materials through the climate system, predicting precipitation and temperature with drivers of atmosphere, oceans, land surface, and ice, whereas the meteorological models mimic air circulation, water vaporization, and vapor movement and condensation, forecasting weather (i.e. precipitation, temperature, wind speed, and humidity) with drivers of vapor content in air, vertical and horizontal air temperature gradients, and topographic altitude and gradient. The climatic models have a modeling domain from continent to globe in size, a spatial resolution larger than 50 km, and a temporal resolution longer than 3 h. In contrast, the meteorological models can have a modeling domain smaller than towns in size, a spatial resolution of less than street districts, and a temporal resolution of seconds. The prediction accuracy of meteorological models is usually much higher than that of climatic models.

The hydrological models simulate the formation and movement of runoff as well as the fate and transport of soluble constitutes through the drainage system (e.g. watershed or soil column), predicting runoff volume and peak and constitutes

doi:10.1088/978-0-7503-4020-5ch4

loadings with drivers of weather, topography, land use and land cover (LULC), soil properties, and management practices. The soil-erosion models simulate the displacement processes of topsoil particles by water and wind, predicting the soil erosion rate with drivers of weather, hydrology, LULC, topography, soil properties, and management practices. These two categories of models share several common drivers, while the runoff predicted by the hydrological models is also a driver of the soil-erosion models. The phenological models mimic the physiological and morphological processes of plant growth, development, and senescence, predict biomass with drivers of plant species, climate, soil properties and nutrients, and management practices. Regardless of the models, management practices can be any human activities that influence the processes of interest, such as cultivation, diversion, drainage, irrigation, land conversion, and pasturing, to name just a few.

To simulate water–soil–vegetation interactions in semi-arid ecosystems, models should include algorithms for the abiotic and biotic dynamic processes presented in chapter 2 and conceptualized in chapter 3. That is, not all models in the above five categories are applicable for the purpose of ecohydrological modeling. Among those that can be adapted as ecohydrological models, HYDRUS-1D, the Soil and Water Assessment Tool (SWAT), and Soil–Water–Atmosphere–Plant (SWAP) have been most often applied around the world.

4.2 Comparisons of existing models

This section describes and compares the three widely used ecohydrological models, namely SWAT, HYDRYS-1D, and SWAP. Also, it analyzes and summarizes their application conditions, strengths, and weaknesses.

4.2.1 HYDRUS-1D

HYDRUS-1D is a finite-element model for simulating one-dimensional movement of water, heat, and multiple solutes in unsaturated, partially saturated, or fully saturated media (Šimůnek and van Genuchten 2008). It solves the Richards equation numerically (equation (3.28)) for saturated–unsaturated water flow and Fickian-based advection dispersion equations for heat (equation (3.52)) and solute transport (equation (3.53)). Its 'freezing module' represents hydrological processes for the times when soil temperature is below the freezing point (Zhao *et al* 2016). HYDRUS-1D simulates root growth by means of a logistic function, with water and salinity stresses defined according to functions proposed by Feddes *et al* (1978) or van Genuchten (1987), and considers water uptake by plant roots in the source term of equation (3.28). As a black-box statistical approach, the logistic function does not have detailed physiological parameters of the plant. HYDRUS-1D assumes solute transport by advection and dispersion in the liquid phase and by diffusion in the gaseous phase. Flow and transport can occur vertically, horizontally, or in a generally inclined direction. The upper boundary can be head, flux, or climate, whereas the lower boundary can be head, flux, or free drainage. The boundary conditions can be constant and/or time varying.

For unsaturated soils, HYDRUS-1D describes the retention curve (θ–ψ) as the function presented by Brooks and Corey (1964), van Genuchten (1980), or Schaap and van Genuchten (2006). The third function (equation (3.29)) is the modified van Genuchten retention curve to better describe hydraulic properties near saturation. The hysteresis is estimated by the empirical formulas proposed by Šimůnek *et al* (1999), which assumes that drying scanning curves are scaled from the main drying curve and wetting scanning curves are scaled from the main wetting curve. HYDRUS-1D computes the soil heat capacity and thermal conductivity using equations (2.7) and (2.8), respectively. It offers a Marquardt–Levenberg technique for inverse estimation of selected parameters from observed water contents, pressure heads, concentrations, and/or boundary fluxes.

4.2.2 SWAT

The Soil and Water Assessment Tool (SWAT) is a physically based, watershed-scale, continuous-time, distributed-parameter hydrological model that uses spatially distributed data on topography, land use, soil, and weather for hydrological modeling and operates on a daily time step (Arnold *et al* 1998, Arnold and Fohrer 2005, Neitsch *et al* 2011). Based on topography, SWAT subdivides a watershed into several subbasins for modeling purposes. Each subbasin delineated within the model is simulated as a homogeneous area in terms of climatic conditions, but additional subdivisions are used within each subbasin to represent different soils and land use types. Each of these individual areas is referred to as a hydrological response unit (HRU) and is assumed to be spatially uniform in terms of soils, land use, and topography.

SWAT is composed of three major components, namely subbasin, reservoir routing, and channel routing, each of which includes several subcomponents. For instance, the subbasin component consists of eight subcomponents: hydrology, weather, sedimentation, soil moisture, crop growth, nutrients, agricultural management, and pesticides. The hydrology subcomponent includes surface runoff, lateral subsurface flow, percolation, ground water flow, snowmelt, evapotranspiration, transmission losses, and ponds. Detailed descriptions of the methods used in modeling these components and subcomponents can be found in (Arnold *et al* 1998, Srinivasan *et al* 1998, Neitsch *et al* 2002). SWAT computes surface runoff using the SCS-CN method (equations (3.15)–(3.17)) with the CN adjusted according to soil moisture conditions (equations (3.18) and (3.19)) or the G–A method (equation (3.26)) (Arnold *et al* 1993), estimates PET using one of the three methods by Priestley–Taylor (Priestley and Taylor 1972), Hargreaves (Hargreaves and Samani 1985), and Penman–Monteith (equations (3.39)–(3.50)) (Allen *et al* 1998), and performs channel flow routing using the Muskingum method (Chow *et al* 1988, Wang 2021). Wang *et al* (2006) found that the Hargreaves method is the most accurate when PET < 4.0 mm d^{-1}, the Priestley method is most appropriate when PET > 8.0 mm d^{-1}, and the three PET methods have a comparable performance when PET $= 4.0$–8.0 mm d^{-1}.

SWAT uses four storage volumes to represent the water balance in each HRU in the watershed: snowpack, vadose zone (0–2 m), shallow aquifer (2–20 m), and deep aquifer (>20 m). The vadose zone can be subdivided into multiple layers. For the

SCS-CN method, the excess water available after accounting for initial abstraction and surface runoff infiltrates into the soil. A storage routing technique is used to simulate the flow through each soil layer. SWAT directly simulates only the saturated flow and assumes that water is uniformly distributed within a given layer. The unsaturated flow between layers is indirectly modeled using depth distribution functions for plant water uptake and soil water evaporation. Downward flow occurs when the soil water in the layer exceeds field capacity and the layer below is not saturated. The rate of downward flow is governed by the saturated hydraulic conductivity. Lateral flow in the vadose zone is simulated using a kinematic storage routing technique that is based on slope, slope length, and saturated conductivity (Wang 2021). Upward flow from a lower layer to the layer above is regulated by the soil water to field capacity ratios of the two layers. Percolation from the bottom of the vadose zone is recharged into the shallow aquifer.

SWAT simulates sediment yield using the modified universal soil loss equation (MUSLE) (Williams 1995). MUSLE predicts the amount of eroded soils in a HRU that would be delivered into the channel of the inclusive subbasin. In the channel network, sediment transport is modeled as a function of two processes, deposition and degradation, operating simultaneously in the reach (Bagnold 1977, Williams 1980). The maximum amount of sediment that can be transported from a reach segment is a function of the peak channel velocity. SWAT provides two options to compute deposition and degradation (Neitsch *et al* 2002). The first option is to use the same channel dimensions for the entire simulation, whereas the second option is to simulate downcutting and widening of the channel and update channel dimensions throughout the simulation.

SWAT simulates phenological plant development based on a daily accumulated heat unit (Neitsch *et al* 2011). A plant grows once the air temperature is above the plant-specific base temperature and goes into dormancy otherwise. The heat unit on a day of interest is the average air temperature of this day minus the base temperature. The plant matures when the summation of the daily heat units exceeds a plant-specific value. The potential biomass is modeled by simulating leaf area development, light interception, and conversion of intercepted light into biomass based on a plant-specific radiation-use efficiency, while the actual yield is estimated using a harvest index. SWAT considers stresses of temperature, water, and nutrients (e.g. nitrogen and phosphorus) on plant growth, but it does not trace root, stem, and storage organ growths, likely misrepresenting roles of vegetation in soil and water dynamics. For a perennial plant (e.g. grass), its root depth is always equal to the allowable maximum of the plant regardless of its current age.

4.2.3 SWAP

Soil–Water–Atmosphere–Plant (SWAP) solves equations (3.28) and (3.52) simultaneously using a finite-difference approach (Kroes *et al* 2017, Kroes 2020). In this regard, the terms in those two equations are determined by equations 2.27), (2.28), and (3.29)–(3.34) and other formulas embedded in the program. As illustrated in figure 3.3, a soil column of interest is subdivided into several sublayers, each of which should be homogeneous in terms of texture and soil–water properties, and in turn, each sublayer can be further discretized into a number of compartments with a smaller thickness to best mimic heat and water dynamics. Driven by climate time series, SWAP continuously simulates the state variables such as $\theta(\psi)$ in terms of specified parameters of management practices (e.g. grazing), physiography, and physiology. Its latest version (i.e. SWAP 4.0.1A) fixes a bug in the grass growth module (Kroes 2020).

SWAP uses a water balance module to calculate water storage change in a ground-surface ponding reservoir (equation (3.7)). The influxes into the reservoir include throughfall, irrigation, snowmelt, run-on, floodwater, and discharge from (or seepage into) the top-most compartment, whereas the effluxes out of the reservoir include evaporation, surface runoff, and water exchange with macropores. The surface runoff occurs when the water storage in the reservoir exceeds its maximum capacity and can be computed as the ratio of an exponential function of the difference of storage and capacity to a resistance parameter for the reservoir attenuation effects (equation (3.51)). In addition, SWAP estimates the interflow from a saturated compartment as an exponential function of the difference between the shallow groundwater level and the hydraulic head of the drain. Further, the direct runoff is computed as the summation of surface runoff and interflows from all saturated compartments. Moreover, SWAP uses either the Penman–Monteith method (Penman 1948, Monteith 1965) or an empirical reference evapotranspiration approach (Allen *et al* 1989, 1998) to determine PET. The actual evaporation is equal to PET if the reservoir is not empty and estimated as the upward Darcy flux out of the top-most compartment if the reservoir is dry. Precipitation is treated as rainfall when air temperature is above a threshold or snowfall when air temperature is below another threshold, and as mixture of thereof when air temperature is between these two thresholds. A temperature-based approach (Wang and Melesse 2005) is used to model snowpack accumulation and snowmelt. The throughfall is computed as the difference of rainfall and interception, which is estimated as a function of LAI (equation (3.13)) (Anten *et al* 1995, Wang *et al* 2012). In the SWAP's finite-difference scheme, the ponding reservoir is taken as the upper boundary.

SWAP simulates plant growth and senescence as conceptualized in section 3.5 by adopting WOFOST, a computer program developed in 1988 (Boogaard *et al* 2014) from the crop models presented by van Ittersum (2003). Since its release, WOFOST has gone through several upgraded versions resulting from its world-wide applications to simulate various agricultural crop systems (Ma *et al* 2011, de Wit *et al* 2012, Cheng *et al* 2016, Hack-ten Broeke *et al* 2016). WOFOST simulates the physiological processes of respiration, assimilation, and organ growing and deceasing to estimate crop yields, subject to field management practices (e.g. irrigation, fertilization, and pesticide control) for alleviating water and nutrient stresses. The WOFOST algorithms can be adapted to model perennial grass systems by using the forcing functions as a function of Julian day number instead of crop development stage as well as by relating root dry matter with rooting depth (Kroes *et al* 2017). Mowing or grazing practices can be triggered either by a minimum amount of aboveground dry matter or a fixed date. For grazing, the consumed biomass depends on the relation between livestock intensity, grazing days, biomass consumption, and dry matter loss due to manure droppings and treading. The dynamics of LAI, stem biomass, and root depth and density can be predicted using corresponding assimilation partitioning functions as a function of Julian day number.

4.2.4 Comparisons

As discussed above, the three ecohydrological models share some aspects in common but are framed differently (table 4.1). HYDRUS-1D and SWAP are applicable for ecosystems where evaporation, transpiration, and infiltration are dominant while runoff may occur sporadically. They model these hydrological processes through a selected soil column (i.e. layer) for a long-term period (e.g. years) at a time step of seconds or longer. The column can consist of several groups or sublayers, each of which is homogeneous in terms of soil properties such as texture, retention curve, porosity, residual soil moisture, field capacity, saturated soil moisture, saturated hydraulic conductivity, heat capacity, and thermal conductivity. To improve the simulation accuracy and increase numerical stability, a group or sublayer is usually further discretized into several thinner sub-sublayers with varying thicknesses. The sub-sublayers at depths where heat and mass fluxes and soil moisture fluctuate frequently and/or have a large vertical gradient, such as those near the ground surface, around root density changing points, and in the vicinity of sublayer interfaces, should have a relatively small (e.g. 1 cm) thickness (figure 4.1). The top-most sublayer may be parameterized to characterize physical or biological crusts if such a crust exists. Both models trace the physical dynamics of heat, water, carbon, and nutrients, including storage changes in a sub-sublayer, movements between adjacent sub-sublayers, and exchanges between the top-most sub-sublayer and upper boundary and between the bottom-most sub-sublayer and lower boundary. The models compute soil evaporation and plant transpiration separately, but they do not explicitly model drainage channels, hydraulic structures (e.g. reservoirs), and soil erosion and sedimentation. In contrast to HYDRUS-1D, SWAP models microtopographical depressions using a ground-surface ponding reservoir, the respiration of soil micro-organisms, the photosynthesis and respiration of plants, the growth and senescence of plant organs (i.e. roots and shoots), and land management practices (e.g. subdrainage, mowing, grazing, irrigation, and fertilization). HYDRUS-1D only models root growth and does not mimic physiological and phenological processes. It has a user-friendly Windows® interface for preparing inputs, estimating parameters, and exploring simulation results. SWAP does not have such an interface: its inputs need to be prepared manually in a text editor (e.g. Notepad®) and outputs explored in a spreadsheet software (e.g. Microsoft Excel®) unless some computer programming efforts are made by the user.

SWAT is applicable for ecosystems where most storm events generate some amount of runoff that routes through overland and streams. It calculates the water balance components, namely runoff, infiltration, evapotranspiration, and storage, on one HRU basis. The runoffs from all HRUs included in a subbasin are convoluted to

Table 4.1. Similarities and differences of the three ecohydrological models.

Item	HYDRUS-1D	SWAT	SWAP
Spatial domain	Column	Watershed	Column
Discretization of soil horizon	Groups and nodes	Rootzone and below-zone	Sublayers and compartments
Temporal domain	Long-term, continuous time	Long-term, continuous time	Long-term, continuous time
Simulation time step	Seconds or longer	Daily	Seconds or longer
Dominant movement direction	Vertical	Horizontal	Vertical
Basic unit for mass balance	Node	Hydrological response unit	Compartment
Drainage channel	No	Yes	No
Subdrain	No	No	Yes
Hydraulic structure	No	Yes	No
Microtopographic depression	No	Yes	Yes
Soil penetration	Yes	No	Yes
Soil water flow	Yes	No	Yes
Soil moisture	Yes	No	Yes
Soil temperature	Yes	No	Yes
Soil erosion and sedimentation	No	Yes	No
Soil respiration	No	No	Yes
Soil evaporation (separate)	Yes	No	Yes
Plant transpiration (separate)	Yes	No	Yes
Plant leaf growth	No	Yes	Yes
Plant stem growth	No	No	Yes
Plant storage organ growth	No	No	Yes
Plant root growth	Yes	No	Yes
Plant senescence	No	No	Yes
Plant photosynthesis	No	Yes	Yes
Plant respiration	No	No	Yes
Mowing and grazing	No	No	Yes
Irrigation	No	Yes	Yes
Fertilization	No	Yes	Yes
Physical crust	Yes	No	Yes
Biocrust	Yes	No	Yes
Preprocessing interface	Yes	Yes	No
Postprocessing interface	Yes	Yes	No

calculate the flows into the stream segment that drains this subbasin; the flows are then combined with those from other subbasins and routed through the downstream stream segments toward the watershed outlet. That is, it models these hydrological processes at three-levels spatially nested areas: HRUs, subbasins, and watershed. SWAT can be run for a long-term period (e.g. years) at daily time steps. For each HRU, its soil layer is subdivided into the rootzone and below-zone, each of which is assumed to be homogeneous in terms of soil properties such as texture, porosity, wilting point, field capacity, saturated soil moisture, and saturated hydraulic conductivity. Given that the soils in the rootzone or below-zone are usually heterogeneous both horizontally and vertically, the soil properties for each of the two zones are the overall averages of the corresponding point values. In this regard, the arithmetic, harmonic, and weighted averages may be used and computed as

soil property at point (x, y, z) in rootzone or below-zone [varying]

volume of soil element around point (x, y, z) [L^3]

arithmetic average soil property of rootzone or below-zone [varying]

$$\varsigma_{zone} = \frac{\iiint_{V_{zone}} \varsigma(x,y,z)\,dV}{V_{zone}} \tag{4.1}$$

gross soil volume of rootzone or below-zone [L^3]

total length of rootzone or below-zone along y-axis [L]

total length of rootzone or below-zone along x-axis [L]

total length of rootzone or below-zone along z-axis [L]

harmonic mean soil property of rootzone or below-zone [varying]

$$\varsigma_{zone} = \frac{X_{zone}Y_{zone}Z_{zone}}{\int_0^{Z_{zone}}\left[\int_0^{Y_{zone}}\left(\int_0^{X_{zone}} \frac{x}{\varsigma(x,y,z)}dx\right)dy\right]dz} \tag{4.2}$$

soil property at point (x, y, z) in rootzone or below-zone [varying]

soil property at point (x, y, z) in rootzone or below-zone [varying]

weight function for the soil property [–]

volume of soil element around point (x, y, z) [L^3]

weighted average soil property of rootzone or below-zone [varying]

$$\varsigma_{zone} = \frac{\iiint_{V_{zone}} \left[w(x,y,z)\varsigma(x,y,z)\right]dV}{\int_0^{V_{zone}} w(x,y,z)\,dV} \tag{4.3}$$

gross soil volume of rootzone or below-zone [L^3]

Given the coarse discretization of the soil layer, it is almost impossible to characterize physical or biological crusts in SWAT. This model does not trace transmissions of heat, water, carbon, or nutrients in soils, but it does calculate soil erosion by overland runoff and subsequent sedimentation by streamflow. SWAT computes soil evaporation and plant transpiration together as evapotranspiration, and explicitly models drainage channels, hydraulic structures (e.g. ponds and diversions), and microtopographic depressions as ground-surface storage. It does not calculate the respiration of soil micro-organisms or plants, or the senescence of plant organs (i.e. roots and shoots). SWAT simulates the growth of plant leaves and photosynthesis but does not mimic the growth of other plant organs, namely roots, stems, and storage organs. That is, SWAT only models leaf growth and does not mimic physiological and phenological processes. It allows agricultural management practices (e.g. subdrainage, irrigation, and fertilization) and cannot simulate grassland management practices (e.g. mowing and grazing). SWAT has a user-friendly ArcGIS® interface for delineating the boundaries of the watershed and its subbasins, defining HRUs, and estimating initial parameter values from three data layers of topography, soils, and LULC. The interface uses climate data and the estimated parameters to create the input files, executes automatic calibration tools, runs the SWAT model, and visualizes the simulation results.

4.3 Model selection

In addition to HYDRUS-1D, SWAT, and SWAP, other models have also been developed for modeling partial processes of water, soil, and vegetation (Krysanova and Arnold 2008, Chen *et al* 2015, Revelli and Porporato 2018). However, most of those models are designed to predict hydrological variables (e.g. flows) and treat soils and vegetation as static factors without considering water–soil–vegetation dynamic interactions. For instance, some models, such as the Hydrologic Engineering Center Hydrologic Modeling System (HEC-HMS) (USACE 2018), do not mimic physiological and phonological processes in plants, whereas other models, such as WOFST (Boogaard *et al* 2014), has very detailed algorithms for crop growth and senescence but oversimplifies hydrological processes and soil water dynamics. In practice, prior to modeling a semi-arid ecosystem of interest, a best model must be selected according to the criteria listed in table 4.2 and elaborated as follows.

Models are approximations of the real world. A model may have physical algorithms for some processes while using statistical relationships or even one or two empirical factors for other processes. To simulate an ecosystem of interest, the selected model must have sufficient details of the processes concerned. That is, the modeler must have a clear idea of what is to be modeled, which is termed the modeling objective. In practice, the modeling objective may predict streamflow, soil water, soil erosion, sedimentation, or vegetation productivity, with and without water and land management practices (e.g. diversion and grazing). First, if the primary concern is to predict streamflow, the candidate models such as HEC-HMS (USACE 2018) need to have accurate algorithms in partitioning rainfall into runoff, evapotranspiration, and infiltration as well as routing flows through overland and

Table 4.2. Selection criteria of models for semi-arid ecosystems.

No.	Item	Concerned variables or processes	Sample model (reference)
1	Modeling objective	• Runoff with soil and plant as factors	HEC-HMS (USACE 2018)
		• Water erosion with rainfall and plant as factors	WEPP (Flanagan and Livingston 1995)
		• Crop growth with climate and soil as factors	WOFST (Boogaard *et al* 2014)
		• Perennial plant with climate and soil as factors	J Model (Jouven *et al* 2006)
		• Integrated aeolian and fluvial erosion with climate and vegetation as factors	IAFP (Wang *et al* 2014)
		• Runoff with partial physiological processes	SWAT (Arnold *et al* 1993)
		• Abiotic and biotic processes except for soil erosion	SWAP (Kroes *et al* 2017)
		• Abiotic and biotic processes including soil erosion	SWAP-E (Wang 2022)
2	Dominant flux direction	• Horizontal	HEC-HMS, WEPP, IAFP, SWAT
		• Vertical	J Model, WOFST, SWAP, SWAP-E
3	Computational time step	• Daily	WEPP, WOFST, J Model, SWAT
		• Sub-daily	HEC-HMS, IAFP, SWAP, SWAP-E
4	Data requirement	• Climate	All
		• Topography	HEC-HMS, WEPP, IAFP, SWAT
		• Soil	WEPP, IAFP, SWAT, SWAP, SWAP-E
		• Land use and land cover (LULC)	SWAT
		• Management practices	HEC-HMS, WEPP, WOFST, SWAT, SWAP, SWAP-E
		• Plant physiology	WOFST, J Model, SWAP, SWAP-E
		• Streamflow	HEC-HMS, SWAT
		• Sediment	WEPP
		• Contaminants	SWAT
		• Soil moisture, temperature, and leachates	SWAP, SWAP-E
5	Graphic user interface	• ArcGIS®	HEC-HMS, SWAT
		• Windows®	HEC-HMS, WEPP, WOFST
		• Plain text editor	J Model, SWAP, SWAP-E
6	Documents	• User's manual	HEC-HMS, WEPP, WOFST, SWAT, SWAP, SWAP-E
		• Technical manual	HEC-HMS, J Model, IAFP, SWAT, SWAP, SWAP-E
		• Tutorial manual and/or video	HEC-HMS, WEPP
7	Technical support	• Access to developer	HEC-HMS, IAFP, SWAT, SWAP, SWAP-E
8	Social media	• Discussion forum for users	HEC-HMS, SWAT
9	Application results	• Compilation available for public	SWAT, SWAP

drainage networks. Second, if the primary concern is to predict the dynamics of soil water, the candidate models such as HYDRUS-1D (Šimůnek and van Genuchten 2008) and SWAP (Kroes *et al* 2017) need to have accurate algorithms in tracing heat and water movements in soils. In this regard, the soil water vaporization and extraction, soil moisture replenishment and redistribution, and pore vapor dispersion and condensation are vital for arid and semi-arid ecosystems (Wang 2015, Pedram *et al* 2018). Third, if the primary concern is to predict soil erosion and/or sedimentation, the candidate models such as WEPP (Flanagan and Livingston 1995) and IAFP (Wang *et al* 2014) need to have accurate algorithms in computing soil shear strengths and erosive forces by water and/or wind. The hydrological processes are simplified to estimate direct runoff with a little interest of infiltration and evapotranspiration, while plant biotic processes are usually not considered. The reinforcement effects of plant roots on soil shear strengths (Bordoni *et al* 2020, Liu *et al* 2021) and the reduction effects of vegetation canopy on erosive forces (Wang *et al* 2014) are represented by empirical multipliers. Fourth, if the primary concern is to predict plant productivity, the candidate models such as WOFST (Boogaard *et al* 2014) and J Model (Jouven *et al* 2006) must be able to mimic plant physiology and phonology with simple formulas for rootzone soil water and nutrient balancing. Finally, if the primary concern is to examine how water, soil, and plants interact in the long run, the candidate models such as SWAP (Kroes *et al* 2017) and SWAP-E (Wang 2022) must have physically based algorithms for the abiotic and biotic processes of water, soil, and vegetation in the ecosystem of interest. Although some degrees of simplifications can be achieved, the algorithms need to reflect the main physical features of the processes.

Once the candidate models are selected in accordance with the modeling objective, the dominant water fluxes in the ecosystem of interest should be elucidated by available data, existing literature, and field reconnaissance. As mentioned above, most semi-arid ecosystems are dominated by vertical water fluxes and may be better modeled using J Model, WOFST, SWAP, and SWAP-E, whereas a few semi-arid ecosystems may frequently incur a measurable amount of runoff and can be more appropriately modeled using HEC-HMS, WEPP, IAFP, and SWAT. The vertical flux-dominant ecosystems have very permeable soils and small topographic gradients, leading to extremely low chances for runoff to be generated by any type of rain, including heavy rains. In contrast, the horizontal flux-important ecosystems have not-very-permeable soils and/or large topographic gradients, inducing runoffs from storms with a rainfall intensity greater than the penetration capacity. On an annual basis, the ratio of runoff to rainfall is negligible (e.g. <1%) for a vertical flux-dominant ecosystem and can be larger than 10% for a horizontal flux-important ecosystem (Glenn *et al* 2015, Wang *et al* 2021). In addition, the allowable modeling time step must be considered in terms of application and climate data. For instance, to predict the runoff peak or diurnal cycle of photosynthesis and respiration, the models (e.g. HEC-HMS or SWAP) should be able to perform computations at a sub-daily (e.g. seconds to hours) time step. If daily climate data are used to drive such a model, a corresponding sub-daily dataset is created by executing a linear or nonlinear algorithm (Kroes *et al* 2017, USACE 2018). Further, the data on topography, soil, and LULC should be obtained from government agencies and other entities. The accuracies and spatial resolutions of these data may vary from country to country and from location to location in one country. For the contiguous USA, topographic data (e.g. the national elevation dataset (NED)) has an average vertical accuracy of 2.44 m (Gesch 2007) and a highest spatial resolution of 3 or 10 m (USGS 2020), the soil data (e.g. soil survey geographic database (SSURGO)) has a spatial scale of 1:12 000–1:63 360 (USDA 2021), and the LULC data (e.g. national land cover dataset (NLCD)) has a spatial resolution of 30 m (USGS 2021). The data on water and land managements, which can be collected from local governments and other relevant agencies, are important inputs into some models. If plant growth and senescence must be modeled in detail, the physiological and phenological parameters should be determined, probably using literature reviews, laboratory analyses, and/or field experiments (Wang *et al* 2021). For model calibration and validation, the horizontal flux-important ecosystems should have observations of streamflow and measured concentrations of sediment and contaminants if they are modeled, whereas the vertical flux-dominant ecosystems need to have observations of soil moisture and soil temperature. If leaching processes must be modeled, concentrations of leachates (e.g. nitrogen) in soil water may need to be measured. Moreover, a model with a graphic user interface (GUI), a user manual, a technical document, and a forum portal is preferred to that without these. Through the portal, modelers can exchange experiences and ask questions. It is also a plus to provide technical support and post a timely updated compilation of the publications based on applications of the model. For instance, HYDRYS-1D, SWAT, and SWAP have excellent user manuals and technical documents (Šimůnek and van Genuchten 2008, Neitsch *et al* 2011, Kroes *et al* 2017) and are fully supported by the developers. HYDRUS-1D has a Windows® interface and SWAT has an ArcGIS® interface, but SWAP does not have a GUI. In terms of collecting model-related publications and making them accessible to the public, the SWAT team has been doing better than the developers of the other two models. SWAP does not have a forum portal, while HYDRUS-1D and SWAT have forum portals with many global registers.

4.4 Development of new algorithms

The existing ecohydrological models, including HYDRUS-1D, SWAT, and SWAP, are developed for generic ecosystems and have no embedded algorithms specific for physical crusts, biocrusts, soils with a very low moisture level, and dry soil layers (DSLs), which are unique features of semi-arid and arid (i.e. dryland) ecosystems. Such algorithms are currently incomplete and need to be developed by carrying out a variety of research.

4.4.1 Physical crusts

Physical crusts result from the mechanical forces of raindrop splashing, livestock trampling, and human disturbances. These external forces harden the previously loose soils with an accumulation of salts and silica (Cantón *et al* 2014). Salts can release sodium (Na^+), calcium (Ca^{2+}), and magnesium (Mg^{2+}) ions, while silica

(SiO_2) is a hard, unreactive, and colorless compound that occurs as the mineral quartz and as a principal constituent of sandstone and other rocks. Once the soil aggregate is broken into particles by external forces, the increasing Na^+ ions disperse into pores, disrupting the forces that bind soil particles and causing soil swelling, whereas the increasing Ca^{2+} and Mg^{2+} ions and SiO_2 compete for the same pores, making soils solidified and almost cemented with little or no structure. These ions exist as various salts in physical crusts. In practice, the amount of Na^+ per unit mass of dry crusts can be termed salinity, while the combined amount of Ca^{2+} and Mg^{2+} per unit mass of dry crusts can be characterized as sodicity. The sodium adsorption ratio (SAR), which is the ratio of salinity to sodicity, determines the soil–water properties of a physical crust. SAR can be computed as

$$\mathrm{SAR} = \sqrt{\frac{2\mathrm{Na}^+}{\mathrm{Ca}^{2+} + \mathrm{Mg}^{2+}}} \tag{4.4}$$

SAR: sodium adsorption ratio of physical crust [–]
Na^+: milliequivalent of sodium ions in a milligram of dry physical crust [mEq mg^{-1}]
Ca^{2+}: milliequivalent of calcium ions in a milligram of dry physical crust [mEq mg^{-1}]
Mg^{2+}: milliequivalent of magnesium ions in a milligram of dry physical crust [mEq mg^{-1}]

Physical crusts have a very small hydraulic conductivity and reduce water penetration into the soils beneath, likely restricting plant emergence and growth. This can be attributed to topsoil swelling and cementing. Laboratory and field experiments need to be carried out to determine the influences of SAR and SiO_2 on residual soil moisture, porosity or saturated soil moisture, hydraulic conductivity, and the retention curve. For instance, equations (3.29) and (3.33) need to be modified to consider the SAR and SiO_2 of the physical crust. Such pending efforts are expected to derive relationships expressed as

$$\theta_{r,pc} = f_{r,pc}\left(\theta_r, \mathrm{SAR}, [\mathrm{SiO_2}]\right) \tag{4.5}$$

$\theta_{r,pc}$: residual moisture of physical crust [–]
$f_{r,pc}$: function to be determined by experiments [–]
θ_r: residual moisture of parent soil of physical crust [–]
SAR: sodium adsorption ratio of physical crust (Eq. 4.4) [–]
$[SiO_2]$: amount of silica in physical crust [mEq mg^{-1}]

$$\theta_{sat,pc} = f_{sat,pc}\left(\theta_{sat}, \mathrm{SAR}, [\mathrm{SiO_2}]\right) \tag{4.6}$$

$\theta_{sat,pc}$: saturated moisture of physical crust [–]
$f_{sat,pc}$: function to be determined by experiments [–]
θ_{sat}: saturated moisture of parent soil of physical crust [–]
SAR: sodium adsorption ratio of physical crust (Eq. 4.4) [–]
$[SiO_2]$: amount of silica in physical crust [mEq mg^{-1}]

unsaturated hydraulic conductivity of parent soil of physical crust (e.g., Eq. 3.33) [L T^-1]
amount of silica in physical crust [mEq mg^-1]
unsaturated hydraulic conductivity of physical crust [L T^-1]
function to be determined by experiments [L T^-1]
sodium adsorption ratio of physical crust (Eq. 4.4) [–]

$$K_{pc} = f_{K,pc}\left(K, SAR, [SiO_2]\right) \tag{4.7}$$

moisture of parent soil of physical crust (e.g., Eq. 3.29) [–]
amount of silica in physical crust [mEq mg^-1]
moisture of physical crust [–]
function to be determined by experiments [–]
sodium adsorption ratio of physical crust (Eq. 4.4) [–]

$$\theta_{pc} = f_{\theta,pc}\left(\theta, SAR, [SiO_2]\right) \tag{4.8}$$

Increasing SiO_2 increases pore size and connectivity, whereas increasing SAR causes more fine materials to disperse into and fill voids. As a result, the residual moisture, saturated moisture, and hydraulic conductivity for a physical crust are generally smaller than those for the parent soil of the crust. These soil–water parameters (equations (4.5)–(4.7)) are positively related to the SiO_2 amount but negatively associated with SAR. At a given moisture, the absolute suction head for a physical crust is greater than that for the crust's parent soil (figure 4.1)—head

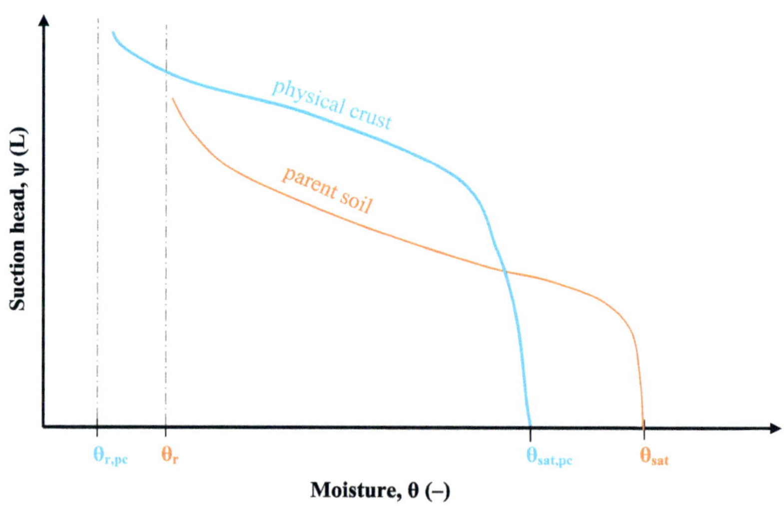

Figure 4.1. Retention curves for a physical crust and its parent soil. Illustration for a hypothetical case. $\theta_{r,pc}$ and $\theta_{sat,pc}$: residual and saturated moisture for the physical crust; and θ_r and θ_{sat}: residual and saturated moisture for the parent soil.

difference increases with an increase of SAR and/or decrease of SiO_2 amount, and vice versa. Equation (4.8) can be experimentally determined and used as the retention curve of the physical crust.

4.4.2 Biocrusts

A biocrust is formed by intimate associations of soil particles with cyanobacteria, algae, fungi, lichens and bryophytes, uncountable bacteria, and mosses. It is omnipresent as a living cover in the open spaces surrounding vascular plants. As with those of a physical crust, the effects of a biocrust on hydrology (figure 3.12) can be reflected by its residual moisture, saturated moisture, hydraulic conductivity, and retention curve, which are changed by living micro-organisms from those of the parent soil. As discussed in section 2.2.3, a biocrust can be characterized well by its level of development (LOD) (Belnap *et al* 2008). The micro-organisms aggregate surface soil particles by releasing extracellular polymeric substances (EPS), making the pores of a biocrust larger in size with better connectivity than those of the biocrust's parent soil. Because adolescent micro-organisms are more likely to clog pores than adult ones (Belnap 2006), a lower LOD biocrust has an overall smaller porosity, and is less permeable, than a higher LOD biocrust, although these two biocrusts share a common parent soil. Living micro-organisms retain and consume water, making the residual and saturated moisture of the biocrust larger than those of the corresponding parent soil. For a given moisture, the biocrust tends to present a smaller negative suction head than the parent soil (figure 4.2). Again, laboratory and field experiments need to be conducted to derive relationships expressed as

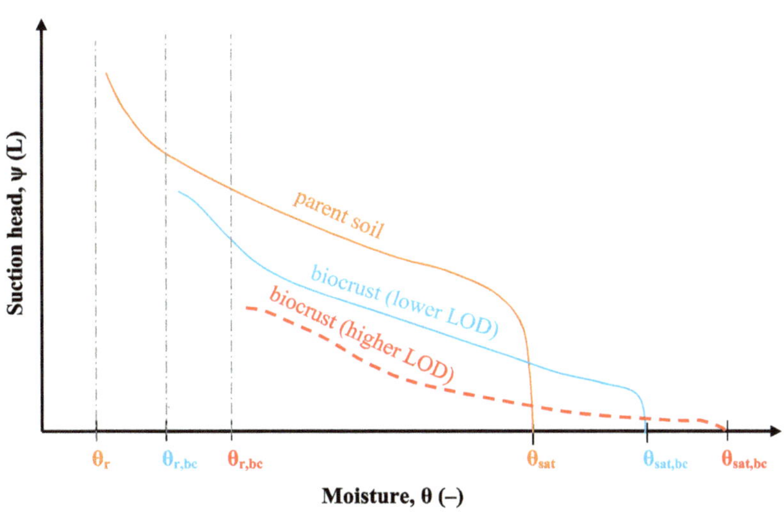

Figure 4.2. Retention curves for a biocrust and its parent soil. Illustration for a hypothetical case. $\theta_{r,bc}$ and $\theta_{sat,bc}$: residual and saturated moisture for the biocrust; θ_r and θ_{sat}: residual and saturated moisture for the parent soil; and LOD: level of development (Belnap *et al* 2008).

residual moisture of parent soil of biocrust [–]
biocrust level of development [1~6 integer]
residual moisture of biocrust [–]
function to be determined by experiments [–]

$$\theta_{r,bc} = f_{r,bc}\left(\theta_r, \mathrm{LOD}\right) \tag{4.9}$$

saturated moisture of parent soil of biocrust [–]
biocrust level of development [1~6 integer]
saturated moisture of biocrust [–]
function to be determined by experiments [–]

$$\theta_{sat,bc} = f_{sat,bc}\left(\theta_{sat}, \mathrm{LOD}\right) \tag{4.10}$$

unsaturated hydraulic conductivity of parent soil of biocrust (e.g., Eq. 3.33) [L T^{-1}]
biocrust level of development [1~6 integer]
unsaturated hydraulic conductivity of biocrust [L T^{-1}]
function to be determined by experiments [L T^{-1}]

$$K_{bc} = f_{K,bc}\left(K, \mathrm{LOD}\right) \tag{4.11}$$

moisture of parent soil of biocrust (e.g., Eq. 3.29) [–]
biocrust level of development [1~6 integer]
moisture of biocrust [–]
function to be determined by experiments [–]

$$\theta_{bc} = f_{\theta,bc}\left(\theta, \mathrm{LOD}\right) \tag{4.12}$$

4.4.3 Low-moisture soils

For a dryland ecosystem, the moisture of the topsoil (e.g. 5–10 cm) can frequently go below the wilting point (θ_ψ) and approach the residual moisture (θ_r), which is particularly true if the ecosystem is degraded and deserted. θ_r is the soil moisture in dry air with a suction head of $\psi = -33$ bar (i.e. -336.52 m water height), while θ_ψ is the soil moisture at $\psi = -15$ bar (i.e. -152.96 m water height). The well-known modified van Genuchen retention curve (equation (3.29)) is problematic when soil moisture $\theta \to \theta_r$ and thus needs somewhat further modification for low-moisture soils. In principle, when $\theta \to \theta_r$, a tiny change in θ corresponds to a very large change in ψ (i.e. $\frac{d\theta}{d\psi} \to 0$). In this regard, equation (3.29) leads to an unrealistic suction head of $\psi \to -\infty$ at θ_r. To solve this issue, one more segment function can be introduced for $\psi = -336.52$ to -152.96 m. Such a function must satisfy $\frac{d\theta}{d\psi} \mid \theta_r = 0$, $\psi(\theta_r) = -336.52$ m, and zero- and first-order geometric continuities at θ_ψ. As derived in appendix C, the further-modified equation is expressed as

$$\theta = \begin{cases} a + b\dfrac{|\psi_b|}{|\psi|} + c\ln\left(\dfrac{|\psi_b|}{|\psi|}\right) & \text{for } 152.96\,\text{m} \le |\psi| \le 336.52\,\text{m} \\ \theta_r + (\theta_{sat} - \theta_r)\left(1 + |\alpha\psi|^n\right)^{-m} & \text{for } |\psi_b| \le |\psi| \le 152.96\,\text{m} \\ \theta_{sat} & \text{for } |\psi| < |\psi_b| \end{cases} \tag{4.13}$$

soil moisture [–]; coefficients [–]; air-entry pressure head [m]; exponent (Eq. 3.31) [–]; exponent (Eq. 3.32) [–]; residual soil moisture [–]; saturated soil moisture [–]; suction head [m]; soil hysteresis factor (Eq. 3.30) [m^{-1}]

$$c = \frac{152.96^2}{183.56}(\alpha mn)(\theta_r - \theta_{sat})\frac{|152.96\alpha|^{n-1}}{\left(1 + |152.96\alpha|^n\right)^{m+1}} \tag{4.14}$$

coefficient in Eq. (4.13) [–]; soil hysteresis factor (Eq. 3.30) [m^{-1}]; residual soil moisture [–]; exponent (Eq. 3.32) [–]; exponent (Eq. 3.31) [–]; saturated soil moisture [–]

$$a = \theta_r + (\theta_{sat} - \theta_r)\left(1 + |152.96\alpha|^n\right)^{-m} + \left(\frac{336.52}{152.96} - \ln\left(\frac{|\psi_b|}{152.96}\right)\right)c \tag{4.15}$$

coefficient in Eq. (4.13) [–]; residual soil moisture [–]; exponent (Eq. 3.31) [–]; exponent (Eq. 3.32) [–]; air-entry pressure head [m]; saturated soil moisture [–]; soil hysteresis factor (Eq. 3.30) [m^{-1}]; Eq. (4.14) [–]

$$b = -\frac{336.52}{|\psi_b|}c \tag{4.16}$$

coefficient in Eq. (4.13) [–]; Eq. (4.14) [–]; air-entry pressure head [m]

As illustrated in figure 4.3, equations (4.13) and (3.29) are exactly the same for a moisture larger than the wilting point; otherwise, the prediction difference of these two equations is greater for a smaller moisture (i.e. larger negative suction head) than for a larger moisture (i.e. smaller negative suction head), and vice versa. For a suction head of $\psi = -336.52$ to -152.96 m, the moisture predicted by equation

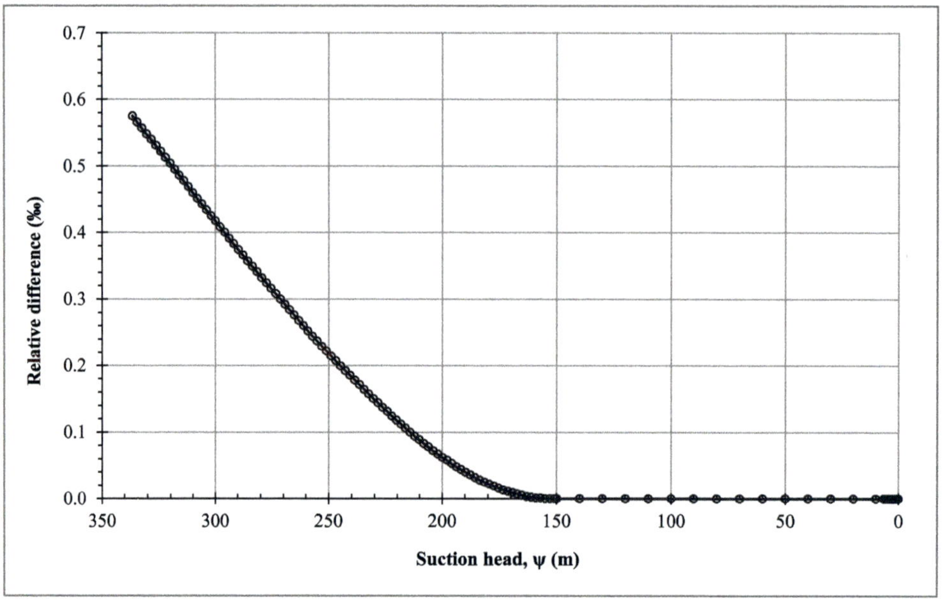

(a)

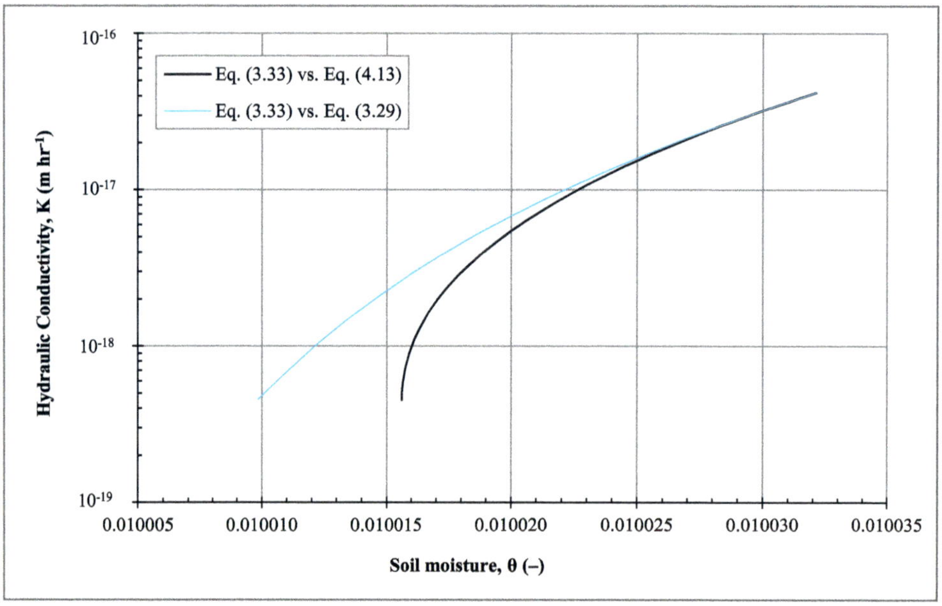

(b)

Figure 4.3. Prediction difference of equations (4.13) and (3.29) for (a) the retention curve and (b) hydraulic conductivity. Illustration for a sandy soil with a residual moisture $\theta_r = 0.01$, saturated soil moisture $\theta_{sat} = 0.38$, air-entry pressure head $\psi_b = -0.3$ m, pore-size distribution index $\lambda = 1.5$, saturated hydraulic conductivity $K_s = 0.05$ m h^{-1}, and pore connectivity parameter $\beta = 0.5$. The absolute suction heads are 152.96 m at the wilting point and 336.52 m at residual moisture.

(4.13) is slightly higher than that by equation (3.29). Given the importance of soil water for arid and semi-arid ecosystems, data on suction head versus moisture should be observed for very dry soils and used to validate which of equations (4.13) and (3.29) is more accurate or to develop a new equation if neither of them is accurate. So far, such data are not available because of technological challenge and historical ignorance. Rationally, equation (3.33) can be used to compute the hydraulic conductivity of dry soils if an accurate retention curve is adopted.

4.4.4 Dry soil layers

As discussed in sections 2.2.1 and 3.3.1, a dry soil layer (DSL) exists most of the time in semi-arid ecosystems (Wang 2015). The pores of the DSL are mainly occupied by vapor (Pedram *et al* 2018), which can move from a soil depth with a higher temperature to another depth with a lower temperature and/or condense into liquid water once the soil temperature is below the dew-point temperature. The condensed liquid water will be vaporized again when soil temperature goes up and the re-vaporized water moves along the soil temperature gradient. DSL can retard underlying soils from further drying. To understand dynamics (i.e. movement, condensation, and re-vaporization) of water within DSL, field observations are needed but very scarce in existing literature (Rossi and Nimmo 1994, Khlosi *et al* 2006). A conceptual observation scheme may be as follows.

For a site of interest, eighteen high-resolution three-needle heat pulse (HRTNHP) sensors (Heitman *et al* 2008) are installed at depths from 0 to 7 cm to capture sharply varying liquid–heat–vapor dynamics, while four three-needle heat pulse (TNHP) sensors (Pedram *et al* 2018, Wang *et al* 2021) are installed at depths from 7 to 16.6 cm, within which a spatial resolution of 6 mm should be sufficient to capture liquid–heat–vapor dynamics. Eighteen SHT75 humidity sensors (Sensirion 2021) and 18 EC-5 soil moisture sensors (Bogena *et al* 2007) are installed at a spacing interval of 4 mm from 0 to 7 cm and of 24 mm from 7 to 16.6 cm. The measured data by the sensors are recorded at a time interval of 5 s by CR3000 data loggers (Campbell 2021) and can be aggregated to have a computational time step (e.g. 15 min).

For each measurement position and each computational time, the SHT75 measured soil temperature and pore air humidity are used to compute the point suction head as (Edlefsen and Anderson 1943)

gas constant of water vapor [= 461.5 J kg^{-1} K^{-1}]
soil temperature [°C]
suction head [m]

$$\psi = \frac{R_{vapor}(T + 273.15)}{g}\ln(RH_{pa}) \qquad (4.17)$$

gravitational acceleration [= 9.81 m s^{-2}]
pore air humidity [–]

The computed suction head (ψ) at a measurement depth is paired with the EC-5 measured volumetric soil moisture (θ) at the same depth. As a result, four datasets of paired (ψ, θ) time series can be obtained. Plotted and analyzed depth by depth as well as for the entire 16.6 cm soil profile, these datasets are used to determine: (i) water retention functions (i.e. $\psi \sim \theta$); (ii) whether the functions are piecewise or continuous with θ; and (iii) whether and how hysteresis influences the retention functions. As elaborated in section 4.4.3, the retention functions for $\psi \leqslant -15$ bar are vital but lacking for dryland ecosystems. In this regard, although equation (4.13) seems theoretically reasonable, it needs to be validated using field observations. The determined water retention functions are substituted into the equation proposed by Mualem (1976) to derive the corresponding hydraulic conductivity functions for water and vapor.

A set of equations governing the heat–water dynamics in DSL are developed by Philip and de Vries (1957). Hereinafter, they are designated the PdV model for description purpose. The PdV model can be expressed as

$$\begin{cases} \dfrac{\partial(\theta_w+\theta_v)}{\partial t} = -\dfrac{\partial(q_w+q_v)}{\partial z} = \dfrac{\partial}{\partial z}\left[K_{L\psi}\left(\dfrac{\partial\psi}{\partial z}+1\right)+K_{LT}\dfrac{\partial T}{\partial z}+K_{v\psi}\dfrac{\partial\psi}{\partial z}+K_{vT}\dfrac{\partial T}{\partial z}\right] \\ \dfrac{\partial S_h}{\partial t} = -\dfrac{\partial q_h}{\partial z} = \dfrac{\partial}{\partial z}\left[\lambda_{sh}\dfrac{\partial T}{\partial z}-C_w(T-T_r)q_w-C_v(T-T_r)\cdot q_v-\lambda q_v\right] \\ S_h=(C_s\theta_s+C_w\theta_w+C_v\theta_v)(T-T_r)+\lambda\theta_v \end{cases} \tag{4.18}$$

volumetric liquid water content [m^3 m^{-3}]
volumetric water vapor content [m^3 m^{-3}]
Darcy flux of liquid water (+ upward) [m s^{-1}]
Darcy flux of water vapor (+ upward) [m s^{-1}]
hydraulic conductivity of liquid water [m s^{-1}]
suction head [m]
thermal hydraulic conductivity of liquid water [m^2 °C^{-1} s^{-1}]
hydraulic conductivity of water vapor [m s^{-1}]
thermal hydraulic conductivity of water vapor [m^2 °C^{-1} s^{-1}]
depth (+ upward) [m]
soil heat storage [J m^{-3}]
soil heat flux [J m^{-2} s^{-1}]
heat capacity of liquid water [= 4.18×10^6 J m^{-3} °C^{-1}]
time [s]
soil thermal conductivity [W m^{-1} °C^{-1}]
heat capacity of water vapor [= 1.8×10^6 J m^{-3} °C^{-1}]
latent heat of water vaporization [= 2.162 × 10^9 J m^{-3}]
soil temperature [°C]
heat capacity of soil particles [= 1.92×10^6 J m^{-3} °C^{-1}]
content of soil particles (1 – θ_w – θ_v) [m^3 m^{-3}]
reference soil temperature below which q_w and q_v contribute no heat flux [°C]

In equation (4.18) θ_v needs to be substituted by an equivalent volumetric liquid water content computed as

$$\theta_v' = \frac{\theta_v \rho_v}{\rho_w} \tag{4.19}$$

equivalent volumetric liquid water content [m^3 m^{-3}]
volumetric water vapor content [m^3 m^{-3}]
density of water vapor (Eq. 4.20) [kg m^{-3}]
density of liquid water [kg m^{-3}]

pore air humidity [–]; atmosphere pressure [kPa]; units conversion factor [–]; density of water vapor [kg m^{-3}]; gas constant of water vapor [= 461.5 J kg^{-1} K^{-1}]; soil temperature [°C]

$$\rho_v = \frac{1000\left(RH_{pa}\right)p_a}{R_{vapor}\left(T+273.15\right)} \tag{4.20}$$

The boundary conditions at ground surface and a depth are (Nassar *et al* 1992)

condensation rate (negative) [m s^{-1}]; water vapor release rate (positive) [m s^{-1}]; net evaporation rate [m s^{-1}]

$$E^t_{sf} = q_w(0,t) + q_v(0,t) \tag{4.21}$$

volumetric water vapor content [m^3 m^{-3}]; Darcy flux of water vapor (+ upward) [m s^{-1}]; depth-dependent vaporization (+) or condensation (-) rate [s^{-1}]; time [s]; depth (+ upward) [m]

$$\frac{\partial\theta_v}{\partial t} = -\frac{\partial q_v}{\partial z} + e \tag{4.22}$$

Fayer and Simmons (1995) proposed a soil water retention curve that accounts for the adsorption of water on soil under dry conditions. The retention curve is expressed as

shape factor [–]; soil moisture at suction head of -10^7 [cm]; exponent (Eq. 4.24) [–]; volumetric liquid water content [–]; saturated soil moisture [–]; soil hysteresis factor (Eq. 3.30) [cm^{-1}]; suction head [cm]; exponent (= 1 – 1/n$_1$) [–]; upper suction head for full saturation [= 10^{-7} to 10^{-20} cm]

$$\theta_w = \begin{cases} \chi\theta_a + \left(\theta_{sat} - \chi\theta_a\right)\left[1+\left|\alpha\psi\right|^{n_1}\right]^{-m} & \text{for } |\psi| > |\psi_c| \\ \theta_{sat} & \text{for } |\psi| \le |\psi_c| \end{cases} \tag{4.23}$$

exponent (Eq. 3.31) [–]; exponent in Eq. 4.23 [–]

$$n_1 = 1.23n - 0.162 \tag{4.24}$$

The shape factor in equation (4.23) can be determined as (Fayer and Simmons 1995)

$$\chi = 1 - \frac{\ln|\psi|}{\ln|\psi_m|} \tag{4.25}$$

shape factor [–]
suction head [cm]
suction head at oven dryness [= -10^7 cm]

Substituting equation (4.23) into the Mualem (1976) equation and then simplifying, yields (Fayer and Simmons 1995)

$$K_{L\psi} = K_s \left(\frac{\theta_w}{\theta_{sat}}\right)^{\beta} \left[\frac{F_s(|\psi|)}{F_{max}}\right]^2 \tag{4.26}$$

hydraulic conductivity of liquid water [L T^-1]
saturated hydraulic conductivity [L T^-1]
volumetric liquid water content [–]
saturated soil moisture [–]
pore connectivity parameter (0 ~ 1) [–]
Mualem integral (Eq. 4.27) [cm^-1]
suction head [cm]
total Mualem integral (Eq. 4.40) [cm^-1]

$$F_s(|\psi|) = I_1(w)\alpha\left[1-(\gamma+1)\frac{\theta_a}{\theta_{sat}}\right] + \frac{\alpha}{\ln|\psi_m|}\frac{\theta_a}{\theta_{sat}}\left[I_2(|\psi|, w) + \frac{I_3(w)}{n_1}\right] + C_0 \tag{4.27}$$

Mualem integral in Eq. 4.26 [cm^-1]
soil hysteresis factor (Eq. 3.30) [cm^-1]
soil moisture at suction head of -10^7 [cm]
function III [–]
suction head [cm]
function I [–]
variable by Eq. 4.28 [–]
parameter by Eq. 4.30 [–]
saturated soil moisture [–]
suction head at oven dryness [= -10^7 cm]
function II [–]
integral constant (assumed to be zero) [cm^-1]
exponent (Eq. 4.24) [–]

$$w = \frac{1}{1 + |\alpha\psi|^{n_1}} \tag{4.28}$$

variable in Eq. 4.27 [–]
exponent (Eq. 4.24) [–]
soil hysteresis factor (Eq. 3.30) [cm^-1]
suction head [cm]

$$w_m = \frac{1}{1 + |\alpha\psi_m|^{n_1}} \tag{4.29}$$

value of Eq. 4.28 at oven dryness [–]
exponent (Eq. 4.24) [–]
soil hysteresis factor (Eq. 3.30) [cm^-1]
suction head at oven dryness [= -10^7 cm]

soil hysteresis factor (Eq. 3.30) [cm^{-1}]

parameter in Eq. 4.27 [–]

$$\gamma = \ln\left(\left|\alpha \psi_m\right|\right) \tag{4.30}$$

suction head at oven dryness [= -10^{7} cm]

exponent in Eq. 4.23 [–]

function I in Eq. 4.27 [–]

$$I_1(w) = (1 - w_m)^m - (1 - w)^m \tag{4.31}$$

value by Eq. 4.29 [–]   variable by Eq. 4.28 [–]

exponent in Eq. 4.23 [–]

function II in Eq. 4.27 [–]

$$I_2(|\psi|, w) = \frac{1}{\alpha}\left(\frac{1}{|\psi_m|} - \frac{1}{|\psi|}\right) + (1 - w_m)^{m-1} - (1 - w)^{m-1} \tag{4.32}$$

value by Eq. 4.29 [–]

soil hysteresis factor (Eq. 3.30) [cm^{-1}]

suction head [cm]

variable by Eq. 4.28 [–]

suction head at oven dryness [= -10^{7} cm]

sub-function a (Eq. 4.34) [–]   sub-function b (Eq. 4.35) [–]

function III in Eq. 4.27 [–]

$$I_3(w) = \begin{cases} I_{3a}(w) + I_{3b}(w_0) & \text{for } w > w_0 \\ I_{3b}(w) & \text{for } w \le w_0 \end{cases} \tag{4.33}$$

variable by Eq. 4.28 [–]

threshold (Eq. 4.34) [–]

threshold in Eq. 4.33 [–]

$$w_0 = \min\left\{w_m, 10^{-10}\right\} \tag{4.34}$$

value by Eq. 4.29 [–]

sub-function a in Eq. 4.33 [–]

variable by Eq. 4.28 [–]

$$I_{3a}(w) = (1 - w_0)^m \left[\ln\left(\frac{1 - w_0}{w_0}\right) - \frac{1}{m}\right] - (1 - w)^m \left[\ln\left(\frac{1 - w}{w}\right) - \frac{1}{m}\right] + G(1 - w) - G(1 - w_0) \tag{4.35}$$

threshold (Eq. 4.34) [–]   exponent in Eq. 4.23 [–]   function (Eq. 4.37) [–]

$$I_{3b}(w) = m\left[(w - w_m) + w_m \ln(w_m) - w\ln(w)\right] \tag{4.36}$$

sub-function b in Eq. 4.33 [–]
variable by Eq. 4.28 [–]
exponent in Eq. 4.23 [–]
value by Eq. 4.29 [–]

$$G(x) = \begin{cases} G_s(0.9) + H(0.1) - H(1-x) & \text{for } x > 0.9 \\ G_s(x) & \text{for } x \le 0.9 \end{cases} \tag{4.37}$$

function in Eq. 4.35 [–]
independent variable [–]
function (Eq. 4.38) [–]
function (Eq. 4.39) [–]

$$G_s(y) = y^{m+1}\sum_{i=0}^{100}\frac{y^i}{m+1+i} \tag{4.38}$$

exponent in Eq. 4.23 [–]
function in Eq. 4.37 [–]
independent variable [–]
counter [–]

$$H(z) = \ln(z) - mz + \frac{m(m-1)}{4}z^2 - \frac{m(m-1)(m-2)}{18}z^3 \tag{4.39}$$

function in Eq. 4.37 [–]
exponent in Eq. 4.23 [–]
independent variable [–]

The total Mualem integral is computed as

$$F_{max} = F_s(|\psi_0|) + \alpha\left[1 - \frac{\theta_a}{\theta_{sat}}\left(\gamma + 1 - \frac{1}{n_1 \ln(|\psi_m|)}\right)\right]\left[|\alpha\psi_0|^{n_1-1} - |\alpha\psi_c|^{n_1-1}\right] + \frac{\alpha}{\ln|\psi_m|}\frac{\theta_a}{\theta_{sat}}\left[\begin{array}{l}|\alpha\psi_0|^{n_1-1}\left(\ln(|\alpha\psi_0|) - \frac{1}{n_1-1}\right) \\ -|\alpha\psi_c|^{n_1-1}\left(\ln(|\alpha\psi_c|) - \frac{1}{n_1-1}\right)\end{array}\right] \tag{4.40}$$

total Mualem integral in Eq. 4.26 [cm^{-1}]
upper suction head for natural saturation [≈ 0.025/α cm]
parameter by Eq. 4.30 [–]
exponent (Eq. 4.24) [–]
computed by Eq. 4.27 [cm^{-1}]
saturated soil moisture [–]
soil moisture at suction head of -1[cm]
suction head at oven dryness [= -10^7 cm]
soil hysteresis factor (Eq. 3.30) [cm^{-1}]
upper suction head for full saturation [= 10^{-7} to 10^{-20} cm]

The thermal hydraulic conductivity of liquid water is determined as (Nimmo and Miller 1986, Noborio *et al* 1996)

$$K_{LT} = K_{L\psi} \frac{|\psi| G_{wT}}{\sigma_0} \frac{d\sigma}{dT} \tag{4.41}$$

suction head [cm]; gain factor [–]; thermal hydraulic conductivity of liquid water [m² °C⁻¹ s⁻¹]; surface tension as function of T [mN m⁻¹]; pore water (i.e., soil) temperature [°C]; hydraulic conductivity of liquid water [m s⁻¹]; surface tension at 25°C [= 71.97 mN m⁻¹]

The surface tension of pure liquid water in contact with its vapor is determined as (IAPWS 2014)

$$\sigma = 1.07399\times10^{-4}\left(373.946 - T\right)^{1.256}\left(413.37975 + 0.625T\right) \tag{4.42}$$

pore water (i.e., soil) temperature [°C]; surface tension [mN m⁻¹]

Differentiating equation (4.42) with respect to T and simplifying, yields

$$\frac{d\sigma}{dT} = -1.07399\times10^{-4}\left(373.946 - T\right)^{0.256}\left(285.488716 + 1.41T\right) \tag{4.43}$$

pore water (i.e., soil) temperature [°C]; change gradient of surface tension curve [mN m⁻¹ °C⁻¹]

The temperature- and moisture-dependent gain factor in equation (4.41) can be determined from Nimmo and Miller (1986) and is determined as

$$G_{wT} = \frac{\psi_{Tj}/\psi_{Ti} - 1}{\sigma_{Tj}/\sigma_{Ti} - 1} \quad \text{for } T_j < T < T_i \tag{4.44}$$

suction heat at temperature Tj [cm]; suction heat at temperature Ti [cm]; gain factor in Eq. (4.41) [–]; surface tension at temperature Tj [mN m⁻¹]; surface tension at temperature Ti [mN m⁻¹]; pore water (i.e., soil) temperature [°C]

The hydraulic conductivities of water vapor are computed as (Sakai *et al* 2009)

$$K_{v\psi} = D\frac{\rho_{vs}}{\rho_w}\frac{g}{R_{vapor}\left(T + 273.15\right)}\left(RH_{pa}\right) \tag{4.45}$$

saturated vapor density (Eq. 4.20 with pore air humidity of 1.0 [kg m⁻³]; vapor diffusivity in soil [m² s⁻¹]; gravitational acceleration [= 9.81 m s⁻²]; hydraulic conductivity of water vapor [m s⁻¹]; density of water [kg m⁻³]; pore air humidity [–]; gas constant of water vapor [= 461.5 J kg⁻¹ K⁻¹]; pore water (i.e., soil) temperature [°C]

$$K_{vT} = D \frac{\eta}{\rho_w} \frac{d\rho_{vs}}{dT} \left(RH_{pa} \right) \tag{4.46}$$

- K_{vT}: thermal hydraulic conductivity of water vapor [m² °C⁻¹ s⁻¹]
- D: vapor diffusivity in soil [m² s⁻¹]
- η: enhancement factor [–]
- ρ_w: density of water [kg m⁻³]
- ρ_{vs}: saturated vapor density (Eq. 4.20 with pore air humidity of 1.0 [kg m⁻³]
- T: pore water (i.e., soil) temperature [°C]
- RH_{pa}: pore air humidity [–]

The vapor diffusivity in soil is determined as

$$D = \frac{\theta_v^{\frac{10}{3}}}{\theta_{sat}^2} D_a \tag{4.47}$$

- D: vapor diffusivity in soil [m² s⁻¹]
- θ_v: volumetric water vapor content [–]
- θ_{sat}: saturated soil moisture [–]
- D_a: vapor diffusivity in air [m² s⁻¹]

The enhancement factor is determined as

$$\eta = a + 3\frac{\theta_w}{\theta_{sat}} - (a-1)e^{-\left[\left(1+\frac{2.6}{\sqrt{f_c}}\right)\frac{\theta_w}{\theta_{sat}}\right]^3} \tag{4.48}$$

- η: enhancement factor [–]
- θ_w: volumetric liquid water content [–]
- θ_{sat}: saturated soil moisture [–]
- a: calibration constant [–]
- f_c: clay content [–]

For equation (4.20), letting RH_{pa} = 1.0, differentiating with respect to T, and simplifying, yields

$$\frac{d\rho_{vs}}{dT} = -\frac{1000 p_a}{R_{vapor} \left(T + 273.15 \right)^2} \tag{4.49}$$

- $\frac{d\rho_{vs}}{dT}$: saturated vapor density gradient [kg m⁻³ °C⁻¹]
- p_a: atmosphere pressure [kPa]
- R_{vapor}: gas constant of water vapor [= 461.5 J kg⁻¹ K⁻¹]
- T: soil temperature [°C]

The vapor diffusivity in air can be determined as (Campbell 1985)

$$D_a = 2.12 \times 10^{-5} \left(\frac{T + 273.15}{273.15} \right)^2 \tag{4.50}$$

- D_a: vapor diffusivity in air [m² s⁻¹]
- T: air temperature equivalent to soil temperature [°C]

Fayer and Simmons (1995) presented hydraulic properties of six soils (table 4.3), namely Palouse, Palouse B, Walla, Salkum, Royal, and L soil, and experimental

Table 4.3. Hydraulic properties of the six soils presented by Fayer and Simmons (1995).

Property	Palouse	Palouse B	Walla	Salkum	Royal	L soil
Residual soil moisture, θ_r [–]	0.070	0.000	0.070	0.000	0.066	0.034
Saturated soil moisture, θ_{sat} [–]	0.442	0.549	0.398	0.479	0.352	0.203
Soil hysteresis factor, α [cm^{-1}]	0.0103	0.0262	0.0100	0.0045	0.0078	0.0307
Exponent by equation (4.24), n_1 [–]	1.6707	1.3017	1.8552	1.4985	2.8884	2.4087
Soil moisture at suction head of 10^{-7} cm in equation (4.23), θ_a [–]	0.228	0.251	0.182	0.122	0.134	0.070

paired data on suction head and liquid water content. For each of these soils, the Fayer–Simmons retention curve (equation (4.23)) fits the data better than the modified van Genuchen retention curve (equation (3.29)), which is particularly true for low and high soil moisture. The latter retention curve is above the former retention curve for low and high moisture while it is below the curve for middle moisture levels. The hydraulic conductivity predicted by the van Genuchen–Mualem formula (equation (3.33)) is higher than that predicted by the Fayer–Simmons–Mualem formula (equation (4.26)) for a moisture of larger than the residual soil moisture and either higher or lower otherwise, depending on the soils. For the Royal and L soil, the Fayer–Simmons–Mualem formula seems to be unstable for moisture of less than 0.05, as evidenced by an unreasonable sudden jump of the predictions. This issue needs to be understood and addressed through future research. For illustrative purpose, figures 4.4–4.6 show the water retention and hydraulic conductivity curves for the three soils Palouse, Salkum, and Royal, respectively.

4.4.5 The SWAP-E model

SWAP-E is developed using SWAP as the general framework with a variety of upgrades while keeping other SWAP algorithms (Kroes *et al* 2017, Kroes 2020) unchanged. The upgrades, which are realized by modifying the source codes of SWAP, include:

- Substituting the first part of the PdV model for the Richards equation. As discussed above, water movements in arid and semi-arid soils may be in liquid, vapor, or mixture thereof. The Richards equation (equation (3.28)) works for liquid water only, whereas the PdV model (equation (4.18)) describes the dynamics of both liquid and vapor.
- Substituting the second part of the PdV model for the one-dimensional convection equation. The heats associated with liquid and vapor are vital for dryland ecosystems and must be considered well. However, such heats are not included in the one-dimensional convection equation (equation (3.52)), which is used by SWAP.
- Substituting the Fayer–Simmons retention curve for the modified van Genuchen retention curve. As shown in figures 4.4–4.6, the Fayer–Simmons retention curve (equation (4.26)) is much more accurate than the

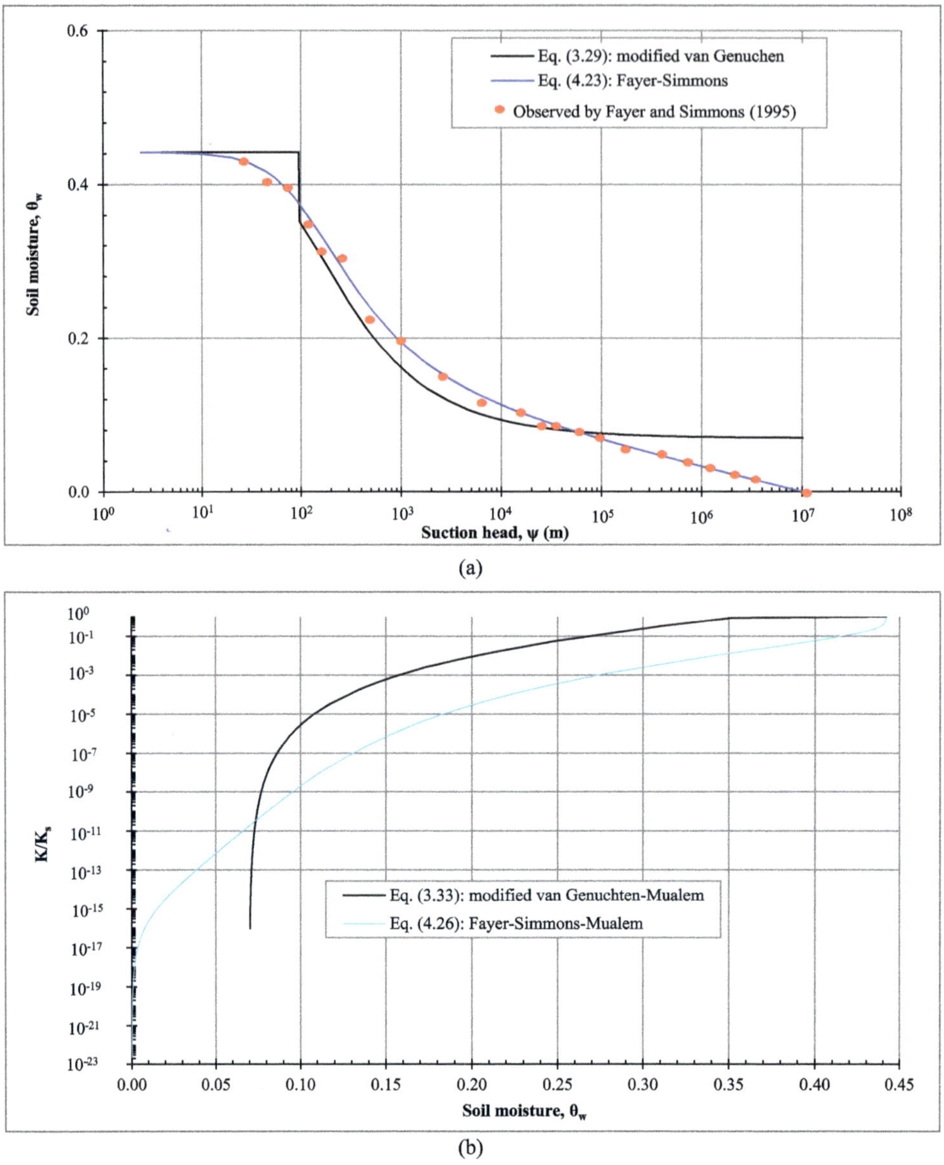

Figure 4.4. Predictions for the Palouse soil. The curves of (a) retention and (b) hydraulic conductivity. K: hydraulic conductivity; and K_s: saturated hydraulic conductivity.

van Genuchen retention curve (equation (3.33)) for low soil moisture, which are the situations of dryland ecosystems most of the time. Equation (3.33) assumes that soil moisture is independent of the suction head in the ranges of smaller than the residual soil moisture and larger than the air-entry soil moisture. This invalid assumption is overcome by equation (4.26).

- Substituting the Fayer–Simmons–Mualem hydraulic conductivity for the modified van Genuchen–Mualem hydraulic conductivity. The PdV model

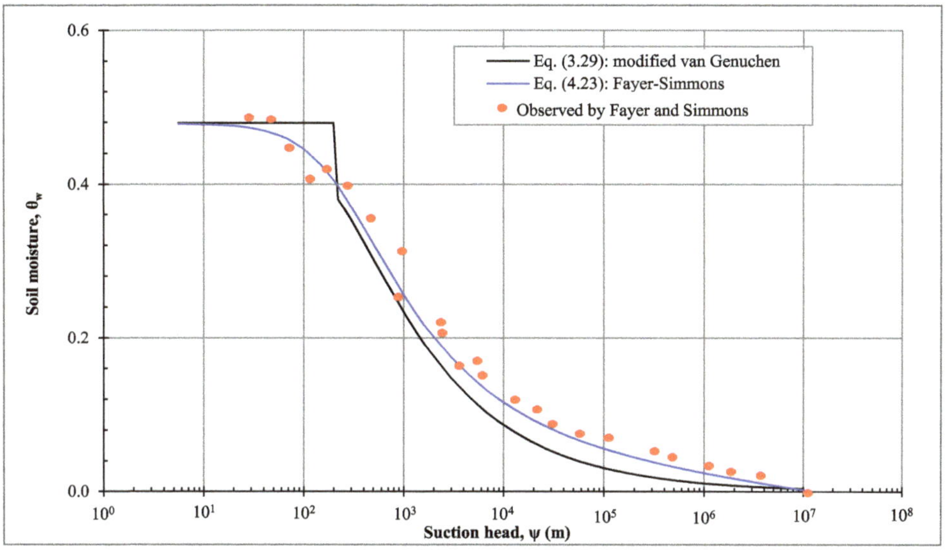

(a)

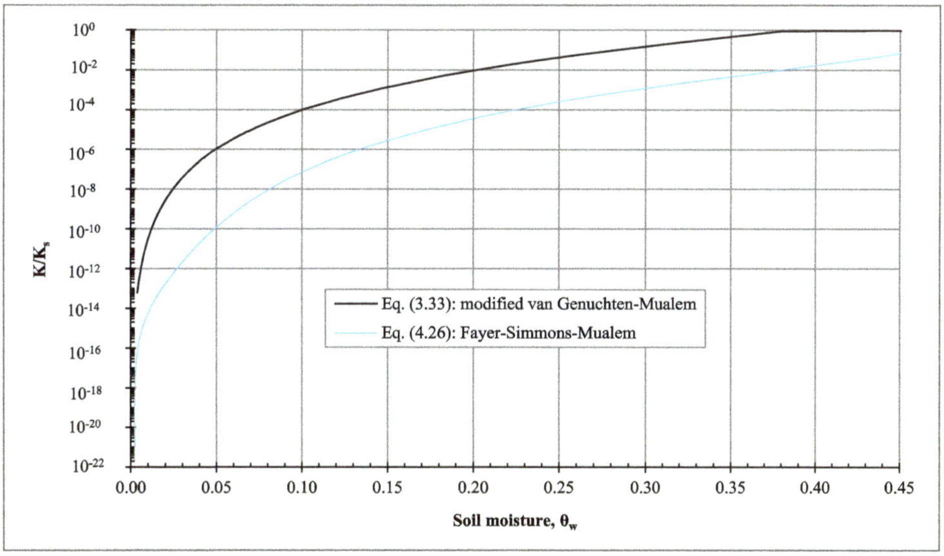

(b)

Figure 4.5. Predictions for the Salkum soil. The curves of (a) retention and (b) hydraulic conductivity. K: hydraulic conductivity; and K_S: saturated hydraulic conductivity.

requires four types of hydraulic conductivities, namely hydraulic conductivity of liquid (equation (4.26)), thermal hydraulic conductivity of liquid (equation (4.41)), hydraulic conductivity of vapor (equation (4.45)), and thermal hydraulic conductivity of vapor (equation (4.46)). As shown in figures 4.4–4.6, the modified van Genuchen–Mualem hydraulic conductivity (equation

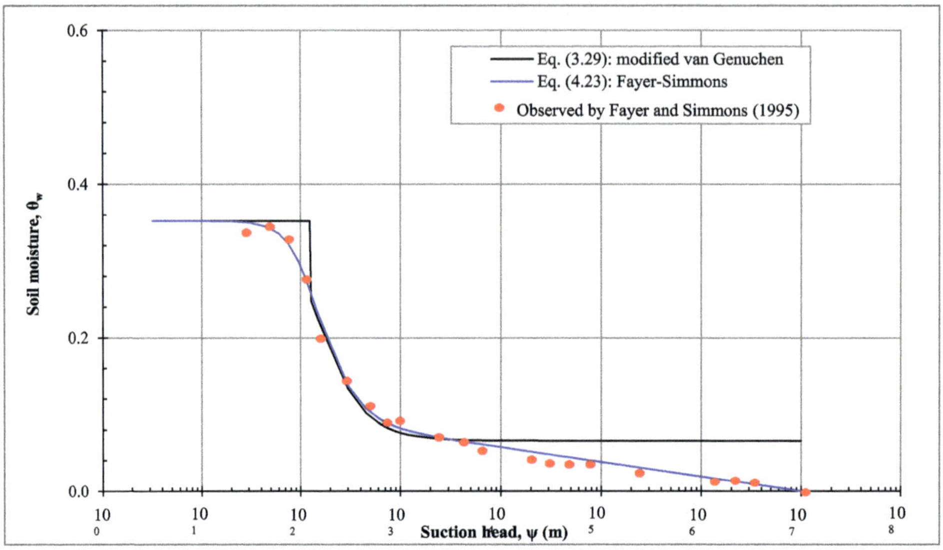

(a)

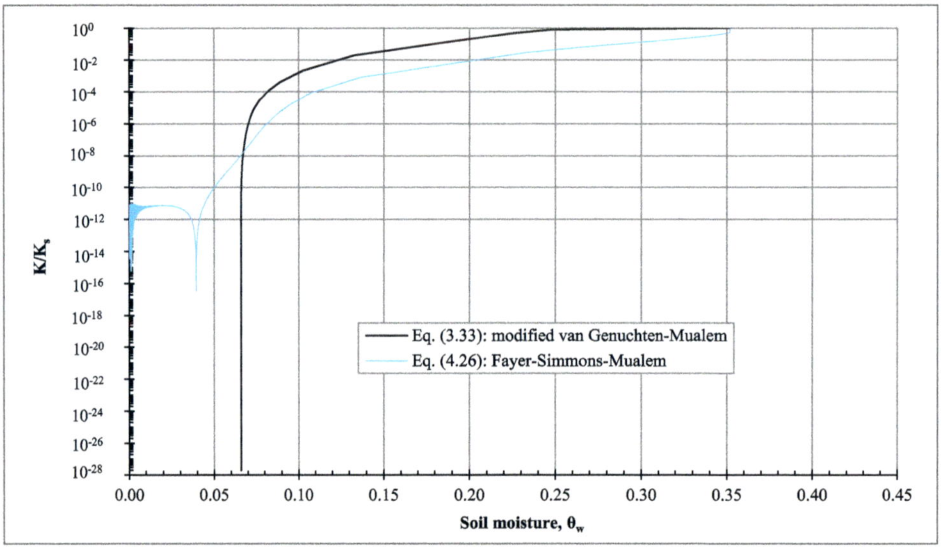

(b)

Figure 4.6. Predictions for the Royal soil. The curves of (a) retention and (b) hydraulic conductivity. K: hydraulic conductivity; and K_S: saturated hydraulic conductivity.

(3.33)) is for liquid only and assumes that hydraulic conductivity is independent of soil moisture in the range of smaller than the residual soil moisture. This assumption can result in overestimation of liquid fluxes while underpredicting vapor fluxes, biasing the dynamics of ecosystems.

- Adding new algorithms of physical and biological crusts. As discussed above, it is not uncommon for a dryland ecosystem of interest to have a large percentage of areas covered by physical crusts and/or biocrusts. The soil–water properties (e.g. residual soil moisture, saturated soil moisture, and saturated hydraulic conductivity) of such a crust are distinctly different from those of its parent soils. The algorithms like equations (4.5)–(4.8) and (4.9)–(4.12) are incorporated into SWAP-E to explicitly model the hydraulic effects of crusts.
- Embedding the IAFP model into SWAP. Dryland ecosystems incur frequent and widespread soil erosion by wind and water. The erosion physical processes are well described by the IAFP model (Wang *et al* 2014). SWAP-E can predict the mutual impacts between soil loss, hydrological alteration, and vegetation degradation.
- Coupling SWAP-E with a stand-alone module for ecosystem equilibrium and state shift. The algorithms like equations (3.1)–(3.6) are coded to create the stand-along module. Taking outputs of SWAP-E, the module predicts possible shifts of an ecosystem of interest and threshold conditions triggering an irreversible degradation of the ecosystem, as illustrated in figure 2.14.

4.5 Measures of model performance

For an ecohydrological model, its prediction accuracy (i.e. performance) is profound and should be assessed in accordance with observations and/or rationales. In this regard, a variety of measures have been developed and used. This section presents these measures in three categories, namely empirical judgment, visualization plot, and statistics.

4.5.1 Empirical judgement

For ecosystems where observed data are not available, the performance of a model cannot be quantitatively assessed. Instead, empirical knowledge or scientific judgements may be used to speculate on the reasonableness of the model's predictions. In generic hydrological principles, runoff and soil moisture tend to be larger for a heavier than a lighter storm, while soil moisture tends to be lower and evapotranspiration higher at a warmer than a cooler temperature. In phenological theory, plant productivity is generally higher in a wetter than a drier year, and vice versa.

Figure 4.7 illustrates predictions by two hypothetical models, designated Model I and Model II for description purpose. Obviously, Model I has a more reasonable performance than Model II. The predictions by Model I are consistent with the hydrological and phenological principles mentioned above, whereas Model II predicts lower runoff and soil moisture (figure 4.7(a)) for a larger than a smaller rainfall, a lower evapotranspiration (figure 4.7(b)), and a higher soil moisture (figure 4.7(c)) for a larger than a smaller air temperature, and a higher plant production (figure 4.7(d)) for a larger than a smaller aridity index. That is, the predictions by Model II are inconsistent with or even contradictory to the relevant

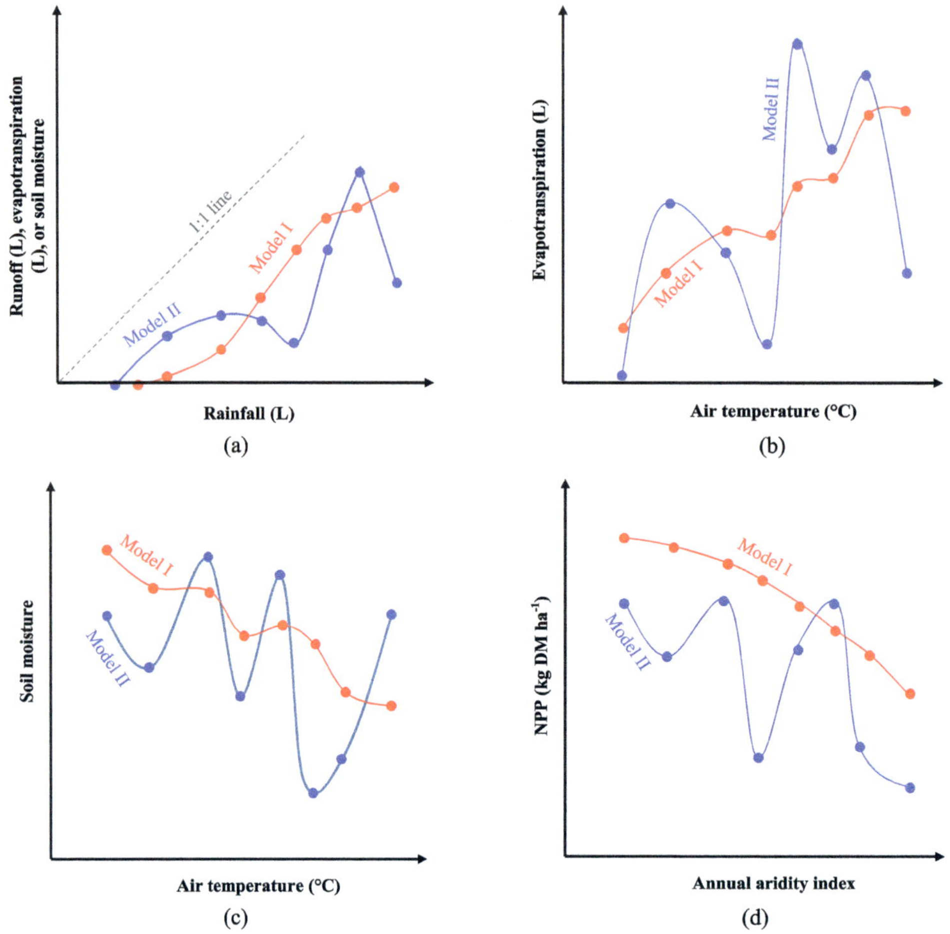

Figure 4.7. Reasonableness of model performance. Illustration of (a) runoff and soil moisture versus rainfall, (b) evaporation versus air temperature, (c) soil moisture versus air temperature, and (d) net primary production (NPP) versus aridity index at the annual scale. DM: dry mass.

rationales. Nonetheless, there is no evidence of prediction accuracy. Model I may or may not be more accurate than Model II because many other factors (e.g. grazing) also affect the assessment variables, which herein are runoff, soil moisture, evapotranspiration, and net primary production.

4.5.2 Visualization plot

For ecosystems where observed data are available, the performance of a model can be visually assessed by plotting modeled versus observed values of a concerned assessment variable (figure 4.8). If the paired dots are on the 1:1 line, the model is ideally accurate and has a perfect performance. However, it is more common for the dots to be scattered. If the paired dots have a distribution (i.e. clustering) pattern, the

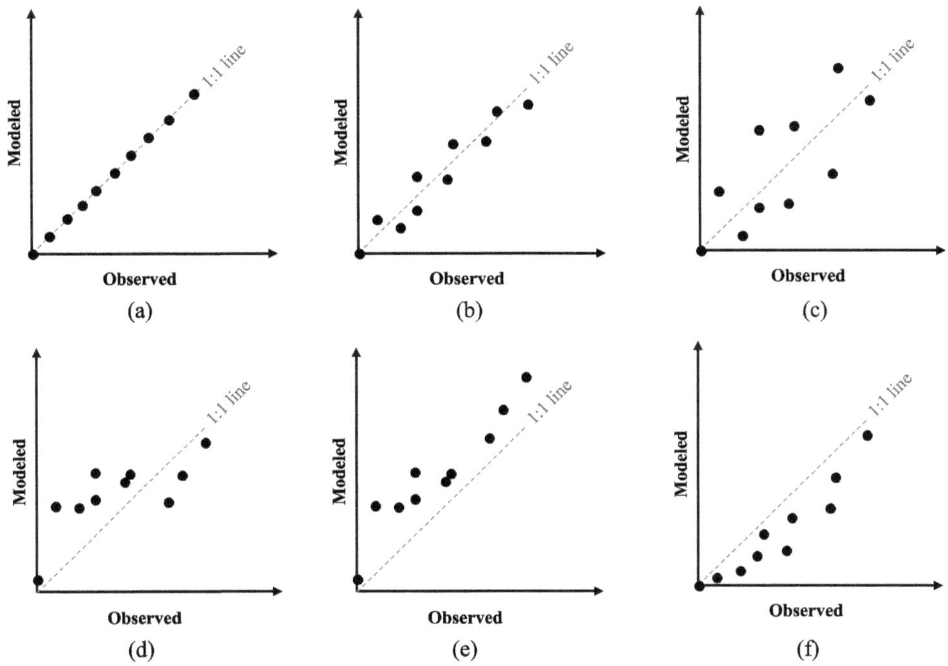

Figure 4.8. Visualization of model performance. Illustration of being: (a) on the 1:1 line; (b) randomly scattered with a small scatteredness; (c) randomly scattered with a large scatteredness; (d) clustered by ranges; (e) consistently above the 1:1 line; and (f) consistently below the 1:1 line.

model can be judged to have a biased performance, whereas if the paired dots are randomly scattered around the 1:1 line, the model can be judged to have a consistent performance, with a smaller scatteredness indicating a higher model accuracy and vice versa.

4.5.3 Statistics

In addition to the visualization plots discussed in section 4.5.2, a variety of statistics have been developed and used to quantitatively measure model performance. Among them, dimensionless statistics are independent of units and more favorable than dimensional ones. Based on the specific performance aspects to be measured, these statistics can be classified into five groups. Table 4.4 presents the widely used statistics, with numerical ranges and empirical thresholds for excellent, good, fair, and poor model performance.

The statistics in group I measure how accurately a model simulates the gross quantity of a concerned variable (e.g. runoff, evaporation, transpiration, or soil water) for a time duration of interest. In practice, five statistics, namely percent error (PE), mean percent error (MPE), relative root mean squared error (RRMSE), root mean squared relative error (RMSRE), and normalized root mean squared error (NRMSE), have been widely used.

Table 4.4. Measurement statistics for model performance.

Group	Performance aspect to be measured	Widely used statistics: possible range	Threshold[a] EP	GP	FP	PP
I	Mass conservation	• Percent error (PE): 0 to +∞	< 5	5–15	15–35	>35
		• Mean percent error (MPE): 0 to +∞	< 0.1	0.1–0.2	0.2–0.3	>0.3
		• Relative root mean squared error (RRMSE): 0 to +∞	< 0.1	0.1–0.2	0.2–0.3	>0.3
		• Root mean squared relative error (RMSRE): 0 to +∞	<0.1	0.1–0.2	0.2–0.3	>0.3
		• Normalized root mean squared error (NRMSE): 0 to +∞				
II	Observed variability	• Coefficient of determination (R^2): 0–1	>0.9	0.8–0.9	0.6–0.8	<0.6
		• Adjusted coefficient of determination (R^2_{adj}): 0–1	>0.9	0.8–0.9	0.6–0.8	<0.6
III	Goodness-of-fit	• Nash–Sutcliffe efficiency (NSE): −∞ to 1	>0.9	0.8–0.9	0.6–0.8	<0.6
		• Normalized Nash–Sutcliffe efficiency (NNSE): 0–1	>0.9	0.8–0.9	0.7–0.8	<0.7
		• Modified Nash–Sutcliffe efficiency (MNSE): −∞ to 0	>0.7	0.6–0.7	0.5–0.6	<0.5
		• Root relative squared error (RRSE): 0 to +∞	<0.3	0.3–0.4	0.4–0.6	>0.6
		• Kling–Gupta efficiency (KGE): −∞ to 1	>0.9	0.8–0.9	0.6–0.8	<0.6
IV	Frequency distribution	• Kolmogorov–Smirnov test (D_n): 0 to +∞	Table 4.5			
		• Cramér–von Mises test (ω^2): 0 to +∞	Table 4.6			
		• Kuiper's test (V): 0–1	Table 4.7			
V	Hydrograph	• Performance virtue (PV_k): −∞ to 1	>0.8	0.7–0.8	0.6–0.7	<0.6

[a] EP: excellent performance; GP: good performance; FP: fair performance; PP: poor performance. The thresholds may vary with actual applications (Wang and Melesse 2005, Ritter and Muñoz-Carpena 2013).

Table 4.5. Critical values of the Kolmogorov distribution.[a]

N	$\alpha = 0.01$	$\alpha = 0.05$	$\alpha = 0.1$
1	0.995	0.975	0.950
5	0.669	0.565	0.510
10	0.490	0.410	0.368
15	0.404	0.338	0.304
20	0.356	0.294	0.264
25	0.320	0.270	0.240
30	0.290	0.240	0.220
35	0.270	0.230	0.210
>35	$1.63/N^{½}$	$1.36/N^{½}$	$1.22/N^{½}$

[a] α: significance level; N: number of observed values.

Table 4.6. Critical values of Cramér–von Mises test.[a]

N	$\alpha = 0.01$	$\alpha = 0.05$	$\alpha = 0.1$
2	0.186	0.175	0.162
5	0.300	0.199	0.160
10	0.320	0.212	0.167
15	0.330	0.215	0.169
20	0.330	0.217	0.172
30	0.330	0.218	0.172
60	0.330	0.220	0.173
$\geqslant 100$	0.340	0.220	0.173

[a] α: significance level; N: number of observed values.

Table 4.7. Critical values of Kuiper's test.[a]

$\alpha = 0.01$	$\alpha = 0.05$	$\alpha = 0.1$
2.001	1.757	1.620

[a] α: significance level.

PE is computed as

$$PE = 100 \frac{\left| \sum_{i=1}^{N} y_{o,i} - \sum_{i=1}^{N} y_{m,i} \right|}{\sum_{i=1}^{N} y_{o,i}} \qquad (4.51)$$

number of observed values [–]
i^{th} modeled value [varying]
percent error [%]
multiplier [–]
i^{th} observed value [varying]

MPE is computed as

$$\text{MPE} = 100\frac{\left|\sum_{i=1}^{N} y_{o,i} - \sum_{i=1}^{N} y_{m,i}\right|}{N\sum_{i=1}^{N} y_{0,i}} \quad (4.52)$$

percent error [%]; multiplier [–]; number of observed values [–]; i^{th} modeled value [varying]; i^{th} observed value [varying]

RRMSE is computed as

$$\text{RRMSE} = \frac{\sqrt{\frac{1}{N}\sum_{i=1}^{N}\left(y_{o,i} - y_{m,i}\right)^2}}{\sum_{i=1}^{N} y_{0,i}} \quad (4.53)$$

relative root mean squared error [–]; number of observed values [–]; i^{th} modeled value [varying]; i^{th} observed value [varying]

RMSRE is computed as

$$\text{RRMSE} = \sqrt{\frac{1}{N}\sum_{i=1}^{N}\left(\frac{y_{o,i} - y_{m,i}}{y_{0,i}}\right)^2} \quad (4.54)$$

root mean squared relative error [–]; number of observed values [–]; i^{th} modeled value [varying]; i^{th} observed value [varying]

NRMSE is computed as

$$\text{NRMSE} = \frac{\sqrt{\frac{1}{N}\sum_{i=1}^{N}\left(y_{o,i} - y_{m,i}\right)^2}}{\max_{i=1}^{N}\left\{\left|y_{o,i} - y_{m,i}\right|\right\} - \min_{i=1}^{N}\left\{\left|y_{o,i} - -y_{m,i}\right|\right\}} \quad (4.55)$$

normalized root mean squared error [–]; number of observed values [–]; i^{th} modeled value [varying]; i^{th} observed value [varying]

The statistics in group II measure the proportion of variance in observations that can be explained by a model; that is, they tell how well the modeled and observed values are associated or correlated. In practice, two statistics, namely coefficient of determination (R^2) and adjusted coefficient of determination (R^2_{adj}), have been widely used. R^2 increases with an increase in the number of observations (Neter *et al* 1996) and thus may be misleading. In contrast, R^2_{adj} is independent of the number of

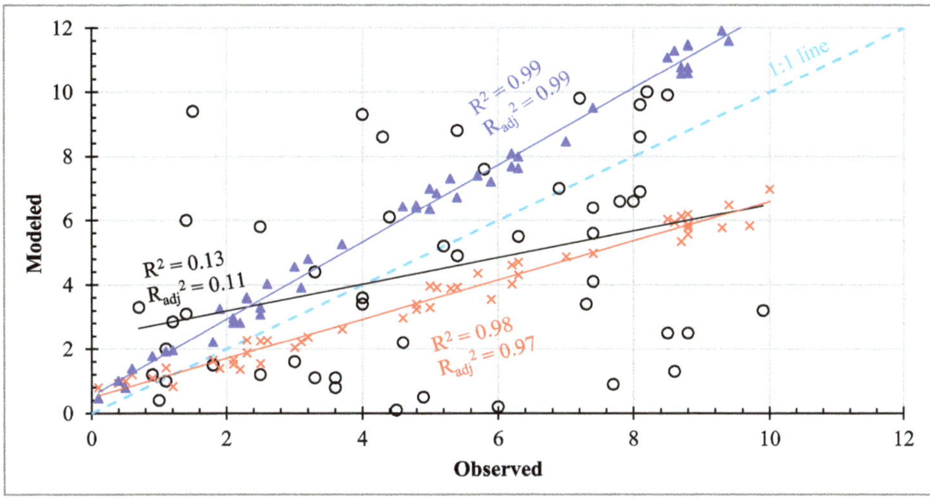

Figure 4.9. Coefficient of determination (R^2) and adjusted R^2 (R^2_{adj}). Hypothetical situations.

observations and relatively more realistic than R^2. Figure 4.9 shows the situations with contrasting values of R^2 and R^2_{adj}.

R^2 is computed as

number of observed values [–] · modeled mean [varying] · coefficient of determination [–]

$$R^2 = \frac{\text{Model-explained variance}}{\text{Total variance in observations}} = \frac{\left[\sum_{i=1}^{N}\left(y_{o,i}-\bar{y}_o\right)\left(y_{m,i}-\bar{y}_m\right)\right]^2}{\sum_{i=1}^{N}\left(y_{o,i}-\bar{y}_o\right)^2\sum_{i=1}^{N}\left(y_{m,i}-\bar{y}_m\right)^2} \tag{4.56}$$

i^{th} observed value [varying] · i^{th} modeled value [varying] · observed mean [varying]

R^2_{adj} is computed as

coefficient of determination (Eq. 4.56) [–]

adjusted coefficient of determination [–]

$$R^2_{adj} = \frac{(N-1)R^2-1}{N-2} \tag{4.57}$$

number of observed values [–]

The statistics in group III measure how well a model fits the observations (i.e. the magnitude closeness of modeled and observed values). That is, such a statistic indicates the overall distance from the points of modeled versus observed values to the 1:1 line. A larger value of the statistic implies a better goodness-of-fit. In practice, five statistics, namely Nash–Sutcliffe efficiency (NSE), normalized Nash–Sutcliffe efficiency (NNSE), modified Nash–Sutcliffe efficiency (MNSE), root relative squared error (RRSE), and Kling–Gupta efficiency (KGE), have been widely used. In comparison to NSE, MNSE is less sensitive to large outliers. Figure 4.10 shows that the model can explain most (or little) variance in the observations while it has a poor (or an excellent) goodness-of-fit.

NSE is computed as (Nash and Sutcliffe 1970)

$$NSE = 1 - \frac{\sum_{i=1}^{N}\left(y_{o,i} - y_{m,i}\right)^2}{\sum_{i=1}^{N}\left(y_{o,i} - \bar{y}_o\right)^2} \tag{4.58}$$

number of observed values [–]; i^{th} modeled value [varying]; Nash-Sutcliffe efficiency [–]; i^{th} observed value [varying]; observed mean [varying]

NNSE is computed as

$$NNSE = \frac{1}{2 - NSE} \tag{4.59}$$

normalized Nash-Sutcliffe efficiency [–]; Nash-Sutcliffe efficiency (Eq. 4.58) [–]

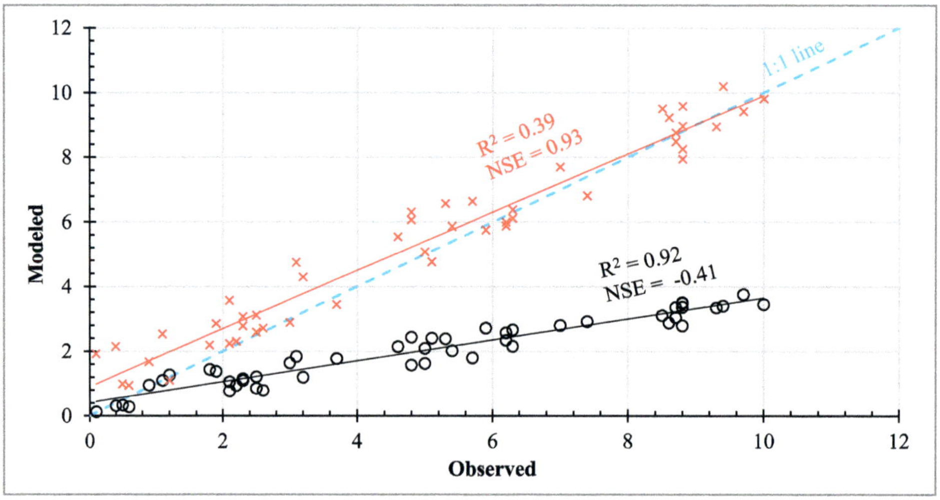

Figure 4.10. Goodness-of-fit versus explanation of variance in observations. Hypothetical situations. R^2: coefficient of determination; NSE: Nash–Sutcliffe efficiency.

MNSE is computed as

number of observed values [–] ; i^{th} modeled value [varying]

modified Nash-Sutcliffe efficiency [–]

$$\mathrm{MNSE} = 1 - \frac{\sum_{i=1}^{N} \left| y_{o,i} - y_{m,i} \right|}{\sum_{i=1}^{N} \left| y_{o,i} - \bar{y}_o \right|} \tag{4.60}$$

i^{th} observed value [varying] ; observed mean [varying]

RRSE is computed as

number of observed values [–] ; i^{th} modeled value [varying]

root relative squared error [–]

$$\mathrm{RRSE} = \sqrt{\frac{\sum_{i=1}^{N} \left(y_{o,i} - y_{m,i} \right)^2}{\sum_{i=1}^{N} \left(y_{o,i} - \bar{y}_o \right)^2}} \tag{4.61}$$

i^{th} observed value [varying] ; observed mean [varying]

KGE decomposes NSE into four components (Gupta *et al* 2009, Kling *et al* 2012). The first component is the Pearson product–moment correlation coefficient (Neter *et al* 1996). The second component is the ratio of the mean of the modeled values to that of the observations. The third component is the ratio of the standard deviation of the modeled values to that of the observations. The fourth component is the ratio of the variation coefficient of the modeled values to that of the observations. The variation coefficient is the division of the standard deviation by the mean. Ideally, the correlation coefficient and three ratios are equal to one. KGE is computed as

Kling-Gupta efficiency [–]

$$\mathrm{KGE} = 1 - \sqrt{\left(r_{mo} - 1 \right)^2 + \left(\mu_{mo} - 1 \right)^2 + \left(\sigma_{mo} - 1 \right)^2} \tag{4.62}$$

correlation coefficient (Eq. 4.63) [–] ; ratio of means (Eq. 4.64) [–] ; ratio of variation coefficients (Eq. 4.65) [–]

number of observed values [–] ; modeled mean [varying]

Pearson product-moment correlation coefficient [–]

$$r_{mo} = \frac{\sum_{i=1}^{N} \left(y_{o,i} - \bar{y}_o \right) \left(y_{m,i} - \bar{y}_m \right)}{\sqrt{\sum_{i=1}^{N} \left(y_{o,i} - \bar{y}_o \right)^2 \sum_{i=1}^{N} \left(y_{m,i} - \bar{y}_m \right)^2}} \tag{4.63}$$

i^{th} observed value [varying] ; i^{th} modeled value [varying] ; observed mean [varying]

modeled mean [varying]
ratio of modeled to observed mean [–]
observed mean [varying]

$$\mu_{mo} = \frac{\bar{y}_m}{\bar{y}_o} \tag{4.64}$$

i[th] modeled value [varying]
ratio of modeled to observed variation coefficients [–]
modeled mean [varying]
observed mean [varying]
i[th] observed value [varying]
number of observed values [–]

$$\sigma_{mo} = \frac{\sqrt{\dfrac{\sum_{i=1}^{N}\left(y_{m,i}-\bar{y}_m\right)^2}{N}}\Bigg/\bar{y}_m}{\sqrt{\dfrac{\sum_{i=1}^{N}\left(y_{o,i}-\bar{y}_o\right)^2}{N}}\Bigg/\bar{y}_o} \tag{4.65}$$

The statistics in group IV are used to test whether the modeled and observed values are from the same population (i.e. have an identical statistical distribution). In this regard, the modeled and observed cumulative density functions (CDFs) (Wang 2021) are determined using the corresponding values. In ascending order, the observations ($y_{o,1}$, $y_{o,2}$, …, $y_{o,i-1}$, $y_{o,i}$, $y_{o,i+1}$, …, $y_{o,N}$) are $\tilde{y}_{o,1} \leqslant \tilde{y}_{o,2} \leqslant \cdots \leqslant \tilde{y}_{o,i-1} \leqslant \tilde{y}_{o,i} \leqslant \tilde{y}_{o,i+1} \leqslant \cdots \leqslant \tilde{y}_{o,N}$ and the modeled values ($y_{m,1}$, $y_{m,2}$, …, $y_{m,i-1}$, $y_{m,i}$, $y_{m,i+1}$, $y_{m,N}$) are $\tilde{y}_{m,1} \leqslant \tilde{y}_{m,2} \leqslant \cdots \leqslant \tilde{y}_{m,i-1} \leqslant \tilde{y}_{m,i} \leqslant \tilde{y}_{m,i+1} \leqslant \cdots \leqslant \tilde{y}_{m,N}$. To determine a CDF, the corresponding ordered values are assigned consecutive ranks, and the cumulative probability of a sorted value can be computed as the division of the value's rank by N+1. If there are duplicates, each duplicate is assigned the mid-rank. For instance, if $\tilde{y}_{m,1} = \tilde{y}_{m,2} = \tilde{y}_{m,3}$ and there are no other duplicates, the rank for each of the three duplicates is (1 + 3)/2 = 2.0, and the subsequent values have ranks of 3, 4, …, N−1. Figure 4.11 illustrates how such CDFs would look like. In practice, three statistical tests, namely Kolmogorov–Smirnov (K–S), Cramér–von Mises (C–M), and Kuiper, are widely used.

The K–S statistic is computed as (Smirnov 1948, Simard and L'Ecuyer 2011)

supremum or maximum [–]
variable of interest [varying]
K-S statistic [–]

$$D_{KS} = \sup_{\tilde{y}}\left|F_o(\tilde{y}) - F_m(\tilde{y})\right| \tag{4.66}$$

observed cumulative density function [–]
modeled cumulative density function [–]

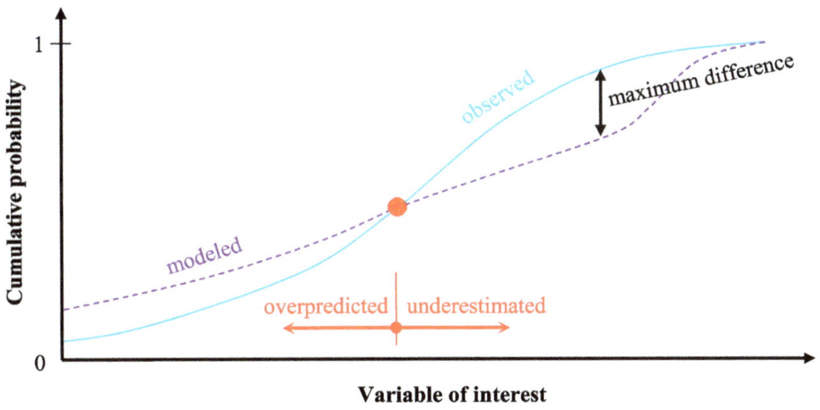

Figure 4.11. Modeled versus observed cumulative density function. Illustration of the concepts.

At a significance level of α, the modeled and observed values are statistically same for the condition defined as

$$D_{KS} \le \frac{K_{KS,\alpha}}{\sqrt{N}} \tag{4.67}$$

where D_{KS} = K-S statistic [–]; $K_{KS,\alpha}$ = critical value of Kolmogorov distribution (Table 4.5) [–]; α = significance level of interest [–]; N = number of observed values [–].

For a C–M test, the sorted observed and the sorted modeled values are pooled together into one dataset, whose members are in turn sorted in ascending order and assigned ranks as discussed above. The C–M statistic is computed as (Cramér 1928, von Mises 1928, Anderson 1962)

$$D_{CM} = \frac{6\left[\sum_{i=1}^{N}\left(r_{o,i}-i\right)^2 + \sum_{i=1}^{N}\left(r_{m,i}-i\right)^2\right] - \left(4N^2-1\right)N}{12N^2} \tag{4.68}$$

where D_{CM} = C-M statistic [–]; $r_{o,i}$ = new rank of ith sorted observed value in pooled dataset [–]; $r_{m,i}$ = new rank of ith sorted modeled value in pooled dataset [–]; N = number of observed values [–].

At a significance level of α, the modeled and observed values are statistically the same for the condition defined as

$$D_{CM} \le K_{CM,\alpha} \tag{4.69}$$

where D_{CM} = C-M statistic [–]; $K_{CM,\alpha}$ = critical value of Cramér–von Mises test (Table 4.6) [–]; α = significance level of interest [–].

The Kuiper's statistic is computed as (Kuiper 1960, Pearson and Hartley 1972)

$$D_{Kuiper} = \sup_{i \subseteq [1,N]} \left| \frac{r_{m,i}}{N} - F_o\left(\tilde{y}_{m,i}\right) \right| + \sup_{i \subseteq [1,N]} \left| F_o\left(\tilde{y}_{m,i}\right) - \frac{r_{m,i} - 1}{N} \right| \quad (4.70)$$

Kuiper's statistic [–]; supremum or maximum [–]; i^{th} sorted modeled value [varying]; rank of ith sorted modeled value [–]; counter [–]; observed cumulative density function [–]; number of observed values [–]

At a significance level of α, the modeled and observed values are statistically same for the condition defined as

$$D_{Kuiper} \leq K_{Kuiper,\alpha} \quad (4.71)$$

Kuiper's statistic [–]; critical value of Kuiper's test (Table 4.7) [–]; significance level of interest [–]

The statistics in group V measure how satisfactorily a model can predict all three of the aspects (silhouette, volume, and peak) of the observed hydrograph of a concerned variable (e.g. runoff, evapotranspiration, soil moisture, or soil temperature). To date, only one statistic, namely performance virtue (PV_k), has been developed and used for model evaluation (Wang and Melesse 2005, Wang *et al* 2006). It is defined as the weighted average of Nash–Sutcliffe efficiency (equation (4.58)), percent error (equation (4.51)), and percent peak deviation (equation (4.73)). The weights reflect the relative contributions of the three components to PV_k and their summation must be equal to one. For equivalent contributions, the weights are equal to 1/3. PV_k is computed as

$$PV_k = w_1\left(NSE\right) + w_2\left(1 - \frac{PE}{100}\right) + w_3\left(1 - \frac{PPD}{100}\right) \quad (4.72)$$

percent virtue [–]; Nash-Sutcliffe efficiency (Eq. 4.58) [–]; percent error (Eq. 4.51) [%]; percent peak deviation (Eq. 4.73) [%]; first weight (≤ 1.0) [–]; second weight (≤ 1.0) [–]; third weight (≤ 1.0) [–]

$$PPD = \sqrt{\left(\frac{Q_{p,o} - Q_{p,m}}{Q_{p,o}}\right)^2 + \left(\frac{t_{p,o} - t_{p,m}}{t_{p,o}}\right)^2} \quad (4.73)$$

percent peak deviation [%]; modeled peak magnitude [varying]; modeled peak timing [T]; observed peak magnitude [varying]; observed peak timing [T]

4.6 Model parameterization and calibration

It is not uncommon for an ecohydrological model to have many parameters. Before the model can be applied to simulate an ecosystem of interest, those parameters must be appropriately determined. This section presents how to estimate initial values of the parameters, identify the sensitive versus insensitive parameters, optimize the sensitive parameters, and analyze model uncertainty and robustness.

4.6.1 Initial parameter values

Depending on the types and details of the processes considered by an ecohydrological model, its parameters can be classified into five categories: climatic, hydrological, physiological, phenological, and anthropogenic. Climatic parameters (e.g. ocean water temperature and atmosphere circulation) are needed if the model is to be used to predict the future impacts of changing climate. Some models (e.g. SWAT) have default values for climatic parameters to generate site-specific data on precipitation, air temperature, humidity, wind speed, and solar radiation. Otherwise, the relevant parameter values can be extracted from various sources (IPCC 2001). Hydrological parameters (e.g. slope, slope length, and soil–water properties) are required by almost all ecohydrological models. Their initial values can be estimated by overlaying data layers on topographic elevation, soil, and LULC in a geographic information system such as ArcGIS®. Physiological (e.g. optimal root depth and leaf area) and phenological parameters (e.g. ideal photosynthesis rate and base temperature) are needed to model plant growth. The initial values for these parameters can be extracted from the existing literature and/or estimated by conducting experiments. Anthropogenic parameters (e.g. livestock grazing intensity) can be determined in terms of governmental documents, interviews of stakeholders, and field reconnaissance.

4.6.2 Sensitivity analysis

Given the large number of model parameters, the model simulation is usually sensitive to a subset of the parameters but least influenced by others (figure 4.12). The main purpose of sensitivity analysis is to differentiate sensitive from insensitive parameters (Forrester *et al* 2001) so that efforts can be made to refine the values of the sensitive parameters to a great deal of accuracy. In contrast, the insensitive parameters can be reasonably assumed to be the corresponding initial values because any changes in such parameters would have a minimal influence on the model performance.

Conventionally, sensitivity analysis is implemented one parameter at a time. That is, when a parameter is selected for sensitivity analysis, it is varied within a plausible range to have different values while all other parameters remain unchanged. Based on the model simulations, the changes of concerned variables (e.g. soil moisture and plant biomass) or model performance statistics discussed in section 4.5.3 in response to the changes in the selected parameter are examined to evaluate the sensitivity of this parameter (figure 4.12). However, there are three strong arguments for such a

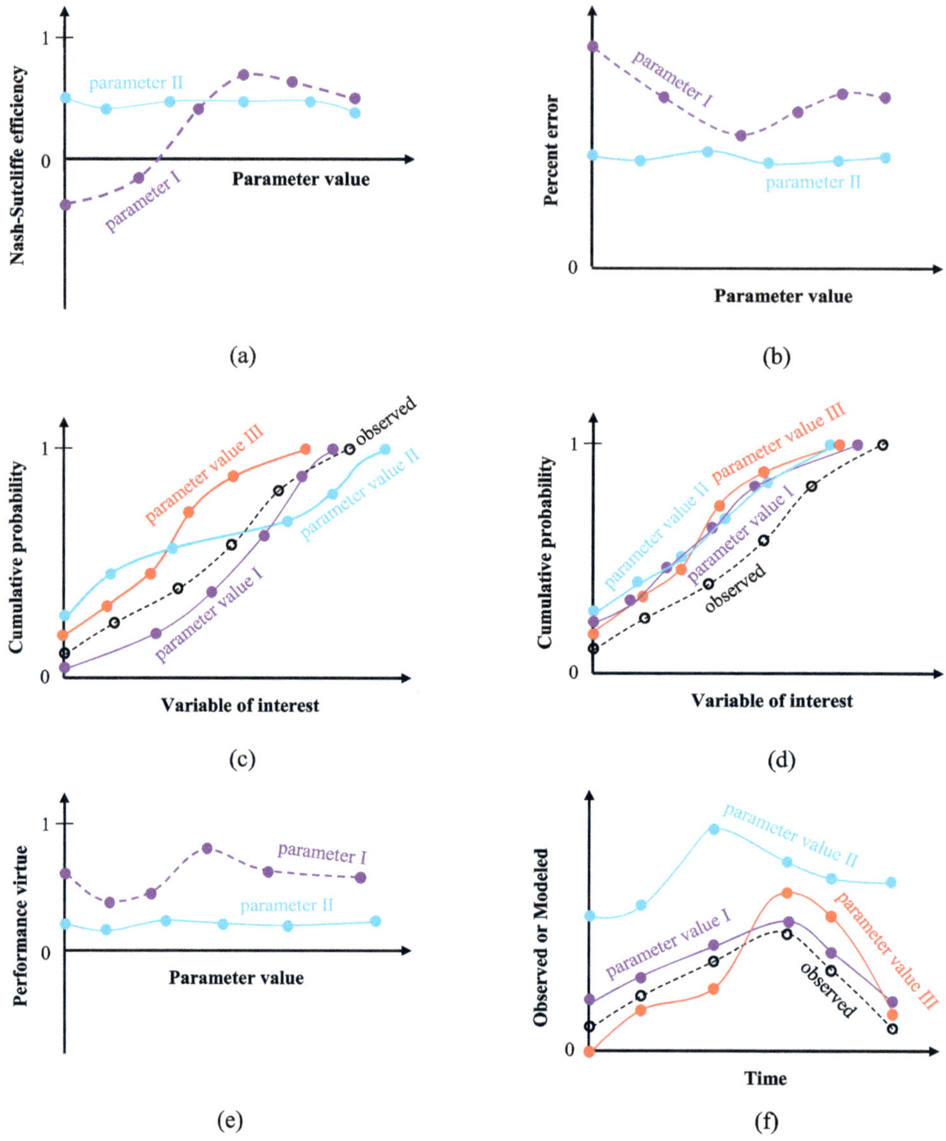

Figure 4.12. Sensitivity analysis results. Illustration in terms of (a) Nash–Sutcliffe efficiency, (b) percent error, (c) and (d) frequency distribution, (e) performance virtue, and (f) visualization plot. In (a), (b), and (e), parameter I is sensitive, whereas parameter II is insensitive. The parameter in (c) and (f) is sensitive and the parameter in (d) is insensitive.

sensitivity analysis approach (figure 4.13). First, the determined sensitivity of a parameter can depend on the start value of the parameter; so, this same parameter may be sensitive for a start value but insensitive for an alternative one. Second, the determined sensitivity of a parameter can depend on the fixed values of other parameters. The combinations of the fixed values are usually myriad and thus

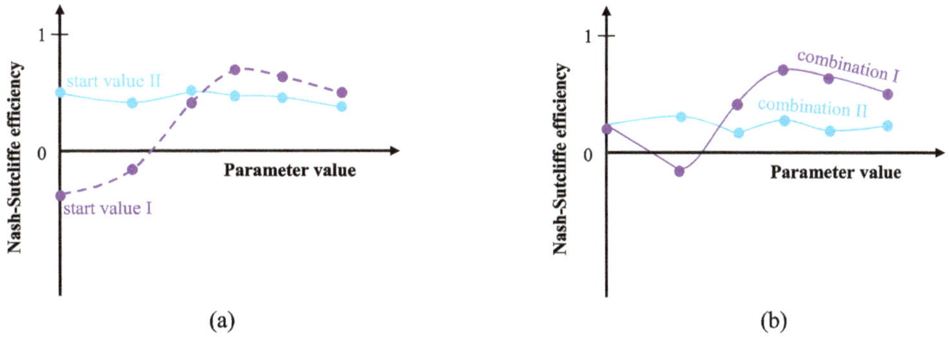

Figure 4.13. Arguments on sensitivity analysis of a concerned parameter. Possible influence of (a) start value of the parameter and (b) combined values of other parameters. The parameter is sensitive for start value I or combination I, whereas it is insensitive for start value II or combination II.

practically impossible to be extensively treated. Third, the determined sensitivity of a parameter can be compounded and/or biased by large errors in observed values, against which the model performance is assessed. A sensitivity analysis is likely limited by all three arguments, making it vital for the modeler to excise empirical speculation and scientific judgement.

To address the first two arguments, Monte Carlo simulations (Hastings 1970, Kroese *et al* 2014) can be carried out to create a CDF of each standardized parameter (figure 4.14), which is the division of this parameter by its maximum and has values of less than or equal to one. For a significance level such as $\alpha = 0.05$, determine the values of the parameters as the multiplications of the readings from the CDFs with the corresponding maximums. The determined value for a concerned parameter is taken as its start value, whereas the determined values for the other parameters remain unchanged while the concerned parameter varies.

To address the third argument, the approach presented by Wang *et al* (2010) can be implemented. The measure statistics discussed in section 4.5.3 assume that the observed data are error free, which is very rare in reality (White *et al* 2009). For instance, Harmel *et al* (2006) examined the approaches currently used to measure streamflow and found that errors in streamflow data can be introduced by four potential sources, including: (i) individual streamflow measurements to establish the stage–discharge relationship; (ii) application of the stage–discharge relationship; (iii) continuous stage measurements; and (iv) effects of streambed condition on stage measurement. The absolute errors vary from 6% to 19% (Harmel *et al* 2006). For a given observation, its error can be described by probable error range, a statistic proposed by Harmel and Smith (2007) and computed as

total number of error sources [–]

probable error range for i^{th} observed value [%]

$$PER_i = \pm 100\sqrt{\sum_{j=1}^{M}\left(E_{i,j}\right)^2} \tag{4.74}$$

error contribution from source j to i^{th} observed value [–]

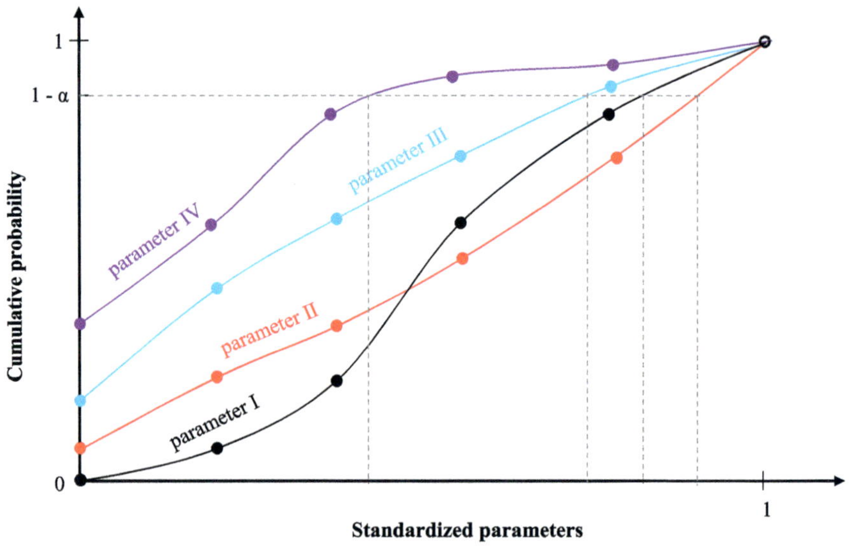

Figure 4.14. Cumulative density functions of standardized parameters based on Monte Carlo simulations. Illustration for four hypothetical parameters, designated parameter I to IV. α: significance level of interest.

The upper and lower limits of an observation are determined as (Wang *et al* 2010)

probable error range for i^{th} observed value [%]
upper limit of ith observed value [varying]
i^{th} observed value [varying]

$$y_{o,i}^{U} = \left(1 + \left|PER_i\right|\right) y_{o,i} \tag{4.75}$$

probable error range for i^{th} observed value [%]
lower limit of ith observed value [varying]
i^{th} observed value [varying]

$$y_{o,i}^{L} = \left(1 - \left|PER_i\right|\right) y_{o,i} \tag{4.76}$$

To be more conservative, the statistics in section 4.5.3 should be calculated by replacing observed values with corresponding corrected values determined as

i^{th} modeled value [varying]
upper limit of ith observed value (Eq. 4.75) [varying]
corrected value of ith observation [varying]

$$y'_{o,i} = \begin{cases} y_{m,i} & \text{for } y_{o,i}^{L} < y_{m,i} < y_{o,i}^{U} \\ y_{o,i}^{U} & \text{for } y_{m,i} \le y_{o,i}^{L} \\ y_{o,i}^{L} & \text{for } y_{m,i} \ge y_{o,i}^{U} \end{cases} \tag{4.77}$$

lower limit of ith observed value (Eq. 4.76) [varying]

4.6.3 Optimization of sensitive parameters

The identified sensitive parameters need to be adjusted to make the modeled values closely match corresponding observations, resulting in a set of adopted parameter values. The adjustment can be performed using three approaches. The first approach (approach I) is to manually change the parameters, run the model, and compare the modeled versus observed values. Because such a three-step process must be repeated until the model performance cannot be further improved, approach I is very tedious and time consuming, which is particularly true when the number of parameters is large. The success or failure of approach I relies heavily on the modelers' experiences, so the adopted parameter values are usually subjective (i.e. different from one modeler to another) and rarely optimal. In contrast, the second approach (approach II) is to execute an optimization algorithm to search the feasible region of the high-dimensional parameter space for the best values. Two algorithms, namely Shuffled Complex Evolution by University of Arizona (SCE-UA) and Gauss–Marquardt–Levenberg (GML), have been used widely. The SCE-UA (Duan *et al* 1992, Duan *et al* 1993, Naeini *et al* 2019) is a global population-evolution based search algorithm, whereas the GML (Levenberg 1944, Marquardt 1963) is a local gradient-based algorithm. For ecohydrological models with many parameters, the SCE-UA is superior to the GML partially because of the nonlinear and multimodal features of the parameter space (Blasone *et al* 2006, Naeini *et al* 2019). Although approach II can automatically execute the above three-step process, some parameters tend to be adjusted to their upper or lower limits, compounding the necessary variations of other parameters. To overcome the above drawbacks of approaches I and II, the third approach (approach III) is to manually adjust the parameters as much as possible and then automatically execute an algorithm to optimize the parameters. Such adjustment and optimization need to be looped interchangeably, ensuring that no parameter reaches its upper or lower limit while the best model performance is achieved.

The SCE-UA algorithm is realized by four steps (figure 4.15) (Naeini *et al* 2019). First, define the population of a parameter in accordance with its physical range. Second, intersect the ranges of the parameters to determine the feasible region (e.g. figure 4.16) of the parameter space, from which the populations are sampled randomly or uniformly to draw several sets of parameter points (i.e. arrays). The sampled sets are clustered into several complexes, each of which has the same number of members. Third, for a given complex, draw a subcomplex with several sets of arrays using a search strategy such as the Downhill Simplex Method (Johnston and Pilgrim 1976) and compute and short the objective functions; this can be repeated more than one time. Repeat such a drawing–computing–sorting process several times for each complex. Finally, once all the complexes are worked out, the optimal set of parameter values can be determined as the one that optimizes (i.e. maximizes or minimizes) the objective function, which can be a single statistic or weighted average of multiple statistics in section 4.5.3 (Wallner *et al* 2012).

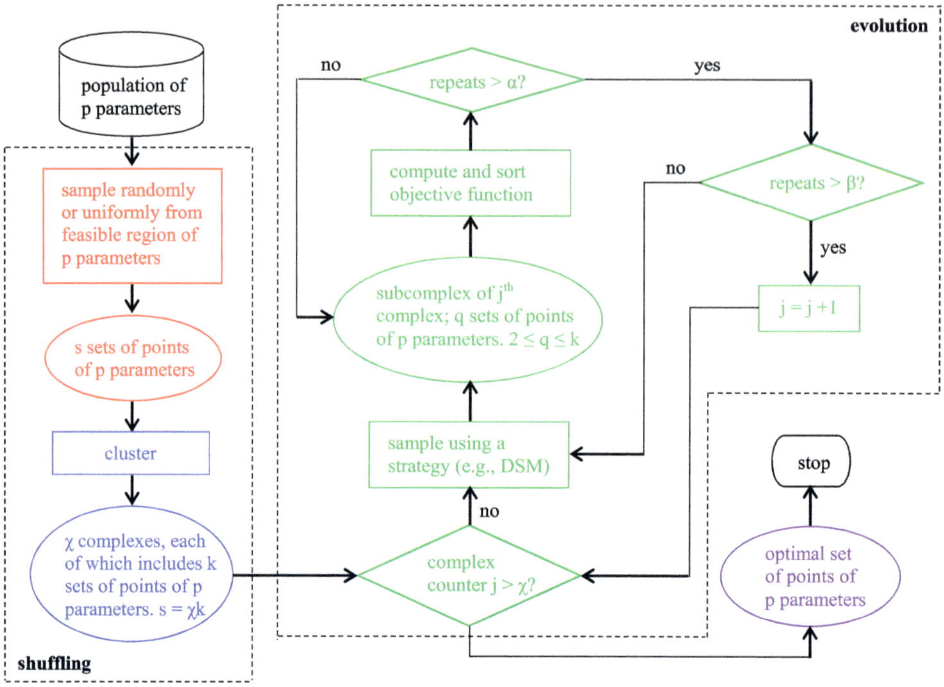

Figure 4.15. Implementation flowchart of the Shuffled Complex Evolution by University of Arizona (SCE-UA) algorithm. DSM: Downhill Simplex Method; p: number of parameters; s: sample size for each parameter; χ: number of clusters or complexes; k: sets of values of the parameters in each complex; q: sets of values of the parameters in each subcomplex; α: repeating times for each subcomplex; β: repeating times for each complex. Naeini *et al* (2019) suggest $\alpha = 1$, $k = 2p + 1$, $\chi = s/k$, $q = p + 1$, and $\beta = 2p + 1$.

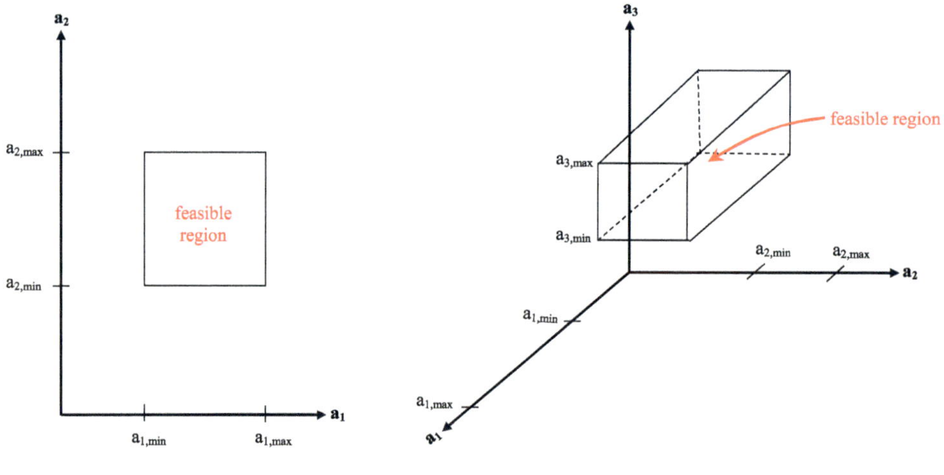

Figure 4.16. Feasible region of parameter space. Illustration for (a) two parameters and (b) three parameters. $a_{1,min}$ and $a_{1,max}$: lower and upper limits of parameter a_1; $a_{2,min}$ and $a_{2,max}$: lower and upper limits of parameter a_2; and $a_{3,min}$ and $a_{3,max}$: lower and upper limits of parameter a_3.

The GML algorithm is a damped least-squares method (Levenberg 1944). First, the initial values of the parameters are specified to obtain the modeled values of a concerned variable. Second, the modeled and observed values of this variable are used to compute the sum squared error (SSE) or other statistics in section 4.5.3. Third, the Gauss–Newton method is used to determine the parameter gradients for adjusting the parameter values within specified ranges. The new values of the parameters are used to obtain the new modeled values of the variable, which in turn are used to compute the new SSE or statistics. Fourth, the new SSE or other statistics are compared with the previous SSE or other statistics to assess whether the model performance is becoming better. Finally, the above second to fourth steps are repeated until the model performance cannot be improved further. Moreover, the convergence of the GML to a local optimum depends on the initial values of the parameters. A damping factor needs to be chosen for a good success rate and efficiency of the algorithm. If a step size is too small for efficient convergence, the damping factor should be increased, whereas if a step size is too large for steady convergence, the damping factor must be decreased. Increasing the damping factor decreases step size, and vice versa. The damping factor may be adjusted up or down by a constant delta or a series of varying deltas (Nielsen 1999).

4.6.4 Parameter uncertainty and model robustness

To set up an ecohydrological model, five procedures (table 4.8) need to be carried out as shown in figure 4.17. The first three procedures are elaborated in sections 4.6.1–4.6.3, while the last two procedures are discussed in this section. In practice, it is not unusual for modelers to misinterpret the third procedure (i.e. sensitivity analysis) and fourth procedure (i.e. uncertainty analysis) as being equivalent without examining possible variations of the adopted values of the model parameters. As a result, the certainty of model predictions tends to be overstated. The purpose of sensitivity analysis is to distinguish sensitive model parameters from insensitive ones,

Table 4.8. Comparison of model set-up procedures.

Procedure	Section	Purpose	Product
Estimate initial values of parameters	4.6.1	Create a preliminary model	Preliminary model
Conduct sensitivity analysis	4.6.2	Discern sensitive parameters	List of sensitive parameters
Optimize sensitive parameters	4.6.3	Calibrate the model against observations	Calibrated model
Analyze parameter uncertainty	4.6.4	Examine deviations from adopted values of sensitive parameters	Variations of sensitive parameters
Evaluate model robustness	4.6.4	Show model applicability for those beyond the calibration condition	Predictability of calibrated model

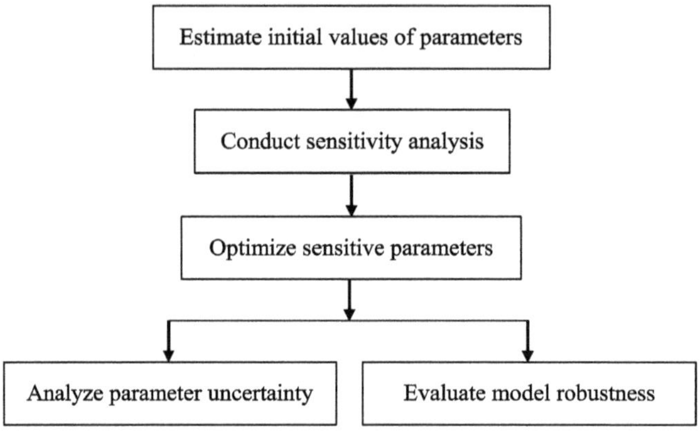

Figure 4.17. Flowchart for model set-up. A comparison is presented in table 4.8.

whereas the purpose of uncertainty analysis is to determine how calibrated parameter values change if different calibration data are used. For sensitivity analysis, the observed data for model assessment (as a whole) are usually used to compute the measure statistics, for parameter optimization (i.e. model calibration), uncertainty analysis, and robustness evaluation (i.e. model validation), however, the assessment data are empirically subdivided into two datasets: one for model calibration and another for model validation.

The procedure of parameter uncertainty analysis is carried out in four steps. First, the summary statistics (e.g. mean, standard deviation, and skewness) of the calibration dataset are computed and used in Monte Carlo simulations to mutate multiple datasets. Second, as with the adopted values from model calibration discussed in section 4.6.3, the optimal values of the sensitive parameters are determined for each mutated dataset. After that, the percent change of the optimal value of each parameter from the corresponding adopted value is computed. Third, for each parameter, the percent changes for the mutated datasets are sorted in ascending order to create a cumulative density function (CDF) for visual exploration of the parameter's uncertainty (figure 4.18), and t-tested for a null hypothesis that the mean change is nil at a selected significance level (e.g. $\alpha = 0.05$). Finally, if the mean change for a parameter is significantly different from zero, the $(1 - \alpha)$ confidence interval is created by assuming that the percent changes follow a Student's t-distribution (Neter *et al* 1996).

The procedure of robustness evaluation is realized in seven steps. First, the calibrated model, which is parameterized by the adopted values, is used to make predictions for the validation period. Second, the predictions are compared with the validation dataset to assess the model performance in accordance with the measure statistics discussed in section 4.5.3. A better performance indicates a greater robustness, and vice versa. The model is judged to be robust if the statistics for the validation dataset are comparable with those for the model calibration. Third, for a significantly uncertain parameter, its adopted value is substituted by a randomly selected value within the confidence interval and the model is run to

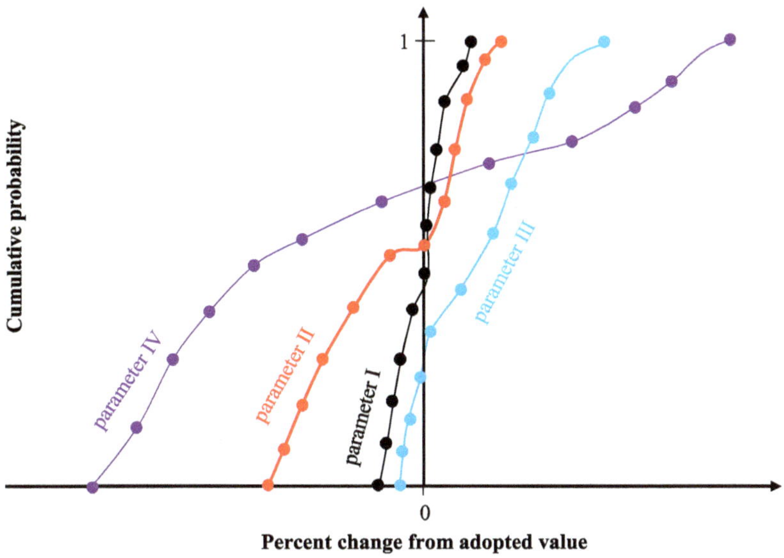

Figure 4.18. Visualization of parameter uncertainty. Illustration of the cumulative density functions for four hypothetical parameters. Parameter I is the least uncertain, whereas parameter IV is most uncertain. The adopted value of parameter II is likely too large, but the adopted value of parameter III may be too small.

make predictions for the validation period again. In this regard, the other parameters remain the corresponding adopted values. Fourth, the second step is repeated. Fifth, the third and fourth steps are repeated for several selected values of the parameter. Sixth, the third to fourth steps are repeated for all significantly uncertain parameters. Finally, in terms of the validation dataset, the probability for the model to be robust is computed as the ratio of the number of robust runs to the number of total runs.

4.6.5 Relevant computer programs

SWAT-CUP, which stands for Soil and Water Assessment Tool (SWAT) Calibration and Uncertainty Program (CUP), is a computer interface program that links the SUFI2, PSO, GLUE, ParaSol, and MCMC algorithms to SWAT for sensitivity analysis, model calibration, uncertainty analysis, and model validation (Abbaspour 2015). These five algorithms have also been coded into stand-alone computer programs that can be applied for any other models. Herein, the uncertainty is analyzed for predictions as influenced by errors in model structure, input data, and observations.

Sequential Uncertainties Fitting ver. 2 (SUFI2) is an algorithm for sensitivity and uncertainty analyses (Khalid *et al* 2016). On the one hand, the algorithm assesses uncertainty by the P-factor and R-factor. The P-factor is the percentage of the observations bracketed by the 95% confidence interval of the predictions, whereas the R-factor is the ratio of the delta of the concerned variable between the 2.5% and 97.5% cumulative probabilities of the predictions to the standard deviation of the observations. The P-factor can range from zero to one and the R-factor from zero to positive infinity. A simulation that exactly reproduces the observations corresponds to P-factor = 1.0 and R-factor = 0.0, implying a zero uncertainty. A higher uncertainty has a smaller P-factor and a larger R-factor than a lower uncertainty, and vice versa. On the other hand, the algorithm determines sensitive parameters using Student's t-tests at a selected significance level.

Particle swarm optimization (PSO) is a population-based stochastic optimization technique developed by Eberhart and Kennedy (1995). It is a pseudo-evolution technique that has no crossover and mutation. PSO is initialized with a group of random particles (solutions) and then searches for optima by updating generations. In every iteration, each particle is updated by two best values: one that has been achieved so far and another that has been obtained so far by any particle in the population.

Generalized likelihood uncertainty estimation (GLUE) allows for the possible non-uniqueness (i.e. equifinality) of parameter sets during the estimation of model parameters in over-parameterized models (Beven and Binley 1992). It is based on the estimation of the weights or probabilities associated with different parameter sets and uses a subjective likelihood measure to derive a posterior probability function, which in turn is used to derive the predictive probability of the output variables. Romanowicz *et al* (1994) propose that the likelihood function be explicitly derived based on the error between the observed outputs and those simulated by the model. GLUE is usually applied by directly likelihood weighting the outputs of multiple model realizations, which can be stochastic or deterministic and are defined by sets of parameter values within one or more model structures, to form a predictive distribution of a variable of interest. The prediction uncertainty of this variable is then related to variation in model outputs, unnecessarily adding an additional explicit error component.

The parameter solution (ParaSol) aggregates objective functions into a global criterion (van Griensven and Meixner 2006), which is optimized using the SCE-UA algorithm (discussed in section 4.6.3) and performs uncertainty analysis. Markov chain Monte Carlo (MCMC) generates samples from a random walk that adapts to the posterior distribution (Kuczera and Parent 1998, Gelman *et al* 2013).

In addition to SWAT-CUP, parameter estimation (PEST) is a model-independent computer program for parameter estimation and sensitivity analysis (Doherty and Johnston 2003). Its model independence relies on the fact that it can communicate with a model through the latter's own input and output files, thus allowing easy calibration set up with an arbitrary model. PEST implements a particularly robust variant of the GML method of nonlinear parameter estimation. It includes SENSAN for model-independent sensitivity analysis.

4.7 Summary and discussion

Although a variety of models have been developed since the 1980s, not all of them can be applied to simulate water–soil–vegetation interactions in dryland ecosystems. Relative to other models, three models, namely SWAT, HYDRUS-1D, and SWAP, include the primary ecohydrological processes and are more widely used across the world. Some processes that are included in one model are not included in another model, making it necessary to develop a new ecohydrological model (e.g. SWAP-E) by overcoming the drawbacks of the existing models. For instance, HYDRUS-1D does not simulate plant growth and carbon dynamics but SWAT and SWAP do, whereas SWAT does not trace the dynamic movements of heat and water in soils but HYDRUS-1D and SWAP do. Also, SWAT simulates water erosion and sedimentation but HYDRUS-1D and SWAT do not. However, none of those models simulate wind erosion, the effects of physical and biological crusts, the resistance of dry soil lays (DSLs), and vapor dynamics (i.e. movement, condensation, and re-vaporization) in DSLs. In dryland ecosystems, wind erosion is very important, DSLs are common, and soil moisture is very low most of the time. The current water retention and hydraulic conductivity functions used by HYDRUS-1D and SWAP can be problematic for low soil moisture and liquid–vapor two-phase flows. In this regard, SWAP-E intends to develop and incorporate new algorithms that can address such limitations.

To set up an ecohydrological model, five procedures need to be carried out. First, the initial values of the model parameters are estimated using physiographical, physiological, phenological, and anthropogenic data. Second, the sensitive parameters are determined by carrying out a sensitivity analysis. Third, the sensitive parameters are manually and/or automatically adjusted to optimize an objective function in accordance with a calibration dataset, yielding the adopted parameter values and a calibrated model. The objective function can be a single statistic or weighted average of multiple statistics. Such a statistic measures model performance in terms of mass conservation, explained percentage of the variability in observations, goodness-of-fit, frequency distribution, or overall hydrograph. Fourth, the uncertainty is analyzed for predictions as influenced by errors in model structure, input data, and observations. The uncertainty resulting from model structure can be reduced by model selection, whereas the uncertainty resulting from errors in input data can be controlled by increasing the data accuracy. The uncertainty resulting from errors in observations can be considered by substituting the corrected observations (equation (4.77)) for the observed values in calculating measurement statistics. The uncertainty analysis yields a significant range for each sensitive parameter. Fifth, the robustness of the calibrated model is evaluated in accordance with a validation dataset. A series of model runs can be done by varying the sensitive parameters within their ranges. The probability for the model to be robust is computed as the ratio of the number of robust runs to the number of total runs. A variety of computer programs (e.g. SWAT-CUP and PEST) are available for executing these five procedures. However, it is always crucial for modelers to exercise the best judgement for successful model applications.

Declaration

Parts of this chapter have been reprinted from (Wang *et al* 2010, Wang *et al* 2021).

References

Abbaspour K C 2015 *SWAT-CUP: SWAT Calibration and Uncertainty Programs* (Dübendorf: Swiss Federal Institute of Aquatic Science and Technology) pp 1–100

Allen R G, Pereira L S, Raes D and Smith M 1998 *Crop Evapotranspiration Guidelines for Computing Crop Water Requirements* (Washington, DC: United Nations (UN) Food and Agriculture Organization (FAO))

Allen R G, Wright M E and Burman R D 1989 Operational estimates of evapotranspiration *Agron. J.* **81** 650–62

Anderson T W 1962 On the distribution of the two-sample Cramer–von Mises criterion *Ann. Math. Stat.* **33** 1148–59

Anten N, Schieving F, Werger M and Schuffelen P 1995 Optimal leaf area indices in C_3 and C_4 mono- and dicotyledonous species at low and high nitrogen availability *Physiol. Plant.* **95** 541–50

Arnold J G, Allen P M and Bernhardt G 1993 A comprehensive surface-groundwater flow model *J. Hydrol.* **142** 47–69

Arnold J G and Fohrer N 2005 SWAT2000: current capabilities and research opportunities in applied watershed modelling *Hydrol. Processes* **19** 563–72

Arnold J G, Srinivasan R, Muttiah R S and Williams J R 1998 Large-area hydrologic modeling and assessment: Part I. Model development *J. Am. Water Resour. Assoc.* **34** 73–89

Bagnold R A 1977 Bedload transport in natural rivers *Water Resour. Res.* **3** 303–12

Belnap J 2006 The potential roles of biological soil crusts in dryland hydrologic cycles *Hydrol. Process.* **20** 3159–78

Belnap J, Phillips S L, Witwicki D L and Miller M E 2008 Visually assessing the level of development and soil surface stability of cyanobacterially dominated biological soil crusts *J. Arid. Environ.* **72** 1257–64

Beven K and Binley A 1992 The future of distributed models: model calibration and uncertainty prediction *Hydrol. Processes* **6** 279–98

Blasone R S, Madsen H and Rosbjerg D 2006 Comparison of parameter estimation algorithms in hydrological modelling *Calibration and Reliability in Groundwater Modelling: From Uncertainty to Decision Making* vol 304 (The Hague: IAHS) pp 67–72

Bogena H R, Weuthen A, Rosenbaum U, Huisman J A and Vereecken H 2007 SoilNet-A Zigbee-based soil moisture sensor network *AGU Fall Meeting* (San Francisco, CA: American Geophysical Union)

Boogaard H L, de Wit A J W, Roller J A d, van Diepen C A, Rötter R P, Cabrera J M C A and van Laar H H 2014 *WOFOST Control Centre 2.1 and WOFOST 7.1.7* (Wageningen: Wageningen University and Research Centre)

Bordoni M, Cislaghi A, Vercesi A, Bischetti G B and Meisina C 2020 Effects of plant roots on soil shear strength and shallow landslide proneness in an area of northern Italian Apennines *Bull. Eng. Geol. Environ.* **79** 3361–81

Brooks R H and Corey A T 1964 Hydraulic properties of porous media and their relation to drainage design *Trans. ASAE* **7** 26–8

Campbell G S 1985 *Soil Physics with BASIC: Transport Models for Soil-Plant Systems Developments in Soil Science* vol 14 (Amsterdam: Elsevier)

Campbell Scientific 2021 CR3000 Measurement and Control Datalogger http://campbellsci.com/cr3000

Cantón Y, Román J R, Chamizo S, Rodríguez-Caballero E and Moro M J 2014 Dynamics of organic carbon losses by water erosion after biocrust removal *J. Hydrol. Hydromech.* **62** 258–68

Chen L, Wang L, Ma Y and Liu P 2015 Overview of ecohydrological models and systems at the watershed scale *IEEE Syst. J.* **9** 1091–9

Cheng Z, Meng J and Wang Y 2016 Improving spring maize yield estimation at field scale by assimilating time-series HJ-1 CCD data into the WOFOST model using a new method with fast algorithms *Remote Sens.* **8** 1–22

Chow V T, Maidment D R and Mays L W 1988 *Applied Hydrology* (New York: McGraw-Hill)

Cramér H 1928 On the composition of elementary errors *Scand. Actuar. J.* **1** 13–74

de Wit A, Duveiller G and Defourny P 2012 Estimating regional winter wheat yield with WOFOST through the assimilation of green area index retrieved from MODIS observations *Agric. For. Meteorol.* **164** 39–52

Doherty J and Johnston J M 2003 Methodologies for calibration and predictive analysis of a watershed model *J. Am. Water Resour. Assoc.* **39** 251–65

Duan Q, Gupta V K and Sorooshian S 1993 Shuffled complex evolution approach for effective and efficient global minimization *J. Optim. Theory Appl.* **76** 501–21

Duan Q, Sorooshian S and Gupta V K 1992 Effective and efficient global optimization for conceptual rainfall–runoff models *Water Resour. Res.* **28** 1015–31

Eberhart R C and Kennedy J 1995 A new optimizer using particle swarm theory *The Sixth International Symposium on Micro Machine and Human Science (Nagoya)* (Piscataway, NJ: IEEE) pp 39–43

Edlefsen N E and Anderson A B C 1943 Thermodynamics of soil moisture *Hilgardia* **15** 31–298

Fayer M H and Simmons C S 1995 Modified soil water retention functions for all matric suctions *Water Resour. Res.* **31** 1233–8

Feddes R A, Kowalik P J and Zaradny H 1978 *Simulation of Field Water Use and Crop Yield. Simulation Monographs* (Wageningen: Centre for Agricultural Publishing and Documentation) (New York: Wiley) p 189

Flanagan D C and Livingston S J 1995 *WEPP User Summary* (West Lafayette, IN: USDA-ARS NSERL)

Forrester J W, Breierova L and Choudhari M 2001 *An Introduction to Sensitivity Analysis* (Cambridge, MA: Massachusetts Institute of Technology)

Gelman A, Carlin J B, Stern H S, Dunson D B, Vehtari A and Rubin D B 2013 *Bayesian Data Analysis* (New York: Chapman and Hall)

Gesch D B (ed) 2007 The national elevation dataset *Digital Elevation Model Technologies and Applications: The DEM Users Manual* (Bethesda, MD: American Society for Photogrammetry and Remote Sensing) ch 4

Glenn E P, Scott R L, Nguyen U and Nagler P L 2015 Wide-area ratios of evapotranspiration to precipitation in monsoon-dependent semiarid vegetation communities *J. Arid. Environ.* **117** 84–95

Gupta H V, Kling H, Yilmaz K K and Martinez G F 2009 Decomposition of the mean squared error and NSE performance criteria: implications for improving hydrological modelling *J. Hydrol.* **377** 80–91

Hack-ten Broeke M J D, Kroes J G, Bartholomeus R P, van Dam J C, de Wilt A J W, Supit I, Walvoort D J J, van Bakel P J T and Ruijtenberg R 2016 Quantification of the impact of hydrology on agricultural production as a result of too dry, too wet or too saline conditions *Soil* **2** 391–402

Hargreaves G H and Samani Z A 1985 Reference crop evapotranspiration from temperature *Appl. Eng. Agric.* **1** 96–9

Harmel R D, Cooper R J, Slade R M, Haney R L and Arnold J G 2006 Cumulative uncertainty in measured streamflow and water quality data for small watersheds *Trans. ASABE* **49** 689–701

Harmel R D and Smith P K 2007 Consideration of measurement uncertainty in the evaluation of goodness-of-fit in hydrologic and water quality modeling *J. Hydrol.* **337** 326–36

Hastings W K 1970 Monte Carlo sampling methods using Markov chains and their applications *Biometrika* **57** 97–109

Heitman J L, Xiao X, Horton R and Sauer T J 2008 Sensible heat measurements indicating depth and magnitude of subsurface soil water evaporation *Water Resour. Res.* **44** W00D05

IAPWS 2014 Revised release on surface tension of ordinary water substance *Report* R1–76 The International Association for the Properties of Water and Steam Moscow, Russia

IPCC 2001 Climate change 2001: the scientific basis *Contribution of Working Group 1 to Third Assessment Report of the IPCC* ed J Houghton and Y Ding (Cambridge: Cambridge University Press) p 881

Johnston P and Pilgrim D 1976 Parameter optimization for watershed models *Water Resour. Res.* **12** 477–86

Jouven M, Carrère P and Baumont R 2006 Model predicting dynamics of biomass, structure and digestibility of herbage in managed permanent pastures. 1. Model description *Grass Forage Sci.* **61** 112–24

Khalid K, Ali M F, Rahman N F A, Mispan M R, Haron S H, Othman Z and Bachok M F 2016 Sensitivity analysis in watershed model using SUFI-2 algorithm *Proced. Eng.* **162** 441–7

Khlosi M, Cornelis W M, Gabriels D and Sin G 2006 Simple modification to describe the soil water retention curve between saturation and oven dryness *Water Resour. Res.* **42** W11501

Kling H, Fuchs M and Paulin M 2012 Runoff conditions in the upper Danube basin under an ensemble of climate change scenarios *J. Hydrol.* **424** 264–77

Kroes J G 2020 Personal Email communication: modeling grass growth. X. Wang

Kroes J G, van Dam J C, Bartholomeus R P, Groenendijk P, Heinen M, Hendriks R F A, Mulder H M, Supit I and van Walsum P E V 2017 *SWAP Version 4: Theory Description and User Manual* (Wageningen: Wageningen Environmental Research)

Kroese D P, Brereton T, Taimre T and Botev Z I 2014 Why the Monte Carlo method is so important today *WIREs Comput. Stat.* **6** 386–92

Krysanova V and Arnold J G 2008 Advances in ecohydrological modelling with SWAT—a review *Hydrol. Sci.* **53** 939–47

Kuczera G and Parent E 1998 Monte Carlo assessment of parameter uncertainty in conceptual catchment models: the Metropolis algorithm *J. Hydrol.* **211** 69–85

Kuiper N H 1960 Tests concerning random points on a circle *Proc. K. Ned. Akad. Wet.* A **63** 38–47

Levenberg K 1944 A method for the solution of certain non-linear problems in least squares *Q. Appl. Math.* **2** 164–8

Liu X *et al* 2021 Effects of vegetation type on soil shear strength in Fengyang Mountain nature reserve, China *Forests* **12** 490

Ma Y, Feng S, Huo Z and Song X 2011 Application of the SWAP model to simulate the field water cycle under deficit irrigation in Beijing, China *Math. Comput. Model.* **54** 1044–52

Marquardt D W 1963 An algorithm for least-squares estimation of non-linear parameters *J. Soc. Indus. Appl. Math.* **11** 431–41

Monteith J L 1965 Evaporation and the environment *The State and Movement of Water in Living Organisms* ed G E Fogg (New York: Cambridge University Press) pp 205–34

Mualem Y 1976 A new model for predicting the hydraulic conductivity of unsaturated porous media *Water Resour. Res.* **12** 513–22

Naeini M R, Analui B, Gupta H V, Duan Q and Sorooshian S 2019 Three decades of the Shuffled Complex Evolution (SCE-UA) optimization algorithm: review and applications *Sci. Iran.* **26** 2015–31

Nash J E and Sutcliffe J V 1970 River flow forecasting through conceptual models—part I: a discussion of principles *J. Hydrol.* **10** 282–90

Nassar I N, Globus A M and Horton R 1992 Simultaneous soil heat and water transfer *Soil Sci.* **154** 465–72

Neitsch S L, Arnold J G, Kiniry J R and Williams J R 2011 Soil and water assessment tool theoretical documentation version 2009 *Technical Report* No. 406 Texas Water Resources Institute, Texas A&M University p 613

Neitsch S L, Arnold J G, Kiniry J R, Williams J R and King K W 2002 Soil and water assessment tool—theoretical documentation (version 2000) *Report* Grassland, Soil and Water Research Laboratory of Agricultural Research Service & Blackland Research Center of Texas Agricultural Experiment Station

Neter J, Kutner M H, Nachtsheim C J and Wasserman W 1996 *Applied Linear Statistical Models* (New York: McGraw-Hill)

Nielsen H B 1999 Damping parameter in Marquardt's method *Technical Report* IMM-REP-1999-05 Technical University of Denmark, Lyngby, Denmark pp 1–29

Nimmo J R and Miller E E 1986 The temperature dependence of isothermal moisture vs potential characteristics of soils *Soil Sci. Soc. Am. J.* **50** 1105–13

Noborio K, McInnes K J and Heilman J L 1996 Two-dimensional model for water, heat, and solute transport in furrow-irrigated soil: I. Theory *Soil Sci. Soc. Am. J.* **60** 1001–9

Pearson E S and Hartley H O 1972 *Biometrika Tables for Statisticians* (Cambridge: Cambridge University Press)

Pedram S, Wang X, Liu T and Duan L 2018 Simulated dynamics of soil water and pore vapor in a semiarid sandy ecosystem *J. Arid. Environ.* **151** 58–82

Penman H L 1948 Natural evaporation from open water, bare soil, and grass *Proc. R. Soc. London* **193** 120–46

Philip J R and de Vries D A 1957 Moisture movement in porous materials under temperature gradients *Trans. Am. Geophys. Union* **38** 222–32

Priestley C H B and Taylor R J 1972 On the assessment of surface heat flux and evaporation using large-scale parameters *Mon. Weather Rev.* **100** 81–92

Revelli R and Porporato A 2018 Ecohydrological model for the quantification of ecosystem services provided by urban street trees *Urban Ecosyst.* **21** 489–504

Ritter A and Muñoz-Carpena R 2013 Performance evaluation of hydrological models: statistical significance for reducing subjectivity in goodness-of-fit assessments *J. Hydrol.* **480** 33–45

Romanowicz R, Beven K and Tawn J 1994 Evaluation of predictive uncertainty in nonlinear hydrological models using a Bayesian approach *Stat. Environ.* **2** 297–315

Rossi C and Nimmo J R 1994 Modeling of soil water retention from saturation to oven dryness *Water Resour. Res.* **30** 701–8

Sakai M, Toride N and Šimůnek J 2009 Water and vapor movement with condensation and evaporation in a sandy column *Soil Sci. Soc. Am. J.* **73** 707–17

Schaap M G and van Genuchten M T 2006 Modified Mualem–van Genuchten formulation for improved description of the hydraulic conductivity near saturation *Vadose Zone J.* **5** 27–34

Sensirion 2021 Digital Humidity Sensor SHT7 https://sensirion.com/en/environmental-sensors/humidity-sensors/pintype-digital-humidity-sensors/

Simard R and L'Ecuyer P 2011 Computing the two-sided Kolmogorov–Smirnov distribution *J. Stat. Softw.* **39** 1–18

Šimůnek J, Kodešová R, Gribb M M and van Genuchten M T 1999 Estimating hysteresis in the soil water retention function from cone permeameter experiments *Water Resour. Res.* **35** 1329–45

Šimůnek J and van Genuchten M T 2008 Modeling nonequilibrium flow and transport with HYDRUS *Vadose Zone J.* **7** 782–97

Smirnov N 1948 Table for estimating the goodness of fit of empirical distributions *Ann. Math. Stat.* **19** 279–81

Srinivasan R, Ramanarayanan T S, Arnold J G and Bednarz S T 1998 Large-area hydrologic modeling and assessment: part II. Model application *J. Am. Water Resour. Assoc.* **34** 91–101

USACE 2018 *Hydrologic Modeling System (HEC-HMS) (Version 4.3) Quick Start Guide* (Davis, CA: US Army Corps of Engineers Institute for Water Resources)

USDA 2021 *Soil Survey Geographic Database (SSURGO)* https://data.nal.usda.gov/dataset/soil-survey-geographic-database-ssurgo

USGS 2020 What types of elevation datasets are available, what formats do they come in, and where can I download them?

USGS 2021 *National Land Cover Database* https://usgs.gov/centers/eros/science/national-land-cover-database?qt-science_center_objects=0#qt-science_center_objects

van Genuchten M T 1980 A closed-form equation for predicting the hydraulic conductivity of unsaturated soils *Soil Sci. Soc. Am. J.* **44** 892–8

van Genuchten M T 1987 *A Numerical Model for Water and Solute Movement in and Below the Root Zone* (Washington, DC: United States Department of Agriculture Agricultural Research Service US Salinity Laboratory)

van Griensven A and Meixner T 2006 Methods to quantify and identify the sources of uncertainty for river basin water quality models *Water Sci. Technol.* **53** 51–9

van Ittersum M K, Leffelaar P A, van Keulen H, Kropff M J, Bastiaans L and Goudriaan J 2003 On approaches and applications of the Wageningen crop models *Eur. J. Agron.* **18** 201–34

von Mises R E 1928 *Wahrscheinlichkeit, Statistik und Wahrheit* (Berlin: Julius Springer)

Wallner M, Haberlandt U and Dietrich J 2012 Evaluation of different calibration strategies for large scale continuous hydrological modelling *Adv. Geosci.* **31** 67–74

Wang X 2015 Vapor flow resistance of dry soil layer to soil water evaporation in arid environment: an overview *Water* **7** 4552–74

Wang X 2021 *Water Resources and Hydraulics* (New York: Cambridge University Press)

Wang X 2022 *Water–soil–vegetation dynamic interrelationships in a typical Eurasian Steppe Watershed in China* Final Report to NSF-IRES (Norfolk, VA: Old Dominion University) p 200

Wang X, Gao R and Yang X 2021 Responses of soil moisture to climate variability and livestock grazing in a semiarid Eurasian steppe *Sci. Total Environ.* **781** 146705

Wang X, Liu T, Li F, Gao R, Yang X, Duan L, Luo Y and Li R 2014 Simulated soil erosion from a semiarid typical steppe watershed using an integrated aeolian and fluvial prediction model *Hydrol. Process.* **28** 325–40

Wang X, Liu T, Shang S, Yang D and Melesse A M 2010 Estimation of design discharge for an ungauged overflow-receiving watershed using one-dimensional hydrodynamic model *Int. J. River Basin Manag.* **8** 79–92

Wang X, Liu T and Yang W 2012 Development of a robust runoff-prediction model by fusing the Rational equation and a modified SCS-CN method *Hydrol. Sci. J.* **57** 1118–40

Wang X and Melesse A M 2005 Evaluation of the SWAT model's snowmelt hydrology in a northwestern Minnesota watershed *Trans. ASAE* **48** 1359–76

Wang X, Melesse A M and Yang W 2006 Influences of potential evaporation estimation methods on SWAT's hydrologic simulation in a northwestern Minnesota watershed *Trans. ASABE* **49** 1755–71

White E D, Feyereisen G W, Veith T L and Bosch D D 2009 Improving daily water yield estimates in the Little River watershed: SWAT adjustments *Trans. ASABE* **52** 69–79

Williams J R 1980 SPNM, a model for predicting sediment, phosphorus, and nitrogen yields from agricultural basins *Water Resour. Bull.* **16** 843–8

Williams J R 1995 The EPIC model *Computer Models of Watershed Hydrology* ed V P Singh (Highlands Ranch, CO: Water Resources) pp 909–1000

Zhao Y, Si B, He H, Xu J, Peth S and Horn R 2016 Modeling of coupled water and heat transfer in freezing and thawing soils, Inner Mongolia *Water* **8** 424

Chapter 5

Case studies

This chapter presents eight case studies to elucidate the ecohydrological processes and concepts presented in chapters 2 and 3 as well as to demonstrate how to apply the models presented in chapter 4 to examine water–soil–vegetation interactions in semi-arid ecosystems. The studies are conducted for two sites in the Mongolian Plateau of north China and for a third site in central Texas in the United States. These meteorologically semi-arid and arid sites are grasslands for livestock production. For each case study, scientific questions and hypotheses are formulated and addressed in terms of the modeling results. Following the site descriptions, each successive section presents one case study. The last section summarizes the findings from the case studies and draws general conclusions.

5.1 Study sites

The case studies are carried out for three dryland ecosystems: two in Inner Mongolia in China and another in Texas in the United States of America (USA). These ecosystems share some common characteristics while have some contrasting features. This section presents the basic information of the ecosystems that is needed to understand the studies.

5.1.1 The Balagaer River watershed in North China

The 5365 km^2 Balagaer River watershed (figure 5.1), located in the northeast Inner Mongolia Autonomous Region (IMAR) of China, is a typical Eurasian steppe grassland. It is almost uniformly covered by *Stipa grandis* and *Leymus chinensis* (Chen *et al* 2005, Wang *et al* 2014a). These two perennial rhizome grass species grow from late April to early October and are grazed by livestock (i.e. sheep and cattle) from late May to late September (Shiyomi *et al* 2011). The watershed has topographic elevations ranging from 857 to 1876 m above mean sea level, leading to a mean topographic gradient of 0.07. The annual precipitation varies from 170 to 615 mm,

doi:10.1088/978-0-7503-4020-5ch5

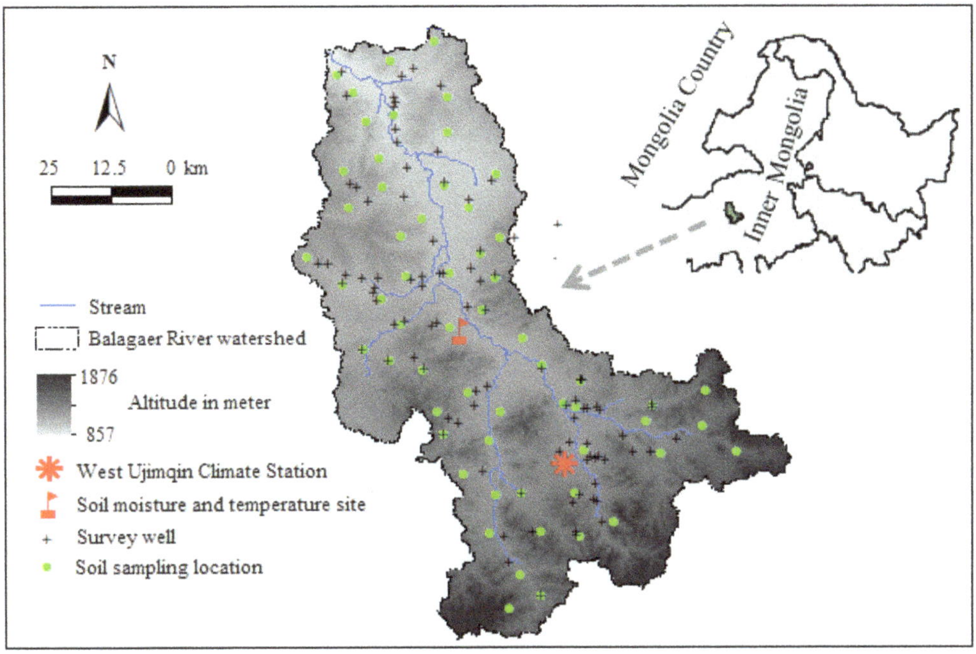

Figure 5.1. The Balagaer River watershed. Topography and data collection locations are superimposed.

with a long-term average of 335 mm. For a given year, the precipitation falls as rain from July to September and as snow from October to January. In comparison, the annual potential evapotranspiration (PET) is usually higher than 1165 mm, resulting in a semi-arid climate. Given that on average less than 0.5% of the precipitation can be converted into runoff, the hydrological cycle of the grassland is dominated by vertical rather than lateral water fluxes. The daily air temperature fluctuates from −38.5 °C in winter to 37.5 °C in summer, with an annual mean of 1.2 °C. The number of windy days in a year of interest varies from 28 to 148, with an annual mean daily average wind speed of about 15 km h^{-1} and a maximum wind gust of 125 km h^{-1}.

The soil samples from 70 locations across the watershed (figure 5.1) indicate two soil layers: the upper (above 30 cm depth) and lower (30–100 cm depth). The sand or sandy loam soils (Saxton and Rawls 2006) consist of 72%–76% sand particles, about 23% silt particles, and 2%–3% clay particles (figure 5.2(a)). The textures of the lower-layer soils have a larger variation than those of the upper-layer soils probably due to the soil geomorphic history of the Mongolia Plateau (Luo *et al* 2012). This is further verified by the larger mean coefficient of uniformity (C_u) and coefficient of curvature (C_c) for the lower-layer soils (figure 5.2(b)). C_u is the ratio of 60%- to 10%-finer diameters, whereas C_c is the ratio of 30%-finer diameter squared to the multiplication of 10%- and 60%-finer diameters (Charbeneau 2006). The lower-layer soils can be more permeable than the upper-layer soils at one site but less permeable at another site. As expected, the upper-layer soils have overall larger values for saturated soil moisture and field capacity than the lower-layer soils (figure 5.2(c)). Moreover, surveys of the 100 domestic wells shown in figure 5.1 reveal that the

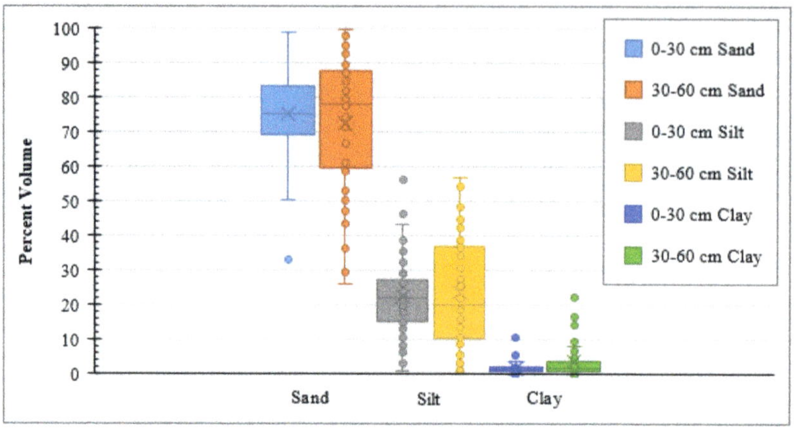

(a)

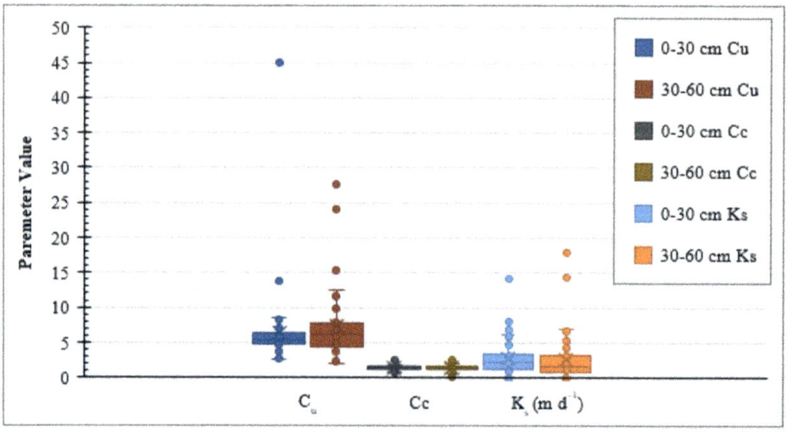

(b)

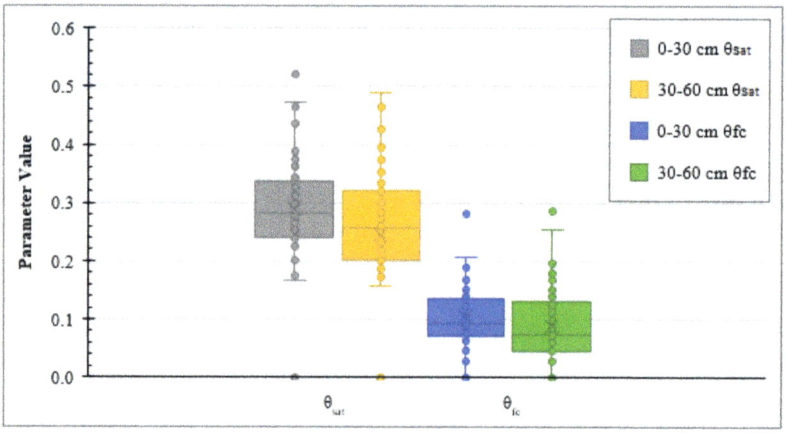

(c)

Figure 5.2. Box–whisker plots of the data on soil texture and soil–water properties. C_u: coefficient of uniformity; C_c: coefficient of curvature; K_s: saturated hydraulic conductivity; θ_{sat}: saturated soil moisture; θ_{fc}: field capacity. (Adapted with permission from figure 2 in Wang *et al* (2021). Copyright 2021 Elsevier.)

watershed, except for some low-lying spots recharged by intermittent streams and ponds, has a water table of deeper than 3.0 m. The data on soil properties and depths to water table can be found in Luo *et al* (2014) and Wang *et al* (2021).

5.1.2 The HSL site in North China

The study site (122°37′30″ E, 43°20′56″ N) is a spot within the 51 700 km^2 Horqin Sandy Land (HSL) (118°35′ to 123°30′ E, 42°41′ to 45°15′ N) (Wang 2015). It is designated the HSL site (HSLS) for description purposes. The HSL is one of the most typical sandy lands in the world and was historically part of the largest and most characteristic Eurasian grassland (Zhu *et al* 2020), but it has been undergoing moderate to severe desertification in recent decades (He *et al* 2008, Wang *et al* 2008c, Duan *et al* 2011, Li *et al* 2017); deserted land has reached 57.8% of the HSL's total area, primarily because of inappropriate reclamation for agriculture (e.g. chisel plow tillage in the autumn) and overgrazing has adversely altered the natural hydrology (Zuo *et al* 2008). As a result, the HSL is losing biome productivity (Christensen *et al* 2005) and incurring environmental problems such as more frequent dust storms with an increasing magnitude (Rooyen 1998, Dey *et al* 2004, Portnov and Safrielb 2004, Takemi 2005, Gao *et al* 2006). Similar desertification-related issues exist all over the world (D'Odorico *et al* 2013), particularly in Africa, Australia, and the southwestern USA (FAO 2021). Currently, the HSL landscape features consist of sand dunes (55%), meadows (27%), farmlands (10%), lakes (5%), and residential areas (3%) (Duan *et al* 2011). The HSL has a temperate and semi-arid continental monsoonal climate. It receives an average annual precipitation of 389 mm, of which 69.3% falls during the growing season (June to August), while it incurs a very high annual PET of 1412 mm. Given that the PET is almost four times larger than the annual precipitation, the HSL is climatically semi-arid. For a given month, the actual evapotranspiration (AET) is almost the same as the precipitation (figure 5.3), that is, most precipitation is lost to evaporation. The HSL has an average annual

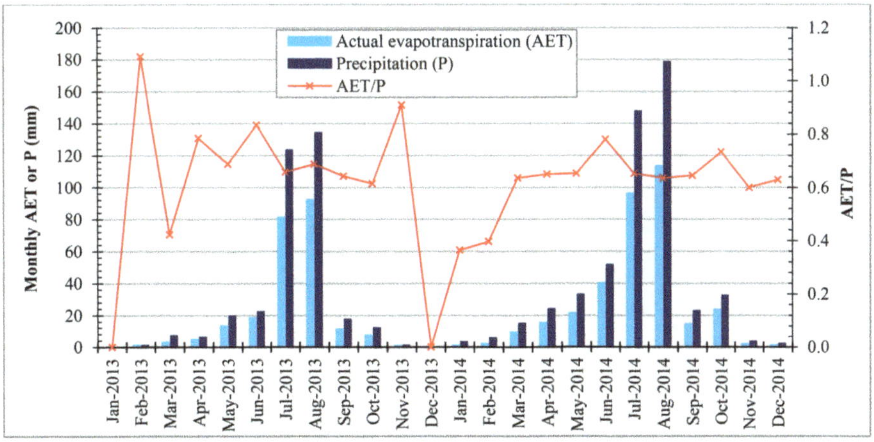

Figure 5.3. Monthly precipitation and actual evapotranspiration. The observed data at the Horqin Sandy Land site (HSLS) in 2013 and 2014.

temperature of around 6.6 °C, with a minimum monthly mean temperature of −13.3 °C in January and a maximum temperature of 23.8 °C in July, and an average annual wind speed of 3.8 m s^{-1}, with a minimum monthly mean wind speed of 3.0 m s^{-1} in August and a maximum of 5.0 m s^{-1} in April. The prevalent wind direction is northwest in winter and spring while it is southwest in summer and autumn.

The HSLS has a sparse vegetation coverage with desert sandy soils (figure 5.4). From the ground surface to a depth of around 80 cm, the soils (median porosity of 0.387) consist of 98.9% sand particles (a diameter of 0.05–2.0 mm) and 1.1% silt particles (a diameter of 0.002–0.05 mm), whereas below the depth of 80 cm, the soils (median porosity of 0.382) consist of 99.7% sand particles and 0.3% silt particles. There are no clay particles (a diameter of less than 0.002 mm) at all. The wilting point of the soils is 0.02 (Dong *et al* 2003) and the depth to the water table is greater than 4.0 m (Duan *et al* 2015).

5.1.3 The upper Cowhouse Creek watershed in USA

The 117 845 ha drainage area upstream of the flow gauging station USGS 08101000 is the upper part of the 188 715 ha Cowhouse Creek watershed in north central Texas in the USA (figure 5.5). It is designated the TXCCW for description purpose. The TXCCW covers portions of five Texas counties: Comanche, Coryell, Hamilton, Lampasas, and Mills. The soils in the TXCCW (figure 5.6 and table 5.1) are dominated by clay and sandy clay, with an infiltration capacity ranging from 0.8 to 332 mm h^{-1} (Mace *et al* 1999, TCEQ 2007). The land use within the TXCCW (figure 5.7) consists of almost entirely range brush (55.0%) and range grasses (40.0%), and the rest is composed of forest (4.4%), water (0.4%), and agriculture (0.1%). The TXCCW has a very low topographic relief; its local relief is less than 5 m and global relief is only up to 340 m, with altitudes from 150 to 490 m. The Cowhouse Creek has an average sinuosity of 1.3 and is characterized by broad flat alluvial floodplains, river terraces, and lower Cretaceous Walnut Clay with adjacent ridges of the Edwards-Duck Creek–Comanche Peak limestone (Nordt 2004).

The TXCCW has an average annual precipitation of 865 mm and an average annual PET of 2315 mm. Its annual average daily temperature ranges from −20 °C in winter to 45 °C in summer, with a mean of 19 °C. For three summer months, the

Figure 5.4. A picture of the HSLS. Photograph taken by the author in June 2010.

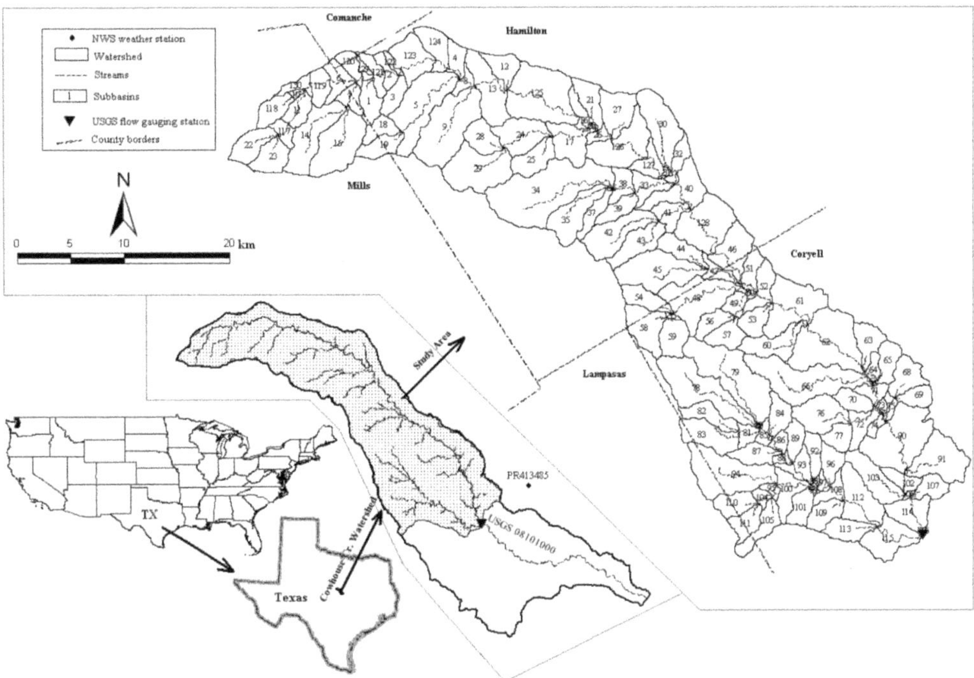

Figure 5.5. The location and boundary of the upper Cowhouse Creek watershed. Its 130 subbasins, the National Weather Service (NWS) weather station, and the US Geological Survey (USGS) flow gauging station are also superimposed. (Adopted with permission from figure 1 in Wang *et al* (2010). Copyright 2010 Elsevier.)

temperature could be 32 °C or higher for 75 days, whereas for three winter months, 30 days might have a temperature below zero. Overall, the TXCCW has a cool, windy, and semi-arid climate (Greer 2005). The Cowhouse Creek is a seasonal stream, with low flows in summer and fall, intermediate flows in winter, and high flows in spring. The daily streamflow varies from 0 to 1000 $m^3 s^{-1}$, with an annual average of 2.8 $m^3 s^{-1}$. The streamflow in spring is predominantly generated from rainfall runoff, but the streamflow in the other three seasons is predominantly discharged from the Edwards–Trinity aquifer (USGS 2005). The daily peaks range from 0.1 to 996.8 $m^3 s^{-1}$ while the instantaneous peaks could be up to 3115 $m^3 s^{-1}$, as occurred on 20 December 1991. The peak for a water year (i.e. from December to November) could occur in any season.

5.2 Long-term soil moisture responses to climate and livestock grazing

Soil moisture plays a pivotal role in regulating the biotic and abiotic processes of semi-arid ecosystems. However, the information is scare regarding how soil moisture responds to climate variations and anthropogenic activities, practically limiting concerted efforts in sustaining the sparse vegetation coverage, fragile hydrology, and

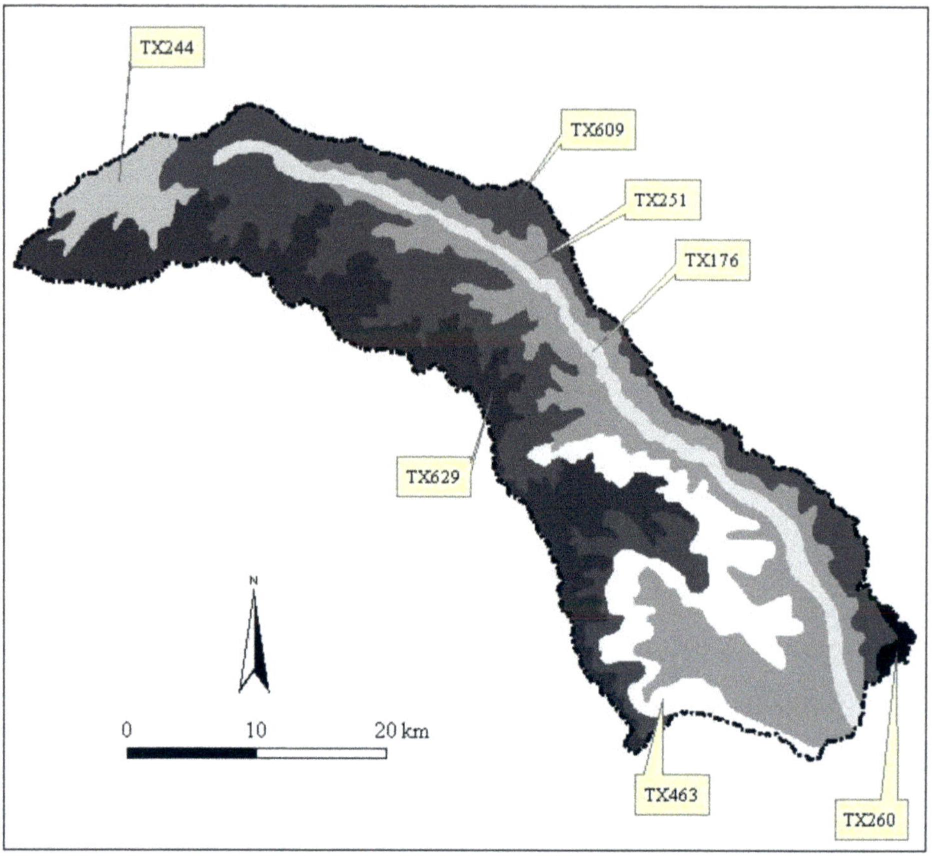

Figure 5.6. Soils in the upper Cowhouse Creek watershed. The soil types are classified in the State Soil Geographic (STATSGO) database (Wang and Melesse 2006) and their properties are presented in table 5.1. (Adopted with permission from figure 2 in Wang *et al* (2010). Copyright 2010 Elsevier.)

fertile topsoil of such ecosystems. This case study aimed to show how to use an ecohydrological model to simulate long-term soil moisture as affected by climate and livestock grazing. It was conducted in a typical Eurasian grassland located in northeast China. The model was run from 1955 to 2017, with the first five years as the warm-up period. Initial values of the model parameters were estimated from literature review, field reconnaissance, and *in situ* laboratory experiments. Subsequently, calibration parameters were selected and manually adjusted to make the simulated daily soil temperatures and daily soil moisture closely match the corresponding observed values at four depths from 2014 to 2017. The model reproduced the observations well, indicating a good performance. Averaged across the 58 assessment years, the daily soil moisture varied from 0.02 to 0.38; the moisture at a shallower depth tended to be smaller, but to have a greater range, than those at a deeper depth. The moisture around 30 cm was predicted to have the largest mean and variance. For a given depth, the daily soil moisture pulsed in the

Table 5.1. Properties of soils in the upper Cowhouse Creek watershed shown in figure 5.6[1].

Soil code	Name	Texture	Hydrological soil group (HSG)	Hydraulic conductivity (mm h^{-1})	USLE erodibility (0.013 t m^2 h m^{-3} t^{-1} cm^{-1})[2]	Remark
TX176	Faddin	Sandy clay, clay	D	1.2–120	0.43	Permeability decreases with depth
TX244	Jalmar–Penwell	Fine sand	A	280–450	0.10	
TX251	Kaufman–Tinn–Gladewater	Clay	D	0.09–0.13	0.32	
TX260	Kimbrough	Loam	D	8.4–11	0.37	
TX463	Reagan–Hodgins–Iraan Permeability increases with depth	Clay	loam	B	1.3–5.7	0.32
TX609	Wilson–Burleson–Mabank	Clay	D	0.23–12	0.43	Permeability decreases with depth
TX629	Tencee	Gravel	D	9.8–10	0.15	Permeability is almost constant throughout the soil profile

[1] Reproduced with permission from table 1 in Wang *et al* (2010). Copyright 2010 Elsevier.
[2] USLE: Universal Soil Loss Equation (Williams 1975).

beginning of March and plateaued from May to September. As a result of livestock grazing, the moisture in the top 1.5 cm soils would be reduced, while the moisture in the soils beneath would be increased. Moisture reduction would not become remarkable until a grazing intensity of 3.0 cattle ha^{-1}, below which grazing would not aggregate wind erosion. The grazing impacts on soil moisture tended to decrease monophonically with an increase of evapotranspiration or precipitation. For the years with an annual precipitation of less than 220 mm, the predicted impacts would either increase or decrease with an increase of precipitation, depending on the ratio of evapotranspiration to precipitation. The soil moisture pulses in early spring will likely be diminished by climate change, resulting in an elevated wind soil erosion in the future.

5.2.1 Background

Soil water is central for semi-arid vegetation (e.g. Eurasian steppe) ecosystems (UNEP 1992, Singh *et al* 1998, Kurc and Small 2007), where the annual potential evapotranspiration (PET) is several times larger than the annual precipitation (UNEP 1991, Luo *et al* 2014, Yang 2014, Davies *et al* 2016). As a result, the shallow soils are dry for most of the time and may only be moistened during sporadic storm events for a short period of time because the infiltrated water quickly

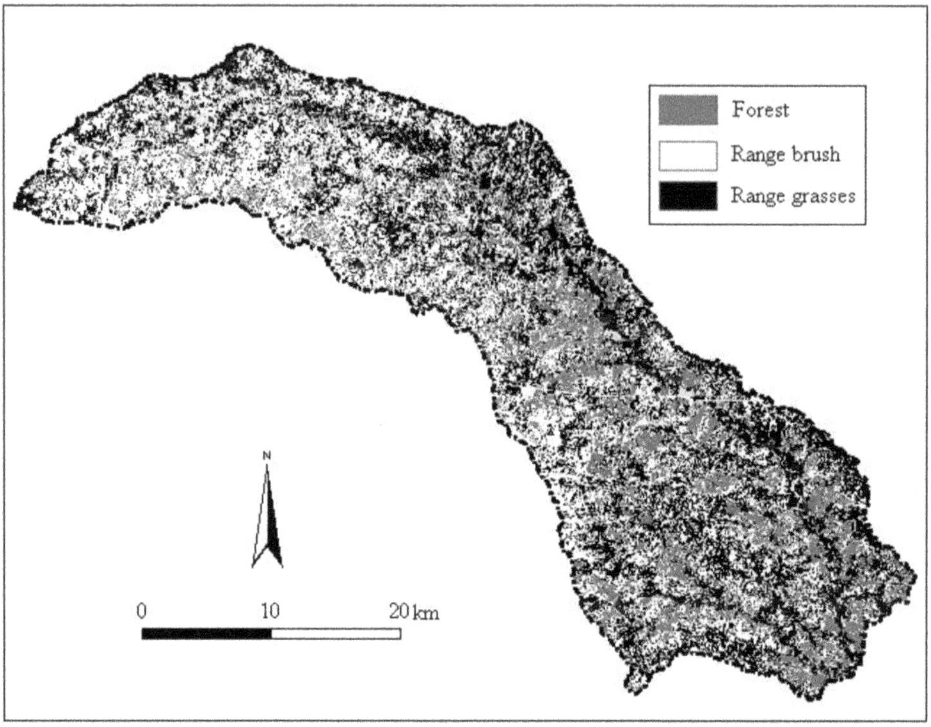

Figure 5.7. Major land uses in the upper Cowhouse Creek watershed. The information is extracted from the 30 m National Land Cover Database 2001 (Wickham *et al* 2021). Range brush accounts for 55% of the watershed area, while range grasses account for 40%. (Adopted with permission from figure 3 in Wang *et al* (2010). Copyright 2010 Elsevier.)

evaporates and/or percolates. That is, the soil moisture (i.e. volumetric water content) is very dynamic, regulating water–heat–carbon cycles and controlling plant respiration and assimilation. It is a determinant of plant biomass because net ecosystem exchange (NEE) increases with an increase of water availability. Herein, NEE is the difference between the fixed carbon dioxide (CO_2) by assimilation and the released CO_2 by respiration. A higher soil moisture leads to larger evaporation (i.e. respiration) and transpiration (i.e. assimilation) (Singh *et al* 1998, Huxman *et al* 2004). For a semi-arid ecosystem, the respiration is mainly affected by soil water in the 0–5 cm layer, within which evaporation dominates the hydrological cycle, whereas the assimilation is primarily controlled by soil water in the rootzone, within which transpiration dominates the hydrological cycle; so, the NEE varies with precipitation fluctuation and soil wetting–drying dynamics. Given that evaporation and transpiration tend to vary interannually more with a decrease of annual precipitation, the ecosystem can switch from a carbon source in a wetter year to a carbon sink in a drier year and vice versa. Also, the NEE can be affected by the intensity and timing of livestock grazing (Wang *et al* 2020). Thus, it is imperative to understand well the long-term responses of soil moisture to climate variations and livestock grazing practices. However, such an understanding is incomplete in the existing literature, limiting the utilization and protection of vulnerable semi-arid ecosystems, while a few studies have been conducted to monitor relevant variables.

Among others, three representative studies were conducted at Sevilleta National Wildlife Refuge (SNWR) located in central New Mexico, USA (Kurc and Small 2007), the Central Plains Experimental Range (CPER) located in north central Colorado, USA (Wang *et al* 2010), and Loess Plateau Grassland Ecosystem Trial Station (LPGETS) located in Gansu Province, northwest China (Wang *et al* 2020). These sites receive an average annual precipitation of 200–350 mm but have an average annual PET of more than 1500 mm. The SNWR has sandy soils and was covered by C_4 perennial black grama (*Bouteloua eripoda*). The soils at the LPGETS can be classified as sandy loam, clay loam, and sandy clay loam (Saxton and Rawls 2006), while the steppe shortgrass was dominated by C_4 perennial bunchgrass (*Bouteloua gracilis*). The LPGETS has sandy free-drainage loess soils and was covered by steppe grass species of *Stipa bungeana*, *Lespedeza davurica*, *Pennisetum flaccidum*, *Artemisie capillaris*, and *Stetarua viridis*.

Using the data collected at the SNWR from 1 June 2002 to 1 January 2005, Kurc and Small (2007) found that soil moisture pulsed at 2.5 cm following almost all storms but only four moisture pulses occurred at 53.5 cm following large storm events or a series of successive small events. The moisture in the top 15 cm soils was mainly lost to evaporation, benefiting respiration, while the moisture in the soils beneath was mostly subject to transpiration, prompting assimilation. NEE and evapotranspiration were negatively related, indicating that the infiltrated water could move downward beyond the 15 cm depth before it was quickly evaporated from the above soils. On the other hand, Wang *et al* (2010) studied three topo-sequences of upland, midland, and lowland of the CPER. They revealed a positive relationship between ANPP and annual precipitation but a negative relationship between ANPP and the moisture of deeper than 30 cm soils. Such superficially contradictory relationships can be explained by the fact that more than 80% of the plant roots reside within the top 30 cm soils. That the deeper soils had higher moisture was because more infiltrated water from rainfall could not be held by the very permeable rootzone loamy soils for plant transpiration. The soil water in the rootzone can be more efficiently ingested by the rich roots and thus plays more important roles in increasing plant biomass than that in the layers beneath (Luo *et al* 2014, Pedram *et al* 2018). Nonetheless, the soil water beneath the rootzone is also beneficial for the ecosystem because it can be transferred upward by capillary action to sustain the rootzone soil moisture to be higher than the permanent wilting point during a drought period, allowing plants to survive and revive afterwards (Wang *et al* 2016a). Moreover, Wang *et al* (2020) assessed the effects of livestock grazing on soil respiration at the LPGETS. They found that a higher grazing intensity resulted in smaller ANPP and below-ground primary production (BNPP), a larger moisture of the top 1.0 cm soils, and a greater respiration rate. However, such a causal relationships could be weakened by climate variability.

Obviously, the existing knowledge is incomplete of how climate and livestock grazing interactively affect the water and carbon cycles in semi-arid steppe grasslands, as regulated by soil moisture dynamics. The objective of this case study was to demonstrate how to simulate the long-term variability of soil moisture profiles as affected by climatic fluctuations and grazing practices. The simulation was carried out in the Balagaer River watershed, which is detailed in section 5.1.1, The simulation was carried out for historical (i.e. 1955–2017) climates and four representative grazing regimes to fill the knowledge gap in sustaining semi-arid grasslands as an important carbon sink to confront climate change (Neumann and Wigen 2013).

5.2.2 Materials and methods

5.2.2.1 Data and grazing culture

This case study used data on soil properties, climate, soil moisture, and soil temperature. The data on soil properties were determined using samples taken by the author and his collaborators (Wang *et al* 2021a) and are described in section 5.1.1. The daily data on climate, namely precipitation, air temperature, relative humidity, wind speed, solar radiation, and reference evapotranspiration, at West Ujimqin Climate Station 54012 (117°36′ E, 44°35′ N) were obtained from the Chinese Meteorological Information Center (CMIC 2020) for the record period of 1 January 1955 to 31 December 2017. The 12 h data on soil moisture and soil temperature were continuously measured from 2014 to 2017 by four EC-5 three-needle sensors (METER Group 2020) installed at the site (117°52′ E, 44°20′ N) (figure 5.1) at 5, 10, 15, and 30 cm depths, respectively. For a given day and given depth, the daily soil moisture was computed as the average of the two corresponding 12 h soil moisture, while the daily soil temperature was computed as the average of the two corresponding 12 h soil temperatures. Moreover, the physiological parameters (table 5.2) for a hybrid of *Stipa grandis* and *Leymus chinensis* (Chen *et al* 2005, Luo *et al* 2014) were estimated from the literature (Kurc and Small 2007, Wang and Zhou 2012, Pereyeara *et al* 2017), field reconnaissance, and laboratory analyzes. The hybrid plant was determined to have the root density distribution shown in figure 5.8.

The grazing culture of the study watershed was determined by extrapolating the data for a 19 700 ha typical pasturing land (table 5.3). From 1985 to 2017 the land was co-grazed by 100 000–260 000 cows and horses as well as 800 000–1 700 000 sheep and goats per year. When the animal unit equivalent (AUE) is measured by cattle (USDA-NRCS 2020), the land was incurred grazing intensities of 11.8–27.6 cattle ha^{-1}, with an average of 19.7 cattle ha^{-1}. Grazing starts once the aboveground dry matter grows to 350 kg ha^{-1} and stops once the aboveground dry matter reduces to 150 kg ha^{-1}. The loss from livestock dropping and treading is assumed to be 0.43–0.5 kg ha^{-1}. The dry matter consumption per AUE, which is estimated to be 8.5–10.0 kg ha^{-1}, is a negative function of actual grazing intensity (AHDB 2020). The rainfed watershed has never been fertilized and mowed.

5.2.2.2 Set-up of the SWAP model

The SWAP model was parameterized for a 200 cm long soil column that can represent the overall characteristics of the Balagaer River watershed. To consider

Table 5.2. List of physiological parameters for the hybrid plant[1].

Class (source)	Parameter	Description	Possible range[2]	Value
Initialization (authors)	LAIEM [m^2 m^{-2}]	Leaf area index (LAI) at emergence	0–10	0.01
	RGRLAI [m^2 m^{-2} d^{-1}]	LAI increasing rate per day	0–1	0.1
Surface area (authors)	SSA [ha kg^{-1}]	Specific stem area	0–1	0.0004
	SPAN [d]	Optimum life span of leaves	0–366	30
	TBASE [°C]	Lowest temperature for leaves to grow	−10–30	0
	SLA [ha kg^{-1}]	Specific leave area	0–1	0.0015–0.002
Assimilation (Wang and Zhou 2012)	KDIF	Extinction coefficient for diffuse visible light	0–2	0.6
	KDIR	Extinction coefficient for direct visible light	0–2	0.75
	EFF [kg ha^{-1} h^{-1} W^{-1} m^2]	Light use efficiency	0–10	0.5
	AMAX [kg ha^{-1} h^{-1}]	Maximum CO_2 assimilation rate	0–100	25–40
Conversion (Wang and Zhou 2012)	CVL [kg kg^{-1}]	Efficiency of conversion into leaves	0–1	0.685
	CVR [kg kg^{-1}]	Efficiency of conversion into roots	0–1	0.694
	CVL [kg kg^{-1}]	Efficiency of conversion into stems	0–1	0.662
Respiration (Wang *et al* 2016a)	Q10 [°C^{-1}]	Respiration rate increase with temperature	0–0.5	0.2
	RML [kg CH_2O kg^{-1} d^{-1}]	Maintenance respiration rate of leaves	0–1	0.03
	RMR [kg CH_2O kg^{-1} d^{-1}]	Maintenance respiration rate of roots	0–1	0.015
	RMS [kg CH_2O kg^{-1} d^{-1}]	Maintenance respiration rate of stems	0–1	0.015
Partitioning (Wang and Zhou 2012)	FRTB [kg kg^{-1}]	Fraction of total dry matter partitioned to roots	0–1	0.3
	FLTB [kg kg^{-1}]	Fraction of aboveground dry matter partitioned to leaves	0–1	0.6
	FSTB [kg kg^{-1}]	Fraction of aboveground dry matter partitioned to stems	0–1	0.4
Death rate (Wang and Zhou 2012)	PERDL [kg kg^{-1} d^{-1}]	Maximum relative death rate of leaves due to water stress	0–3	0.05
	RDRRTB [kg kg^{-1} d^{-1}]	Maximum relative death rate of roots	0–3	0.02
	RDRSTB [kg kg^{-1} d^{-1}]	Maximum relative death rate of stems	0–3	0.02

Oxygen stress (Feddes *et al* 1978, Chen *et al* 2005, Kroes *et al* 2017)	HLIM1 [cm]	Potential head above which no water extraction	−100–100	−2
	HLIM2U [cm]	Potential head above which optimum water extraction starts for top layer	−100–100	−16.1
	HLIM2L [cm]	Potential head above which optimum water extraction starts for sublayer	−100–100	−44.4
Drought stress (Feddes *et al* 1978, Chen *et al* 2005, Kroes *et al* 2017)	HLIM3H [cm]	Potential head below which water uptake reduction at high T_{pot}	<100	−44.4
	HLIM3L [cm]	Potential head below which water uptake reduction at low T_{pot}	<100	−81.2
	HLIM4 [cm]	Potential head below which no water extraction	<100	−170.7
	ADCRH [cm d^{-1}]	Level of high atmospheric demand	0–5	5
	ADCRL [cm d^{-1}]	Level of low atmospheric demand	0–5	5
	ALPHACRIT [–]	Critical stress index for compensation of root water uptake	0.2–1	0.95
Interception (authors)	COFAB [cm]	Maximum interception amount	0–1	0.01
Root growth (authors)	RDI [cm]	Initial rooting depth	>0	20
	RRI [cm d^{-1}]	Maximum daily increase in rooting depth	>0	0.01
	RDC [cm]	Maximum rooting depth of grass cultivar	>0	40
	WRTMAX [kg DM ha^{-1}]	Maximum root dry mass	>0	20 000

[1] Reproduced with permission from table 1 in Wang *et al* (2021). Copyright 2021 Elsevier. The hybrid plant represents a mixture of *Stipa grandis* and *Leymus chinensis*.
[2] Source: Kroes *et al* (2017).

the vertical heterogeneity of soil properties, from the top to the bottom, the column was subdivided into four sublayers of 20, 10, 30, and 140 cm thick, respectively. The first and second sublayers are parts of the upper (i.e. 0–30 cm) layer or rootzone, while the third and fourth sublayers are the lower (i.e. 30–60 cm) and deeper layers. In addition, to better mimic the nonlinearity of heat–water–carbon processes, the first or third sublayer was subdivided into 20 compartments, the second sublayer into 5 compartments, and the fourth sublayer into 14 compartments. For a given sublayer, the compartment thickness is uniform and equal to the ratio of the sublayer thickness to the number of compartments.

The lower and deeper layers were determined to be identical in terms of soil–water properties, namely θ_r, θ_s, K_s, α, n, and β (Wang *et al* 2014b). The initial values

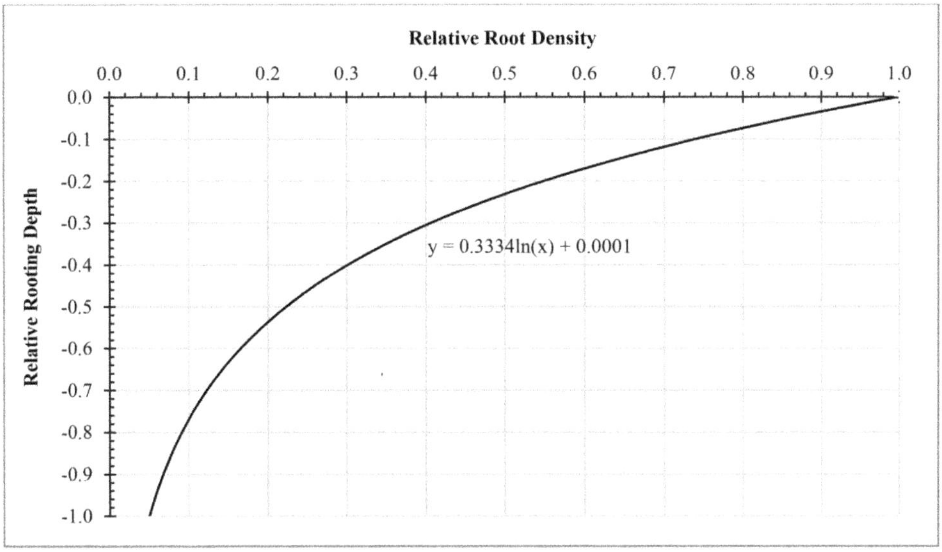

Figure 5.8. Vertical distribution of the hybrid plant. The plant represents a mixture of *Stipa grandis* and *Leymus chinensis*. At a root depth of interest, the relative root density is the ratio of the root density at this depth to the maximum root density of the whole rootzone, while the relative rooting depth is the ratio of this depth to the rootzone thickness (i.e. 40 cm). (Adopted with permission from figure 3 in Wang *et al* (2021). Copyright 2021 Elsevier.)

Table 5.3. Statistics of livestock grazing from 1985 to 2017[1].

Statistics	Number of cattle and horses	Number of sheep	AUE [cattle][2]	Grazing area [ha]	Grazing intensity [cattle ha^{-1}]
Min	100 000	800 000	233 333	19 700	11.8
Max	260 000	1 700 000	543 333	19 700	27.6
Average	180 000	1 250 000	388 333	19 700	19.7

[1] Reproduced with permission from table 2 in Wang *et al* (2021). Copyright 2021 Elsevier. The grazing season is from 20 May to 30 September.
[2] AUE: animal unit equivalent to cattle; AUE = Number of cattle and horses + Number of sheep/6 (AHDB 2020, USDA-NRCS 2020).

for these properties were estimated from laboratory experiments (figure 5.2) and existing literature (Charbeneau 2006), whereas the final values (table 5.4 and figure 5.9) were determined by model calibration from 2014 to 2017, which automatically adjusted the relevant model parameters using parameter estimation (PEST) (Doherty 2004, Wang *et al* 2008b) to make simulated soil moisture and soil temperatures closely match the corresponding observations. As stated in section 4.6.5, PEST is a model-independent parameter estimator with advanced predictive analysis and regularization features. The model's upper and lower boundary conditions were defined as the time series of air temperature and free drainage, respectively. The free drainage out of the column bottom is applicable because of the

Table 5.4. The calibrated parameters of the SWAP model[1].

Hydrological process	Parameter	Description	Possible range[2]	Calibrated value
Snow and frost	TePrRain [°C]	Temperature above which all precipitation is rain	0–5	2.0
	TePrSnow [°C]	Temperature below which all precipitation is snow	−5–0	0.0
	SNOWCOEF [–]	Calibration factor for snowmelt	0–10	0.3
Surface runoff	PONDMX [cm]	Maximum storage depth of ground-surface ponding reservoir	>0	0.01
	RSROEXP [–]	Exponent of water depth above PONDMX	0.1–10	1.0
	RSRO [$cm^{RSROEXP-1}$ d]	Drainage resistance	<1	0.95
Evapotranspiration	CF [–]	Adjustment factor of reference evapotranspiration	0–2	figure 5.9
Soil water[3]	θ_r [$cm^3\ cm^{-3}$]	Residual soil moisture	0–0.4	UL: 0.02 LL: 0.02
	θ_{sat} [$cm^3\ cm^{-3}$]	Saturated soil moisture	0–0.95	UL: 0.35 LL: 0.40
	K_s [$cm\ d^{-1}$]	Saturated vertical hydraulic conductivity	>0	UL: 286.98 LL: 312.78
	α [cm^{-1}]	Retention curve shape factor (equation (5.4))	0.0001–1	UL: 0.05 LL: 0.05
	n [–]	Retention curve shape factor (equation (5.5))	1–4	UL: 2.5 LL: 2.5
	β [–]	Empirical pore-connectivity parameter (equation (5.7))	−25–25	UL: 2.5 LL: 2.5
	ψ_b [cm]	Air-entry potential head	<0	UL: −2.0 LL: −2.0

[1] Reproduced with permission from table 3 in Wang *et al* (2021). Copyright 2021 Elsevier.
[2] Source: Kroes *et al* (2017).
UL: upper (i.e. 0–30 cm) soil layer; LL: lower (i.e. below-30 cm) soil layer.

high permeability of the soils as well as the large depth to water table. Table 5.2 lists the physiological parameters for the SWAP model's grass growth module.

In addition to visualization plots, three statistics, namely the coefficient of determination (R^2), Nash–Sutcliffe efficiency (NSE), and mean percent error (MPE), were used to assess how the model reproduces the sensor-measured soil moisture and soil temperatures. The detailed formulizations of these statistics can be found in section 4.5.3 (Wang and Melesse 2005). R^2 measures percent of the variance presented in the observations that can be accounted for by the model: the larger R^2, the better the model, and vice versa. NSE measures the goodness-of-fit of the model to the observations, higher values indicating a better overall fit and 1.0 indicating a perfect fit. A negative NSE implies that the simulation results are less

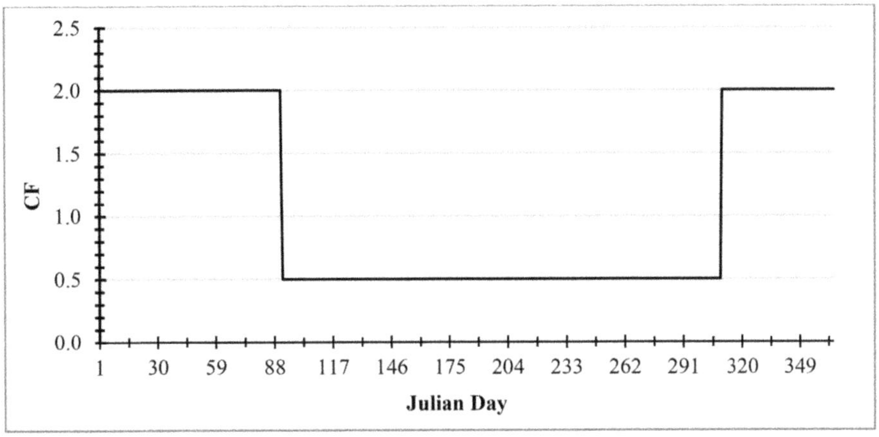

Figure 5.9. Adjustment factor (CF) for reference evapotranspiration. The values were determined by model calibration. (Adopted with permission from figure 4 in Wang *et al* (2021). Copyright 2021 Elsevier.)

reliable than if one had used the average of the observed values, while a positive value indicates that the simulation results are more reliable than using this average. MPE measures the overall closeness of predictions to observations and is computed as the arithmetic average of the paired differences.

5.2.2.3 Model runs

Continuous simulations were performed at a daily time steps for the historical period of 1 January 1955 to 31 December 2017. The initial five years were taken as the warm-up period to stabilize the model's initial parameters (e.g. soil water storage), while the other 58 years (i.e. 1 January 1960 to 31 December 2017) were used to assess soil moisture dynamics. This study ran the model for several trials. The first trail used the average grazing intensity of 19.7 cattle ha^{-1} to calibrate the model and establish the baseline against which assessments were done, whereas the second and third trials used grazing intensities of 0.0 and 27.6 cattle ha^{-1}, respectively, to assess the zero and most-intensive grazing scenarios. Additional trials were carried out for grazing intensities of 0.0 to 27.6 cattle ha^{-1} to determine the threshold above which soil moisture becomes independent of livestock grazing. In principle, such a threshold does exist because the grazing impact on soil moisture diminishes with removing aboveground biomass and depends on grazing–climate interactions.

For a given trial, the simulated daily soil moisture during the 58 assessment years was summarized as three statistics, including average, minimum, and maximum, by soil depth, grazing season (20 May–30 September), and non-grazing season (1 October–19 May). The values of a statistic (e.g. average) for all trials were plotted versus soil depth to visualize how livestock grazing impacts soil moisture. On the other hand, for a given soil depth, a time series plot of the simulated daily soil moisture was created and compared virtually with the corresponding plots of daily precipitation and daily reference ET to scrutinize how climate influences soil moisture. Moreover, the interactive grazing–climate effects on soil moisture were examined by cross comparing the statistical plots with the time series plots, which were created as discussed above.

5.2.3 Results

5.2.3.1 The model parameters and performance

Based on the model calibration, the lowest air temperature for rain was determined to be 2.0 °C, whereas the highest air temperature for snow was determined to be 0.0 °C; if the air temperature is between these two thresholds, precipitation falls in a mixture of rain and snow (table 5.4). The factor for snowpack melting rate was determined to be 0.3. In addition, the storage capacity of surface depressions was determined to be 0.01 cm, below which all precipitation would be absorbed. The surface runoff was estimated as 1.053 times rainfall excess (i.e. the difference of total rainfall and 0.01 cm). Further, the reference evapotranspiration was adjusted by the factor shown in figure 5.9. The lower-layer soils were determined to be slightly more permeable than the upper-layer soils, but the other soil–water parameters were determined to be identical between both layers and are as shown in figure 5.2.

The model performed very well in reproducing the observed daily soil temperatures (table 5.5). The average daily mean temperatures were predicted accurately, as indicated by small prediction errors of −0.72 °C to 0.70 °C, $R^2 > 75\%$, and NSE > 0.9. From 1 January to 5 March 2017, the observed soil temperatures at a depth of interest had a very small variability, which made NSE less meaningful. The model's prediction ability was verified by the visual plots (e.g. figure 5.10) showing the predicted versus observed daily mean soil temperatures. The plots for other years showed a similar pattern and thus are not shown to reduce redundancy. Similarly, the model had a very good performance in reproducing the observed daily soil moisture (table 5.5)—it had mean percent errors of less than 0.04 and captured the observed variations well (figure 5.11). However, the model could explain less than 30% of the variances presented in the observations of 2014, 2015, and 2017.

Relative to the long-term average annual precipitation, the year 2014 was wetter, the year 2015 was normal, and the years 2016 and 2017 were drier. The SWAP model performed very well for all three hydroclimatic conditions, indicating that the model was sufficiently calibrated for simulating the overall responses of soil moisture to climate variations and grazing practices.

5.2.3.2 The predicted soil moisture

As expected, at a depth of concern, while the grazing season had a higher average soil moisture than the non-grazing season (figure 5.12), the daily soil moisture of the non-grazing season fluctuated more widely than those of the grazing season. For either season, the soil moisture was larger at a deeper depth than at a shallower depth and approached an asymptote with increasing depth. The moisture of the 20–60 cm soils had a steeper vertical gradient than those of the above and beneath soils. The soils in the top centimeters might be moistened on a rainy day and either quickly or gradually dried on following sunny days. The model showed the highest soil moisture around 32 cm below ground surface.

Table 5.5. The SWAP model's performance statistics[1].

	Soil temperature				Soil moisture				Soil temperature				Soil moisture			
Calibration period	Mean [°C]	Sd. [°C]	R^2	NSE	Mean	Sd.	MPE	R^2	Mean [°C]	Sd. [°C]	R^2	NSE	Mean	Sd.	MPE	R^2
	at 5 cm soil depth								at 10 cm soil depth							
31 May to	**7.83**	**13.15**	0.97	0.95	**0.10**	**0.04**	0.03	0.15	**7.87**	**12.77**	0.97	0.96	**0.10**	**0.04**	0.03	0.09
31 Dec 2014	7.81	12.51			0.08	0.02			7.92	12.29			0.09	0.02		
01 Jan to	**3.53**	**13.20**	0.96	0.94	**0.09**	**0.04**	0.03	0.13	**3.56**	**12.91**	0.97	0.95	**0.09**	**0.04**	0.03	0.10
31 Dec 2015	3.21	13.13			0.07	0.02			3.19	12.97			0.08	0.02		
01 Jan to	**−7.19**	**10.65**	0.91	0.85	**0.07**	**0.03**	0.04	0.84	**−7.19**	**10.30**	0.91	0.85	**0.08**	**0.03**	0.04	0.83
10 May 2016	−7.63	11.63			0.05	0.01			−7.89	11.22			0.06	0.02		
30 Aug to	**−1.20**	**10.04**	0.95	0.94	**0.10**	**0.05**	0.04	0.45	**−0.77**	**9.82**	0.96	0.94	**0.11**	**0.05**	0.04	0.46
31 Dec 2016	−0.69	10.46			0.08	0.02			−0.42	10.37			0.08	0.02		
01 Jan to	**−15.21**	**3.06**	0.76	0.33	**0.05**	**0.00**	0.02	0.36	**−14.89**	**2.87**	0.76	0.23	**0.05**	**0.00**	0.02	0.00
05 Mar 2017	−14.49	4.55			0.06	0.00			−14.48	4.25			0.07	0.00		
	at 15 cm soil depth								at 30 cm soil depth							
31 May to	**7.83**	**12.29**	0.97	0.96	**0.09**	**0.03**	0.03	0.05	**7.91**	**10.99**	0.98	0.97	**0.10**	**0.04**	0.03	0.08
31 Dec 2014	8.03	12.09			0.09	0.02			8.34	11.59			0.10	0.02		
01 Jan to	**3.52**	**12.53**	0.97	0.95	**0.08**	**0.03**	0.03	0.06	**3.46**	**11.55**	0.98	0.95	**0.08**	**0.03**	0.02	0.29
31 Dec 2015	**3.18**	**12.83**			0.08	0.02			3.19	12.54			0.09	0.01		
01 Jan to 10	**−7.20**	**9.89**	0.92	0.84	**0.07**	**0.03**	0.04	0.83	**−6.95**	**9.16**	0.92	0.78	**0.08**	**0.04**	0.04	0.82
May 2016	−8.13	10.85			0.06	0.02			−8.58	10.15			0.08	0.01		
30 Aug to	**−0.40**	**9.57**	0.96	0.94	**0.10**	**0.04**	0.03	0.47	**0.56**	**9.02**	0.96	0.91	**0.12**	**0.05**	0.04	0.44
31 Dec 2016	−0.16	10.29			0.09	0.02			0.56	10.13			0.10	0.02		
01 Jan to	**−14.58**	**2.68**	0.75	0.10	**0.05**	**0.00**	0.03	0.27	**−13.75**	**2.31**	0.77	−0.48	**0.06**	**0.00**	0.02	0.00
05 Mar 2017	−14.48	3.99			0.07	0.00			−14.46	3.34			0.08	0.00		

[1] Reproduced with permission from table 4 in Wang *et al* (2021). Copyright 2021 Elsevier. S_d.: standard deviation; R^2: coefficient of determination; NSE: Nash–Sutcliffe efficiency; MPE: mean percent error; the bold numbers are based on the daily measurements.

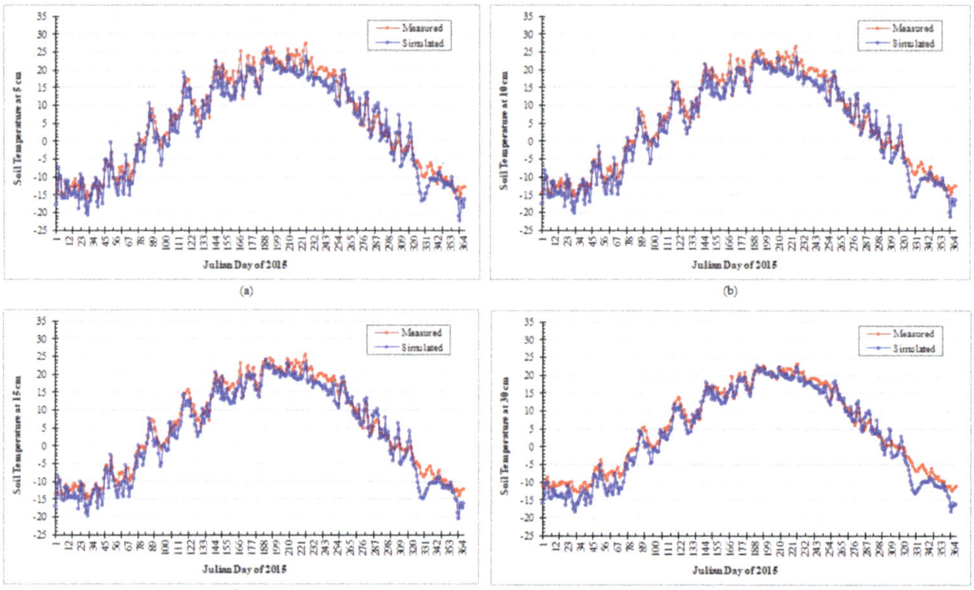

Figure 5.10. The SWAP model's performance in predicting soil temperatures. Plots showing the simulated versus measured daily mean soil temperatures in 2015 at a depth of (a) 5 cm, (b) 10 cm, (c) 15 cm, and (d) 30 cm. (Adopted with permission from figure 5 in Wang *et al* (2021). Copyright 2021 Elsevier.)

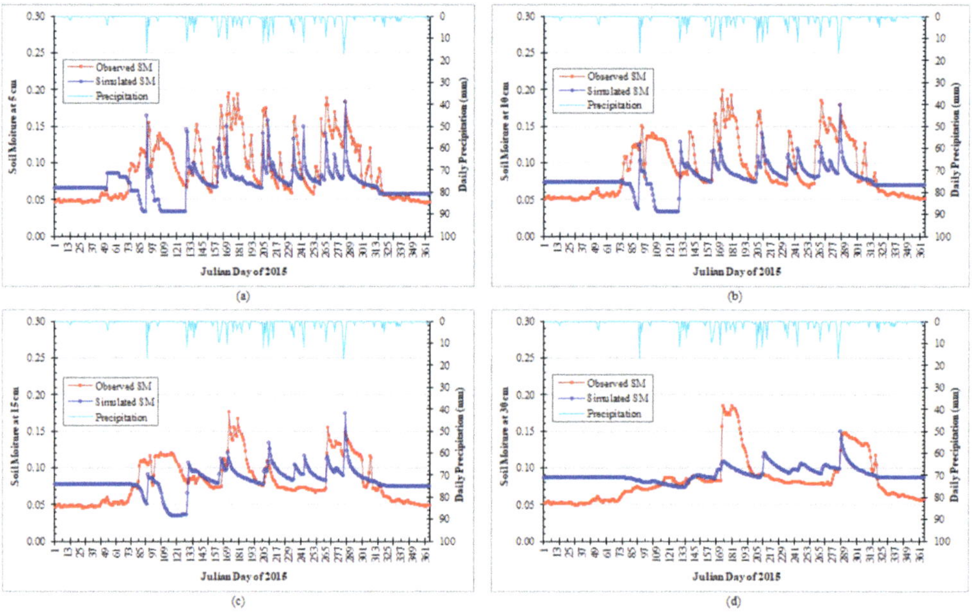

Figure 5.11. The SWAP model's performance in predicting soil moisture. Plot showing the simulated versus measured daily mean soil moisture in 2015 at a depth of (a) 5 cm, (b) 10 cm, (c) 15 cm, and (d) 30 cm. (Adopted with permission from figure 6 in Wang *et al* (2021). Copyright 2021 Elsevier.)

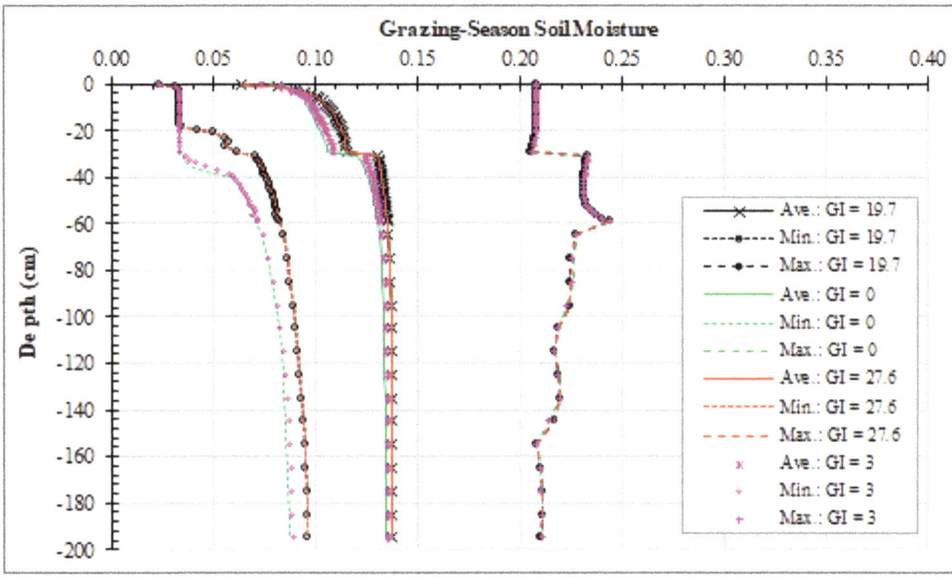

(a)

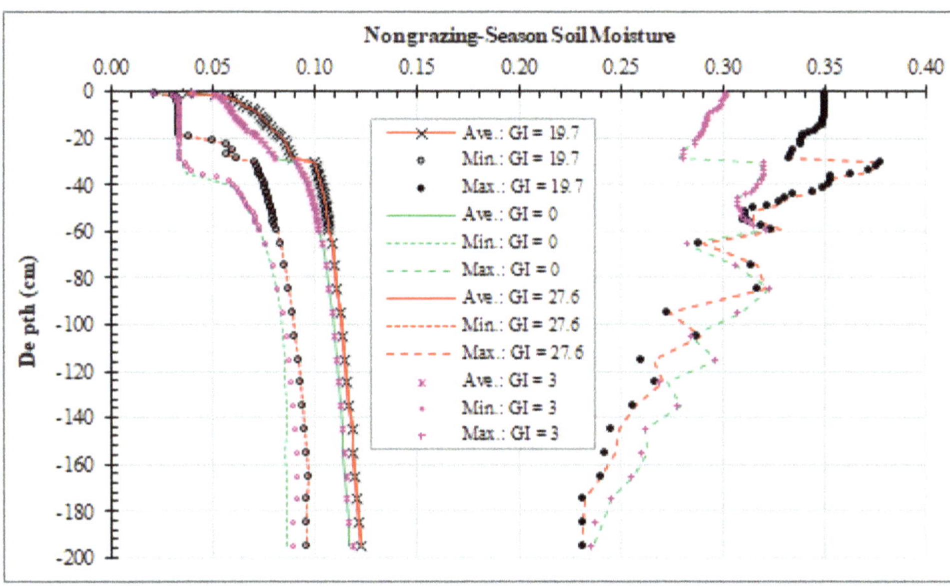

(b)

Figure 5.12. The predicted impacts of livestock grazing on soil moisture. Plots showing the averages and ranges of the daily mean soil moisture from 1960 to 2017 for selected grazing intensities during: (a) the grazing season (20 May–30 September); and (b) the non-grazing season (1 October–19 May). Ave.: average; Min.: minimum; Max.: maximum; and GI: grazing intensity, in cattle ha^{-1}. (Adopted with permission from figure 7 in Wang *et al* (2021). Copyright 2021 Elsevier.)

At a given depth and on an average daily basis, the mean soil moisture fluctuated synchronically with precipitation and reference evapotranspiration (e.g. figure 5.13). The average soil moisture exhibited a seasonal pattern: staggering in January and February, pulsing from March to April, increasing steadily in May, plateauing from late May to September, and decreasing gradually from October to December. From

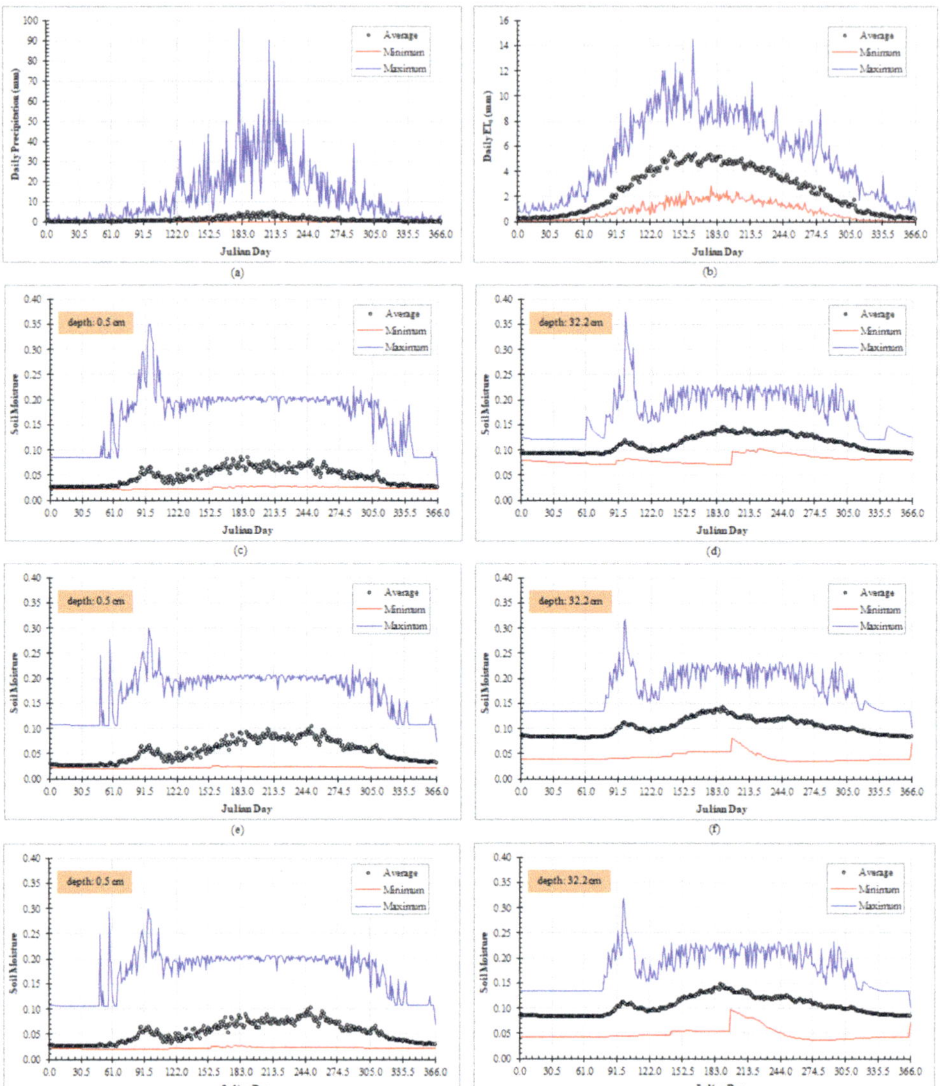

Figure 5.13. The predicted climate–grazing interactive effects on soil moisture. Plots showing the average and range from 1960 to 2017 for the daily (a) precipitation, (b) reference evapotranspiration, ET_0, (c) and (d) soil moisture for grazing intensity GI = 19.7 cattle ha^{-1}, (e) and (f) soil moisture for GI = 0.0, and (g) and (h) soil moisture for GI = 3.0 cattle ha^{-1}. The soil moisture at 0.5 and 32.2 cm had the largest fluctuations and are shown for explanation purposes. (Adopted with permission from figure 8 in Wang *et al* (2021). Copyright 2021 Elsevier.)

late February to early March, the interannual variability of daily soil moisture might be very large on a day of interest and was largest on the day of pulsing peak. In contrast, the interannual variability was gentle from April to September, moderate from October and November, and minimal from December to mid-February. Such interannual patterns coincided with those of daily precipitation and daily reference evapotranspiration (ET_0). For instance, from April to September the large variabilities of daily precipitation and ET_0 were asynchronous, offsetting each other. As a result, the interannual variability of the daily soil moisture was not large during these six months.

Grazing could alter soil moisture in multifaceted ways relative to the baseline (i.e. zero grazing) condition. On the one hand, the moisture of the top 1.5 cm soils was slightly lowered, while the moisture of the soils beneath was elevated (figure 5.12). For instance, if the grassland were grazed at a stress of 4.0 cattle ha^{-1}, the model predicted a reduction of up to 0.025 in the moisture of the top 0.5 cm soils while an increase of up to 0.04 in the moisture of the soils around the 10.5 cm depth. For February and March, grazing could make the top 1.5 cm soils have fewer days with a high moisture (e.g. figure 5.13(e) versus figure 5.13(c)) while the soils beneath have more days with a high moisture (e.g. figure 5.13(f) versus figure 5.13(d)). For the other ten months of April to January, grazing did not have similar impacts. On the other hand, grazing could reduce the interannual variability of daily soil moisture. Such a reduction effect was smaller for a lower grazing intensity than for a higher grazing intensity. Moreover, grazing would lower the maximum daily soil moisture by 0.05 but did not affect the minimum daily soil moisture. These grazing impacts on soil moisture would not stop escalating until a threshold grazing intensity of 3.0 cattle ha^{-1} (figures 5.13(g) and (h)). The current average grazing intensity of 19.7 cattle ha^{-1} was much higher than this threshold, indicating that the grassland has been being overgrazed.

The soil moisture change was found to be related to precipitation (figure 5.14). For the years with an annual precipitation of less than 220 mm or 0.6 mm d^{-1}, the changes could decrease or increase with increase of precipitation (figure 5.14(a)), whereas for the years with an annual precipitation of more than 0.6 mm d^{-1}, the changes tended to decrease with increase of precipitation and become apparently zero around 0.9 mm d^{-1} (figure 5.14(a)). Also, the soil moisture changes monotonically decreased with increase of ET_0 and tended to become apparently zero at $ET_0 \geqslant 2.5$ mm d^{-1} (figure 5.14(b)).

When the annual precipitation was less than 0.6 mm d^{-1}, all three factors of grazing, precipitation, and ET_0 were found to interactively affect soil moisture change, when the annual precipitation was more than 0.6 mm d^{-1}, however, only grazing and precipitation would interactively affect, while ET_0 might individually influence, soil moisture change.

5.2.4 Discussion

The measured soil temperatures were reproduced more accurately than the measured soil moisture by the model. A possible reason is because the

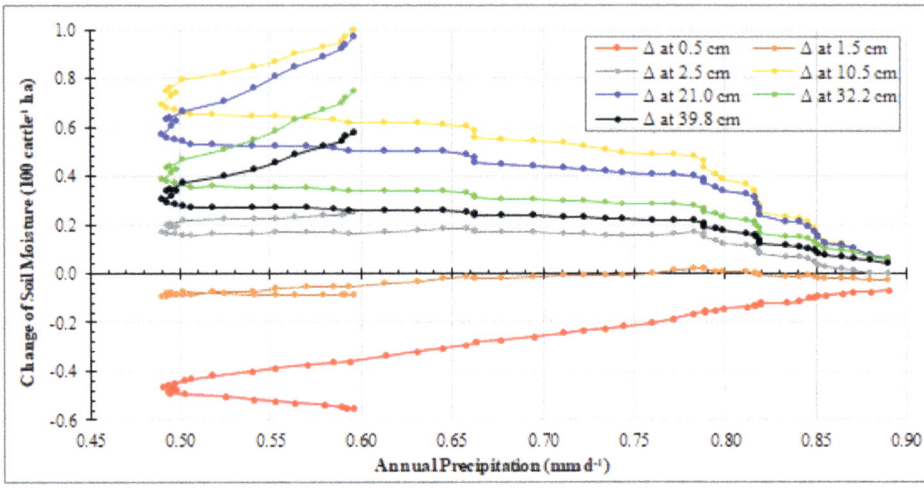

(a)

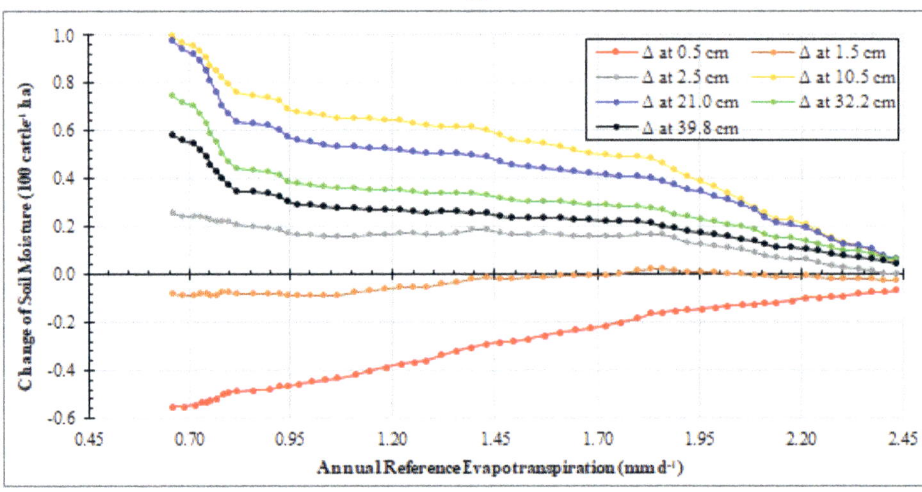

(b)

Figure 5.14. The predicted soil moisture changes relative to the non-grazing condition. Plots showing average annual change of soil moisture per unit increase of grazing intensity (GI) versus annually standardized: (a) precipitation and (b) reference evapotranspiration. For a given year between 1960 and 2017, the change for a given day was computed as the soil moisture for GI = 4.0 cattle ha^{-1} minus the soil moisture for GI = 0.0, and then the arithmetic average of the computed daily changes was divided by 4.0 to obtain the y-axis value of this year. The x-axis value was standardized as the ratio of the annual total to the corresponding number of days, leading to a units of mm d^{-1}, for the convenience of plotting. (Adopted with permission from figure 9 in Wang *et al* (2021). Copyright 2021 Elsevier.)

instrumented location is about 35 km distant from the climate station (figure 5.1), where the weather data were used to drive the model. Rationally, soil temperature is positively correlated with air temperature and solar radiation; these two climatic variables do not have a spatial variability within a 35 km distance that is large enough to differentiate soil heat absorption, emission, and reflection. In contrast, soil moisture is a positive function of precipitation; this climatic variable can vary drastically within a short distance (figure 5.11), which is particularly true for the watershed studied (Gao *et al* 2017a). On some days, the predicted soil moisture was high, but the corresponding observed values were low; this is probably because it rained at the climate station while did not rain at the instrumented location. On other days, the predicted soil moisture was low, but the corresponding observed values were high; this is probably because it did not rain at the climate station while rained at the instrumented location. On the days when the climate station and the instrumented location received similar storms in terms of rain amount and duration, the predicted soil moisture matched the corresponding observed values. The performance statistics (table 5.5) show the model's predictability of soil moisture dynamics.

The grazing season (20 May–30 September) usually has more storm events and larger rainfall as well as greater ET_0 than the non-grazing season (figure 5.13). As a result, the soil moisture during the grazing season is higher but has a smaller intraday variability than those during non-grazing season. The higher soil moisture is caused by the larger rainfall, whereas the smaller intraday variability may be due to the wetting effects of the larger rainfall being offset by the greater ET_0. On the other hand, the soil moisture during the non-grazing season is lower because of the smaller precipitation, snow packing, and soil freezing from late autumn to early spring. The pulsing soil moisture is caused by infiltration of melting snow water into thawing soils in mid- to late spring when air temperature steadily increases (figure 5.13). Controlled by air temperature fluctuations, the snowpack can go through multiple melting–packing loops, while the shallow soils can incur several thawing–freezing cycles. In response to the loops and cycles, the soil moisture pulses up and down multiple times synchronically, increasing the soil moisture variability of the non-grazing season.

The moisture of the top 30 cm soils is quickly evaporated and transpired to dry up this layer of soil (Wang 2015, Pedram *et al* 2018). This soil layer has an interface with the atmosphere and receives direct sunshine, providing sufficient heat to quickly vaporize soil water and efficiently transfer the pore vapor upward into the ambient environment. In addition, this soil layer has the densest roots (figure 5.8) to effectively absorb the soil water to meet the plant water consumption. The combination of quick evaporation and effective transpiration is responsible for the large variability of moisture of this layer of soil (figures 5.12 and 5.13). The lower-layer soils are more permeable than the upper-layer soils, thus the soil water just beneath the rootzone (i.e. 30–40 cm) tends to seep downward into deeper soils, leading to a large variability of soil moisture of this 10 cm soil layer (figure 5.2(b) and table 5.4). The soil moisture vertical distributions, which are consistent with those presented by Kurc and Small (2007) and Pereyra *et al* (2017), are similar for the grazing and non-grazing seasons, although the non-grazing season is dominated by cold-climate processes (e.g. snowfall, snow accumulating and melting, and soil freezing and thawing) but the grazing season is dominated by conventional rainfall-runoff and biotic processes.

Grazing removes plant stems and leaves, reducing vegetation coverage and increasing the bare-surface areas for soil water evaporation. This is the primary reason for the lower topsoil moisture from grazing (figure 5.14). However, the dried topsoil layer can inhibit the moisture of the soils beneath from being vaporized, which is beneficial for vegetation regrowth, and less soil water will be transpired because of the reduced plants (Wang 2015). As a result, the soil moisture below 1.5 cm is higher than that of the topsoil. These impacts on soil moisture are larger for a higher than a lower grazing intensity. For grazing stresses of larger than 4.0 cattle ha^{-1}, the grazing impacts do not measurably increase with an increase of grazing intensity (figure 5.13), which can be attributed to that above this stress threshold, primary aboveground biomasses may have been removed and thus grazing will not influence transpiration anymore.

As expected, the grazing impacts in a wetter year tend to be lighter than those in a drier year (figure 5.14(a)), although such impacts depend on the relative magnitude of ET_0 versus precipitation in years with an annual precipitation of less than 0.6 mm d^{-1}. The grazing impacts are a positive function of precipitation if ET_0 is large enough to vaporize most of the precipitation within a short period of time and are a negative function of precipitation otherwise (Charbeneau 2006). The canopy interception tends to increase with an increase of precipitation and the intercepted rainwater can substitute soil water for atmospheric evaporation demand, alleviating the grazing impacts. On days with a very high ET_0, the precipitation may be vaporized before it reaches the ground, leaving no chance for replenishing soil moisture (figure 5.14(b)). In contrast, when ET_0 is not large enough to quickly vaporize rainwater, the grazing impacts depend on vegetation coverage. Normally, the vegetation coverage under a lower grazing stress is smaller than that under a higher grazing stress. A higher vegetation coverage can intercept and absorb solar radiation, transmitting less energy for water evaporation and thus allowing more rainwater to seep into soils, whereas a lower vegetation coverage allows more solar radiation to warm the ground surface and soils beneath, facilitating rainwater evaporation rather than infiltration. A consequence is that transpiration is largely reduced while evaporation is greatly increased, causing the deteriorating impacts of grazing on soil moisture.

Semi-arid regions are challenged by wind erosion (Wang *et al* 2014b, Potter 2019), which can be very severe in late spring because of the combined effects of minimal vegetation coverage, low topsoil moisture, and high wind speed. In this regard, the moisture pulses in late spring play a vital role in stabilizing fertile-fine soil particles (figure 5.13). At low soil moisture, the wind can easily detach these fine particles, which in turn are suspended in the air to form dust storm (Duan *et al* 2011). For the Balagaer River watershed, if the topsoil moisture can be maintained at 0.15 or above, wind erosion will be under control (Wang *et al* 2014b, Potter 2019). Across the assessment years, relative to the non-grazing condition, grazing practices with a stress level of above 3.0 cattle ha^{-1} will result in 3–24 more days in late spring and 9–78 more days in the grazing season of a topsoil moisture of less than 0.15. To maintain sustainability and resilience, a grazing intensity of lower than 3.0 cattle ha^{-1} should be practiced for the semi-arid steppe ecosystem.

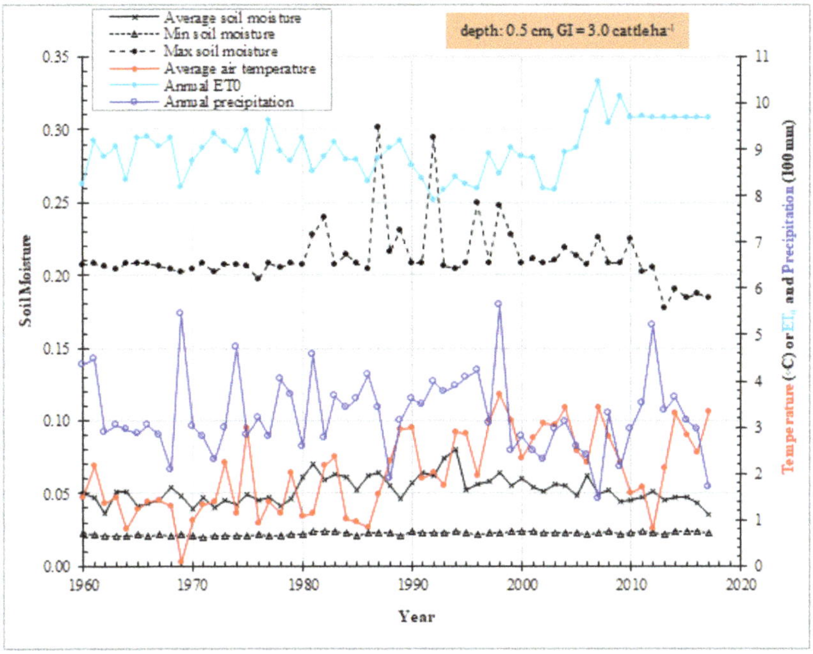

(a)

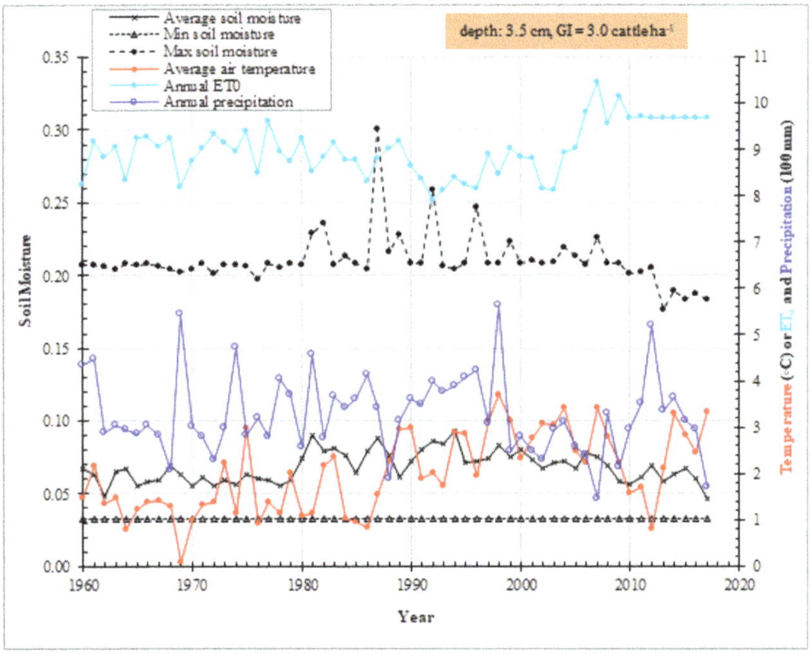

(b)

Figure 5.15. The modeled soil moisture. Time series plots for: (a) a soil depth of 0.5 cm and grazing intensity (GI) of 3.0 cattle ha^{-1}; (b) a soil depth of 3.5 cm and GI = 3.0 cattle ha^{-1}; and (c) a soil depth of 0.5 cm and GI = 19.7 cattle ha^{-1}. The results are for historical (i.e. 1960–2017) climatic conditions, which are surrogated by the average annual air temperature, annual reference evapotranspiration (ET_0), and annual precipitation. (Adopted with permission from figure 10 in Wang *et al* (2021). Copyright 2021 Elsevier.)

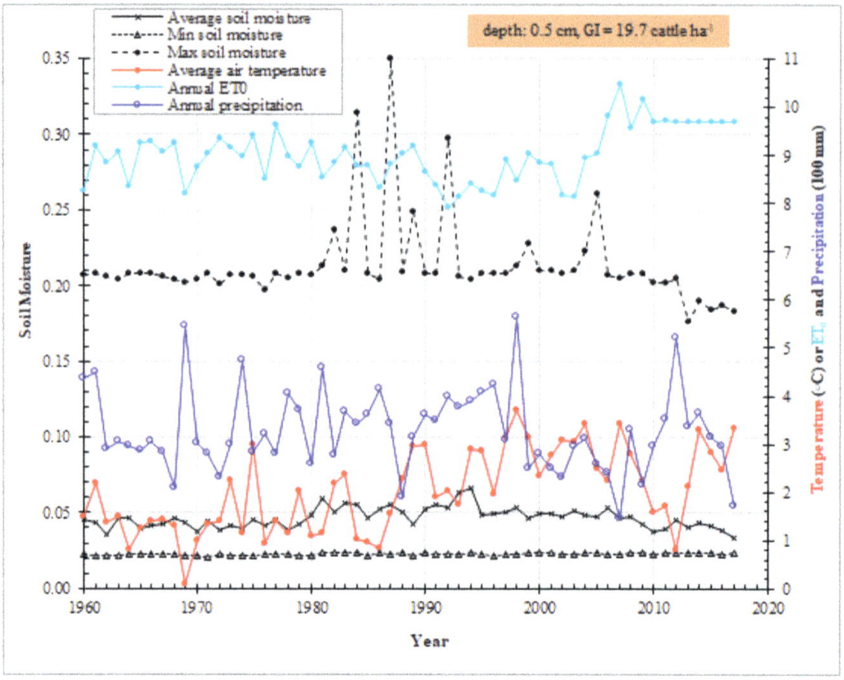

(c)

Figure 5.15. (Continued.)

Across the assessment years, the predicted annual minimum daily soil moisture is the same, but the annual maximum daily and the average annual soil moisture, however, exhibit a clear temporal pattern: slightly varying from 1960 to 1980, largely fluctuating from 1980 to 2000, and gradually decreasing since 2000 (figure 5.15); this is true for all grazing stresses and soil depths. As expected, for a grazing intensity of interest, soil moisture tends to vary more greatly at a shallower than a deeper depth (figure 5.15(a) versus figure 5.15(b)), and for a depth of interest, soil moisture fluctuates less at a grazing stress of below 3.0 cattle ha^{-1} than at a grazing stress of above 4.0 cattle ha^{-1} (figure 5.15(a) versus figure 5.15(c)). Precipitation and ET_0 directly influence the moisture of the topsoil and indirectly affect the moisture of the soils beneath (figure 5.14). ET_0 reflects the overall atmospheric water demand and is dependent on multiple climatic variables, including air temperature, vapor content, wind speed, and solar radiation. The post-2000 soil moisture shows a declining trend, likely caused by varying climate and livestock grazing.

Other parts of the Eurasian steppe grasslands share major characteristics with the Balagaer River watershed; hence the soil moisture responses discussed above to climate and livestock grazing can be reasonably extrapolated beyond the study watershed. It has been predicted that climate change will reduce snowfall and snowpack in winter. As a result, the soil moisture pulses in early spring would gradually diminish and ultimately disappear in the future, making the ecosystem more susceptible towards degradation and desertification. Policies should be made to regulate and guide ranchers to graze at stress levels below a threshold livestock intensity such as 3.0 cattle ha^{-1}.

5.2.5 Conclusions

This case study demonstrates how to use a SWAP model to examine the responses of soil moisture to climate and livestock grazing. The model was calibrated to have a good performance in reproducing observed soil temperatures and soil moisture in a semi-arid steppe watershed. The predicted daily soil moisture varies from 0.02 to 0.38 and tend to be lower at a smaller than a larger depth. The soils around the 30 cm depth were found to be the most moist on average with a largest temporal variability. The model shows soil moisture pulses in late spring and an asymptote between summer and early autumn. Livestock grazing will lower the moisture of the top 1.5 cm soils while elevating the moisture of the soils beneath. To prevent wind soil erosion, a topsoil moisture of 0.15 needs to be maintained by regulating the grazing stresses to be below 3.0 cattle ha^{-1}. When the annual precipitation is larger than 0.6 mm d^{-1}, grazing was predicted to reduce soil moisture with an increases of atmospheric water demand (ET_0) and precipitation; otherwise, grazing was found to increase or decrease soil moisture with an increase of precipitation, depending on whether ET_0 is large enough to quickly vaporize rainwater. The case study results can be logically extrapolated to other parts of the semi-arid Eurasian grasslands. The takeaway is that grazing must be practiced below a threshold livestock intensity.

5.3 The effects of climate and livestock grazing on precipitation consumption

In semi-arid ecoregions, because precipitation is far smaller than climatic water demand and tends to directly evaporate without benefiting the hydro-ecosystems, the relative proportion of non-beneficial (NBPC) over beneficial (BPC) precipitation consumption controls the water–carbon and assimilation–respiration processes. Increasing NBPC over BPC likely triggers irreversible degradation or even desertification of such a hydro-ecosystem; the relevant information, however, is missing in the existing literature. The objective of this study is to assess the effects of climate variability and livestock grazing on changes of the NBPC–BPC ratio, taking the Balagaer River watershed, a typical Eurasian steppe grassland in northeast China, as the testbed. The model is initially parameterized using literature values and experimental data and subsequently calibrated using *in situ* daily soil temperatures and soil moisture. The results reveal a threshold grazing intensity of around 3.0 cattle ha^{-1}, above which the NBPC–BPC ratio will be significantly elevated at a family significance level of 0.05. The ratio is predicted to be higher in a drier than a wetter year and to sharply increase with an increase of grazing intensity in years with an annual precipitation of less than 60 cm or an aridity index larger than 1.5. In the future, climate change will result in decreasing precipitation, mandating policies to balance livestock productivity and eco-environmental sustainability across the Eurasian Steppe.

5.3.1 Introduction

Water is vital for sustaining the fragile vegetation ecosystems of semi-arid regions (Singh *et al* 1998, Austin *et al* 2004, Kurc and Small 2007), such as the Eurasian Steppe (also called the Great Steppe), which is the vast steppe ecoregion of Eurasia in the temperate grasslands, savannas, and shrublands biome (Wang *et al* 2014a). On average, most parts of the ecoregion receive an annual precipitation of less than 600 mm while they incur an annual potential evapotranspiration (PET) greater than 1200 mm (Luo *et al* 2014, Yang 2014). For such an ecosystem, precipitation is subject to non-beneficial and beneficial consumption. The non-beneficial precipitation consumption (NBPC), which consists of canopy interception and soil evaporation, has a minimal contribution to vegetation growth and water availability, whereas the beneficial precipitation consumption (BPC), which consists of vegetation transpiration, surface runoff, and groundwater recharge, prompts vegetation growth and/or sustains water resources. The ratio of NBPC to BPC can be a handy indicator of whether the hydrological cycle of such an ecosystem has been shifted away from its pristine condition due to climate change and/or livestock grazing. Herein, the null hypothesis is that the NBPC–BPC ratio is smallest for the pristine condition and tends to increase as the ecosystem degrades and vice versa. It is BPC that drives abiotic (i.e. water and carbon) cycles that are directly related to the biotic (i.e. physiological) processes of respiration and assimilation. The difference between the CO_2 released by respiration and that fixed by assimilation is the determinant of plant biomass. If the NBPC–BPC ratio increases, the ecosystem's productivity will decrease while its fragility will increase, likely raising eco-environmental concerns.

Previous studies (Singh *et al* 1998, Huxman *et al* 2004, Kurc and Small 2007) indicate that respiration and assimilation are positively affected by evaporation and transpiration and that plant biomass tends to increase with an increase of soil moistness. In semi-arid environments, the respiration depends on water evaporation from surficial soils, whereas the assimilation is positively related to water transpiration from rootzone soils. As a result, these physiological processes are very sensitive to precipitation fluctuation and consumption. The interannual variabilities of evaporation and transpiration increase with a decrease of precipitation, possibly shifting a semi-arid ecosystem from a carbon source in a wet year to a carbon sink in a dry year. Moreover, for a semi-arid steppe grassland, in addition to precipitation, the grazing intensity and timing can also influence evapotranspiration (ET) and thus affect the physiological processes (Wang *et al* 2020).

Steppe grasslands occupy about 8% of the Earth's terrestrial surface and are suffering from moderate to severe degradation (Henwood 1998, Bai *et al* 2004). This has become a key constraint for the sustainability of 'grassland agriculture', a system of agriculture in which the major emphasis is on grasses, legumes, and other fodder or soil-building crops (Barnes *et al* 2003). The main factors for degradation have been widely cited as overgrazing, cultivation, overdevelopment, and climate change

(Chepil 1946, Li *et al* 1995, White *et al* 2000, IPCC 2001, Peart 2008), which has altered natural hydrology (Li *et al* 2007) and caused loss of top (i.e. 10–20 cm) calcic castanozem soils. This topsoil plays a key role in sustaining grasslands because it is loose and has a plentiful supply of humus and organic matter. Policy choices to reduce or reverse grassland degradation are often made with only a vague understanding of the causative complexity linking steppe hydrology, topsoil erosion, and grassland degradation, limiting efforts to manage, protect, and/or restore steppe grasslands.

Some studies (Peart 2008, Liu *et al* 2019) examined the policy and economic reasons for grassland degradation, while others (Gőri *et al* 2005, Kwon *et al* 2016) assessed the impacts of grassland degradation on wildlife species. In addition, some researchers (Christensen *et al* 2004, Endo *et al* 2006) analyzed climate change and its effects on the vulnerability of steppe grasslands. The climate of the Inner Mongolia steppe, for instance, has been shown to experience a significant spatiotemporal trend, with increasing air temperature during the winter and more frequent drought in the spring (Shao *et al* 2008, Wang *et al* 2014a). Other researchers (Christensen *et al* 2003, Christensen *et al* 2005, Li *et al* 2007, Blank *et al* 2009) have investigated the ecological functions of steppe grasslands and how these functions are affected by grazing and overgrazing, demonstrating that a grazing removal index (i.e. the ratio of the aboveground biomass reduction in a grazed area to the aboveground biomass in the same area if non-grazed) of 0.3 or below may not lead to a sustainability problem for steppe grasslands. Further, a few studies (Zhou *et al* 2002, Zhang *et al* 2005, Li *et al* 2007, Hao *et al* 2008) assessed the hydrological processes of steppe grasslands. Those few studies mainly focus on how ET varies with precipitation and/or soil water content, revealing that ET is limited by the availability of soil water (i.e. atmospheric heat stress is weaker than soil water stress and soil evaporation is the primary contributor to ET), and that the maximum daily ET varies from 3.0 to 6.2 mm d^{-1} in those grasslands. Daily ET rates tend to increase with increasing soil water content, which in turn is highly dependent upon the drying–wetting fluctuations experienced by steppe grasslands. Overall, soil water content is higher when grass coverage is denser (Liu *et al* 2006). Those studies also indicate that the aboveground biomass of grassland communities has a positive relation with precipitation and that precipitation in the first half of the year (i.e. January to July) is the primary climatic factor causing fluctuations in community biomass production (Bai *et al* 2004, Luo *et al* 2014, Wang *et al* 2014a, Yang 2014). Moreover, several recent studies (Zhang *et al* 2017, Zhai *et al* 2018, Song *et al* 2020, Wang *et al* 2020) examined the effects of grazing regimes (i.e. intensities and seasons) on plant physiology and biomass production. For instance, Wang *et al* (2020) collected soil respiration data in 2010 and 2011 at the Loess Plateau Grassland Ecosystem Trial Station located in Gansu Province of northwest China. The station has sandy free-drainage loess soils and is a typical temperate steppe grassland with dominant species of *Stipa bungeana*, *Lespedeza davurica*, *Pennisetum flaccidum*, *Artemisia capillaris*, and *Stetarua viridis*. The data indicate that for a given grazing season, a higher grazing intensity tends to result in smaller above- and below-ground biomasses but a larger soil respiration rate, and vice versa.

Such causal relationships, however, can be compounded by interannual variations of precipitation, air temperature, and radiation.

In the existing literature, including those cited above, the information is lacking on how the relative magnitude of NBPC over BPC can be affected by climate variability and livestock grazing, while such information is needed for developing practically effective measures to sustain semi-arid steppe ecosystems as important carbon sinks to confront global climate change (Neumann and Wigen 2013). The objective of this study was to fill this information gap by simulating the effects of climate variability and grazing intensity on changes of the NBPC–BPC ratio in semi-arid steppe environments. The testbed is the Balagaer River watershed, a typical Eurasian steppe grassland located in the northeast Inner Mongolia Autonomous Region (IMAR) of China. The descriptions of the watershed are presented in section 5.1.1. The simulation is carried out for combinations of retrospective (i.e. 1955–2017) climates and seven representative grazing intensities.

5.3.2 Materials and methods

5.3.2.1 Data and preprocessing

In addition to the soil data (Wang *et al* 2021a), this study also used historical (i.e. 1 January 1955 to 31 December 2017) daily data on precipitation, air temperature, relative humidity, wind speed, solar radiation, and reference evapotranspiration at West Ujimqin Climate Station 54012 (117°36′ E, 44°35′ N), acquired from the Chinese Meteorological Information Center (CMIC 2020). In addition, from 2014 to 2017, the data on soil moisture and soil temperature were continuously recorded at a 12 h time interval using four EC-5 three-needle sensors (METER Group 2020) installed, respectively, at 5, 10, 15, and 30 cm depths beneath the site (117°52′ E, 44° 20′ N). The sensors have a detection range of zero to one at an accuracy of ±1% for soil moisture and a measurement range of –40 °C to 50 °C at an accuracy of ±0.3 °C. The recorded values were processed in Microsoft® Excel into the corresponding daily time series, which are complete from 31 May to 31 December 2014, 1 January to 31 December 2015, 1 January to 10 May 2016, 30 August to 31 December 2016, and 1 January to 5 March 2017 (Wang *et al* 2021a). Further, the literature and experimental data (Kurc and Small 2007, Wang and Zhou 2012, Luo *et al* 2014, Pereyra *et al* 2017) were used to estimate the physiological parameters (table 5.2) for a hybrid of *Stipa grandis* and *Leymus chinensis* (Wang *et al* 2021a) because they intermingle across the watershed (Luo *et al* 2014) and share primary physiological features (Chen *et al* 2005). Moreover, in reference to cattle, the animal unit equivalent (AUE) (USDA-NRCS 2020) grazing intensities across the watershed range from 11.8 to 27.6 cattle ha^{-1}, with a mean of 19.7 cattle ha^{-1} (table 5.3). The livestock might start grazing once the aboveground dry matter is more than 350 kg ha^{-1}, with the after-grazing dry matter of less than 150 kg ha^{-1} and a dropping and treading loss of 0.43–0.5 kg ha^{-1}. The dry matter consumption per AUE was estimated to be 8.5–10.0 kg ha^{-1} (AHDB 2020). Other typical grassland management practices (e.g. irrigation, fertilization, and mowing) were not implemented in the past.

5.3.2.2 The calibrated SWAP model

Wang *et al* (2021) parameterized a SWAP model for a 200 cm long soil column of the synthesized study site. As discussed in sections 5.2.2.2 and 5.2.3.1, the model subdivides the soil column into four sublayers of 20, 10, 30, and 140 cm thick, respectively, in the downward direction. The first or third sublayer was discretized into 20 compartments, while the second sublayer was discretized into 5 compartments and the fourth sublayer into 14 compartments. The compartments of a given sublayer have the same thickness, which is computed as the division of the sublayer thickness by the number of compartments. The first and second sublayers belong to the upper (i.e. 0–30 cm) layer and the third sublayer corresponds to the lower (i.e. 30–60 cm) layer. The fourth sublayer can be considered as the deeper layer. The soil–water properties were assumed to be the same for the lower and deeper layers (Wang *et al* 2014a).

The SWAP model was sufficiently calibrated (Wang *et al* 2021a), as evidenced by table 5.5 and figures 5.10 and 5.11. It could accurately reproduce the measured daily soil temperatures, as indicated by the small random prediction errors of −0.72 °C to 0.70 °C, a strong capability in capturing the variances of the observations (coefficient of determination $R^2 > 0.75$), and the excellent tracing of the fluctuations of the observed values (Nash–Sutcliffe efficiency NSE > 0.9). In addition, the model performed well in reproducing the average daily mean soil moisture, as indicated by the compatible standard deviations and small values of mean percent errors (MPE < 0.04). It traced the overall fluctuation patterns of the observations fairly.

5.3.2.3 Simulation and assessment

The calibrated SWAP model was run from 1 January 1955 to 31 December 2017: the first five years were used as the warm-up period to stabilize the model parameters and the successive 58 years were taken for assessments. Model runs were executed for seven AUE grazing intensities, including 0, 3, 5, 10, 15, 19.7, and 27.6 cattle ha^{-1}. The non-grazing might surrogate the pristine condition, while the grazing intensity of 27.6 cattle ha^{-1} was the highest that has been practiced in the study watershed. The results from these model runs were examined to determine the threshold grazing intensity above which further increasing intensity would not affect the NBPC–BPC ratio anymore. Herein, the null hypothesis was that grazing influences would become minimal once the aboveground biomass was too small to be further grazed. The different grazing intensities were used to assess individual and interactive effects of grazing and climate on the ratio.

To verify this null hypothesis, a Bonferroni pairwise comparison test (Dunn 1961, Demidenko 2019) was carried out using the simulated NBPC–BPC ratios across the seven grazing intensities. The family comparison had a significance level of $\alpha = 0.05$ and thus each paired comparison had a significance level of $\alpha/(2 \times 7) = 1/280$. In addition, for a given grazing intensity, the NBPC–BPC ratios were plotted versus responding annual precipitations, annual PETs, or aridity indices to visually examine how the ratios would vary with climates. Further, for each component of the NBPC or BPC, the simulated values were linearly regressed on annual precipitations and aridity indices. The BPC consists of bottom seepage and transpiration, whereas the NBPC includes interception and soil evaporation. For each regression, its intercept and slope were statistically tested to be zero or not at $\alpha = 0.05$ and the value of R^2 was computed. The slopes and R^2 values were used to evaluate how each of the four BPC and NBPC components would be related to climates for a given grazing intensity. Moreover, the interactive climate–grazing effects were visually assessed by surveying the previously generated plots across the seven grazing intensities as well as evaluated quantitatively by comparing the regression intercept and slope for a grazing intensity with the corresponding values for another grazing intensity.

5.3.3 Results

5.3.3.1 The livestock grazing impacts

The relative magnitude of NBPC versus BPC was predicted to be affected by livestock grazing (figure 5.16). The NBPC–BPC ratios for a higher grazing intensity tended to have a larger interannual variability than those for a lower grazing intensity; for instance, for the non-grazing condition, the ratios varied from 0.11 to 0.65, whereas for GI = 27.6 cattle ha^{-1}, the ratios ranged from 0.40 to 3.31. In addition, the annual average of the ratios for a larger grazing intensity tended to be greater than that for a lower grazing intensity, indicating that the hydro-ecological benefits of precipitation could be lessened by livestock grazing. The Bonferoni test for multiple comparisons at a family significance level of $\alpha = 0.05$ indicated that the NBPC–BPC ratios for GI = 3.0 cattle ha^{-1} were statistically the same as those for the non-grazing condition (p-value = 0.30) and that the ratios for GI = 5.0 cattle ha^{-1} were statistically same as those for the higher grazing intensities (p-values >0.93). However, the ratios for a GI $\geqslant$ 5.0 cattle ha^{-1} were significantly different from those for

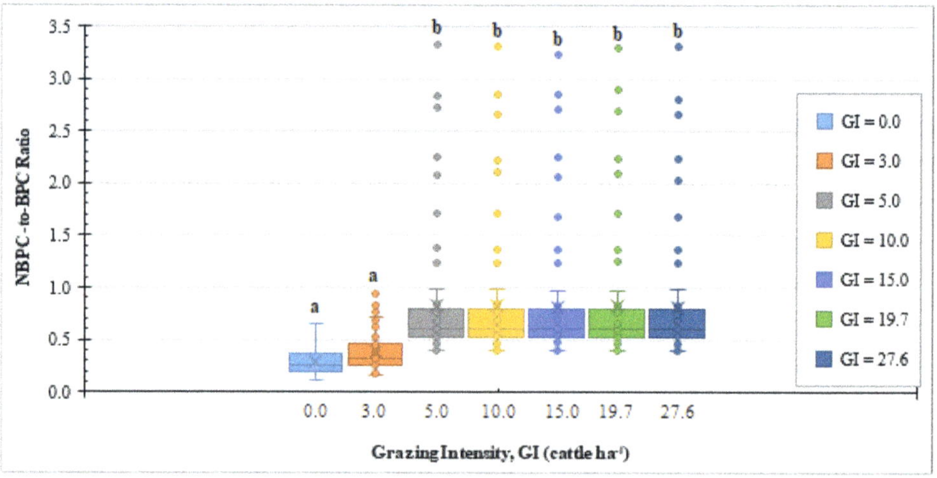

Figure 5.16. Ratios of annual non-beneficial (NBPC) to beneficial (BPC) precipitation consumptions for selected livestock grazing intensities (GIs). The ratios for GIs labeled by the same letter a or b are statistically identical at a family significance level of 0.05, whereas the ratios for GIs labeled by different letters are statistically different.

a GI $\leqslant$ 3.0 cattle ha^{-1} (p-values < 1.67×10^{-6}), revealing a threshold grazing intensity of around 3.0 cattle ha^{-1} in terms of balancing livestock productivity and ecosystem protection.

5.3.3.2 The climatic effects

Regardless of the grazing intensities, the NBPC–BPC ratios tended to decrease with an increase of annual precipitation (figure 5.17(a)). Such a negative correlation was visually strong when the precipitation was less than 60 cm, but it became very weak when the annual precipitation was larger, as indicated by the ratios that randomly fluctuated between 0.13 and 0.98, with an approximate mean of 0.5. In contrast, the NBPC–BPC ratios were virtually independent of PET (figure 5.17(b)), which reflects the climatic water demand as determined by solar radiation, air temperature, humidity, vapor pressure, and wind speed. However, the ratios exhibited a weak positive relationship with aridity index (figure 5.17(c)), which is defined as division of annual PET by annual precipitation (Middleton *et al* 1997, Greve *et al* 2019). The NBPC–BPC ratios randomly fluctuated between 0.16 and 0.98, with an approximate mean of 0.5, for aridity indices of less than 1.5, whereas the ratios tended to increase with an increase of aridity index otherwise. Overall, the relative magnitude of NBPC versus BPC was more closely related to the climatic factors in a drier than a wetter year.

Although the precipitation amount had a positive effect on its overall use, its specific effect on each consumption component would vary (e.g. figure 5.18(a)). On the one hand, among the three BPC components, the surface runoff was predicted to be zero for all simulation years, which was expected for semi-arid steppe grasslands with very permeable soils. The bottom seepage and transpiration were found to be positively related with precipitation, as indicated by the statistically significant positive slopes (p-values $\ll$ 0.05) (table 5.6). The significant (p-values <0.05) negative intercepts from regressing bottom seepage on precipitation implied that for a year with a small (<8.5 cm) annual precipitation, the shallow groundwater would rise by capillary suction to replenish above-bottom soil moisture for plant transpiration and/or soil evaporation; otherwise, the above-bottom soil water would recharge the shallow aquifer. For a year with zero annual precipitation, the transpiration would be either dependent on or independent of grazing intensity, as indicated by the significant (p-values $\ll$ 0.05) positive intercepts for GI $\leqslant$ 3.0 cattle ha^{-1} but the insignificant (p-values $\approx$ 0.90) intercepts for GI $\geqslant$ 5.0 cattle ha^{-1}. The correlation between bottom seepage and precipitation (adjusted R^2 = 0.84–0.85) was stronger than that between transpiration and precipitation (adjusted R^2 = 0.32–0.39). Per each centimeter increase of precipitation, the bottom seepage and transpiration would increase by about 0.64–0.68 cm and 0.042–0.085 cm, respectively. On the other hand, among the two NBPC components, the interception was statistically independent of precipitation because of the insignificant slopes and very small values of adjusted R^2. The interceptions were predicted to be about 0.12–0.18 cm for GI $\leqslant$ 3.0 cattle ha^{-1} and zeros for GI $\geqslant$ 5.0 cattle ha^{-1}. In contrast, the soil evaporation was found to be positively related with precipitation, as indicated by the statistically significant positive slopes (p-values $\ll$ 0.05). The soil evaporation would increase by about 0.18–0.28 cm

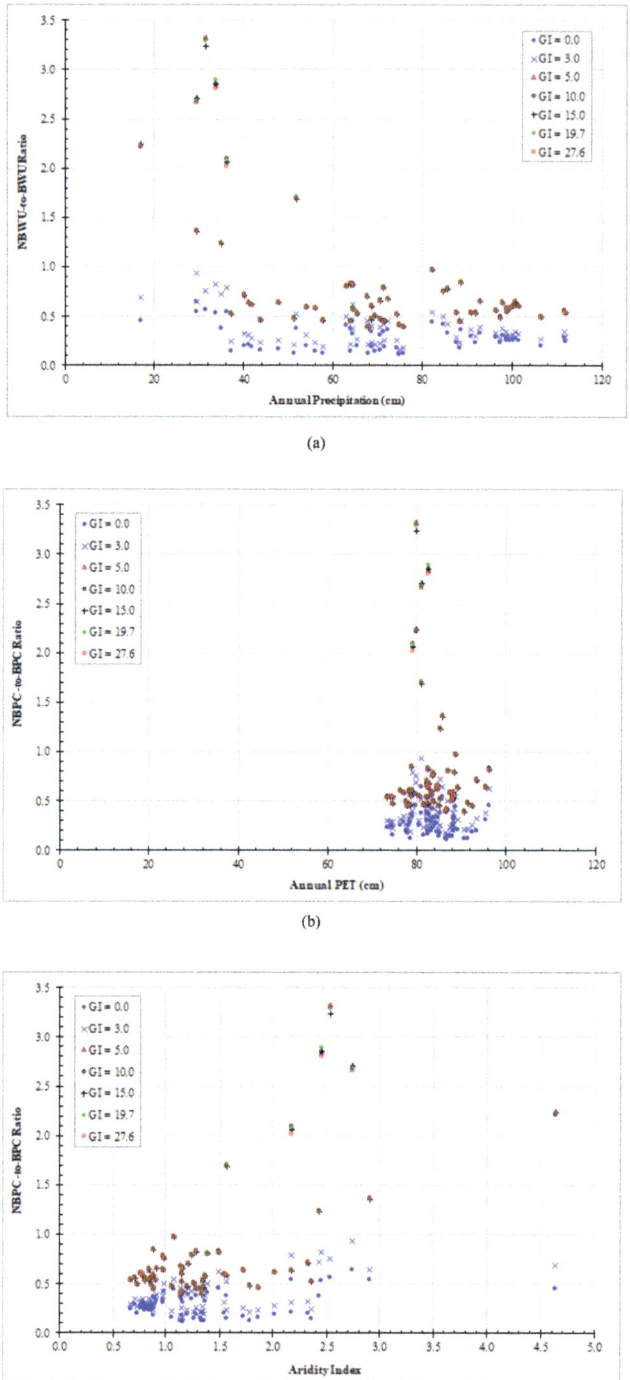

Figure 5.17. Simulated ratios of annual non-beneficial (NBPC) to beneficial (BPC) precipitation consumption for variations of livestock grazing intensity (GI). Plot of the ratios versus the (a) annual precipitation and (b) annual potential evapotranspiration (PET); (c) aridity index (i.e. PET–precipitation quotient).

for each centimeter increase of precipitation. For a year with zero annual precipitation, the soil evaporation would be statistically zero under non-grazing condition and was predicted to be dependent upon grazing intensity.

In comparison, the effect of climatic aridity on overall precipitation use was negative, while the specific effect on each consumption component was variable (figure 5.18(b)). On the one hand, the bottom seepage and transpiration were found to be negatively related to the aridity index, as indicated by the statistically significant negative slopes (p-values $\ll 0.05$) (table 5.6). The significant (p-values < 0.05) positive intercepts implied that for a year with a small (<0.3) aridity index, most of the annual precipitation would recharge the shallow aquifer. The correlation between bottom seepage and aridity index (adjusted $R^2 = 0.68$–0.72) was stronger than that between transpiration and aridity index (adjusted $R^2 = 0.29$–0.39). For a unit increase of aridity index, the bottom seepage and transpiration would decrease by about 19.9–20.7 cm and 1.34–2.85 cm, respectively. On the other hand, the interception was statistically independent of aridity index because of the insignificant slopes and very small values of adjusted R^2. The interceptions were predicted to be about 0.13–0.21 cm for GI $\leqslant 3.0$ cattle ha^{-1} and zeros for GI $\geqslant 5.0$ cattle ha^{-1}. In contrast, the soil evaporation was found to be negatively related with aridity index, as indicated by the statistically significant negative slopes (p-values $\ll 0.05$). The soil evaporation would decrease by about 5.15–7.72 cm for a unit increase of aridity index. For a year with a small (<0.3) aridity index, the soil evaporation was predicted to vary from 21.88–38.31 cm.

5.3.3.3 The interactive influences of climate and livestock grazing

The impact of livestock grazing tended to be larger in a drier than a wetter year (figure 5.17 and table 5.6). Grazing would increase the proportion of annual precipitation for non-beneficial consumption regardless of the precipitation amount, and in a dry year, such a non-beneficial proportion could become greater than the proportion for beneficial consumption, reducing the available water resources for sustaining the semi-arid steppe ecosystem. For a year with an annual precipitation of less than 60 cm or an aridity index larger than 1.5, the NBPC–BPC ratio could be increased by several times with an increase of grazing intensity. For instance, provided an annual precipitation of about 30 cm, the ratio would be increased from 0.6 for the non-grazing condition to about 3.3 for the current maximum grazing intensity of 27.6 cattle ha^{-1}; implying that grazing would more greatly reduce the eco-environmental benefits of limited water resources in drier years. In contrast, for a year with an annual precipitation of more than 60 cm or an aridity index of less than 1.5, the NBPC–BPC ratio was predicted to increase because of grazing, but the NBPC would not exceed the BPC regardless of the grazing intensities, that is, a larger proportion of the precipitation could still benefit the vegetation ecosystem.

Livestock grazing was predicted to have a minor influence on bottom seepage, but it could greatly affect transpiration, interception, and soil evaporation (table 5.6). The increasing rate of bottom seepage with an increase of precipitation was independent of grazing intensity, as indicated by the nearly constant slopes of

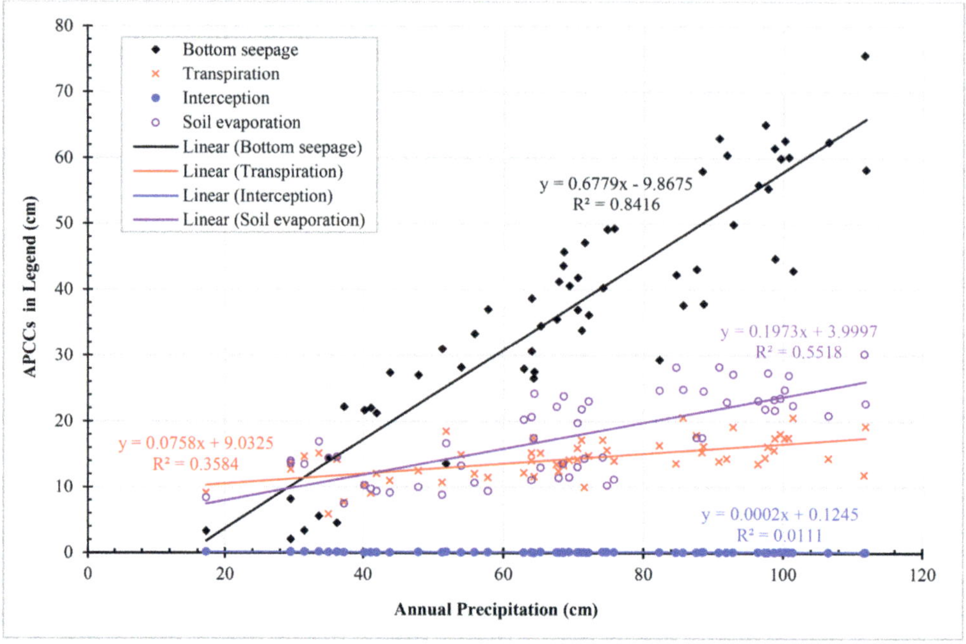

(a)

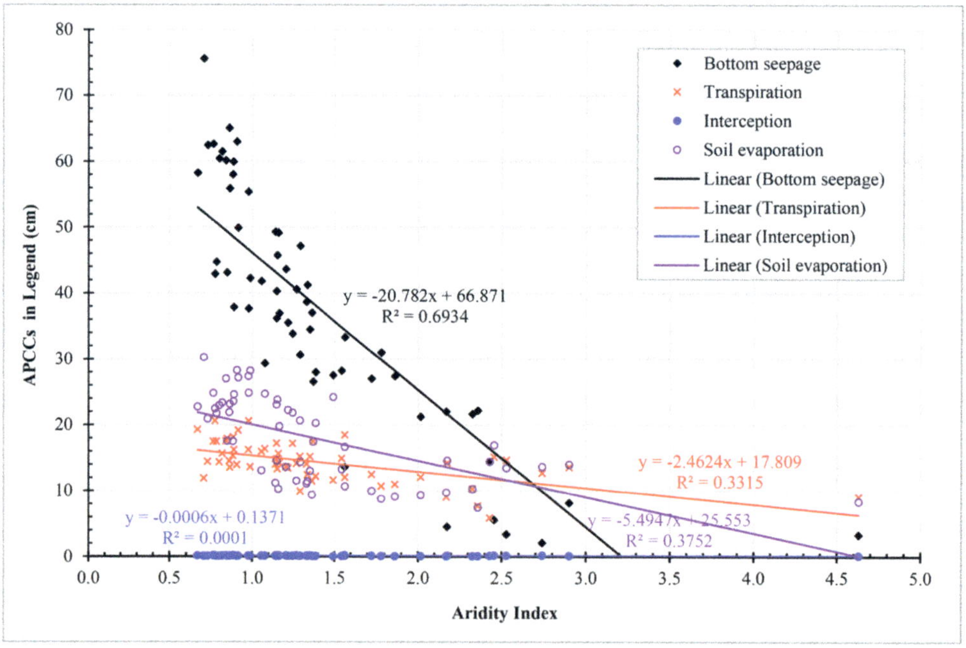

(b)

Figure 5.18. Simulated annual precipitation consumption components (APCCs). Plot of the APCCs versus (a) annual precipitation and (b) aridity index. The results are for a livestock grazing intensity of GI = 3.0 cattle ha^{-1}, below which grazing impacts on the natural hydrology of the study steppe grassland will be statistically insignificant.

Table 5.6. Regressions of precipitation consumption components on precipitation and aridity index[1].

		Linear regression on annual precipitation by				Linear regression on aridity index by			
GI (cattle ha^{-1})	Item	Bottom seepage	Transpiration	Interception	Soil evaporation	Bottom seepage	Transpiration	Interception	Soil evaporation
0.0	Intercept [cm]	−11.0626	12.7534	0.1779	1.5183	65.5303	22.6851	0.2080	21.8869
	p-value	3.99×10^{-4}	6.69×10^{-18}	3.30×10^{-13}	3.90×10^{-1}	2.13×10^{-29}	7.55×10^{-37}	1.69×10^{-21}	5.96×10^{-21}
	Slope [cm cm^{-1}]	0.6777	0.084 58	0.000 324	0.1874	−20.6885	−2.8457	−0.005 29	−5.1454
	p-value	7.66×10^{-24}	8.69×10^{-8}	2.09×10^{-1}	1.11×10^{-10}	1.13×10^{-15}	1.01×10^{-7}	5.45×10^{-1}	1.08×10^{-6}
	Adjusted R^2	0.84	0.39	0.011	0.52	0.68	0.39	0.0066	0.34
3.0	Intercept [cm]	−9.8675	9.0325	0.1245	3.9997	66.8711	17.8089	0.1371	25.5531
	p-value	1.26×10^{-3}	1.70×10^{-12}	8.15×10^{-11}	2.66×10^{-2}	3.62×10^{-30}	2.66×10^{-31}	2.68×10^{-17}	6.91×10^{-24}
	Slope [cm cm^{-1}]	0.6779	0.075 80	0.000 167	0.1973	−20.7820	−2.4624	−0.0006	−5.4948
	p-value	4.54×10^{-24}	6.91×10^{-7}	4.32×10^{-1}	2.46×10^{-11}	5.32×10^{-16}	2.26×10^{-6}	9.33×10^{-1}	3.22×10^{-7}
	Adjusted R^2	0.84	0.35	0.011	0.54	0.69	0.32	0.000 13	0.36
5.0	Intercept [cm]	−5.3080	0.069 01		8.3755	67.2481	4.8742		38.3155
	p-value	4.64×10^{-2}	9.06×10^{-1}		2.64×10^{-5}	5.17×10^{-33}	4.19×10^{-16}		6.20×10^{-30}
	Slope [cm cm^{-1}]	0.6373	0.041 73		0.2723	−19.8279	−1.3367		−7.7222
	p-value	4.81×10^{-25}	2.07×10^{-6}		1.28×10^{-15}	2.64×10^{-17}	8.60×10^{-6}		1.49×10^{-9}
	Adjusted R^2	0.85	0.32		0.68	0.72	0.29		0.47
10.0	Intercept [cm]	−5.4536	0.066 44		8.3431	67.4669	4.8811		38.2577
	p-value	4.07×10^{-2}	9.10×10^{-1}		2.74×10^{-5}	5.94×10^{-33}	4.66×10^{-16}		6.70×10^{-30}
	Slope [cm cm^{-1}]	0.6409	0.041 82		0.2721	−19.9092	−1.3392		−7.7097
	p-value	3.48×10^{-25}	2.16×10^{-6}		1.22×10^{-15}	2.84×10^{-17}	8.96×10^{-6}		1.56×10^{-9}
	Adjusted R^2	0.85	0.32		0.68	0.72	0.29		0.47
15.0	Intercept [cm]	−5.3828	0.077 18		8.3352	67.5322	4.8824		38.1113
	p-value	4.32×10^{-2}	8.96×10^{-1}		2.92×10^{-5}	4.96×10^{-33}	4.79×10^{-16}		8.77×10^{-30}
	Slope [cm cm^{-1}]	0.6406	0.041 73		0.2710	−19.9176	−1.3368		−7.6670
	p-value	3.48×10^{-25}	2.30×10^{-6}		1.61×10^{-15}	2.52×10^{-17}	9.39×10^{-6}		1.89×10^{-9}
	Adjusted R^2	0.85	0.32		0.68	0.72	0.29		0.47

19.7	Intercept [cm]	−5.4757	0.066 65		8.3767	67.4461	4.8842		38.2603
	p-value	3.96×10^{-2}	9.10×10^{-1}		2.64×10^{-5}	5.84×10^{-33}	4.57×10^{-16}		6.87×10^{-30}
	Slope [cm cm^{-1}]	0.6409	0.041 85		0.2719	−19.9097	−1.3368		−7.7019
	p-value	3.28×10^{-25}	2.13×10^{-6}		1.36×10^{-15}	2.76×10^{-17}	8.90×10^{-6}		1.63×10^{-9}
	Adjusted R^2	0.85	0.32		0.68	0.72	0.29		0.47
27.6	Intercept [cm]	−5.4239	0.074 65		8.3277	67.5821	4.8792		38.1404
	p-value	4.17×10^{-2}	8.99×10^{-1}		2.77×10^{-5}	5.41×10^{-33}	4.88×10^{-16}		7.10×10^{-30}
	Slope [cm cm^{-1}]	0.6416	0.041 74		0.2712	−19.9353	−1.3357		−7.6826
	p-value	3.29×10^{-25}	2.26×10^{-6}		1.33×10^{-15}	2.68×10^{-17}	9.49×10^{-6}		1.63×10^{-9}
	Adjusted R^2	0.85	0.32		0.68	0.72	0.29		0.47

[1] GI: livestock grazing intensity; blank cell: not applicable because of zero interception; red cell: statistically insignificant at a significance level of $\alpha = 0.05$; and adjusted R^2: coefficient of determination adjusted by sample size and number of independent variables (Demidenko 2019). In this study, the aridity index is defined as the quotient of annual potential evapotranspiration and annual precipitation.

0.64–0.68 for the varying grazing intensities. Such a positive correlation could explain about 85% of the variance presented in the interannual bottom seepages. In contrast, if GI $\leqslant$ 3.0 cattle ha^{-1}, the increasing rate of transpiration with an increase of precipitation would decrease by 0.002 93 per unit increase of grazing intensity; if GI $\geqslant$ 5.0 cattle ha^{-1}, the increasing rate became independent of grazing intensity. If GI = 3.0–5.0 cattle ha^{-1}, the increasing rate would decrease nonlinearly, with an average decreasing rate of 0.017 per unit increase of grazing intensity, which is about 5.8 times larger than that for GI $\leqslant$ 3.0 cattle ha^{-1}. In addition, the interceptions were predicted to be about 0.13–0.21 cm for GI $\leqslant$ 3.0 cattle ha^{-1} and zeros for GI $\geqslant$ 5.0 cattle ha^{-1}. Further, if GI $\leqslant$ 3.0 cattle ha^{-1}, the increasing rate of soil evaporation with increase of precipitation would increase by 0.0033 per unit increase of grazing intensity; if GI $\geqslant$ 5.0 cattle ha^{-1}, the soil evaporation was barely affected by grazing. If GI = 3.0–5.0 cattle ha^{-1}, the increasing rate would increase nonlinearly, with an average of 0.0375 per unit increase of grazing intensity, which is about 11.5 times larger than that for GI $\leqslant$ 3.0 cattle ha^{-1}. Overall, in terms of varying precipitation, the grazing intensity of around 3.0 cattle ha^{-1} might be considered as the threshold, above which some BPC would be reallocated to NBPC, making the ecosystem more fragile.

Similarly, the decreasing rate of bottom seepage with increase of aridity index gently varied with grazing intensity; the steepest decreasing rate was determined to be 20.78 at GI = 3.0 cattle ha^{-1} (table 5.6). In addition, if GI $\leqslant$ 3.0 cattle ha^{-1}, the decreasing rate of transpiration with increase of aridity index would decrease by about 0.13 per unit increase of aridity index; if GI $\geqslant$ 5.0 cattle ha^{-1}, the decreasing rate became independent of grazing intensity. If GI = 3.0–5.0 cattle ha^{-1}, the decreasing rate would decrease nonlinearly, with an average of 0.56 per unit increase of aridity index, which is about 4.3 times larger than that for GI $\leqslant$ 3.0 cattle ha^{-1}. Further, the interceptions were predicted to be about 0.13–0.21 cm for GI $\leqslant$ 3.0 cattle ha^{-1} and zeros for GI $\geqslant$ 5.0 cattle ha^{-1}. Moreover, if GI $\leqslant$ 3.0 cattle ha^{-1}, the decreasing rate of soil evaporation with increase of aridity index would increase by 0.12 per unit increase of aridity index; if GI $\geqslant$ 5.0 cattle ha^{-1}, the soil evaporation was barely affected by grazing. If GI = 3.0–5.0 cattle ha^{-1}, the decreasing rate would increase nonlinearly, with an average of 1.11 per unit increase of aridity index, which is about 9.3 times larger than that for GI $\leqslant$ 3.0 cattle ha^{-1}. In terms of aridity index, the grazing intensity of around 3.0 cattle ha^{-1} could again be judged as the threshold for the study watershed.

5.3.4 Discussion

In the long run, livestock grazing would reallocate some BPC to NBPC, reducing the proportion of precipitation that could be used to sustain the fragile semi-arid hydro-ecosystem. As expected, because grazing removes aboveground canopy and biomass, more percent bare soil surfaces will be exposed directly to the ambient atmosphere, leading to two consequences in favor of NBPC. First, due to the large PET (i.e. climatic water demand), throughfall water will likely be evaporated before it can infiltrate into the ground to replenish soil moisture and/or recharge shallow aquifer. Second, due to receiving solar radiation directly, bare soils can absorb more heat, increasing soil temperature and thus soil water evaporation. For the study watershed, pasturing practices with a grazing intensity of above 3.0 cattle ha^{-1} were predicted to significantly increase the NBPC–BPC ratio (figure 5.16), which may be attributed to significant reductions of aboveground canopy due to intensely grazing. The current grazing intensities (11.8–19.7 cattle ha^{-1}) are much higher than the threshold value of around 3.0 cattle ha^{-1}, mandating new policies and regulations to restrict the maximum number of livestock animals per hectare grassland to increase the sustainability and resilience of the hydro-ecosystem.

For a given grazing intensity, the NBPC–BPC ratio was generally larger in a drier than a wetter year (figure 5.17). This can be attributed to the fact that vegetation growth in drier years can incur more severe drought stress, exposing a greater percent of bare soil surfaces directly to solar radiation in favor of the evaporation of throughfall and soil water. Overall, the NBPC–BPC ratio tended to exponentially decrease with an increase of precipitation and increase with an increase of aridity index. Luo *et al* (2014) found that the aboveground biomass was positively related to precipitation, with 12%–65% of the net primary production variances explained by the May to July precipitation. The increasing precipitation can prompt vegetation growth to have a larger leaf area, shading more soil surfaces from direct solar radiation and thus reducing soil water evaporation. In contrast, the aridity index indicates the degree of climatic dryness or how much climatic water demand can be met by precipitation. The increasing aridity will increase the drought stress on vegetation growth, leading to more bare soil patches from which throughfall and soil water can be efficiently evaporated. For years with an annual precipitation of less than 60 cm or an aridity index of larger than 1.5, more than 50% of the precipitation would be lost to evaporation without benefiting vegetation growth. Such climatic conditions were true for one third of the 58 assessment years, indicating the natural frangibility of the hydro-ecosystem in the study watershed. However, the NBPC–BPC ratio was independent of PET, which can be attributed to that actual ET was controlled by precipitation for most years. The aridity index can be used as a surrogate of climatic evaporation demand and available water resources.

The NBPC–BPC ratio was affected interactively by climate and grazing (figure 5.17). Such interactive effects tended to become stronger with a drying climate. One possible reason is that the degraded vegetation growth due to drought stress in a drier year can be further deteriorated by grazing. For a year with an annual precipitation of less than 60 cm or an aridity index of larger than 1.5, the NBPC–BPC ratio would increase by several times with an increase of grazing intensity, with the steepest increase when the grazing intensity increases from 3.0–5.0 cattle ha^{-1}. Regardless of the climatic conditions, pasturing practices with a grazing intensity of below 3.0 cattle ha^{-1} would have much smaller hydro-ecological impacts than those with a grazing intensity of above 5.0 cattle ha^{-1}. With increasing precipitation or decreasing aridity index, the NBPC–BPC ratios for different grazing intensities converge into the asymptote of 0.1, indicating that grazing would exert a minimal impact in wet years.

Among the four precipitation consumption components (table 5.6), the bottom seepage was most sensitive to climatic fluctuations, followed by the soil evaporation and then transpiration. The interception was statistically independent of climate. The bottom seepage or transpiration increased with an increase of precipitation, whereas it decreased with an increase of aridity. A possible reason is because the highly permeable sandy-loam soils have an infiltration capacity (285–315 cm d^{-1}) that is much larger than prevalent rainfall intensities, allowing throughfall water to quickly penetrate through the soil surface and move downward to replenish soil moisture and/or recharge the shallow aquifer. However, some throughfall water that was supposed to infiltrate under a lower grazing intensity would be reallocated to evaporate under a higher grazing intensity, with a maximum reallocation under a grazing intensity between 3.0 and 5.0 cattle ha^{-1}. As discussed above, the percent bare soil surfaces tend to increase with an increase of grazing intensity, decrease of precipitation, and increase of aridity, causing transpiration to decrease and soil evaporation to increase. Grazing would remove the aboveground grass stems and leaves to a threshold level below which canopy interception capacity became negligible. For the studied watershed, such a threshold level was predicted to occur at a grazing intensity of around 3.0 cattle ha^{-1}.

Grazing reduces the aboveground biomass and grass leaf areas. On the one hand, this can create more bare ground surfaces directly exposed to solar radiations, increasing soil water evaporation. On the other hand, once a top dry soil layer is formed, it will retard the moist soils beneath from being dried out further (Wang 2015, Pedram *et al* 2018). Also, the grazing removal of grass stems and leaves will reduce the transpiration rate and thus water uptake from the rootzone. The retardation and reduction effects are added up to alleviate soil water loss to ET. Once the grazing intensity is greater than 3.0 cattle ha^{-1}, no more grass stems and leaves may be available for grazing to further reduce transpiration.

For semi-arid steppe grasslands, including the Balagaer River watershed, precipitation is the sole source of water in sustaining the fragile vegetation ecosystems (Penman 1948, Wang *et al* 2014a, Wang 2015). If some BPC is reallocated into NBPC because of livestock grazing, the proportion of precipitation transpired by plants will be reduced, increasing the hydro-ecological frangibility. This is particularly true for years with a precipitation of less than 60 cm or an aridity index of larger than 1.5. Wang *et al* (2014) and Gao *et al* (2017a) found that precipitation of the study watershed exhibited a significant (at $\alpha = 0.05$) downward trend. Thus, future climates tend to become drier, leading to a higher occurrence of the abovementioned critical climatic condition. To balance livestock productivity and eco-environmental sustainability, policies on pasturing practices should be made to ensure not to exceed the threshold grazing intensity of around 3.0 cattle ha^{-1}. If the current grazing intensity of 11.8–27.6 cattle ha^{-1} would be continued in the future, the steppe ecosystem of the studied watershed will likely incur irreversible degradation, triggering inevitable desertification. As a logical generalization, it is imperative for the Great Steppe to practice livestock production below a threshold grazing intensity (e.g. 3.0 cattle ha^{-1}); otherwise, the aboveground grass canopies could be removed so much that a significant amount of BPC will be reallocated into NBPC, reducing topsoil moisture and vegetation cover as a protective armor for fertile-fine topsoil particles. This could become an eco-environmental concern in the foreseeable future because climate change is leading to more frequent and severe droughts across the vast steppe ecoregion of Eurasia (Wen *et al* 2014, Gao *et al* 2017a), as foretold by China's historically largest dust storm on 15 March 2021 that originated from the deserted steppe grasslands in Mongolia (Imster 2021).

5.3.5 Conclusions

This study put forward notions of non-beneficial (NBPC) and beneficial (BPC) precipitation consumption to evaluate the long-term effects of climate variability and livestock grazing intensity (GI) on a semi-arid steppe hydro-ecosystem. Averaged across the 58 assessment years, the NBPC–BPC ratios for GI $\geqslant$ 5.0 cattle ha^{-1} were significantly larger than those for GI $\leqslant$ 3.0 cattle ha^{-1}. A grazing intensity of around 3.0 cattle ha^{-1} can be considered as the threshold, above which livestock grazing will likely trigger irreversible degradation or even desertification of the steppe grassland. For a given grazing intensity, the NBPC–BPC ratio tended to be higher in a drier than a wetter year. Such a negative correlation was stronger for 'critical' years, which have an annual precipitation of less than 60 cm or an aridity index of larger than 1.5. The NBPC–BPC ratios were predicted to increase with increase of grazing intensity regardless of the amounts of annual precipitation. The ratio could become larger than one in a 'critical' year with GI $\geqslant$ 5.0 cattle ha^{-1}, implying that more percent precipitation will be reallocated to be non-beneficial to hydro-ecological sustainability and resilience. It is recommended that policies on pasturing practices be made to ensure grazing intensities less than an appropriate threshold (e.g. 3.0 cattle ha^{-1}) to sustain the hydro-ecosystems of the study watershed and the Great Steppe.

5.4 Topsoil erosion as influenced by climate and vegetation coverage

Semi-arid steppe grasslands are susceptible to topsoil erosion induced by wind and water shears. In the real world, these two erosion processes usually coexist, although one process may dominate another process depending on antecedent soil moisture conditions. The wind erosion process is dominant when the topsoil is relatively dry, whereas the water erosion process is dominant when the topsoil is relatively moist. However, these two processes have rarely been modeled simultaneously, neglecting their interactions and thus biasing erosion predictions. This case study puts forward and uses an integrated aeolian and fluvial prediction (IAFP) model to examine how these two processes interact under different hydroclimatic conditions and respond to variations in soil moisture and plant coverage. The model performed very well in tracing aeolian and fluvial dynamics. Across the assessment period and for the studies grassland, the model revealed that fluvial erosion occurred more often than aeolian erosion, with an average total erosion of 0.07 mm year^{-1}. Overall, the contributions from these two processes to the total erosion were comparable. The model showed that at the status quo of the grassland can be protected well from erosion by the plant canopy. The canopy was found to play a larger role in alleviating wind than water erosion and be more important for a drier than a wetter year. Relative to the canopy, soil moisture was predicted to be more influential on soil erosion; the dominant process can be switched from aeolian to fluvial erosion when soil moisture reaches above 0.23–0.25, while aeolian erosion will become predominant and may raise serious eco-environmental concerns once soil moisture decreases to below 0.15.

5.4.1 Background

Soil erosion, which is the detachment and movement of soil particles, can be induced by wind and/or water shears. In the literature, wind-induced erosion is also termed as the aeolian process, while water-induced erosion is also called the fluvial process. These two interrelated processes are cited as the basic reasons for land degradation and desertification across the world (Valentin *et al* 2005, Belnap *et al* 2011). The convolution is much greater than the linear addition of the individual impacts of these two processes, which is particularly true for dryland ecosystems (Bullard and Livingstone 2002). For most of the world, roughly 42% of the total erosion is caused by each of these two processes and the remaining 6% can be attributed to human activities (e.g. construction) (UNEP 1992). In Australia, the fluvial process dominates, while in Africa the aeolian process dominates. For an area of interest, it is imperative to convolute rather than simply add these two processes to accurately assess and solve the eco-environmental problems resulting from soil erosion (Flanagan *et al* 2006, Field *et al* 2009, Ravi *et al* 2010, Belnap *et al* 2011).

Field *et al* (2009) presented a modeling framework by assuming that total erosion is simply the addition of wind erosion and water erosion. Herein, wind erosion is a negative function of precipitation, whereas water erosion is a positive function of precipitation. At a specific amount of precipitation, aeolian and fluvial processes can have a 'maximum interaction' at which they contribute equally to the total erosion. Given that the relative dominance of one process over another depends on soil moisture, erosion is expected to have a better correlation with soil moisture than precipitation (Rodriguez-Iturbe 2000, Field *et al* 2011). In the framework, inappropriate land practices (e.g. overgrazing) are assumed to cause more erosion by water than wind, with most interactive impacts at the watershed scale and daily time step. Those authors acknowledged the importance of simultaneously considering both processes in addressing soil erosion resulting from climate variability and human activity. However, the linearly additive algorithm of the framework is insufficient to mimic the coexisting features of these two erosion processes, limiting concerted efforts in sustaining dryland ecosystems.

Ravi *et al* (2010) suggested a holistic approach for considering the aeolian–fluvial interactions in semi-arid ecosystems as influenced by climate and vegetation. In such an ecosystem, the aeolian and fluvial processes coexist and can be equally important. This is different from arid ecosystems, where the aeolian process dominates, and humid ecosystems, where the fluvial process dominates. The relative importance of the aeolian over fluvial process can be augmented because of climate change by extending drought duration, increasing wind speed and air temperature, and stressing growth of native plant species. The soil erodibility or critical shear strength can be enhanced by soil water, which bounds and aggregates individual soil particles together, and plants, whose roots hold soil in place and improve soil structure. On the other hand, the plant canopy shelters bare topsoil from wind and water shears, reducing erosive forces. For some areas, the aeolian process used to be predominant, but wind erosion gradually altered the landscape and topography to be more favorable for the fluvial process, whereas for other areas, the fluvial process used to be predominant, but water erosion gradually reduced vegetation coverage and exposed fine soil particles to the aeolian process. The holistic approach recognized that the aeolian and fluvial processes can enhance each other and evolve as time passes, but it needs to be improved by including vegetation dynamics to examine how climate change, soil loss, and human activity interactively degrade semi-arid ecosystems.

Belnap *et al* (2011) presented a conceptual model to seamlessly couple aeolian with fluvial processes. This model allows eroded soils by wind (or water) to be transported, deposited, and then re-eroded again by water and/or wind. Such dynamic aeolian–fluvial linkages will likely intensify because of climate change and human activity. Previous studies (Eichelberger *et al* 2008) have shown that climate change will increase wind speed, likely exceeding soil critical shear strength more often and thus notably increasing wind erosion. In contrast, the effects of climate change on water erosion are due to more frequent storms with an elevated rainfall intensity. Overall, climate change would have a larger impact on wind erosion than water erosion. The model revealed the importance of accurately predicting the erosion from these two processes individually and aggregately.

Although a few studies, including those cited above, have somewhat elaborated on aeolian–fluvial interactions, the relevant understanding and quantitative information are still incomplete (Bullard and Livingstone 2002). For instance, the semi-empirical model developed by Gao *et al* (2002) can predict wind and water erosions individually without considering the interactions: the aeolian process is described by four parameters, namely soil wind erodibility, critical wind velocity for bare and dry soil, soil–water reduction factor, and vegetation sheltering factor, while the fluvial process is described by another four parameters, namely soil water erodibility, runoff enhancement factor, topographic gradient, and vegetation cover index. The model's applicability and predictability are hardly extrapolated because the adopted values of those eight parameters were not clearly explained. Field *et al* (2011) conducted *in situ* experiments in three 50 m by 50 m plots within a least-disturbed semi-arid grassland. The experiments revealed that climate change would notably increase wind over water soil erosion loss.

This case study puts forward and uses an integrated aeolian and fluvial prediction (IAFP) model to examine how these two processes interact under different hydro-climatic conditions and respond to variations in soil moisture and plant coverage. The testbed was the Balagaer River watershed shown in figure 5.1. Climate change was represented hypothetically by varying the soil moisture and runoff coefficient, while human activity was arbitrarily reflected by varying the vegetation cover index. For simplicity, plant physiological parameters, namely population density, stem height, and leaf area, were not varied although they are dependent on hydroclimatic conditions and change with time.

5.4.2 Model formulization

As detailed in section 3.6, the IAFP model was devised to have three modules of hydrological simulation, aeolian process, and fluvial process (figure 5.19). The hydrological simulation module can be parameterized using data on topography, soils, and land cover (LC), and is driven by the climatic variables of air temperature, wind speed, and precipitation to estimate ET, runoff, and infiltration. It also traces water movement and redistributes soil water in the unsaturated zone (Brutsaert 2005).

The aeolian process module calculates the soil wind shear strength in terms of soil–water properties and LC. The shear strength is the maximum resultant force per unit surface area, above which topsoil particles can be mobilized, detached, and moved (Chepil 1956, Marshall 1971, Wolfe and Nickling 1993, Dong *et al* 2002). Typically, the component forces originate from gravitation, cohesion, and adhesion. On the other hand, the module determines the effective wind shear stress using data on wind speed and LC. Moreover, if the shear stress is larger than the shear strength,

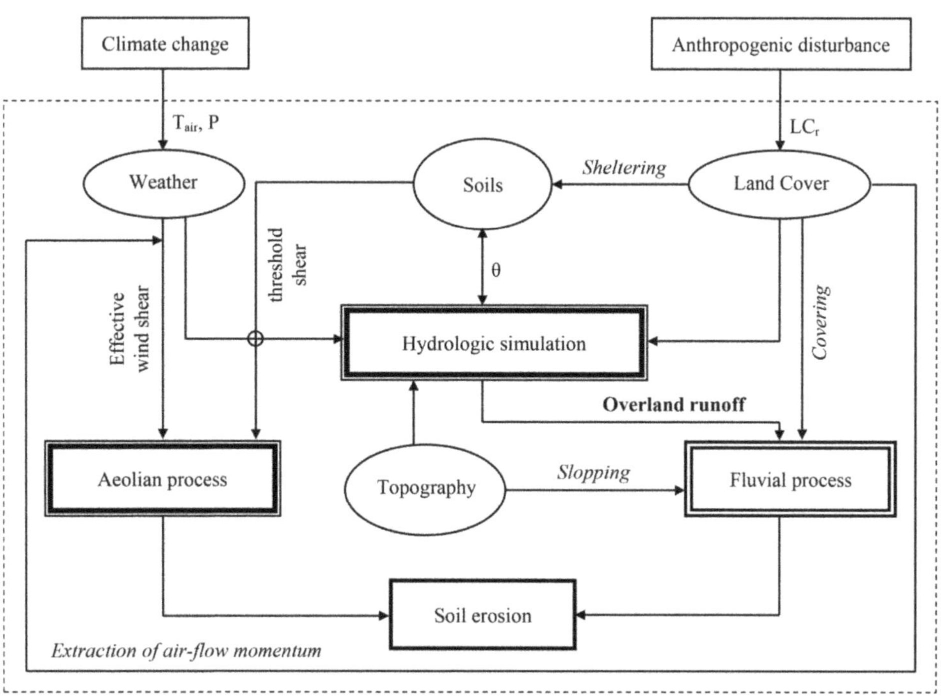

Figure 5.19. The integrated aeolian and fluvial process (IAFP) model. The flowchart shows the interrelations among the three modules and primary variables. T_{air}: air temperature; P: precipitation; LC_r: relative vegetation coverage; θ: volumetric surface soil moisture. (Reproduced with permission from figure 2 in Wang *et al* (2014a). Copyright 2014 Wiley.)

the module uses a physically based equation to estimate the wind erosion modulus or rate; otherwise, the module assumes no wind erosion.

The fluvial process module assumes that the soil water shear strength is same as the soil erodibility, which is the maximum resistance before topsoil particles just start to be mobilized. On the other hand, the module calculates the effective water shear stress in terms of rainfall and runoff (Wang *et al* 2012b), topographic gradient (Nearing 1997), and relative land coverage (Wang *et al* 2012b). Moreover, if the shear stress is larger than the shear strength, the module uses a semi-empirical equation to estimate the fluvial erosion modulus or rate; otherwise, the module assumes no fluvial erosion.

The IAFP model switches the dominance between the aeolian and fluvial process as soil moisture varies. That is, the aeolian and fluvial process modules are interrelated or coupled by soil moisture, which in turn is an independent variable of the hydrological simulation module. The total soil erosion is computed as the summation of the modules' estimated wind and water erosions. Moreover, the model can predict the effects of climate change and human activity (e.g. grazing) on soil erosion in two ways. The first way is to impose *a priori* positive or negative shifts from the module-calculated soil moisture and runoff as well as from the prescribed vegetation coverage by assuming that the shifts can be attributed to climate change and/or human activity. The second way is to use predicted climates by climate models (IPCC 2011) and hypothetical vegetation coverages.

5.4.3 Materials and methods

5.4.3.1 Data

On average, the grasses had a population density of 4500 m^{-2}, a stem height of 20 cm, and a stem diameter of 0.1 cm (Zhao *et al* 2006, Reiche *et al* 2011). These physiological parameters of plants were determined by field reconnaissance and literature review. In addition, other physiological parameters, such as LAI_{min} and LAI_{max} in equation (3.103), were determined from moderate resolution imaging spectroradiometer (MODIS) imagery. Further, the topographic gradient of the watershed was determined from a 30 m digital elevation model (DEM). The MODIS imagery and DEM were downloaded from the International Scientific Data Service Platform website http://datamirror.csdb.cn.

The soils are classified as I-Mo-2c, Kh1–2b, I-K-2c, and Zm2–2/3a (figure 5.20) (Allen *et al* 1998). All four types of soils exist in the upper (<30 cm) and lower (30–100 cm) layers. Relative to the first three types of soils, the fourth type is slightly less permeable. The loamy soils (McCarthy 2002) have saturated hydraulic conductivities of 1.5–107.0 mm h^{-1}, with a mean of 28.5 mm h^{-1}, and a moderately high potential for water erosion (Stone and Hilborn 2000, USDA-NRCS 2004), with an erodibility factor of 0.0011–0.0025 t m^2 h m^{-3} t^{-1} cm^{-1} (Wischmeier and Smith 1978). The soil properties, namely texture, hydrological soil group (HSG) (USDA-NRCS 2004), available water content (AWC) (Viessman and Lewis 2003), organic matter content, pH, and bulk density, were extracted from a World Food and Agriculture Organization (FAO) soil map, available at http://www.iiasa.ac.at/Research/LUC/External-World-Soil-database/HTML. The map consists of

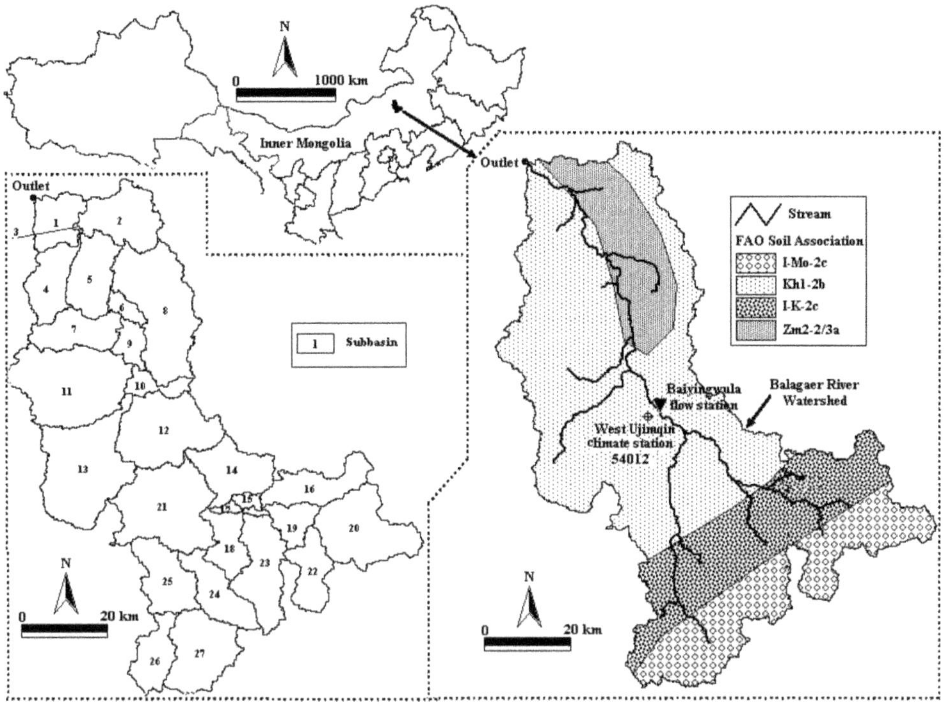

Figure 5.20. Physiographical characteristics of the Balagaer River watershed. Maps showing the location, topographic subdivisions, and soil polygons for the four soil types classified by the World Food and Agriculture Organization (FAO, 2008). (Adopted with permission from figure 1 in Wang *et al* (2014a). Copyright 2014 Wiley.)

22 polygons for the four types of soils (figure 5.20). Subsequently, those soil properties were used to compute soil–water properties (Saxton *et al* 1986, Saxton and Rawls 2006), namely wilting point (θ_ψ), saturated soil moisture (θ_{sat}), and saturated hydraulic conductivity (K_s).

The National Meteorological Information Center of China compiles and publishes (www.nmic.gov.cn/web/index.htm) climatic observations at the West Ujimqin Climate Station 54012 (117°36′ E, 44°35′ N; figure 5.1). This case study obtained and used daily data on precipitation, air temperature, relative humidity, wind speed, solar radiation, and evapotranspiration from 1 November 1954 to 31 March 2011. Also, the Inner Mongolia Agricultural University (IMAU) Cold-Arid Region Water Resources Utilization Laboratory provided data on monthly discharge from January 1959 to December 2006. The discharges were measured at the Baiyingwula flow gauging station (117°37′ E, 44°36′ N; figure 5.1), which is the outlet of the upper 53.6% of the watershed (i.e. 2866 km^2 drainage area). The corresponding monthly runoffs were derived from the discharges as follows: the product of a monthly discharge and the number of days within this month was divided by 2866 km^2. The derived data on monthly runoff were used to calibrate (1959–1982) and validate (1983–2006) the hydrological simulation module.

5.4.3.2 Model set-up

The IAFP model was parameterized to represent the lumped condition of the watershed. In this regard, the composite parameters were determined and used to set up the model. For the soil–water parameters, namely θ_ψ, θ_{fc}, θ_s, and K_s, the polygon-scale values were computed using the soil–plant–air–water (SPAW) model (Saxton *et al* 1986, Saxton and Rawls 2006) in terms of the contents of sand, clay, and organic matter in this polygon. The composite value of a parameter was determined as the area-weighted average of the corresponding polygon-scale values. In addition, the grid-scale values for α in equation (3.104) were computed in ArcGIS® using the elevations presented in the DEM as the steepest downward slopes. The composite value for α was determined as the geometric mean of the grid-scale values. Further, the cell-scale values for LAI, LAI_{min}, and LAI_{max} were extracted in ArcGIS® from the MODIS imagery (Zang 2011). The composite value for each physiological parameter was determined as the median of the corresponding cell-scale values. Moreover, the MODIS imagery and the FAO soil map were overlain in ArcGIS® to delineate hydrological response units (HRUs), each of which is a unique combination of a land cover and an HSG. The initial values of curve number for the normal antecedent moisture condition (CN_{II}) (Hjelmfelt 1991) for the HRUs were extracted from table 3.2. The initial composite value of CN_{II} was determined as the area-weighted average of the corresponding values for the HRUs. Subsequently, this initial composite value was adjusted, as discussed in section 5.4.3.3, to yield the CN_{II} adopted in the calibrated model. Field reconnaissance was conducted to determine the following parameters (Zhao *et al* 2006): P_{clay} (equation (3.96)), ρ_s (equation (3.98)), and d_s (equation (3.102)), plant stem height $h = 0.2$ m, plant diameter $d = 0.1$ cm, and plant population density $D_v = 4500\ m^{-2}$. For an individual plant, the upwind projected and stem cross-sectional areas were calculated as $A_v = 0.2\ m \times 0.001\ m = 0.0002\ m^2$ and $A_h = \pi/4 \times (0.001\ m)^2 = 7.85 \times 10^{-7}\ m^2$, respectively.

In this case study C_a was determined as follows. Equation (3.89) was separately fitted by the method of least squares to the wind-tunnel data presented by Chen *et al* (1996) and the *in situ* data reported by Lancaster and Baas (1998), yielding two values for C_a. In addition, these two values were divided by an area-influence factor $A_i = \sqrt{DA_s/DA_m}$, where DA_s is the area of the Balagaer River watershed and DA_m is the surface area of the experimental soils, to be scaled up to the study watershed. A_i reflects overall area–size effects on physiographic heterogeneity, erosion mechanism, and deposition process (Loughran 1989, Breshears *et al* 2003). Given that eroded soils tend to deposit more over a larger than a smaller area, the net soil erosion tends to increase with a decrease of study area (Reid *et al* 1999, Lal 2001). Further, the final value for C_a was determined as the geometric mean of those two scaled-up values.

Chen *et al* (1996) carried out wind-tunnel experiments to measure the critical wind velocity and wind erosion modulus of least-disturbed loessal bare sandy-loam soils in the northern Loess Plateau of China. The experiments used soil columns with a cross-sectional area of $DA_m = 0.3\ m^2$, subject to nine water contents of $\omega = 2.67\%$–9.52% and wind blowing at 30 cm above soil surface at different speeds. The soils had 72.5% large ($\geqslant 0.05$ mm) particles, 23.9% medium (0.05–0.005 mm) particles, and 3.6% fine ($\leqslant 0.005$ mm) particles, leading to a median particle diameter of 0.114 mm. The least-squares fitting of equation (3.89) to the data was done by assuming $d_s = 0.114$ mm, $D_v = 0.4$, $P_{clay} = 0.0\%$, $d = 0.0$ m, $\rho_s = 2650\ kg\ m^{-3}$, $\rho_a = 1.21\ kg\ m^{-3}$, $\lambda = 0.0$, and $z = 0.3$ m. On the other hand, Lancaster and Baas (1998) conducted field experiments in a 600 m^2 bare plot to directly measure $u_{*\theta\lambda}$ and q_a for 14 different shear velocities (u_*). The plot, located in Owens Lake of California, USA, has sand soils with a low water content of $\omega < 2\%$. The least-squares fitting of equation (3.89) to the data was done by assuming $\rho_a = 1.21\ kg\ m^{-3}$. Similarly, equation (3.103) was least-squares fitted to the field data presented by Fullen and Reed (1986) to obtain an initial value for C_f, which in turn was divided by A_i to derive the final value for the fluvial process module.

5.4.3.3 Model calibration and validation

The IAFP model was set up as a lumped mode because of insufficient data for considering any spatial heterogeneity of the watershed. It was calibrated from 1 January 1959 to 31 December 1982 and validated from 1 January 1983 to 31 December 2006. The soil erosion was assessed using the predictions of the 48 simulation years.

The hydrological simulation module was calibrated by adjusting CN_{II} to best reproduce the observed monthly runoffs at the Baiyingwula station. The corresponding CN_I and CN_{III} were computed using equations (3.18) and (3.19), respectively, in terms of the calibrated CN_{II}. The calibrated module was used to predict the monthly runoffs for the validation period and then validated by directly comparing the predictions with the corresponding observed values. The performance of the hydrological simulation module was measured using visualization plots and Nash–Sutcliffe efficiency (NSE) (equation (4.58)). NSE has been applied widely in assessing hydrological models (Motovilov *et al* 1999, Wang *et al* 2012b). The range of NSE can vary from $-\infty$ to 1.0, with a higher value indicating a better model performance. The predicted runoffs are less reliable than using the average of the observed values if NSE is negative, whereas they are more reliable than using this average if NSE is positive.

However, the aeolian and fluvial process modules could not be calibrated because of no available soil erosion data. These two modules used the parameters of C_a and C_f as determined in section 5.4.3.2 and thus were assumed to be acceptable for this case study.

5.4.3.4 Simulation scenarios

For the entire simulation period (1 November 1954 to 31 March 2011), LAI_{min} and LAI_{max} were taken as constants because these two parameters reflect physiological features of the steppe grasses under extremely- and least-stressed conditions, respectively. That is, LAI_{min} and LAI_{max} are plant specific and independent of exotic (e.g. hydroclimatic and nourishing) factors. For the study watershed, the phenological characteristics of the steppe grasses dictated a growing season of 1 June to 30 September and a decease date before 31 October (Bai *et al* 2004, Zhao *et al* 2004). The daily LAI was assumed to follow the stage curve presented in table 5.7, while the daily θ and Q/P were computed using equation (3.24). In terms of the physiological and phenological parameters and the hydrological variables, the IFAP model was run to predict daily q_a and q_w, which in turn were summed to obtain the corresponding daily total soil erosion.

To assess the impacts of climate change on wind and water erosion, θ was varied from θ_ψ to θ_s (Gao *et al* 2002), while Q/P was multiplied by $e^{-\eta_\theta}$ (Wang *et al* 2012b), where η_θ is the inverse of percent variation in θ. That is, if θ is reduced by 10%, $\eta_\theta = 0.1$, whereas if θ is increased by 10%, $\eta_\theta = -0.1$. For simplicity, LAI was not varied with θ and Q/P (Zhao *et al* 2004). The impacts were examined at an annual scale using visual plots showing how the predicted erosions fluctuate with soil moisture. Second, to assess the impacts of livestock grazing and grassland degradation, LAI was varied from LAI_{min} to LAI_{max} (Tong *et al* 2004). Third, to examine the interactive impacts of climate change and livestock grazing, LAI, θ, and Q/P were synchronically varied by corresponding approaches mentioned above, yielding sets of these three variables. For the second and third assessments, the impacts were examined at annual scale using visual plots showing vegetation-influencing indices versus LAI, with θ as a categorical variable.

The vegetation-influencing index on total erosion is computed as

annual mean total erosion rate at soil moisture θ and scenario leaf area index LAI

vegetation-influencing index on total erosion [–]

$$R_{aw} = \frac{q_{aw}^{\theta,LAI}}{q_{aw}^{\theta,LAI_0}} \tag{5.1}$$

annual mean total erosion rate at soil moisture θ and reference leaf area index LAI_0

The vegetation-influencing index on wind versus water erosion (i.e. the ratio of wind erosion to water erosion) is computed as

Table 5.7. The stage curve of leaf area index[1].

Day in	Leaf area index (LAI) [–]
30 November to 1 June	LAI_{min}
1 June to 30 September	Linear increase from LAI_{min} to LAI_{max}
30 September to 31 October	Linear decrease from LAI_{max} to LAI_{min}

[1] Reproduced with permission from table 1 in Wang *et al* (2014a). Copyright 2014 Wiley. LAI_{min} and LAI_{max} are the minimum and maximum LAI, respectively.

annual mean aeolian erosion rate at soil moisture θ and scenario leaf area index LAI

annual mean aeolian erosion rate at soil moisture θ and reference leaf area index LAI_0

vegetation-influencing index on aeolian versus fluvial erosion [–]

$$R_{a\rightarrow w} = \begin{cases} \dfrac{q_a^{\theta,LAI} - q_a^{\theta,LAI_0}}{q_w^{\theta,LAI} - q_w^{\theta,LAI_0}} & \text{for } LAI \neq LAI_0 \\ \dfrac{q_a^{\theta,LAI_0}}{q_w^{\theta,LAI_0}} & \text{for } LAI = LAI_0 \end{cases} \tag{5.2}$$

annual mean fluvial erosion rate at soil moisture θ and scenario leaf area index LAI

annual mean fluvial erosion rate at soil moisture θ and reference leaf area index LAI_0

The indices are computed for an annual average condition. R_{aw} measures the long-term impact of a change in vegetation coverage on total soil erosion for a given soil moisture (θ). R_{aw} can be from a decimal to a very large real number—a larger value indicating a greater impact. If $R_{aw} = 1.0$, the impact is negligible. In contrast, $R_{a\rightarrow w}$ measures the relative impact of a change in vegetation coverage on wind over water erosion for a given θ. $R_{a\rightarrow w}$ can be from zero to infinite. If $R_{a\rightarrow w} = 0.0$, the impact on wind erosion is minimal, whereas if $R_{a\rightarrow w} = +\infty$, the impact on water erosion is negligible. If $R_{a\rightarrow w} < 1.0$, the impact on wind erosion is smaller than that on water erosion; otherwise, the impact on wind erosion is larger than that on water erosion. If $R_{a\rightarrow w} = 1.0$, the impact on wind erosion is same as that on water erosion. For an LAI same as the reference leaf area index (i.e. $LAI = LAI_0$), $R_{a\rightarrow w}$ is the relative magnitude of wind over water erosion.

5.4.4 Results and discussion

5.4.4.1 The calibrated model

The calibrated parameters for the modules of hydrological simulation, aeolian process, and fluvial process are listed in table 5.8. The seven soil–water parameters, including θ_ψ, θ_{fc}, θ_s, K_s, d_s, ρ_s, and P_{clay}, are in agreement with those presented by Zhao *et al* (2004) and Duan *et al* (2011) for geographic regions that share a variety of physiographical features with the studied watershed. The slope angles of the DEM grids were determined to be from 4.0° to 8.5°, leading to an overall slope angle of 6.3° for the watershed. In equation (3.103), $LAI_{min} = b_3 = 0.03$, as these two parameters are physiologically close and have a physical range of 0.02–0.9 (Zhao *et al* 2004). LAI_{min} occurs at plant wilting or just germinating when the leaf tissue is minimal; it is the minimum total one-sided leaf area per unit-area ground surface. b_3 is the horizontal projected stem area per unit-area ground surface.

The least-squares fittings of equation (3.89) to the data presented by Chen *et al* (1996) and Lancaster and Baas (1998) yielded C_a values of 10.1 g m^{-5} s^2 ($R^2 = 0.6$) and 8.5 g m^{-5} s^2 ($R^2 = 0.6$), respectively. The fitting errors were relatively large for a large net effective wind shear stress (i.e. $\tau_* - \tau_{*c}$). This can be attributed to few (<20%) points having a large $\tau_* - \tau_{*c}$ and to τ_{*c} maybe having a large measurement error due to the soil moisture possibly having been reduced by a high wind speed

Table 5.8. The calibrated parameters of the IAFP model[1].

Module in figure 5.19	Parameter	Description	Value
Hydrological simulation	θ_ψ [%]	Permanent wilting point	13.2
Aeolian process	θ_{fc} [%]	Field capacity	27.0
	θ_{sat} [%]	Saturated soil moisture	41.7
	K_s [mm h^{-1}]	Saturated hydraulic conductivity	7.8
Aeolian process	C_a [g m^{-5} s^2]	See equation (3.89)	4.6×10^{-4}
	d_s [mm]	Mean soil particle diameter	0.15
	ρ_s [kg m^{-3}]	Soil particle density	2650
	P_{clay} [%]	Clay content in soil	8
	λ [–]	Vegetation roughness density (see equation (3.91))	0.9
Aeolian process Fluvial process	LAI_{min} [–]	Lowest value for leaf area index (LAI)	0.03
	LAI_{max} [–]	Highest value for LAI	2.5
Fluvial process	C_f [g m^{-2} s^{-1}]	See equation (3.103)	4.4×10^{-5}
	b_1 [–]	See equation (3.103)	0.7
	b_2 [–]	See equation (3.103)	0.8
	b_3 [–]	See equations (3.103) and (3.105)	0.03
	α [°]	Soil surface slope angle (see equation (3.104))	6.3
Hydrological simulation	CN_{II} [–]	SCS curve number for AMC II (table 3.2)	61

[1] Reproduced with permission from table 2 in Wang *et al* (2014a). Copyright 2014 Wiley.

when τ_{*c} was measured (Fécan *et al* 1999). The area-influence factors are computed as $\sqrt{\frac{(5350\ \text{km}^2) \times (10^6\ \text{m}^2\ \text{km}^{-2})}{0.3\ \text{m}^2}} = 133542$ and $\sqrt{\frac{(5350\ \text{km}^2) \times (10^6\ \text{m}^2\ \text{km}^{-2})}{600\ \text{m}^2}} = 2986$ for these two sites, respectively. For the studied watershed, $C_a = \sqrt{\frac{10.1\ \text{gm}^{-5}\text{s}^2}{133\ 542} \times \frac{8.5\ \text{gm}^{-5}\ \text{s}^2}{2986}} = 4.6 \times 10^{-4}\ \text{gm}^{-5}\ \text{s}^2$.

Similarly, equation (3.103) was least-squares fitted to the data presented by Fullen and Reed (1986), giving an experiment-scale $C_f = 0.64$ g m^{-2} s^{-1}. The fluvial process module explained about 20% of the variance presented in the data, probably because the large erosions from heavy storms were not measured accurately. Nonetheless, for the study watershed, $C_f = (0.64\ \text{gm}^{-2}\ \text{s}^{-1}) \div \sqrt{\frac{5350\ \text{km}^2 \times 10^6\ \text{m}^2\ \text{km}^{-2}}{25\ \text{m}^2}} = 4.4 \times 10^{-5}\ \text{gm}^{-2}\ \text{s}^{-1}$.

With the calibrated $CN_{II} = 61$ (table 5.8), the hydrological simulation module reproduced the annual runoffs well (table 5.9 and figure 5.21), as indicated by $R^2 > 0.7$, NSE > 0.7, and small prediction errors of −2.3% to −4.2%. As expected,

Table 5.9. Predicted versus observed (in bold font) runoff from the study watershed[1].

	Runoff (mm)													NSE[3]	
Period[2]	Jan	Feb	Mar	Apr	May	Jun	Jul	Aug	Sep	Oct	Nov	Dec	Annual	Monthly	Yearly
Calibration	0.00	0.00	0.40	1.49	1.05	1.30	2.19	2.76	1.64	0.76	0.06	0.00	11.66	0.11	0.89
	0.00	**0.00**	**0.24**	**3.05**	**1.40**	**1.28**	**1.56**	**1.83**	**1.07**	**0.77**	**0.18**	**0.00**	**11.39**		
Validation	0.00	0.00	0.04	2.48	0.96	1.16	1.70	1.28	1.02	0.50	0.18	0.00	9.33	−0.18	0.70
	0.00	**0.00**	**0.18**	**2.51**	**0.93**	**0.86**	**1.53**	**1.47**	**0.74**	**0.56**	**0.17**	**0.00**	**8.94**		

[1] Reproduced with permission from table 3 in Wang *et al* (2014a). Copyright 2014 Wiley.
[2] The calibration period is January 1959 to December 1982, while the validation period is January 1983 to December 2006.
[3] NSE: Nash–Sutcliffe efficiency (equation (4.58)).

overpredictions and underestimates did occur (table 5.9). During the calibration period, for March and the growing season (June to September), the observed monthly runoffs were overpredicted by up to 0.94 mm, whereas for April, May, October, and November the observed monthly runoffs were underestimated by up to 1.56 mm. However, during the validation period, the observed monthly runoffs were overpredicted and underestimated in different months from those of the calibration period, indicating some reliance of the module's predictability on hydroclimatic conditions. Nonetheless, the measure statistics and visualization plots revealed that the module was sufficiently calibrated for long-term simulations at the watershed scale.

5.4.4.2 The modeled soil erosion

On an annual basis, the predicted aeolian erosion modulus ranged from 0.0 to 2.3×10^{-3} g m^{-2} s^{-1}, with a mean of 2.0×10^{-6} g m^{-2} s^{-1} (table 5.10). Based on *in situ* measurements along a Tariat–Xilingol transect, Liu *et al* (2008) and Qi *et al* (2010) showed a long-term average modulus range of 1.7×10^{-6} to 1.3×10^{-5} g m^{-2} s^{-1}. Because the Balagaer River watershed is adjacent to the transect and because the predicted mean aeolian erosion modulus falls within the measured range, the aeolian process module was judged to have a reasonably good performance. The module predicted that 179 out of 20 605 simulation days would have a non-zero wind erosion, which is equivalent to an occurrence of 0.9%. Similarly, the predicted fluvial erosion modulus varied from 0.0 to 9.5×10^{-5} g m^{-2} s^{-1}, with a mean of 2.0×10^{-6} g m^{-2} s^{-1} (table 5.10). Based on field experiments, Yan and Li (2002) showed an Inner Mongolian water erosion modulus range of 6.3×10^{-5} to 4.4×10^{-4} g m^{-2} s^{-1}. Because the predicted maximum fluvial erosion modulus falls within the measured range, the fluvial process module was judged to have a good performance. The module predicted that 4532 out of 20 605 simulation days would have a non-zero water erosion, which is equivalent to an occurrence of 22%.

Although the predicted annual mean aeolian and fluvial erosions were comparable (table 5.10), aeolian erosion would be much more severe than fluvial erosion: the aeolian erosion per occurrence was about 25 times greater than the fluvial

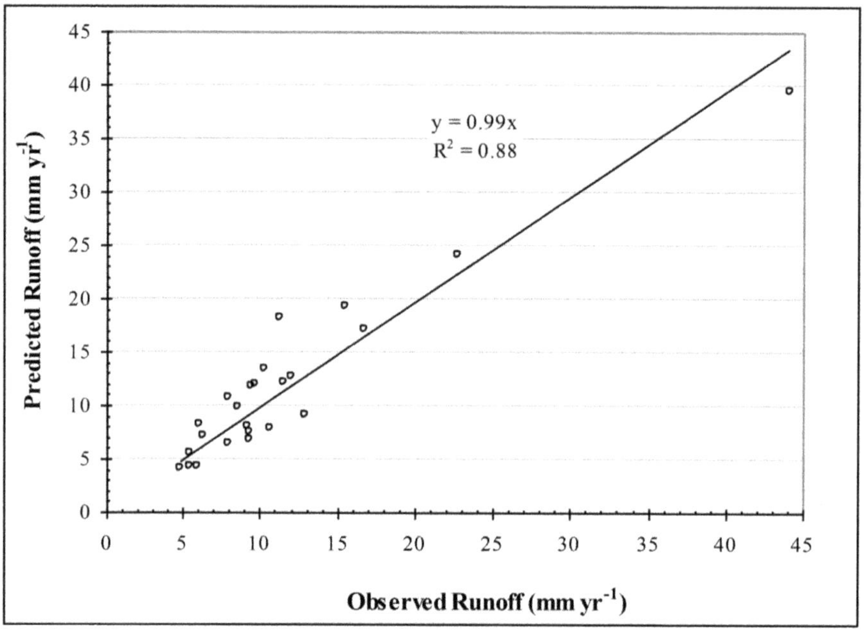

(a)

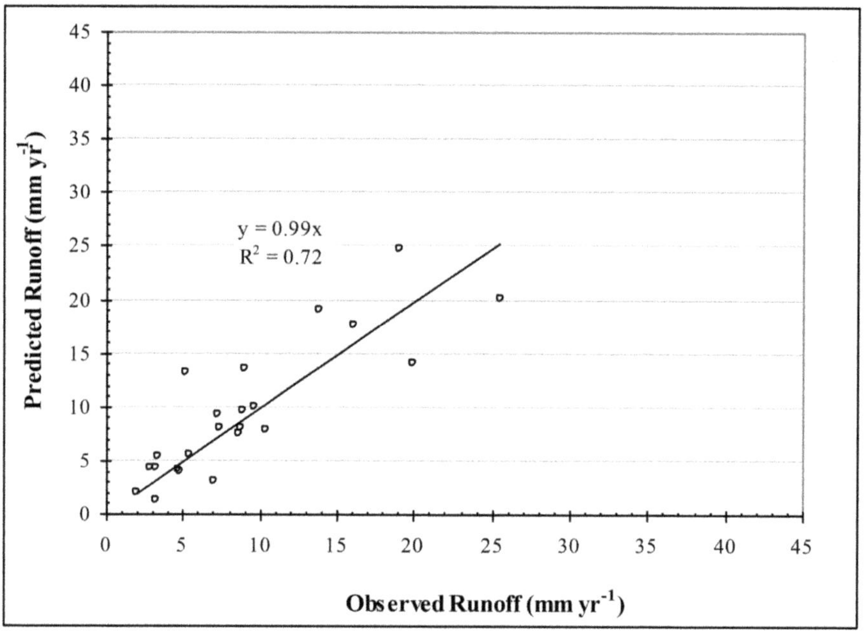

(b)

Figure 5.21. Performance of the hydrological simulation module for the Balagaer River watershed. Plot showing the predicted versus observed annual runoff for the (a) calibration years of 1959–1982 and (b) validation years of 1983–2006. (Adopted with permission from figure 4 in Wang *et al* (2014a). Copyright 2014 Wiley.)

Table 5.10. Summary statistics of the predicted daily soil erosion in the Balagaer River watershed[1].

Item	Aeolian erosion			Fluvial erosion		
	Modulus [g m^{-2} s^{-1}]	Rate [t $year^{-1}$]	Depth [mm $year^{-1}$]	Modulus [g m^{-2} s^{-1}]	Rate [t $year^{-1}$]	Depth [mm $year^{-1}$]
Minimum	0.00	0.00	0.00	0.00	0.00	0.00
Mean (simulation days)[2]	2.0×10^{-6}	3.3×10^{5}	0.04	2.0×10^{-6}	3.3×10^{5}	0.04
Maximum	2.3×10^{-3}	3.9×10^{8}	40.13	9.5×10^{-5}	1.6×10^{7}	1.52
Average (erosion days)[3]	2.3×10^{-4}	3.8×10^{7}	4.10	8.9×10^{-6}	1.5×10^{6}	0.16
Number of erosion days[4]	179			4532		
Number of simulation days	20 605			20 605		
Frequency (%)[5]	0.9			22.0		

[1] Reproduced with permission from table 4 in Wang *et al* (2014a). Copyright 2014 Wiley. The daily values on rate and depth are reported on an annual basis.
[2] Computed by dividing by the number of simulation days.
[3] Computed by dividing by the number of erosion days.
[4] The days with non-zero predicted erosion.
[5] Quotient of the number of erosion days and the number of simulation days.

erosion per occurrence. This is consistent with previous studies (Gao *et al* 2002, Field *et al* 2011) conducted in regions that have similar hydroclimatic, physiographical, and physiological conditions to the study watershed.

Rationally, aeolian erosion process is positively related to soil moisture deficit, wind speed, and vegetation spareness, which explained that the predicted monthly wind erosion was higher when the watershed was drier, windier, and less vegetated than that when the watershed was moister, calmer, and more vegetated (figure 5.22(a)). For April 1980, because of a high wind speed of 4.9 m s^{-1}, a low soil moisture of 0.16, and a nearly bare ground surface (LAI = 0.03), that month was predicted to have the highest aeolian erosion modulus of 9.0×10^{-5} g m^{-2} s^{-1}. In contrast, the fluvial erosion process is positively related to soil moisture and vegetation spareness—the predicted water erosion was higher when the watershed was more moist than that when the watershed was drier (figure 5.22(b)). Vegetation coverage can play some role in reducing water erosion, but such a role may become minimal when soils are very moist and/or subject to heavy storms. For September 1959, because of the almost saturated soils ($\theta = 0.41$), that month was predicted to

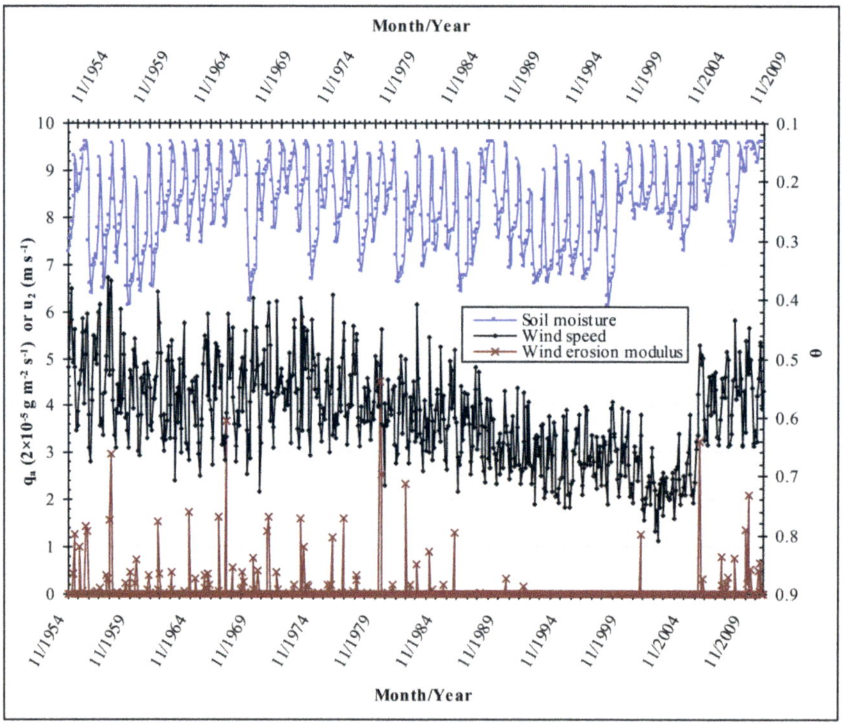

(a)

Figure 5.22. The predicted monthly erosion modulus and influencing variables. Plots showing the results for (a) aeolian erosion q_a and (b) fluvial erosion q_w. u_2: wind speed at 2 m aboveground surface; θ: soil moisture; Q: overland runoff; P: throughfall; P_t: total precipitation; and LAI: leaf area index. (Adopted with permission from figure 5 in Wang *et al* (2014a). Copyright 2014 Wiley.)

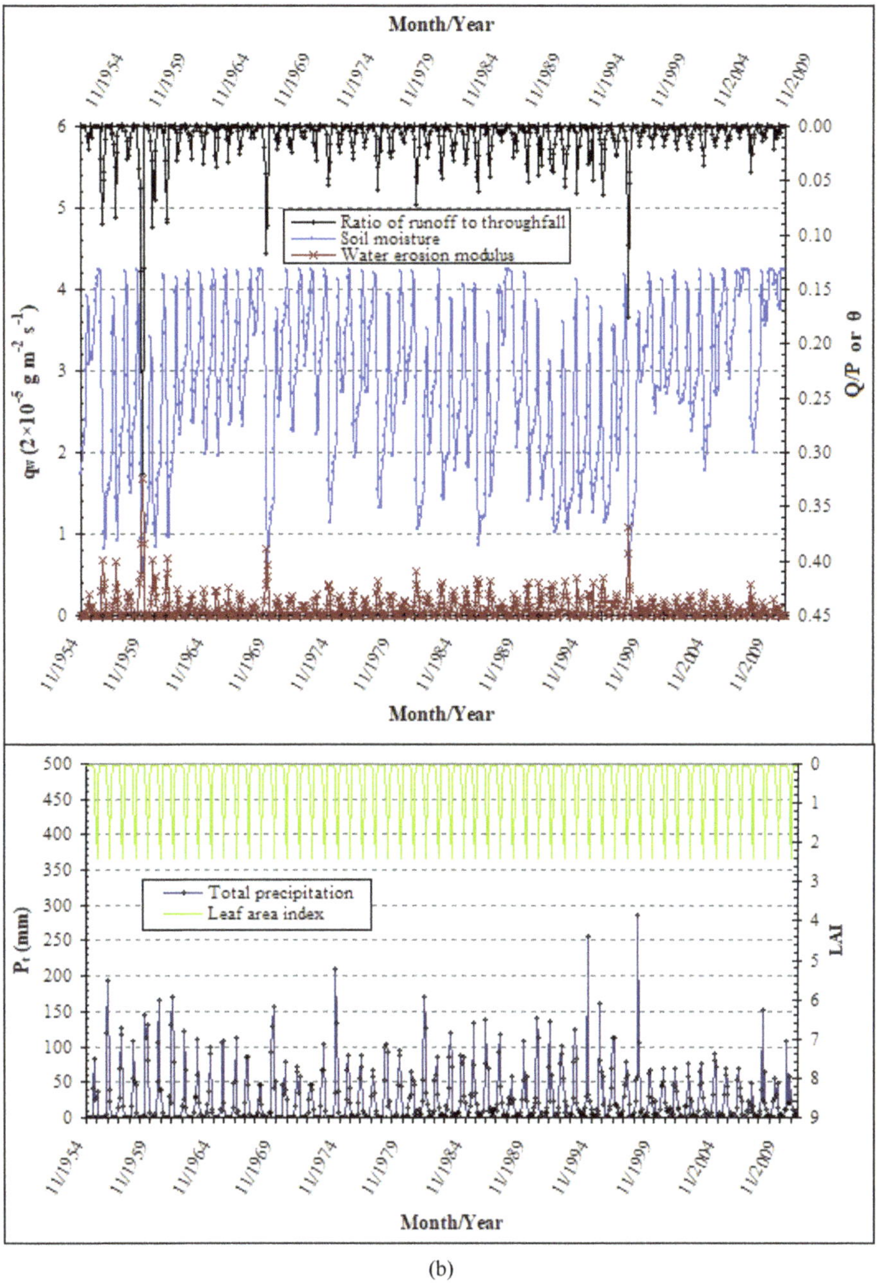

(b)

Figure 5.22. (Continued.)

have the highest fluvial erosion modulus of 3.4×10^{-5} g m^{-2} s^{-1} despite there being the most dense vegetation coverage (LAI = 2.4). For a month of interest, if LAI $\geqslant$ 0.8, the aeolian erosion was predicted to be very weakly related to wind speed, otherwise, the aeolian erosion tended to increase with an increase of wind speed and vice versa.

In contrast, for any LAI, the fluvial erosion was predicted to fluctuate with Q/P while it tended to be smaller for a higher than a lower vegetation coverage (not shown). The reduction of the role of vegetation coverage in soil erosion can be attributed to plant roots being able to bind soil particles together, forming aggregates and enhancing soil erodibility, and that plant stems and leaves can protect soil particles from direct shear by wind and water (Musick and Gillette 1990, Stockton and Gillette 1990, Fullen 1991, Simanton *et al* 1991, Wolfe and Nickling 1993). Generically, plants can improve the physical, chemical, and biological properties of soils, all of which contribute to soil stability and reducing soil erosion.

This study showed that aeolian erosion in a drier (windier) year was greater than that in a wetter (calmer) year (figure 5.23(a)). Relative to water erosion, aeolian erosion was predicted to be more closely related to soil moisture (figure 5.23(a) versus figure 5.24(a)). Across the assessment years, aeolian erosion was low for soil moisture of higher than 0.3. The predicted aeolian erosion events mainly occurred in March, April, May, and October (figure 5.23(b)), when soil moisture was lower than 0.20 and LAI was less than 0.03 for most of the days. Few aeolian erosion events occurred in the other months because June and September had a daily median LAI of 1.27 and a daily median soil moisture of 0.23, while November and February had a daily median soil moisture of 0.25.

In contrast, the fluvial erosion modulus was predicted to be larger in a year with higher soil moisture and more precipitation (e.g. 1959) than a year with lower soil moisture (e.g. 2007) and/or less precipitation (e.g. 1968) (figure 5.24(a)). One explanation is that more overland runoff can be generated in a wetter than a drier year and that the fluvial erosive force is a positive function of runoff peak and volume (Belnap *et al* 2011). Averaged across the assessment period, the fluvial erosion modulus peaked at 5.7×10^{-7} g m^{-2} s^{-1} in August due to a relatively high soil moisture, the largest runoff coefficient (i.e. Q/P), and a relatively low vegetation coverage (figure 5.24(b)). Compared to August, July had a smaller fluvial erosion modulus of 5.4×10^{-7} g m^{-2} s^{-1} because the soil moisture in July was lower than that in August despite more precipitation in July. This is consistent with the generic hydrological principle that the antecedent soil moisture condition is a determinant factor of the rainfall-runoff process. A month with more precipitation may generate less runoff than a month with less precipitation if the former month has lower antecedent soil moisture than the latter month, which may be true for July and August, respectively.

On an annual average basis, the fluvial erosion modulus was predicted to be a positive function of soil moisture, while the aeolian erosion modulus was predicted to be a negative function of soil moisture; both curves have an inflection at $\theta \approx 0.25$ (figure 5.25). On the one hand, the years with $\theta \geqslant 0.25$ were predicted to have a mean fluvial erosion modulus of 2.6×10^{-6} g m^{-2} s^{-1}, with a maximum of 7.2×10^{-6} g m^{-2} s^{-1} and a standard deviation of 1.4×10^{-6} g m^{-2} s^{-1}; such years were predicted to have a mean aeolian erosion modulus of 1.2×10^{-6} g m^{-2} s^{-1}, with a maximum of 8.2×10^{-6} g m^{-2} s^{-1} and a standard deviation of 2.0×10^{-6} g m^{-2} s^{-1}. On the other hand, the years with $\theta < 0.25$ were predicted to have a mean fluvial erosion modulus of 1.6×10^{-6} g m^{-2} s^{-1}, with a maximum of 4.2×10^{-6} g m^{-2} s^{-1} and a standard

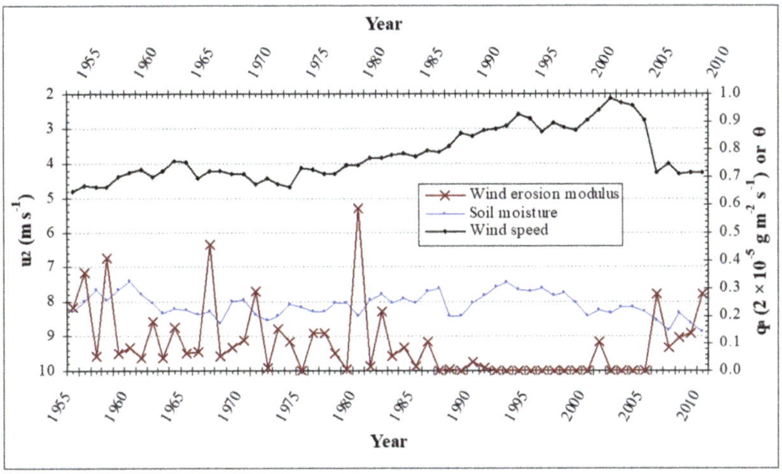

(a)

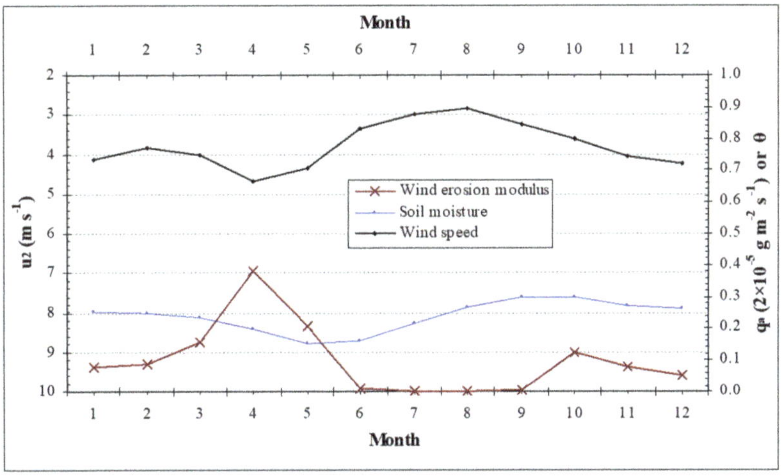

(b)

Figure 5.23. The predicted aeolian erosion modulus q_a and influencing variables. Plots showing the (a) annual values and (b) annual mean monthly values. u_2: wind speed at 2 m above the ground surface; θ: soil moisture. (Adopted with permission from figure 6 in Wang *et al* (2014a). Copyright 2014 Wiley.)

deviation of 0.6×10^{-6} g m^{-2} s^{-1}; such years were predicted to have a mean aeolian erosion modulus of 2.4×10^{-6} g m^{-2} s^{-1}, with a maximum of 11.8×10^{-6} g m^{-2} s^{-1} and a standard deviation of 2.8×10^{-6} g m^{-2} s^{-1}. These statistics show that fluvial erosion tended to be larger and fluctuate more widely in a moister than a drier year but that aeolian erosion tended to be larger and fluctuate more widely in a drier than a moister year. Around the inflection point, the aeolian and fluvial erosion processes had the 'maximum interaction' and contributed equivalently to the total soil erosion (Kirkby 1978, Ravi *et al* 2010).

5.4.4.3 Erosion response to climate change and human disturbance

As mentioned above, fluvial erosion tended to increase, but aeolian erosion tended to decrease, with an increase of soil moisture (figure 5.26). On the one hand, for a year with $\theta < 0.24$, the predicted aeolian erosion was larger than the predicted fluvial erosion; the maximum difference of 206 145 t year^{-1} (or 0.02 mm year^{-1}) was predicted to occur at the permanent wilting point of 0.132. On the other hand, for a

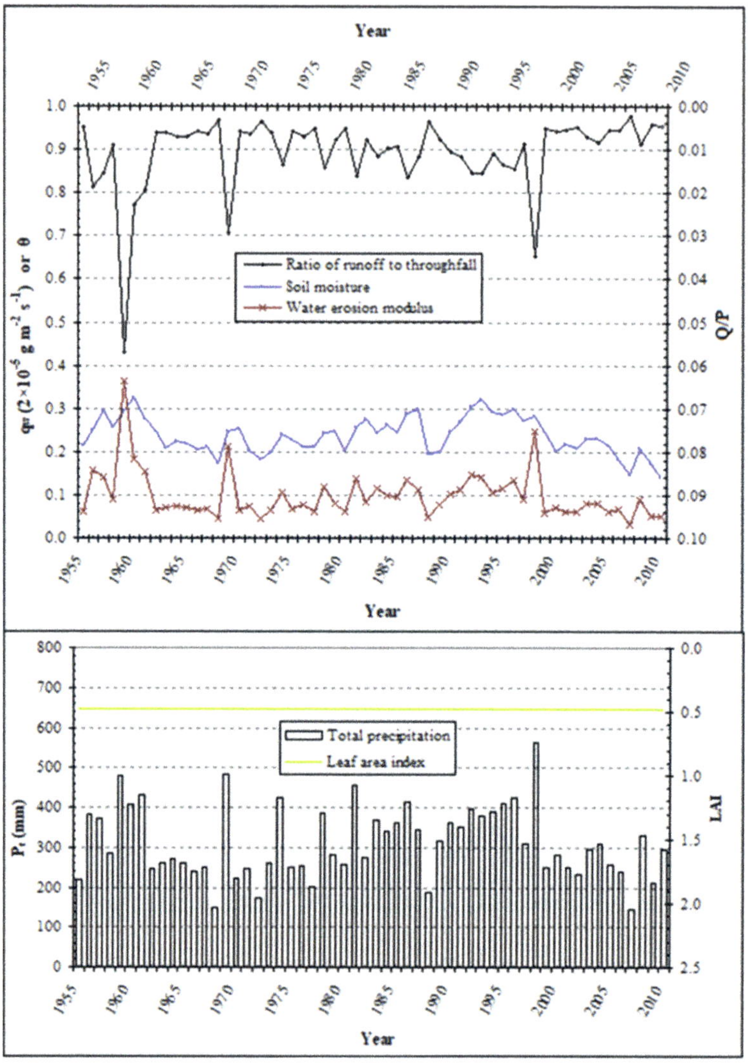

(a)

Figure 5.24. The predicted fluvial modulus q_w and influencing variables. Plots showing the (a) annual values and (b) annual mean monthly values. θ: soil moisture; Q: overland runoff; P: throughfall; P_t: total precipitation; and LAI: leaf area index. (Adopted with permission from figure 7 in Wang *et al* (2014a). Copyright 2014 Wiley.)

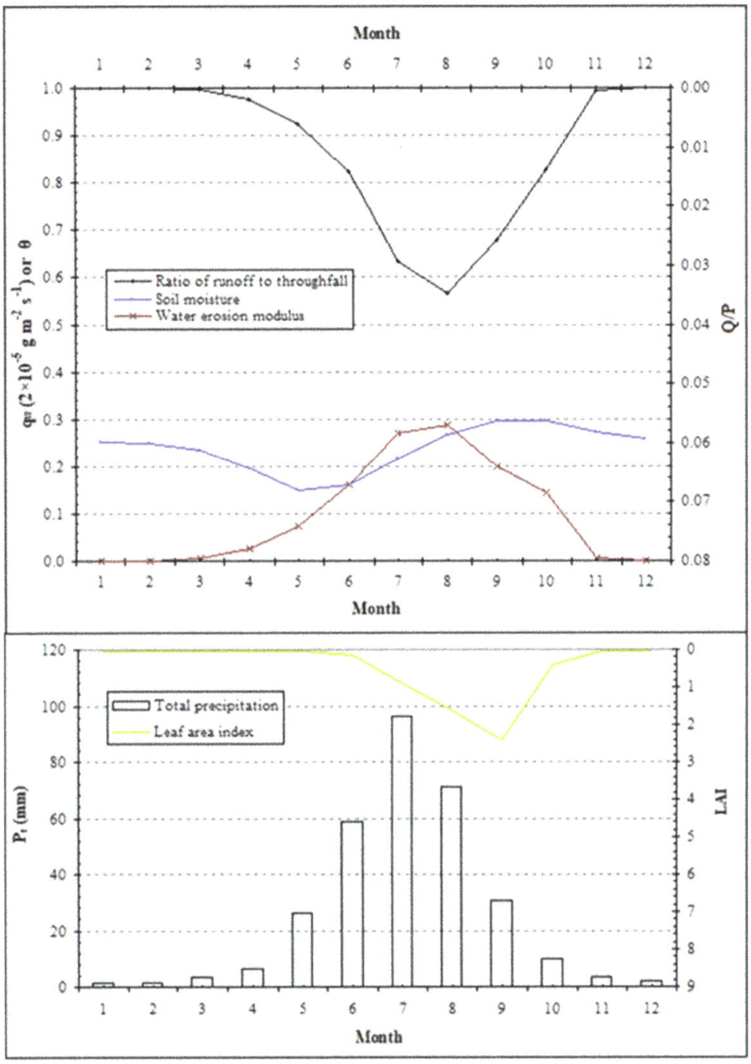

(b)

Figure 5.24. (Continued.)

year with θ ⩾ 0.24, the predicted aeolian erosion was less than the predicted fluvial erosion; the maximum difference of 79 727 t year^{-1} (or 0.01 mm year^{-1}) was predicted to occur at the saturated soil moisture of 0.417. Historically, because the studied watershed had an annual mean soil moisture of 0.24, the predicted mean annual total soil erosion of 660 375 t year^{-1} (or 0.08 mm year^{-1}) was equivalently contributed by aeolian and fluvial processes (table 5.10 and figures 5.25 and 5.26).

For a given soil moisture, the total soil erosion tended to decrease with an increase of vegetation coverage (figure 5.27(a)). The impact of a change in vegetation coverage on total soil erosion was predicted to be larger for a smaller

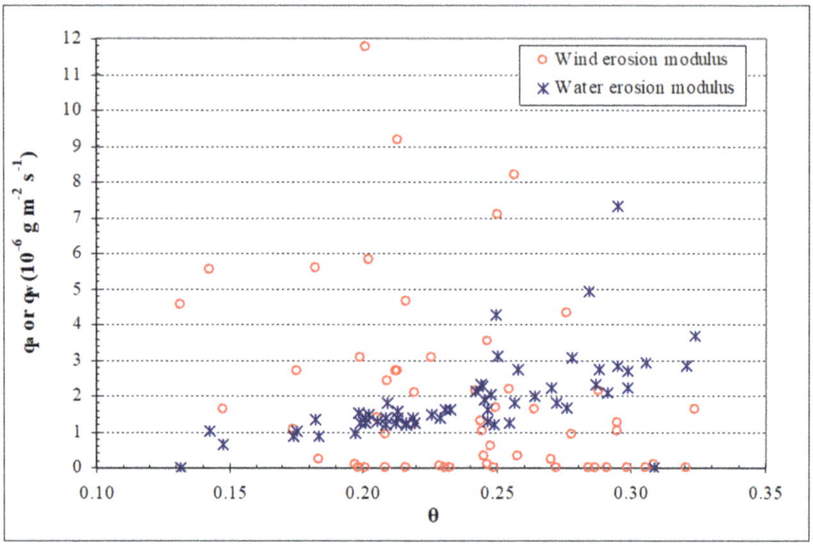

Figure 5.25. Influences of soil moisture (θ) on soil erosion. The annual aeolian (q_a) and fluvial (q_w) erosion modulus versus annual mean θ. (Adopted with permission from figure 8 in Wang *et al* (2014a). Copyright 2014 Wiley.)

than a larger base LAI because the curve of total soil erosion versus LAI has a steeper slope at a smaller than a larger base LAI. For a given LAI, the impact of a change in vegetation coverage was larger for a lower than a higher soil moisture, as indicated by the greater value of R_{aw} for the lower soil moisture. A change in vegetation coverage would have a more severe impact on soil erosion for a lower than a higher soil moisture. Relative to fluvial erosion, aeolian erosion would be more effectively protected by vegetation cover, as indicated by $R_{a \to w} > 1$ (figure 5.27(b)); human disturbance was predicted to result in a higher risk in triggering aeolian than fluvial erosion, particularly for years with a low soil moisture. One reason is that plants can increase resistance to air flows, reducing air momentum (i.e. shear stress) on soils (Wolfe and Nickling 1993, Zhao *et al* 2004, Reiche *et al* 2011). Another reason is that plant roots can aggregate soil particles and enhance soil structures, increasing the critical shear strength of soils (τ_{*c} in equation (3.89)) (Lancaster and Baas 1998). Field *et al* (2009) reported the opposite finding that vegetation cover could more effectively protect soils from fluvial than aeolian erosion. This contradiction can be attributed to their study being conducted in a relatively humid area with minimal wind erosion. For the Balagaer River watershed, its current vegetation coverage (LAI = 0.47) can protect the soils well from aeolian and fluvial processes, as indicated by the small deviations of R_{aw} from one for a wide range of LAI (figure 5.27(a)).

5.4.5 Conclusions

This case study put forward an integrated aeolian and fluvial prediction (IAFP) model that can simulate aeolian, fluvial, and hydrological processes in one domain.

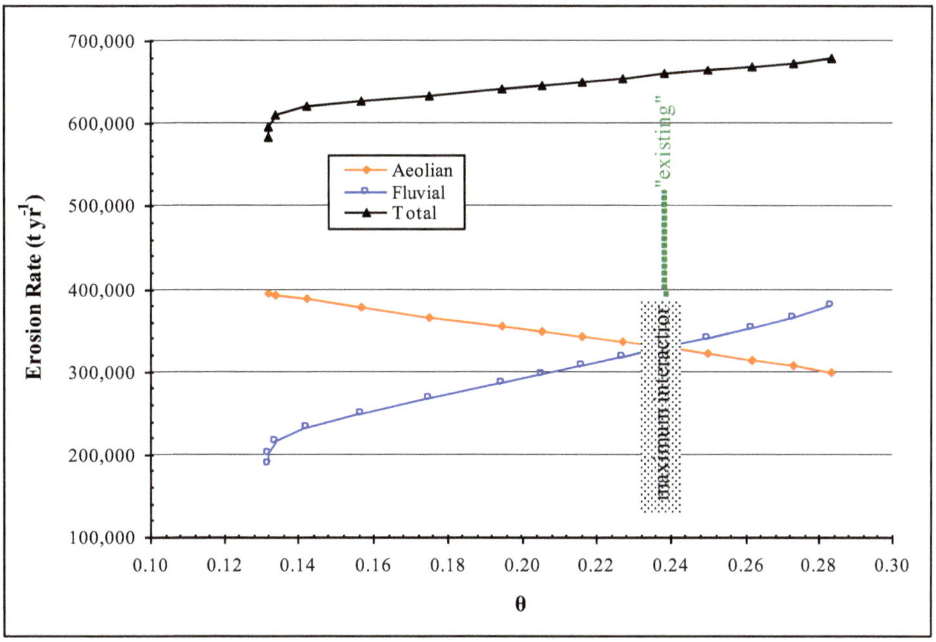

(a)

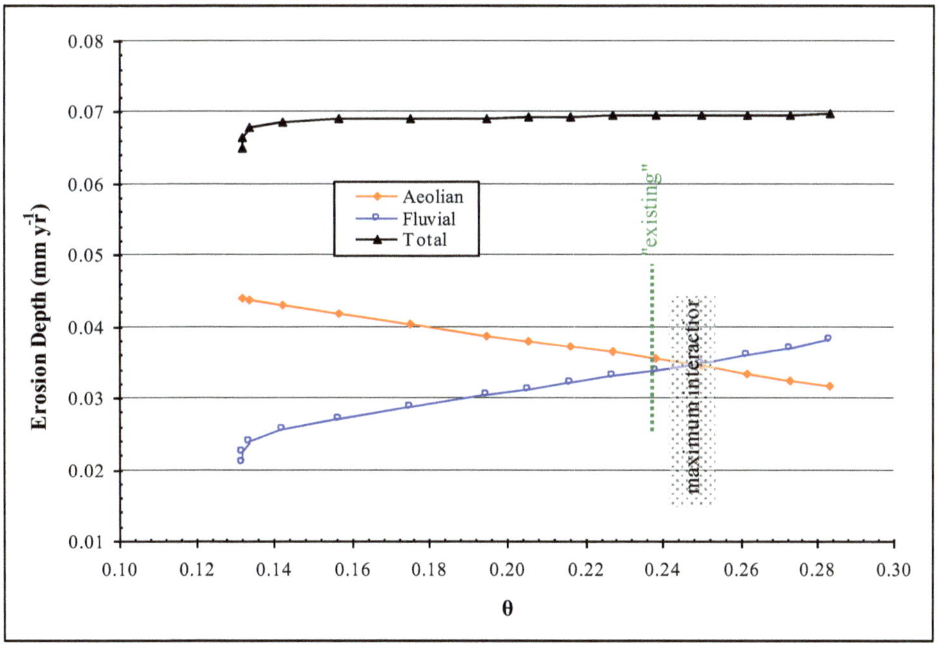

(b)

Figure 5.26. Erosion response to soil moisture. Plots showing influences of annual mean soil moisture (θ) on (a) erosion rate and (b) erosion depth. The results assumed the 'existing' vegetation coverage for the Balagaer River watershed. (Adopted with permission from figure 9 in Wang *et al* (2014a). Copyright 2014 Wiley.)

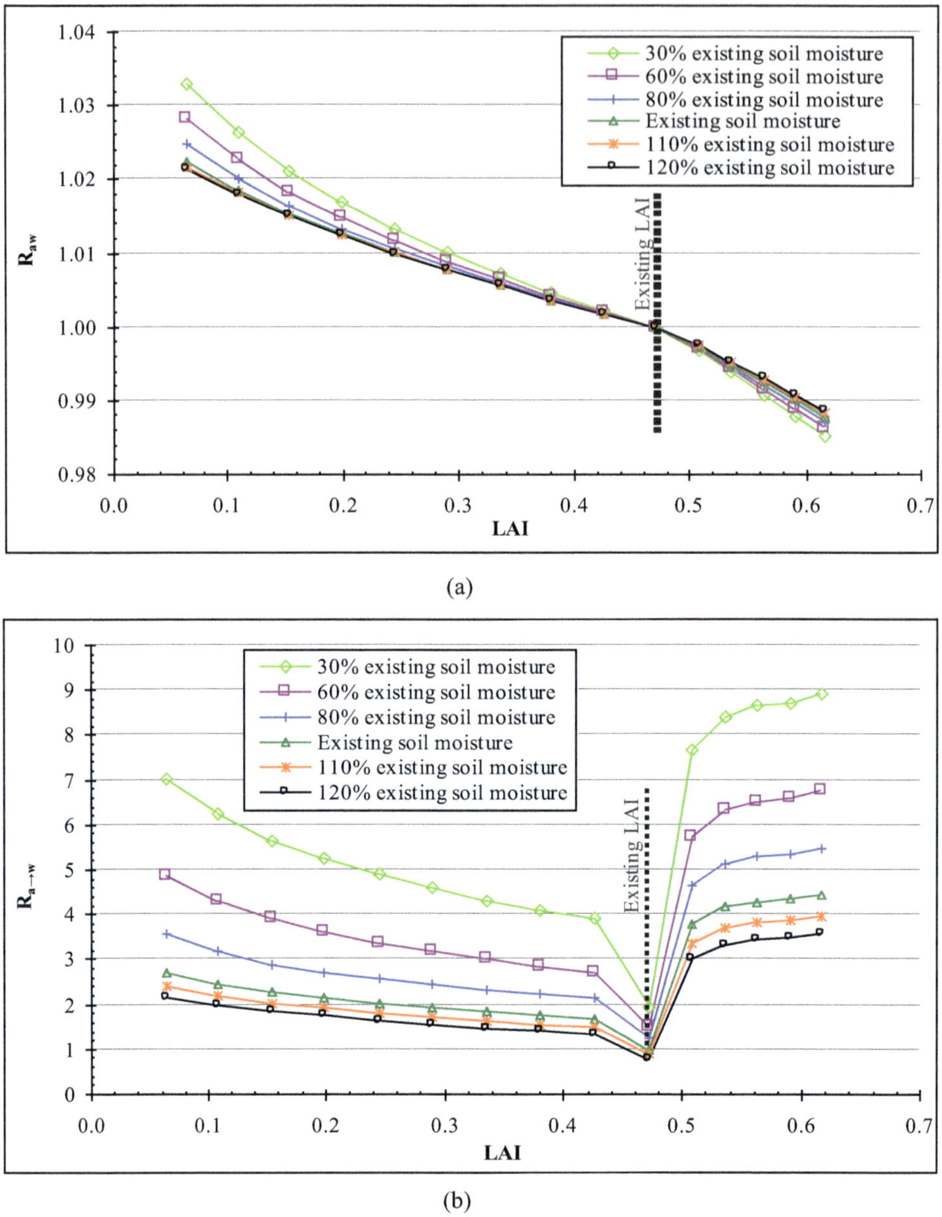

Figure 5.27. Erosion response to vegetation coverage. Plots showing annual mean leaf area index (LAI) versus (a) R_{aw} (equation (5.1)) and (b) $R_{a\to w}$ (equation (5.2)). (Adopted with permission from figure 10 in Wang *et al* (2014a). Copyright 2014 Wiley.)

The hydrological simulation module, driven by precipitation, computes soil moisture and runoff, which in turn are taken as inputs into the interrelated aeolian and fluvial process modules. As a novelty, the IAFP model integrated aeolian and fluvial erosion processes.

The well-calibrated IAFP model for the Balagaer River watershed predicted 0.9% and 22.0% occurrences of aeolian and fluvial erosion, respectively. Across the days with a non-zero erosion, the mean aeolian erosion was predicted to be 3.8×10^7 t year^{-1} or 4.1 mm year^{-1}, while the mean fluvial erosion was predicted to be 1.5×10^6 t year^{-1} or 0.2 mm year^{-1}. This implies that per erosion event, aeolian erosion was almost twenty times more severe than fluvial erosion. However, across the assessment period, either the mean aeolian erosion or the mean fluvial erosion was predicted to be 3.3×10^5 t year^{-1}, indicating that at the annual mean soil moisture of 0.24, these two erosion processes had a 'maximum interaction' and contributed equally to the total soil erosion of 660 375 t year^{-1} or 0.07 mm year^{-1}. Aeolian erosion would dominate if soil moisture was below 0.25, but fluvial erosion would dominate otherwise. Vegetation cover was predicted to play a greater role in reducing aeolian than fluvial erosion, particularly for years with a low soil moisture. Both erosion processes were predicted to be equally related to soil moisture, but aeolian erosion was found to be more closely related to vegetation coverage than fluvial erosion. Currently, the soils in the study watershed seem to be protected well by the existing vegetation cover from aeolian and fluvial shear stresses. However, Wang *et al* (2013) showed that due to climate change, precipitation in the watershed exhibited a significant downward trend, implying that future soil moisture will likely decrease to below 0.24. As a result, aeolian erosion could become a serious concern in the foreseeable future.

5.5 Retardation effects of a dry soil layer on soil water evaporation

This section is reproduced from Pedram (2017) and Pedram *et al* (2018) with extensive reorganizations and revisions. The soil water content (SWC) and pore air relative humidity (RH_{pa}) of desert land is influenced by wetting–drying cycles and the top dry soil layer (DSL) dynamics. However, the understanding of such influences is still incomplete. The objective of this study was to examine such influences at a typical desert site within the Horqin Sandy Land (HSL), located in the Mongolian Plateau of northern China. The results indicated that vaporization mainly occurred at a depth of around 10 cm below the ground surface. The diurnal variations of the SWC and RH_{pa} in the top 10 cm soils were much larger than those in the soils beneath. For a non-rainy day, the SWC and RH_{pa} tended to depend upon the relative magnitude of atmospheric temperature over soil temperature, whereas for a rainy day, the SWC and RH_{pa} were primarily controlled by the rainfall pattern and amount. The retardation role of the top DSL, which is about 10 cm thick and exists most of the time at the study site, might effectively prevent the moist soils beneath from being dried up further, and is beneficial for sustaining the desert ecosystem.

5.5.1 Introduction

In arid and semi-arid regions, where the hydrological cycle is dominated by vertical water movement (i.e. soil water evaporation) as driven by heat transport (Dong *et al* 2003, Wang 2015), soil water content (SWC) and pore air humidity (RH_{pa}) are two important indicators of water availability for sustaining the fragile sparse vegetation ecosystems (Goss and Madliger 2007, Duan *et al* 2011, Duan *et al* 2015). Hereinafter, SWC is defined as the ratio of soil water volume to bulk soil volume, while RH_{pa} is defined as the ratio of the partial pressure of water vapor to the total pressure of air in soil pores (Farouki 1981). When transported into soils, heat will cause an increase of soil temperature, vaporizing soil water into pore vapor; in contrast, when the soil temperature cools down, pore vapor can be condensed back into soil water (Wang 2015). The heat emission out of soils will not only lower the soil temperature but also cause a loss of pore vapor to the ambient atmosphere (i.e. soil evaporation). For an area of interest, when such a vaporization–condensation dynamic process becomes insufficient to meet the water demand of sparse vegetation, this area will likely be subject to degradation or even desertification. A major reason for land deterioration in the arid regions of the world (Kassas 1995, Zucca *et al* 2011), including the Horqin Sandy Land (HSL) of the Eurasian Grassland, desertification is closely related to increasing soil evaporation as a result of inappropriate land management practices (e.g. removal of native grasses and the planting of deep-rooted trees) (Zhao *et al* 2010, Smits *et al* 2012, Wang 2015, Li *et al* 2016a). To develop practical solutions for desertification-related eco-environmental problems (e.g. dust storms and land degradation), a good understanding of soil water and pore vapor dynamics is essential.

As noted by previous studies (Wang *et al* 2006, Liu *et al* 2013), such an understanding can be challenging because of the two-phase (i.e. liquid–vapor) fluid condition and the vapor flow resistance (VFR) effect of a thin (5–10 cm) top dry soil layer (DSL). The semi-arid regions are usually dominated by bare sandy soils with such a DSL most of the time, within which soil moisture is mainly in the vapor phase with a very large capillary suction head (>15 000 cm water height) (Goss and Madliger 2007). In the past two decades, various laboratory and field studies (Daamen and Simmonds 1996, Yamanaka and Yonetani 1999, Saravanapavan and Salvucci 2000, Wang *et al* 2006, Goss and Madliger 2007, Novak 2010, Sakai *et al* 2011, Liu *et al* 2013) have been conducted to measure soil evaporation in a water-limited environment. However, most of the studies were for short-term periods (i.e. from several days to a few months) and rarely measured the vertical profiles of SWC and soil temperature, making it difficult to determine the dynamics of soil water and pore vapor as influenced by wetting–drying cycles. On the other hand, a few studies such as Yamanaka *et al* (1998) and Sakai *et al* (2011) used mathematical models to reproduce the vaporization–condensation–movement dynamics observed in laboratory soil columns, without determining the dynamics of soil water and pore vapor within different soil layers. Wang (2015) reviewed existing studies that aimed to measure and/or model soil water evaporation in arid and/or semi-arid environments, concluding that to fill the knowledge gap of soil water and pore vapor dynamics, field measurements need to be extrapolated using mathematical models.

The liquid–heat–vapor processes in a soil profile can be described by the PdV model (Philip and de Vries 1957) formulized in section 4.4.4. In the real world, these processes interactively determine the soil evaporation phase at any time of interest. Philip (1958) classifies the desiccation of a soil profile into three phases (i.e. phases I to III). Phase I occurs when the soil is sufficiently moist and thus has a soil water evaporation (E) indistinguishable from that of a saturated surface (E_s), whereas phase III occurs when the soil surface layers are very dry and E is sensitive to, and may be negatively correlated with, the heat flux into the soil. Phase II occurs when the soil has an intermediate moisture content and thus E is independent of E_s and depends only on the soil moisture distribution. During phase I, liquid water is supplied from the lower layers and vaporized at the soil surface at a rate E_S, and then the vapor is transferred into the ambient atmosphere. On the other hand, during phases II and III, vaporization takes place not at the surface but within the soil mass (e.g. the evaporation zones or EZs), and the vapor diffuses upward through a top DSL to the surface and then into the atmosphere. Such a classification of evaporation phases is verified by field observations presented by Goss and Madliger (2007).

Based on a few previous studies such as Yamanaka and Yonetani (1999) and Goss and Madliger (2007), during phases I and II, E tends to be maximal when total heat flux into the soil is greatest (e.g. at noon), whereas during phase III, E tends to be maximal when the total heat flux out of the soil is largest (e.g. at midnight) and vice versa. This is because for phases I and II, liquid water vaporizes at or near the soil surface at a rate that is positively related to the solar radiation (i.e. heat) input into the soil. A larger heat input will prompt the desiccation of the soil surface layer, increasing the upward suction gradient and thus facilitating the upward transport of liquid water. When the heat flux into the soil is greatest, the rates of desiccation and liquid upward movement will become fastest, leading to a maximal evaporation rate. In contrast, for phase III, the vaporization occurs within the soil mass and then the vapor is transported upward through the above dry soils into the ambient atmosphere. The transport of vapor is primarily driven by an upward temperature gradient between a deeper and upper layer as well as between the topsoil and the ambient atmosphere. When the heat flux out of the soil is largest, the upward gradient will be the greatest and thus the E rate can reach a maximum. However, because of incorrect modeling premises (e.g. E monotonically increases with heat input) and/or insufficient and nonspecific experimental (mostly laboratory soil column) data (Goss and Madliger 2007, Wang 2015), previous studies (van Bavel and Hillel 1976, Mahfouf and Noilhan 1991, Kondo *et al* 1992, Bristow and Horton 1996, Yamanaka *et al* 1998, Yamanaka and Yonetani 1999, Saravanapavan and Salvucci 2000, Mori *et al* 2003, Gowing *et al* 2006, Heitman *et al* 2008, Saito *et al* 2008, Novak 2010, Liu 2011, Sakai *et al* 2011, Xiao *et al* 2011, Shahraeeni *et al* 2012) primarily examined and reported the dynamics of phase I and II evaporation. For this reason, the features presented by those studies are unlikely to be found at any sites in naturally drylands (e.g. deserts), where most of the time a top DSL overlies moist soils and phase III dominates soil evaporation process.

The advancement of heat pulse sensors in the past decade has made it possible for the PdV to be widely applied in practice to determine soil evaporation (Bristow *et al* 1993, Kamai 2013, Liu *et al* 2013). Although the PdV has been programmed into a family of HYDRUS computer software packages (Šejna *et al* 2022, Šimůnek *et al* 2022), few applications have been reported in the existing literature to very dry regions, where phase III soil evaporation dominates the hydrological cycle most of the time (Zeng *et al* 2009, Zeng *et al* 2011a, Zeng *et al* 2011b). The objective of this study was to examine the SWC and RH_{pa} dynamics in a desert sandy soil, as influenced by heating–cooling and wetting–drying cycles as well as DSL retardation effects, by extrapolating field data on SWC and soil temperature using a HYDRUS-1D model. Such information is needed but incomplete in the existing literature for developing practical measures to protect degrading and/or restore degraded sandy ecosystems in arid and semi-arid environments.

5.5.2 Data and methods

This study was conducted at a site within the HSL described in section 5.1.2. The site is covered by bare desert sandy soils.

5.5.2.1 Data

A sensor-based weather station at the study site was installed to continuously record data on precipitation, air temperature, relative humidity of the ambient air, wind speed and direction, solar radiation, and atmosphere vapor pressure. In parallel, relevant sensors were installed to continuously measure soil moisture and soil temperature at seven depths of 10, 20, 40, 80, 120, 160, and 200 cm. In addition, a three-needle heat pulse (TNHP) sensor was used to continuously measure the temperature and water contents of the surface (i.e. the top 1 cm) soil. All data were automatically stored by a data logger at time intervals of 10 min. For the weather station, the precipitation sensor has an accuracy of ±0.1 mm, the temperature sensor has an accuracy of ±0.1 °C, and the other sensors have accuracies of ±0.5%. The soil moisture and soil temperature sensors have accuracies of ±0.4% and ±0.1 °C, respectively, while the TNHP sensor for surface soil temperature and water content has accuracies of ±0.1 °C and ±0.4%, respectively. The details of the sensors and their installations can be found in Duan *et al* (2011) and Duan *et al* (2015).

This study used the data from 1 January 2013 to 31 December 2014. The data of the year 2013 were used to calibrate the HYDRUS-1D model, while the data of the year 2014 were used to validate the model. Subsequently, the simulated results for both years were used to examine the dynamics of SWC and RH_{pa}. The annual precipitation in 2013 was smaller than the average annual precipitation (344.9 versus 389 mm), whereas the annual precipitation in 2014 was larger than the average annual precipitation (520.6 versus 389 mm) (figure 5.3). This indicates that the year 2013 was relatively drier and the year 2014 was relatively wetter. However, both years shared a similar temporal variation pattern in seasonal precipitation: they were wettest in June, July, and August. For these three wettest months, there were fewer rainy days but higher rainfall intensities in 2014 than 2013, while for the other months, there were more rainy days and higher rainfall intensities in 2014 than 2013 (figure 5.28). Overall, regardless of the year, more than 85% of the days had no precipitation and might be subject to net losses of soil water.

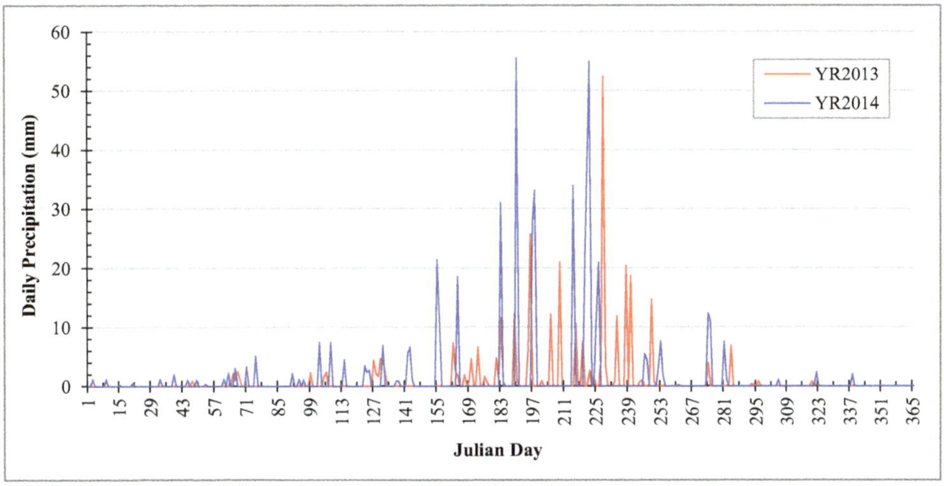

Figure 5.28. The daily precipitation in 2013 and 2014.

5.5.2.2 Model set-up

The HYDRUS-1D model was set up for a 120 cm deep soil profile with a uniform spatial discretization of 1 cm, leading to 120 layers and 121 nodes for mass balances. The sensors at 0, 10, 20, 40, 80, and 120 cm depths correspond to six of the 121 nodes (hereafter designated as 'sensor nodes' for description purpose). For each layer, its preliminary soil–water parameters (table 5.11) were determined by laboratory experiments (Duan *et al* 2011) as well as from the literature (Ghanbarian-Alavijeh *et al* 2010). Herein, the soil water retention function was defined by the modified van Genuchten equation (equation (3.29)). In addition, for each sensor node, the initial condition was specified as the measured value at 0:10 1 January 2013, while for a non-sensor node, the initial condition was determined to be the linear interpolation of the measured values at the two adjacent sensor nodes. For instance, the initial condition at the 1 cm depth node was determined to be $0.9\theta_{0,0} + 0.1\theta_{10,0}$, and the initial condition at the 18 cm depth node was determined to be $0.2\theta_{10,0} + 0.8\theta_{20,0}$, where $\theta_{0,0}$, $\theta_{10,0}$, and $\theta_{20,0}$ are the initial water contents at the surface, 10 cm depth, and 20 cm depth nodes, respectively. Further, the upper boundary conditions were defined as the time series of precipitation, surface soil temperature, and E_s (potential evapotranspiration). E_s was estimated by the Penman–Monteith method (equation (3.39)) using the measured data on air temperature, atmosphere relative humidity, solar radiation, and wind speed. Moreover, the lower boundary condition was specified to be free drainage because of the deep (>4 m) water table as mentioned above.

In addition to visualization plots showing the simulated versus measured time series of soil temperatures and soil moisture, the statistics of the coefficient of determination (R^2) (equation (4.56)) and coefficient of variation (C_v) were also computed. R^2 measures the percentage of the variations presented in a measured time series that can be explained by the model, whereas C_v measures the spread (i.e. amount of variability) relative to the mean of the time series. A good model is

Table 5.11. The calibration parameters in this study[1].

Parameter	Definition	Range	Adopted value
α [cm^{-1}]	Soil hysteresis factor (equation (3.29))	0.029–0.70	0.147
n [–]	Exponent (equation (3.29))	2.0–6.95	2.34
θ_r [–]	Residual soil moisture (equation (3.29))	0.0–0.1	0.02
θ_{sat} [–]	Saturated soil moisture (equation (3.29))	0.35–0.45	0.38
K_s [cm d^{-1}]	Saturated hydraulic conductivity (equation (3.33))	200–1520	650

[1] Reproduced with permission from table 2 in Pedram *et al* (2018). Copyright 2018 Elsevier.

expected to have a relatively large R^2 value and give a C_v value close to that of the responding measured time series. Further, the relative root mean square error (RRSME) (equation (4.53)) was also computed and used to measure the model performance. RRSME can vary from zero to $+\infty$. A zero value indicates that the model can reproduce the measurements perfectly, whereas a greater positive value indicates a poorer model performance.

5.5.2.3 Examination of the SWC and RH_{pa} dynamics

For each simulation year, at the annual or monthly time scale, the means and standard deviations of the simulated SWC and RH_{pa} time series were computed for eight soil layers, including the surface, 0–5 cm (layer 1), 5–10 cm (layer 2), 10–20 cm (layer 3), 20–30 cm (layer 4), 30–40 cm (layer 5), 40–50 cm (layer 6), and 50–60 cm (layer 7). The computations were done separately for dry (i.e. non-rainy) and wet (i.e. rainy) days. Subsequently, the computed means and standard deviations were compared to examine the relative magnitudes and vertical distributions of the simulated values for SWC and RH_{pa}. In addition, the diurnal variations of the SWC and RH_{pa} were scrutinized by plotting the simulated values for selected dry and wet days. Herein, because it is still an unresolved scientific challenge to accurately measure RH_{pa} in sandy soils (Goss and Madliger 2007), this study could not use data on RH_{pa} to calibrate and validate the HYDRUS-1D model. However, it was empirically judged that the simulated values of RH_{pa} were probably more appropriate for assessing the relative diurnal variations of pore vapor, which is the interest of this study, rather than determining the absolute magnitudes of RH_{pa}.

5.5.2.4 Determination of DSL retardation effects

For the two study years (i.e. 2013 and 2014), the daily PET (DPET) was computed using the Penman–Monteith equation (equations (3.39)–(3.50)). The daily data on solar radiation (R_{sl}), heat exchange with rootzone ($g_{gs\leftrightarrow r}$), mean air temperature (T_a), wind speed (u_{zm}), atmosphere pressure (p_a), and air relative humidity (RH) collected at the weather station were used to compute DPET. On the other hand, the daily actual evaporation of soil water (DAE), which is the soil water lost from the soil layers to the ambient atmosphere, was simulated by the calibrated and validated HYDRUS-1D model. Visualization plots showing the DAE versus DPET were used to examine how much the atmospheric evaporation demand was satisfied.

In addition, for a given month in a study year, the summation of the DAE was computed and taken as the monthly actual evaporation of soil water (MAE) of this month, while the summation of the DPET was computed and taken as the monthly potential evapotranspiration (MPET) of this month. At the study site, because the soils are highly permeable and have a very low moisture most of the time (Duan *et al* 2011), overland runoff can be neglected. Thus, the monthly water balance equation can be expressed as

monthly precipitation [mm]
monthly bottom seepage out of soil profile [mm]
change of soil water in entire soil profile [mm]
monthly actual evaporation [mm]

$$\Delta S_m = P_m - MAE - I_{bm} \tag{5.3}$$

In addition, the sub-daily data on R_{sl}, $g_{gs\leftrightarrow r}$, T_a, u_{zm}, p_a, and RH collected at the weather station were also used in the Penman–Monteith equation to compute the instantaneous potential evapotranspiration (IPET), which in turn was plotted along with the instantaneous actual evaporation of soil water (IAE) simulated by the HYDRUS-1D model to examine the diurnal variations of the IPET and IAE. The plots were generated for selected non-rainy and rainy days. For each plot, the responding simulated water moisture in the top three layers (i.e. layer 1 at 5 cm depth, layer 2 at 10 cm depth, and layer 3 at 15 cm depth) were superimposed to examine which layer(s) contribute most to the IAE at a given time.

Further, the thermal liquid and vapor fluxes at different depths (i.e. layers) were analyzed to determine the variations of DSL thickness as affected by the drying–wetting cycle. Herein, the DSL was defined as the soil layers that have a soil moisture of less than $\theta_r = 0.02$ (table 5.11). The variations of DSL and MAE were compared to assess the repellent (i.e. retardation) effects of DSL on soil water evaporation. Such a retardation effect is defined as

monthly actual soil water evaporation without DSL
monthly actual soil water evaporation with DSL
retardation effect of DSL on soil water evaporation [%]

$$RE_{vfr} = \frac{MAE_{noDSL} - MAE_{DSL}}{MAE_{noDSL}} \tag{5.4}$$

The assessment was performed using an empirical formula expressed as Wang (2015)

linear slope (9.743 to 26.004)
intercept (-7.810 to 51.598)
retardation effect of DSL on soil water evaporation [%]
DSL thickness [cm]

$$RE_{vfr} = a_3 \ln(\delta) + a_4 \tag{5.5}$$

5.5.3 Results

5.5.3.1 The calibrated HYDRUS-1D model

The adopted model parameters are presented in table 5.11. During the calibration period (1 January to 31 December 2013), the model reproduced well the soil temperatures measured at the 5, 10, and 20 cm depths below the ground surface, as indicated by $R^2 > 0.79$ and that few of the values fall out of the 95% confidence limits of the simulated–measured regression lines. For a given soil layer, the model tended to underestimate the mean of the measured soil temperatures by about 0.96 °C, whereas it tended to overestimate the temporal variation in the measured soil temperatures, as indicated by the larger coefficient of variation (C_v) for the simulated soil temperatures than that for the measured soil temperatures (table 5.12).

The model had a similar performance for all soil layers, as indicated by the similar values of RRSME or R^2 between the layers. On the other hand, the model also reproduced well the soil moisture measured at the same depths, as indicated by $R^2 > 0.78$ and that few of the values fall out of the 95% confidence limits of the simulated-observed regression lines. Overall, the model performed slightly better in simulating the soil moisture than the soil temperatures. For a given soil layer, the model tended to underestimate the mean of the observed soil moisture by 22%, while it also tended to underestimate the temporal variation of the observed soil moisture, as indicated by the smaller C_v for the simulated soil moisture than that for the measured soil moisture (table 5.12). Again, the model had a similar performance for all soil layers in reproducing the measured soil moisture. Moreover, during the validation period (1 January to 31 December 2014), the model also did a good job ($R^2 \geqslant 0.75$; RRSME $\leqslant 0.215$) (table 5.12). The similar performance for both periods indicates that the model is very robust and can be applicable for different hydro-climatic conditions. Hence, the calibrated model was judged to be accurate enough to mimic the physical processes of the two-phase fluids (i.e. water vapor and liquid water) for the study site.

5.5.3.2 The simulated soil moisture

Figure 5.29 shows the variation of the simulated soil moisture in the 5, 10, and 20 cm layers for 2013 and 2014. Regardless of the layers, during the dry months with a minimal amount of precipitation, the soil moisture was predicted to be very low, whereas during the wet (i.e. summer) months with a high-intensity precipitation, the soil moisture was predicted to be relatively high but to decline very quickly once the rain stopped, which can be attributed to the large evaporation rate of up to 178.5 mm d^{-1} in summer.

At the annual and monthly time scales, regardless of dry or wet days, the average soil moisture of non-rainy days increased with soil depth (tables 5.13(a) and (b)). The increasing gradient within the top 10 cm was generally smaller than that within a lower soil depth, in particularly for the dry days. As expected, for a given time scale, the average soil moisture of a dry day was smaller than that of a wet day. However, the soil moisture within an upper soil layer had a greater temporal variation

Table 5.12. Performance measures of the HYDRUS-1D model[1].

	Soil temperature					Soil moisture				
Soil depth (cm)	Mean (°C)	Std. (°C)	C_v	RRSME	R^2	Mean (–)	Std. (–)	C_v	RRSME	R^2
5 cal.	6.55 (7.45)	2.33 (3.44)	0.35 (0.46)	0.112	0.91	0.0104 (0.0098)	0.001 23 (0.001 22)	0.11 (0.12)	0.113	0.90
val.	4.33 (4.77)	3.17 (3.21)	0.73 (0.65)	0.124	0.92	0.011 (0.013)	0.0014 (0.0015)	0.13 (0.11)	0.222	0.76
10 cal.	5.45 (6.66)	1.88 (2.04)	0.34 (0.30)	0.129	0.89	0.0255 (0.0202)	0.001 03 (0.001 08)	0.066 (0.10)	0.128	0.89
val.	5.63 (6.56)	2.12 (2.44)	0.37 (0.37)	0.147	0.84	0.029 (0.026)	0.0013 (0.0011)	0.044 (0.042)	0.205	0.78
20 cal.	7.55 (8.34)	1.02 (1.13)	0.13 (0.13)	0.133	0.87	0.056 (0.0435)	0.001 01 (0.001 11)	0.017 (0.025)	0.106	0.93
val.	7.48 (8.99)	2.08 (2.87)	0.27 (0.30)	0.199	0.78	0.059 (0.069)	0.0011 (0.0013)	0.018 (0.018)	0.187	0.80
30 cal.	8.01 (9.13)	3.13 (2.23)	0.39 (0.24)	0.130	0.88	0.076 (0.069)	0.000 55 (0.001 01)	0.0072 (0.014)	0.142	0.84
val.	8.12 (9.87)	3.05 (3.16)	0.37 (0.32)	0.215	0.75	0.066 (0.071)	0.0019 (0.0018)	0.028 (0.025)	0.176	0.92
40 cal.	8.83 (9.44)	3.44 (2.62)	0.38 (0.27)	0.202	0.79	0.094 (0.105)	0.002 04 (0.003 45)	0.021 (0.032)	0.204	0.78
val.	9.57 (10.11)	3.14 (3.18)	0.32 (0.31)	0.183	0.80	0.084 (0.096)	0.0021 (0.0027)	0.025 (0.028)	0.172	0.93
50 cal.	9.02 (10.15)	3.02 (3.19)	0.33 (0.31)	0.145	0.83	0.106 (0.113)	0.0045 (0.0048)	0.042 (0.043)	0.135	0.86
val.	10.11 (11.22)	1.22 (1.67)	0.12 (0.14)	0.117	0.93	0.106 (0.115)	0.0031 (0.004)	0.029 (0.034)	0.153	0.98

[1] Reproduced with permission from table 4 in Pedram *et al* (2018). Copyright 2018 Elsevier. The measured value is in brackets. cal., calibration; val., validation; Std., standard deviation; C_V, coefficient of validation, defined as the ratio of Std. to mean; RRSME, relative root mean square error (equation (4.53)); and R^2, coefficient of determination.

(i.e. standard deviation) than that within a lower soil layer. That is, the soil moisture within an upper soil layer varied more greatly from time to time. For a given dry day, the soil moisture within the surface layer had a much greater temporal variation than that within a deeper layer (figures 5.30(a) and 5.30(b)). Overall, the soil moisture within the surface layer reached its minimum around 2:00 pm, whereas the soil moisture within a deeper layer was almost constant throughout the day. At a given time during the day, the soil moisture within a deeper layer tended to be larger than that within a shallower layer. In contrast, for a wet day, the soil moisture within the surface layer was increased because of rainfall (figures 5.30(c) and 5.30(d)). If the rainfall was large, the water content within the 10 cm or a deeper layer could also be increased. The maximum soil moisture within the surface or 10 cm layer occurred during the rainfall event, while the maximum soil moisture within a deeper layer

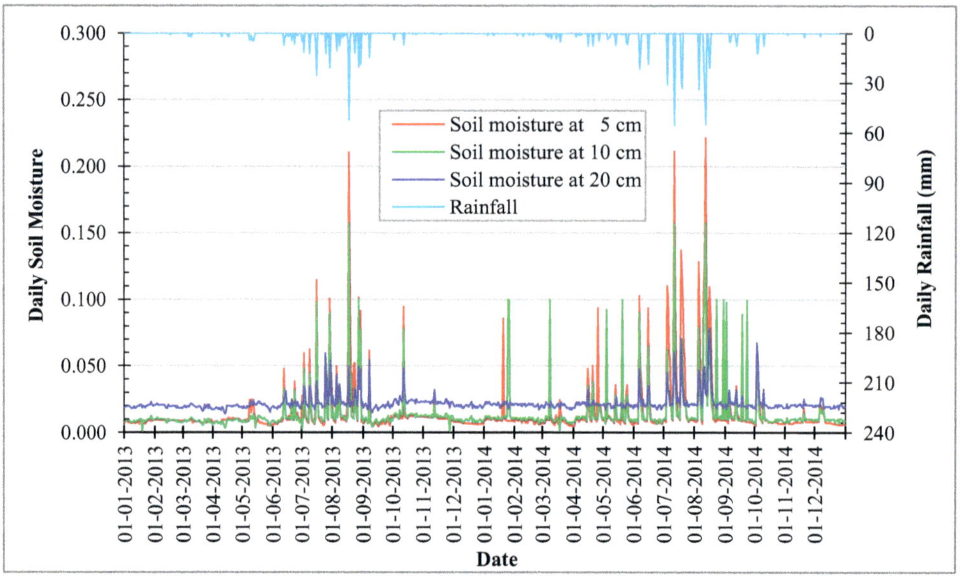

Figure 5.29. The simulated diurnal variations of soil moisture for the two study years.

appeared sometime after the cease of the rain probably due to the redistribution of soil water of infiltration.

5.5.3.3 The simulated pore air relative humidity

At the annual and monthly time scales, regardless of dry or wet days, the average pore air relative humidity (RH_{pa}) increased with soil depth (table 5.14). The increasing gradient within the top 10 cm layer was much larger than that within the lower soil layers, which is particularly true for the dry days. As expected, for a given time scale, the average RH_{pa} of a dry day was smaller than that of a wet day. However, the RH_{pa} within an upper soil layer had a greater temporal variation than that within a lower soil layer. That is, the RH_{pa} within an upper soil layer varied more greatly at different times. Nevertheless, at a given time, the RH_{pa} in 2013 was same as that in 2014 (i.e. invariant from year to year).

For a given dry day, the RH_{pa} within the surface layer had a much greater temporal variation than that within a deeper layer (figures 5.31(a) and (b)). Overall, the RH_{pa} within the surface layer reached its minimum around 2:00 pm, while the RH_{pa} within the 5 or 10 cm layer approached its maximum around this same time. The RH_{pa} within the layers below the depth of 10 cm was almost constant throughout the day. At a given time during the day, the RH_{pa} within a deeper layer tended to be larger than that within a shallower layer. In contrast, for a wet day, the RH_{pa} within the surface layer could be larger than that within a deeper layer, and the timing of its minimum was dependent on the timing and amount of rainfall (figures 5.31(c) and (d)). When rain fell in the morning or the afternoon, the RH_{pa} within the surface layer occurred around 2:00 pm. The RH_{pa} within a layer below the surface was almost constant throughout the day regardless of the rain

Table 5.13(a). Means and standard deviations (in brackets) of the soil moisture within different layers for year 2013[1]

Layer	Days	Annual	Jan	Feb	Mar	Apr	May	Jun	Jul	Aug	Sep	Oct	Nov	Dec
Surface	Dry	0.0085 (0.000 65)	0.0041 (0.0002)	0.0058 (0.0001)	0.0071 (0.0005)	0.0082 (0.000 23)	0.0101 (0.0001)	0.0108 (0.0009)	0.0123 (0.000 16)	0.0132 (0.000 98)	0.0111 (0.000 11)	0.0117 (0.000 87)	0.0093 (0.000 13)	0.0092 (0.000 12)
	Wet	0.0687 (0.0012)	–	0.0132 (0.001)	0.0198 (0.0002)	0.0241 (0.000 44)	0.0275 (0.0007)	0.0855 (0.0011)	0.114 (0.001 06)	0.210 (0.001 66)	0.0850 (0.000 32)	0.0932 (0.001 42)	0.0154 (0.0015)	–
0–5 cm	Dry	0.0107 (0.0006)	0.0062 (0.0003)	0.0062 (0.000 86)	0.0082 (0.000 42)	0.0091 (0.000 12)	0.0121 (0.000 19)	0.0121 (0.000 44)	0.0139 (0.0007)	0.0129 (0.000 83)	0.0119 (0.0001)	0.0121 (0.000 62)	0.0099 (0.000 10)	0.0098 (0.0001)
	Wet	0.0459 (0.000 88)	–	0.0114 (0.000 42)	0.0192 (0.000 18)	0.0223 (0.000 37)	0.0265 (0.000 67)	0.0766 (0.0009)	0.102 (0.001 01)	0.185 (0.001 42)	0.0801 (0.001 42)	0.0905 (0.001 33)	0.0666 (0.001 43)	–
10–20 cm	Dry	0.0131 (0.000 55)	0.0099 (0.000 19)	0.0091 (0.000 54)	0.0121 (0.001 26)	0.0118 (0.000 10)	0.0162 (0.000 12)	0.0176 (0.000 33)	0.0181 (0.000 45)	0.0191 (0.000 94)	0.0155 (0.001 07)	0.0153 (0.000 51)	0.0136 (0.000 93)	0.0112 (0.000 89)
	Wet	0.0476 (0.001 12)	–	0.0131 (0.000 212)	0.0182 (0.001 78)	0.0201 (0.000 19)	0.0259 (0.001 44)	0.0612 (0.000 49)	0.0823 (0.000 978)	0.0860 (0.000 62)	0.0701 (0.001 33)	0.0612 (0.001 19)	0.0388 (0.001 29)	–
20–30 cm	Dry	0.0141 (0.0009)	0.0111 (0.000 16)	0.0108 (0.000 49)	0.0123 (0.001 19)	0.0121 (0.000 93)	0.0171 (0.000 10)	0.0186 (0.000 81)	0.0196 (0.000 34)	0.0193 (0.000 89)	0.0167 (0.001 04)	0.0162 (0.000 59)	0.0144 (0.000 88)	0.0133 (0.0007)
	Wet	0.0439 (0.0012)	–	0.0127 (0.002 08)	0.0166 (0.001 69)	0.0199 (0.001 57)	0.0232 (0.001 41)	0.0557 (0.001 93)	0.0755 (0.000 72)	0.0763 (0.001 69)	0.0693 (0.001 29)	0.0603 (0.001 91)	0.0332 (0.001 21)	–
30–40 cm	Dry	0.0194 (0.0008)	0.0166 (0.000 12)	0.0166 (0.001 38)	0.0181 (0.001 11)	0.0202 (0.000 88)	0.0253 (0.000 88)	0.0254 (0.000 79)	0.0271 (0.000 79)	0.0241 (0.000 82)	0.0255 (0.000 99)	0.0241 (0.000 55)	0.0266 (0.000 82)	0.0201 (0.0008)
	Wet	0.0455 (0.0012)	–	0.0288 (0.001 99)	0.0201 (0.001 58)	0.0312 (0.001 44)	0.0312 (0.001 39)	0.0598 (0.000 77)	0.0689 (0.001 69)	0.0688 (0.000 79)	0.0555 (0.001 11)	0.0566 (0.001 88)	0.0344 (0.001 19)	–
40–50 cm	Dry	0.0293 (0.001 12)	0.0208 (0.000 14)	0.0283 (0.001 29)	0.0293 (0.001 05)	0.0301 (0.000 79)	0.0322 (0.000 81)	0.0366 (0.000 66)	0.0413 (0.000 71)	0.0455 (0.000 75)	0.0367 (0.000 910	0.0266 (0.000 54)	0.0278 (0.000 81)	0.0251 (0.000 65)
	Wet	0.0453 (0.0012)	–	0.0312 (0.001 87)	0.0391 (0.001 44)	0.0433 (0.001 21)	0.0478 (0.001 22)	0.0514 (0.000 69)	0.0532 (0.001 55)	0.0623 (0.000 71)	0.0523 (0.001 08)	0.0412 (0.001 77)	0.0312 (0.001 01)	–
50–60 cm	Dry	0.0305 (0.000 65)	0.0360 (0.000 12)	0.0388 (0.001 23)	0.0452 (0.001 01)	0.0411 (0.000 65)	0.0444 (0.000 77)	0.0479 (0.000 55)	0.469 (0.000 69)	0.0469 (0.000 71)	0.0479 (0.000 88)	0.0459 (0.000 51)	0.0385 (0.000 79)	0.0322 (0.000 55)
	Wet	0.0432 (0.0015)	–	0.0442 (0.001 77)	0.0413 (0.001 32)	0.0488 (0.001 11)	0.0484 (0.001 09)	0.0493 (0.000 59)	0.0519 (0.001 43)	0.0518 (0.000 66)	0.0484 (0.001 04)	0.0499 (0.001 71)	0.0395 (0.000 99)	–

[1] Reproduced with permission from table 6a in Pedram *et al* (2018). Copyright 2018 Elsevier.

Table 5.13(b). Means and standard deviations (in brackets) of the soil moisture within different layers for the year 2014[1].

Layer	Days	Annual	Jan	Feb	Mar	Apr	May	Jun	Jul	Aug	Sep	Oct	Nov	Dec
Surface	Dry	0.0131	0.0055	0.0069	0.0086	0.0099	0.0128	0.0142	0.0153	0.0177	0.0166	0.0133	0.0129	0.0107
		(0.0012)	(0.000 98)	(0.001 32)	(0.001 22)	(0.001 15)	(0.001 45)	(0.001 38)	(0.001 28)	(0.001 44)	(0.001 73)	(0.001 68)	(0.001 73)	(0.001 66)
	Wet	0.0598	0.0116	0.0144	0.0182	0.0232	0.0298	0.101	0.121	0.231	0.0544	0.0596	0.0229	0.0218
		(0.0018)	(0.001 97)	(0.001 66)	(0.001 10)	(0.001 97)	(0.001 97)	(0.001 66)	(0.001 56)	(0.001 94)	(0.001 88)	(0.002 30)	(0.001 88)	(0.001 78)
0–5 cm	Dry	0.015	0.0076	0.0081	0.0104	0.0128	0.0155	0.0245	0.0257	0.0182	0.0169	0.0154	0.0134	0.0122
		(0.0012)	(0.001 54)	(0.001 29)	(0.001 19)	(0.001 04)	(0.001 32)	(0.001 32)	(0.001 22)	(0.001 41)	(0.001 61)	(0.001 55)	(0.001 66)	(0.001 61)
	Wet	0.0536	0.0148	0.0166	0.0199	0.0244	0.0223	0.0956	0.104	0.203	0.0507	0.0443	0.0266	0.0210
		(0.001 82)	(0.001 87)	(0.001 59)	(0.002 08)	(0.001 81)	(0.001 92)	(0.001 62)	(0.001 53)	(0.001 88)	(0.001 85)	(0.002 26)	(0.001 82)	(0.001 70)
5–10 cm	Dry	0.0134	0.0105	0.0116	0.0118	0.0133	0.0162	0.0257	0.0268	0.0202	0.0175	0.0167	0.0148	0.0134
		(0.0011)	(0.001 42)	(0.001 22)	(0.000 98)	(0.001 02)	(0.001 24)	(0.001 22)	(0.001 19)	(0.001 39)	(0.001 55)	(0.001 49)	(0.001 61)	(0.001 59)
	Wet	0.0482	0.0143	0.0161	0.0195	0.0232	0.0218	0.0907	0.0998	0.1552	0.0492	0.0434	0.0255	0.0203
		(0.0015)	(0.001 72)	(0.001 32)	(0.001 97)	(0.001 98)	(0.001 88)	(0.001 57)	(0.001 44)	(0.001 82)	(0.001 79)	(0.002 19)	(0.001 79)	(0.001 59)
10–20 cm	Dry	0.0193	0.0163	0.0177	0.0181	0.0188	0.0194	0.0284	0.0284	0.0224	0.0202	0.0184	0.0155	0.0129
		(0.0012)	(0.001 34)	(0.001 19)	(0.000 91)	(0.000 99)	(0.001 18)	(0.001 15)	(0.000 97)	(0.001 12)	(0.001 43)	(0.001 44)	(0.001 55)	(0.001 46)
	Wet	0.0427	0.0205	0.0222	0.0252	0.0277	0.0208	0.0823	0.0898	0.0999	0.0402	0.0412	0.0233	0.0202
		(0.0015)	(0.001 66)	(0.001 25)	(0.001 88)	(0.001 81)	(0.001 72)	(0.001 42)	(0.001 32)	(0.001 75)	(0.001 71)	(0.001 98)	(0.001 65)	(0.001 38)
20–30 cm	Dry	0.0212	0.0173	0.0182	0.0198	0.0198	0.0202	0.0295	0.0296	0.0252	0.0223	0.0217	0.0164	0.0155
		(0.0011)	(0.001 29)	(0.001 06)	(0.000 89)	(0.000 91)	(0.001 07)	(0.001 04)	(0.000 88)	(0.001 02)	(0.001 21)	(0.001 38)	(0.001 43)	(0.001 42)
	Wet	0.0395	0.0203	0.0205	0.0233	0.0266	0.0209	0.0817	0.0743	0.0823	0.0397	0.0393	0.0226	0.0200
		(0.0013)	(0.001 62)	(0.001 12)	(0.001 75)	(0.001 76)	(0.001 66)	(0.001 34)	(0.001 21)	(0.001 62)	(0.001 65)	(0.001 87)	(0.001 54)	(0.001 22)
30–40 cm	Dry	0.024	0.0204	0.0208	0.0219	0.0239	0.0232	0.0301	0.0332	0.0304	0.0244	0.0267	0.0184	0.0161
		(0.0009)	(0.001 12)	(0.000 98)	(0.000 82)	(0.000 87)	(0.000 95)	(0.000 95)	(0.000 76)	(0.001 23)	(0.001 09)	(0.001 24)	(0.001 32)	(0.001 39)
	Wet	0.0558	0.0245	0.0229	0.224	0.0255	0.0254	0.0801	0.0681	0.0788	0.0429	0.0387	0.0211	0.0185
		(0.0012)	(0.001 54)	(0.001 01)	(0.001 66)	(0.001 52)	(0.001 61)	(0.001 21)	(0.001 15)	(0.001 57)	(0.001 55)	(0.001 65)	(0.001 44)	(0.001 05)
40–50 cm	Dry	0.0323	0.0223	0.0294	0.0237	0.0284	0.0292	0.0452	0.0499	0.0453	0.0323	0.0366	0.0245	0.0212
		(0.0006)	(0.000 98)	(0.000 87)	(0.000 76)	(0.000 77)	(0.000 83)	(0.0078)	(0.000 69)	(0.001 15)	(0.000 88)	(0.001 33)	(0.001 29)	(0.001 22)
	Wet	0.0439	0.0387	0.0298	0.0316	0.0381	0.0344	0.0754	0.0632	0.062	0.0402	0.0432	0.0386	0.0275
		(0.0012)	(0.001 34)	(0.000 96)	(0.001 46)	(0.001 34)	(0.001 55)	(0.001 19)	(0.001 04)	(0.001 08)	(0.001 49)	(0.001 44)	(0.001 39)	(0.001 16)

[1] Reproduced with permission from table 6b in Pedram *et al* (2018). Copyright 2018 Elsevier.

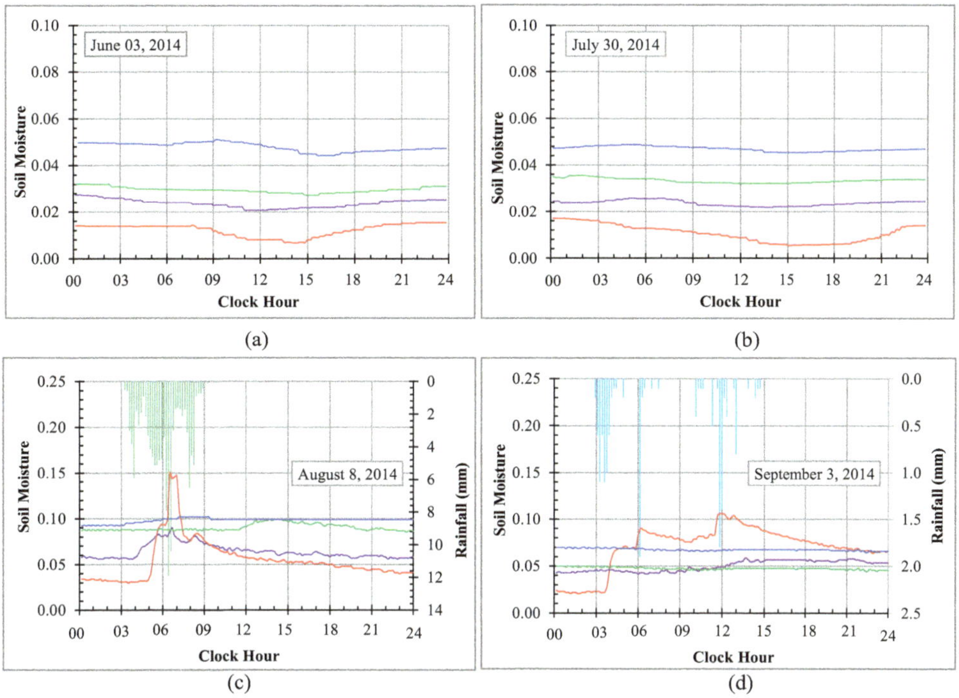

Figure 5.30. The simulated diurnal variations of soil moisture. Illustrations for selected (a) and (b) dry days, and (c) and (d) wet days. Red: soil moisture of surface layer; purple: soil moisture of 5 cm layer; green: soil moisture of 10 cm layer; blue: soil moisture of 20 cm layer; light blue: rainfall. (Reproduced with permission from figures 9 and 10 in Pedram *et al* (2018). Copyright Elsevier 2018.)

pattern. The RH_{pa} within a deeper layer tended to be larger than that within a shallower layer, which is consistent with that for a dry day.

5.5.3.4 The simulated soil evaporation

As expected, for all simulation days, the daily actual evaporation (DAE) was always smaller than the daily potential evapotranspiration (DPET) (figure 5.32(a)). For 75% of the simulation days, the DAE accounted for less than 15% of the DPET, while for the remaining 25% of the simulation days, the DAE accounted for less than 46% of the DPET. For a given day, as indicated by the ratio of DAE to DPET, the available water from rainfall and/or soil water might meet 0%–46% of the atmospheric evaporation demand, with a mean of 14%. For a month in summer (June to August), the atmospheric evaporation demand might be satisfied by 15%–36%, while for a month in other three seasons, the atmospheric evaporation demand could be satisfied by less than 15%. For a month in fall (September to November) and in spring (December to February), the atmospheric evaporation demand was satisfied by 5%–10%. Overall, the daily actual evaporation (MAE) and the monthly potential evapotranspiration (MPET) were highly correlated (figure 5.32(b)), as indicated by the large determination of coefficient $R^2 > 0.95$, and the MAE tended to increase exponentially with the MPET, which indicates that at a monthly scale,

Table 5.14. Means and standard deviations (in brackets) of the pore air relative humidity, in percent, within different layers for the year 2013 or 2014[1].

Layer	Days	Annual	Jan	Feb	Mar	Apr	May	Jun	Jul	Aug	Sep	Oct	Nov	Dec
Surface	Dry	48.88	54.46	56.66	55.34	51.12	35.55	40.77	75.12	73.22	70.13	78.12	71.48	38.66
		(8.66)	(11.61)	(7.76)	(10.45)	(11.21)	(8.69)	(12.34)	(10.11)	(11.65)	(9.99)	(11.12)	(11.43)	(10.44)
	Wet	79.66	78.66	78.89	68.34	55.22	59.99	52.23	75.12	75.44	76.54	81.65	75.12	44.12
		(15.77)	(17.57)	(16.78)	(13.23)	(16.78)	(10.78)	(16.67)	(14.44)	(17.72)	(13.23)	(15.81)	(12.71)	(15.71)
0–5 cm	Dry	65.67	71.32	63.22	69.34	61.47	45.87	55.34	81.23	78.23	71.65	84.34	79.13	44.13
		(7.46)	(4.44)	(6.54)	(5.34)	(6.66)	(5.23)	(4.23)	(5.23)	(4.12)	(5.78)	(6.09)	(6.83)	(5.18)
	Wet	84.34	78.33	89.34	72.22	70.33	51.39	60.12	86.00	81.12	76.44	85.22	81.23	52.27
		(13.65)	(11.22)	(14.33)	(9.34)	(14.44)	(8.76)	(5.67)	(10.12)	(9.23)	(11.66)	(12.66)	(10.24)	(13.24)
5–10 cm	Dry	77.34	79.12	78.88	72.33	66.87	48.67	70.32	83.23	88.23	83.29	85.93	80.43	46.90
		(5.22)	(4.11)	(5.24)	(5.01)	(5.34)	(5.23)	(4.12)	(5.12)	(3.08)	(4.76)	(5.69)	(5.99)	(4.12)
	Wet	88.66	83.32	85.66	78.45	71.12	51.09	75.24	85.33	89.09	86.44	88.44	85.23	56.66
		(12.56)	(10.37)	(11.33)	(8.12)	(9.99)	(6.57)	(5.23)	(8.23)	(6.12)	(8.87)	(9.11)	(9.12)	(10.91)
10–20 cm	Dry	79.34	80.33	80.11	85.44	78.89	52.76	80.13	84.44	89.87	85.65	88.12	82.12	48.71
		(4.33)	(4.01)	(5.09)	(4.65)	(4.44)	(4.23)	(3.43)	(4.23)	(3.04)	(4.34)	(5.12)	(4.81)	(3.12)
	Wet	89.45	88.32	86.32	88.91	81.66	60.12	82.22	87.43	90.12	88.43	86.66	84.28	55.77
		(11.12)	(9.77)	(8.49)	(7.23)	(8.78)	(8.33)	(4.23)	(7.32)	(5.19)	(9.98)	(8.18)	(8.35)	(8.13)
20–30 cm	Dry	80.12	82.22	82.21	88.23	81.21	50.21	85.22	86.34	90.22	87.65	89.32	85.22	55.11
		(4.12)	(3.32)	(4.44)	(4.12)	(3.33)	(3.33)	(2.32)	(4.12)	(2.23)	(3.59)	(4.11)	(3.22)	(3.11)
	Wet	90.13	87.99	87.19	90.12	84.56	61.11	87.21	88.34	91.23	91.79	90.77	88.99	60.12
		(9.55)	(6.76)	(7.45)	(7.09)	(7.99)	(6.23)	(3.39)	(6.21)	(4.44)	(6.98)	(6.18)	(7.08)	(7.06)
30–40 cm	Dry	82.32	88.55	83.33	85.55	85.34	52.11	89.76	87.12	91.11	89.48	90.41	88.71	58.38
		(3.22)	(3.08)	(3.34)	(4.32)	(3.09)	(2.22)	(1.21)	(3.21)	(2.09)	(3.21)	(4.09)	(3.61)	(2.69)
	Wet	89.89	89.99	89.76	91.13	86.77	68.12	90.12	88.21	91.32	93.23	91.77	93.09	65.23
		(8.67)	(5.23)	(6.32)	(6.87)	(6.54)	(5.12)	(3.32)	(5.21)	(4.12)	(5.71)	(5.23)	(6.92)	(8.34)
40–50 cm	Dry	85.12	90.44	85.55	90.09	86.66	55.65	92.12	89.11	92.22	89.99	92.32	90.32	59.12
		(3.12)	(2.24)	(3.12)	(2.89)	(2.12)	(2.34)	(1.09)	(3.11)	(1.98)	(3.13)	(3.18)	(3.14)	(2.13)
	Wet	90.08	93.33	90.22	91.12	92.22	67.32	93.30	90.76	93.21	92.43	93.55	92.56	61.99
		(7.99)	(4.45)	(5.12)	(5.34)	(5.34)	(4.12)	(2.12)	(4.22)	(3.21)	(4.77)	(4.78)	(5.49)	(5.12)
50–60 cm	Dry	86.32	91.55	91.21	92.12	91.21	59.11	94.44	90.59	93.87	90.11	95.21	92.45	61.77
		(2.11)	(2.32)	(2.23)	(1.09)	(1.78)	(1.99)	(1.02)	(2.25)	(1.42)	(2.12)	(2.21)	(2.56)	(2.09)
	Wet	90.33	93.37	94.11	92.89	94.34	69.32	95.22	92.24	94.22	91.12	98.45	93.13	65.33
		(6.67)	(4.12)	(5.09)	(4.44)	(4.32)	(3.33)	(2.09)	(4.12)	(2.73)	(4.22)	(4.45)	(4.34)	(4.99)

[1] Reproduced with permission from table 5a and 5b in Pedram *et al* (2018). Copyright 2018 Elsevier.

the actual evaporation might be primarily controlled by atmospheric evaporation demand rather than precipitation. However, the satisfied proportion of the annual atmospheric evaporation demand was 17% in 2013 and 19% in 2014, which were almost same and independent of climate conditions and precipitation amounts. The annual precipitation (P) in 2014 was more than that in 2013 by $\Delta P = 175.7$ mm (figure 5.28), while the annual PET and actual evaporation (AE) in 2014 were higher than those in 2013 by $\Delta PET = 149.5$ mm and $\Delta AE = 62.6$ mm, respectively. The ratio of the ΔPET to ΔP was much larger than that of ΔAE to ΔP (0.85 versus 0.35), indicating a nonlinear increase of AE with increase of PET.

In both study years, the precipitation resulted in net replenishment of soil water and net recharge into the shallow aquifer (table 5.15). In 2013,the replenishment and recharge were 71.92 and 54.96 mm, respectively, while in 2014 the replenishment and recharge were 141.64 and 98.34 mm, respectively. This is because the precipitation in 2014 was larger than that in 2013. However, for a given year, some months could have a net loss of soil water, as indicated by the negative values of ΔS_m, while other months could have a net replenishment of soil water, as indicated by the positive values of ΔS_m. Similarly, for a given year, some months could have a net recharge of the shallow aquifer, as indicated by the positive values of I_{bm}, while other months could have a net capillary rise of groundwater, as indicated by the negative values of I_{bm}. In addition, for a given year and a given month, some soil layers could have a net loss of soil water, as indicated by the negative change of soil water, while other soil layers could have a net replenishment of soil water, as indicated by the positive change of soil water. As expected, the temporal variation of soil water in an upper layer was generally larger than that in a lower layer.

In both study years, regardless of the month, the liquid water in the soils above the 15 cm depth moved downward because of a downward thermal gradient, whereas the liquid water in the soils below the 15 cm depth moved upward because of an upward thermal gradient. Similarly, the pore vapor in the soils above the 8 cm depth moved downward because of a downward thermal gradient, whereas the pore vapor in the soils below an 8 cm depth moved upward because of an upward thermal gradient. For downward movement, the most active zone of liquid water was around the 5 cm depth, while the most active zone of pore vapor was at the ground surface; for upward movement, on the other hand, the most active zone of liquid water was around the 22 cm depth, while the most active zone of pore water was around the 10 cm depth, which indicates that the evaporation zone was primarily located at the 5 cm depth, at which the liquid water was vaporized. Subsequently, the vapor might be transferred either upward into the ambient atmosphere or downward into the soils, depending on the direction of the thermal gradient. Also, the vapor could be trapped in the soils between the zero and 8 cm depths. Below the 5 cm depth, most of the liquid water was moved upward into the evaporation zone, while only a little of the liquid water was vaporized and transferred upward in the form of vapor. For a given soil depth, the liquid and vapor fluxes in the spring and summer months (March to August) tended to be larger than those in the autumn and winter months (September to February).

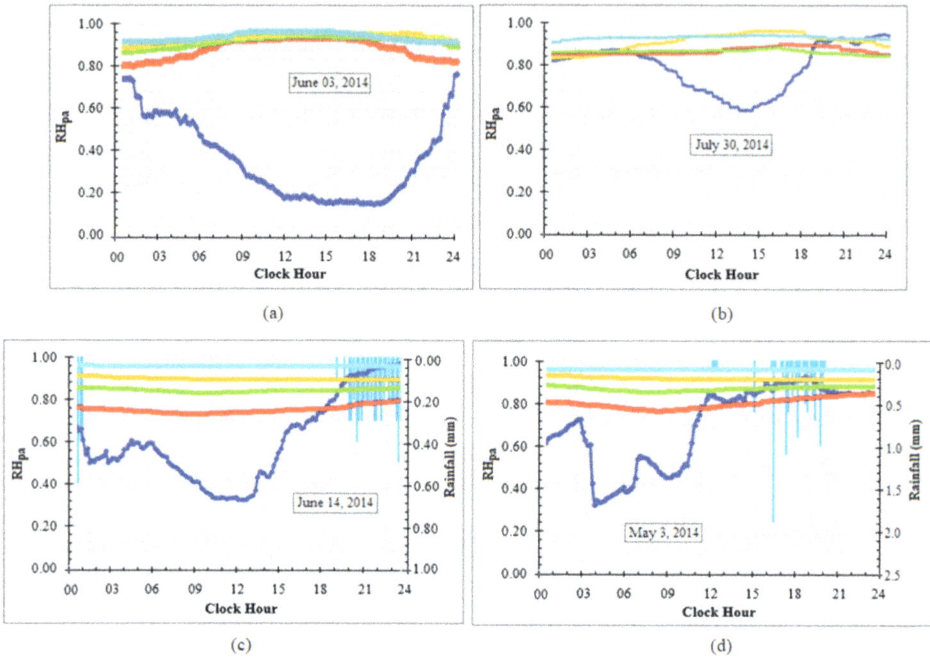

Figure 5.31. The simulated diurnal variations of pore air relative humidity (RH_{pa}). Illustrations for selected (a) and (b) dry days, and (c) and (d) wet days. Dark blue: RH_{pa} of surface layer; red: RH_{pa} of 5 cm layer; green: RH_{pa} of 10 cm layer; orange: RH_{pa} of 20 cm layer; aqua: RH_{pa} of 30 cm layer; light blue: rainfall. (Adopted with permission from figures 7 and 8 in Pedram *et al* (2018). Copyright 2018 Elsevier.)

The pore air relative humidity (RH_{pa}) was maximal around the 5 cm depth, at which the primary evaporation zone was located, but it was minimal at the soil surface, at which the downward flux of vapor was largest. Below 20 cm, the RH_{pa} did not vary with soil depth for a given month. In contrast, the soil water contents in the summer months (June to August) had a primary peak at the 30 cm depth and a secondary peak at the 10 cm depth, whereas the soil moisture in the other months might either gradually increase with soil depth or had a peak at the 30 cm depth. As expected, for a given soil depth, the RH_{pa} and soil moisture in a wetter month tended to be larger than those in a drier month, while their relative magnitudes between such two months could be switched because of the upward and downward movements of vapor and liquid water. In the soils between the zero and 10 cm depth, the RH_{pa} and soil moisture in the winter months tended to be larger than those in other months, which indicates that the vaporization process and liquid and vapor movement might become less active in winter, when the air and shallow-soil temperatures were below zero.

Averaged across the two study years, the thermal liquid flux movement can be classified into three stages. The first stage was from 0:00 to 6:00 am, during which the liquid flux was upward within the zero to 20 cm deep soils while it became downward within the soils below the 20 cm depth. The second stage was from 7:00 am to 6:00 pm, during which the thermal liquid flux was downward within the

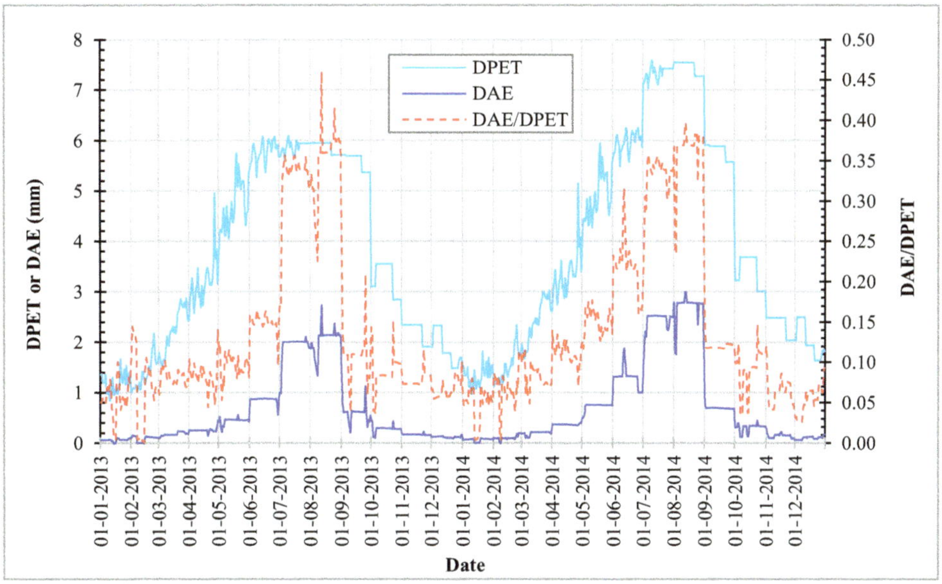

(a)

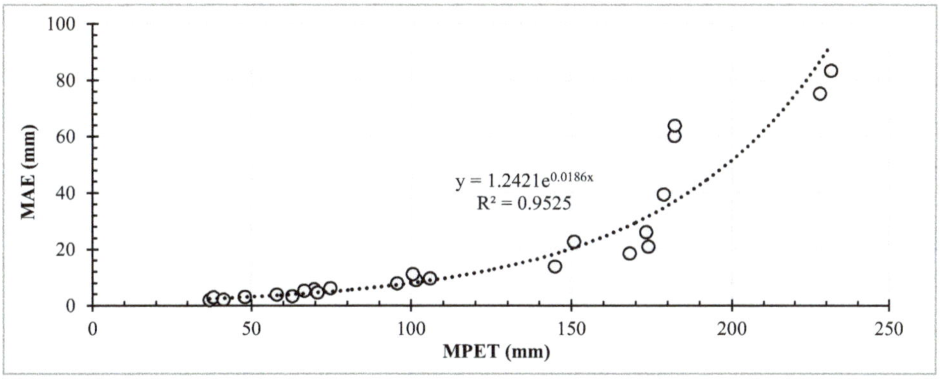

(b)

Figure 5.32. The simulated actual evaporation (AE) and potential evapotranspiration (PET). Plots for (a) daily time interval and (b) monthly time scale. DAE: daily AE; DPET: daily PET; MAE: monthly AE; and MPET: monthly PET. (Reproduced with permission from figure 4-1 in Pedram (2017). Copyright 2017 Old Dominion University.)

zero to 10 cm deep soils while it was upward within the soils below the 10 cm depth. The third stage was from 6:00 pm to 0:00 am, during which the liquid flux was upward within the zero to 20 cm deep soils while it tended to be downward within the soils below the 20 cm depth. The liquid flux provided water for vaporization at different depths, depending on the time of interest, and subsequent evaporation at the soil surface. Similarly, the thermal vapor flux movement can also be classified into three stages. The first stage was from 0:00 to 9:00 am, during which the vapor flux tended to be upward regardless of the soil depth. The second stage was from

Table 5.15. The monthly water balance components in equation (5.3)[1].

		P_m	MPET	MAE	Soil water change (mm) of cm depth						Component (mm)		
Year	Month	(mm)	(mm)	(mm)	0–5	5–10	10–20	20–30	30–40	40–50	ΔS_m	$P_m - MAE$	I_{bm}
2013	Jan	0.0	37.07	1.85	−0.44	−0.31	−0.35	−0.02	−0.01	−0.01	−1.14	−1.85	−0.72
	Feb	1.1	38.26	2.79	−0.46	−0.65	−0.03	−0.05	−0.21	−0.19	−1.59	−1.69	−0.10
	Mar	7.1	69.63	5.64	0.36	0.39	0.42	0.05	0.03	0.02	1.27	1.46	0.19
	Apr	6.2	95.64	7.75	−0.26	−0.61	−0.09	−0.09	−0.04	−0.03	−1.13	−1.55	−0.42
	May	19.7	144.88	13.76	1.30	1.04	1.32	0.81	0.62	0.34	5.44	5.94	0.50
	Jun	22.3	173.35	26.00	−0.93	−0.88	−0.53	−0.41	−0.31	−0.46	−3.52	−3.70	−0.18
	Jul	123.4	182.14	60.11	9.14	7.92	7.33	5.77	4.43	3.66	38.25	63.29	25.04
	Aug	134.2	182.24	63.78	10.11	8.83	6.43	4.11	4.09	3.98	37.55	70.42	32.87
	Sep	17.6	168.17	18.50	−0.33	−0.11	−0.12	−0.12	−0.12	−0.10	−0.89	−0.90	−0.01
	Oct	12.2	101.61	8.94	0.95	0.88	0.67	0.55	0.46	0.34	3.85	3.26	−0.59
	Nov	1.1	66.56	5.13	−1.19	−1.26	−0.34	−0.21	−0.31	−0.27	−3.58	−4.03	−0.45
	Dec	0.0	57.91	3.76	−1.04	−1.20	−0.23	−0.04	−0.04	−0.03	−2.58	−3.76	−1.18
	Annual	344.9	1317.46	218.02	17.22	14.04	14.48	10.34	8.59	7.26	71.92	126.88	54.96
2014	Jan	3.3	41.37	2.03	0.50	0.49	0.05	0.04	0.03	0.02	1.12	1.27	0.15
	Feb	5.8	47.97	2.93	1.10	1.15	0.22	0.18	0.11	0.10	2.85	2.87	0.02
	Mar	15.1	74.70	6.05	1.23	1.76	1.33	1.01	1.00	0.07	6.41	9.05	2.64
	Apr	24.0	100.62	11.07	2.34	2.01	2.88	1.09	1.01	0.95	10.28	12.93	2.65
	May	33.7	150.85	22.63	2.06	2.16	1.43	1.88	1.13	0.82	9.48	11.07	1.59
	Jun	51.6	178.80	39.34	2.61	3.30	2.58	1.32	1.22	0.99	12.02	12.26	0.25
	Jul	147.7	227.81	75.18	10.44	8.44	6.17	5.56	3.21	2.22	36.04	72.52	36.48
	Aug	178.5	231.29	83.26	15.89	10.48	8.32	6.99	2.90	2.11	46.69	95.24	48.54
	Sep	22.9	174.03	20.88	0.90	0.76	0.77	0.32	0.12	0.09	2.96	2.02	−0.95
	Oct	32.4	105.97	9.54	4.57	3.33	2.21	2.13	1.31	1.95	15.51	22.86	7.35
	Nov	3.5	70.66	4.52	−0.18	−0.11	−0.32	−0.12	−0.10	−0.08	−0.92	−1.02	−0.10
	Dec	2.1	62.81	3.20	−0.10	−0.11	−0.15	−0.22	−0.12	−0.11	−0.81	−1.10	−0.29
	Annual	520.6	1466.88	280.62	41.34	33.66	25.49	20.17	11.83	9.15	141.64	239.98	98.34

[1] Adapted with permission from table 4-1 in Pedram (2017). Copyright 2017 Elsevier. Pm: monthly precipitation; MPET: monthly potential evapotranspiration; MAE: monthly actual evaporation; ΔSm: monthly soil water change; and Ibm: monthly bottom seepage.

10:00 am to 3:00 pm, during which the vapor flux was downward within the zero to 10 cm deep soils while it became upward within the soils below the 10 cm depth. The third stage was from 4:00 pm to 0:00 am, during which the vapor flux tended to be upward regardless of the soil depth. The upward vapor flux at the soil surface was the actual evaporation rate, while the downward vapor flux at the soil surface and the downward and upward vapor fluxes below the soil surface contributed to RH_{pa}, which indicates that the study site might have a downward thermal gradient (i.e. air temperature higher than soil temperature) between late morning and early afternoon, which tended to slow down the movement of vapor from the soils into the ambient atmosphere. In early morning and late afternoon and at night, the study site probably had an upward thermal gradient (i.e. air temperature lower than soil temperature), which tended to prompt the movement of vapor from the soils into the ambient atmosphere.

For a given non-rainy day, the peak evaporation rate occurred around 9:30 am in January to May and September to November (e.g. figure 5.33(a)), while it occurred around 1:00 pm in June to August and December (e.g. figure 5.33(b)). The evaporation primarily caused a decrease of soil moisture in layer 1 (i.e. the top 5 cm soils), indicating that this soil layer was the major evaporation zone before the peak time. After the peak time, the soil moisture in layer 1 kept on decreasing until it reached a minimum in late afternoon (i.e. around 2:00–3:00 pm), after which the soil moisture started to increase. This indicates that layer 1 might function as an evaporation zone before 9:30 am or 1:00 pm, depending on the month of interest, while it could function as a condensation and storage zone after 3:00 pm, regardless of the month of interest. In late afternoon, the thermal gradient might become upward positive, causing upward liquid or vapor fluxes from the deeper layers to replenish soil water in layer 1. The upward vapor from the deeper soil layers might roughly compensate the evaporation at a given time, while some of the upward vapor could be condensed into liquid water. The condensed liquid water and the upward liquid water from the deeper layers might be stored in layer 1, causing the increase of soil moisture in this layer. Such a process of evaporation, thermal upward fluxing, condensation, and storing might be the major characteristics of the hydrological cycle at the study site. The negligible fluctuation of soil moisture in layer 2 (i.e. the 5–10 cm soils) and layer 3 (i.e., the 10–15 cm soils) indicates that the upward thermal liquid flux might be almost same as the vaporization rate in these two layers. In late afternoon, the vaporization probably occurred around the 10 cm depth, below which the vapor flux was very small. For a given time, the evaporation rate was much higher on a day in June to September than in the other months.

For a given rainy day, the evaporation rate started to decrease from the inception of rain onward, reached a minimum value at the end of the rain, and then started to increase gradually (e.g. figure 5.33(c)). This indicates that during a storm event, the PET might be greatly lowered, which in turn limited the actual evaporation although relatively more water was available for evaporation. Because the soils at the study site are highly permeable (K_s = 650 cm d^{-1} in table 5.11), the rainwater could quickly infiltrate into layers 1 and 2, as indicated by the synchronic increase of soil moisture in these two layers as it was raining. However, the soil moisture in layer 3

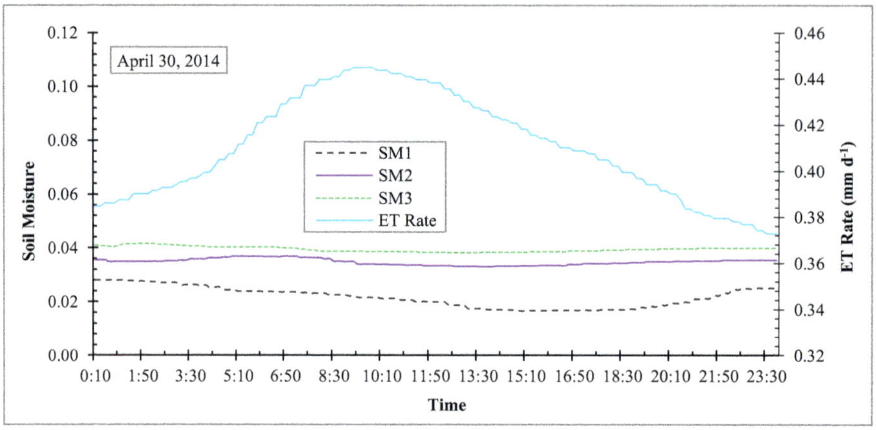

(a)

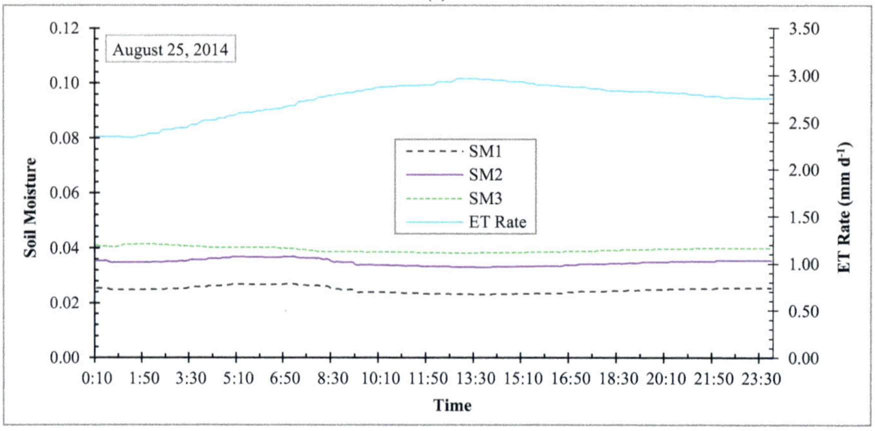

(b)

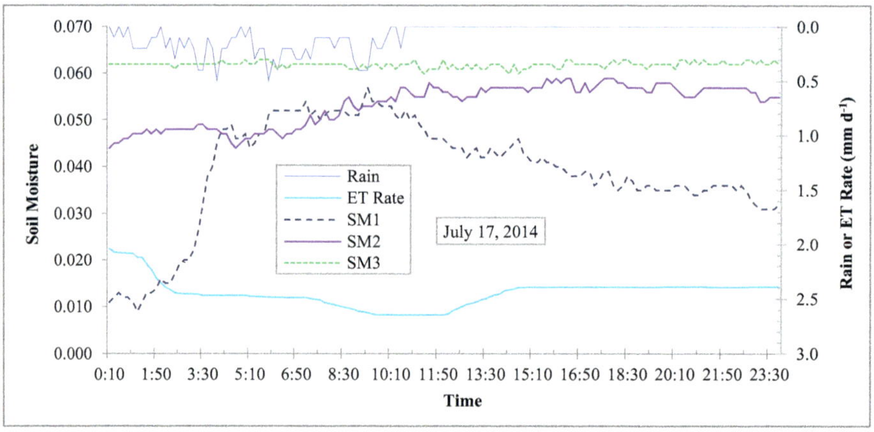

(c)

Figure 5.33. The simulated soil evaporation (ET) rate and soil moisture. Results for selected (a) and (b) dry days, and (c) wet days. SM1: soil moisture in the 0–5 cm layer; SM2: soil moisture in the 5–10 cm layer; and SM3: soil moisture in the 10–20 cm layer. (Reproduced with permission from figures 11 and 12 in Pedram *et al* (2018). Copyright 2018 Elsevier.)

was almost constant and seemed not to be influenced by rain. That is, for most of the rainy days, the rainfall might more often replenish the top 10 cm soils most of the time, and less often wet the deeper soils and recharge the shallow groundwater, which can be further verified by the negative or small positive changes of soil moisture in the deeper layers as well as the negative or small positive values of deep percolation. The diurnal variations of evaporation rate and soil moisture in a rainy day were mainly controlled by the rain pattern (i.e. timing, amount, and duration), which is distinctly different from those in a non-rainy day.

5.5.3.5 The DSL retardation effect

As shown in figure 5.28, the number of rainy days (i.e. precipitation events) at the study site was very small. In 2013 the 47 rainy days had a total precipitation of 344.9 mm, while in 2014 the 52 rainy days had a total precipitation of 520.6 mm. Regardless of the year, most of the events generated a very small (<5 mm) amount of precipitation (table 5.16), while a few of the events had a precipitation of larger than 20 mm. These few precipitation events might be able to replenish the soils below the 10 cm depth and contribute to the percolation in July and August as indicated by the relatively large (25–49 mm) positive values of I_{bm} (table 5.15). The heaviest rainfall occurred on 9 July 2014 with a total rainfall of 55.6 mm, an average intensity of 1.22 mm h^{-1}, and a maximum 1 h intensity of 10.53 mm h^{-1}, whereas the lightest rainfall occurred on 16 January 2014 with a total rainfall of only 0.2 mm and a minimum 1 h intensity of 0.1 mm h^{-1}. Given that the rainfall intensity was much smaller than the soils' saturated hydraulic conductivity of K_s = 650 cm d^{-1} = 271 mm h^{-1}, no runoff was generated. Most of the precipitation events could slightly increase the soil moisture of layers 1 and 2 (i.e. the zero to 20 cm deep soils) (figure 5.29), which might be evaporate quickly once the event ended.

As expected, the soil moisture increased with precipitation, particularly following a heavy rain event and in the top two layers. The increase of soil moisture was dependent on antecedent wetness condition, the number of non-rainy days between two events, and rainfall intensity. The increase of soil moisture in the top two layers was relatively larger when the rainfall was greater than 20 mm and/or when two or

Table 5.16. Summary statistics of the precipitation events in the simulation years[1].

	2013		2014	
Precipitation range (mm)	Rainy days	Precipitation (mm)	Rainy days	Precipitation (mm)
<5	34	141.6	28	59.5
5–10	5	35.0	9	60.0
10–20	4	47.1	5	65.6
20–30	3	67.1	4	91.7
30–40	0	0.0	4	133.3
40–50	0	0.0	0	0.0
>50	1	52.3	2	110.6

[1] Adapted with permission from table 4-4 in Pedram (2017). Copyright 2017 Elsevier.

more rainfall events occurred with a few non-rainy days between them. On the other hand, the small rainfall events might only be able to wet the upper soil layers when a large portion of the soil moisture was lost to evaporation. Thus, the soil water in the deeper soils was mostly replenished by the few large precipitation events.

Correspondingly, the thickness (δ) and vertical position of DSL were dependent on the antecedent soil moisture condition, the number of non-rainy days between two events, and rainfall intensity (figure 5.34). When the number of non-rainy days was larger, a top DSL might be formed and its thickness was also larger, whereas when an event had a total rainfall of greater than 20 mm, the DSL was eliminated, leading to a near-zero value for δ. For the small events with a total rainfall of less than 20 mm, at an hourly or a smaller time scale, a top wet soil layer (WSL) might be formed, and a DSL was sandwiched by the WSL and deeper moist soils; at a daily scale, however, the top WSL disappeared and thus a top DSL was formed probably because the atmospheric evaporation demand (figure 5.32(a)) was much larger than the available water for evaporation. The thickness of the DSL varied from zero to 30 cm, with a mean of 17 cm and a median of 20 cm. Because of the DSL, vaporization primarily occurred in the 10–20 cm deep soils. Using the mean values of $a_3 = 17.874$ and $a_4 = 21.894$ in equation (5.5), the retardation effect of DSL on soil water evaporation was estimated to be $RE_{vfr} = 28.1\%–83.3\%$, with a mean of 77.7% and a median of 78.5%. Based on equation (5.4), for the study site and if the DSL did not exist, the evaporation in 2013 could be 978 mm (= (218.02 mm) ÷ (1 − 77.7%)), which is 760 mm (≈978 mm − 218.02 mm) more, while the evaporation in 2014 could

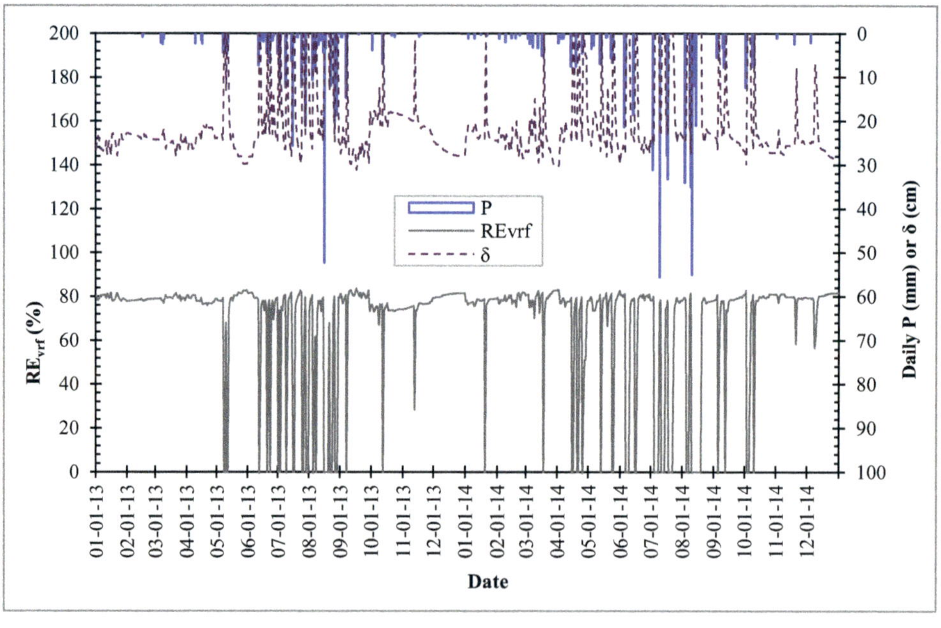

Figure 5.34. The simulated thickness (δ) and retardation effect (RE_{vrf}) of the dry soil layer (DSL). The daily precipitation (P) is superimposed to visualize the influences of P on δ and RE_{vrf}.

be 1258 mm (= (280.62 mm) ÷ (1 − 77.7%)), which is 977 mm (≈1258 mm − 280.62 mm) more. It can be concluded that the DSL was beneficial for the fragile ecosystem because it retarded the moist soils beneath from being dried out further. For the hypothetical situation that all precipitation was evaporated and the entire vertical soil profile saturated, without the retardation effect of the DSL, the depth of dry soils would reach about 120 cm or deeper.

5.5.4 Discussion

Generally, water vapor flow plays a critical role in the assessment of the physical processes related to soil heating, distribution of water, and water vapor exchange between the soil and its ambient atmosphere. In a desert environment, because the evaporation demand is much larger than the amount of available water, soil water vaporization often occurs at some depth (>10 cm in this study) below the ground surface and thus the pore vapor needs to be transported by an upward positive heat gradient through the upper dry soils into the ambient atmosphere (Wang 2015). Such a soil water evaporation process has yet to be fully understood and is barely described by existing evaporation models, including the most widely used Penman–Monteith model (Monteith 1965). This knowledge gap can limit the human capability to sustain fragile desert ecosystems and can likely bias the prediction of global climate change because such ecosystems account for more than 20% of the Earth's terrestrial surface and are distributed across the continents. This study aims to understand the dynamics of soil water and pore vapor using the measured data and the HYDRUS-1D model at a site located within the typical desert environment of HSL.

This study used the latest HYDRUS-1D code to produce the diurnal profiles of pore relative humidity (RH_{pa}) and soil moisture (SM). The results show that the fluctuation of soil temperature tended to decrease with depth and that the surface soil temperature had a maximum diurnal variation. At the depth of 60 cm, the variation of soil temperature was less than 2 °C. The soil temperature gradient, defined as the derivative of soil temperature with respect to vertical distance, was steepest at a depth of around 10 cm, at which the thermal vapor flux was maximal, the heat exchange was most active, and the primary vaporization occurred (Wang 2015). In addition, for dry days, the relative magnitude of atmosphere temperature (T_a) versus soil temperature determines the diurnal variations of RH_{pa} and SM in the top 10 cm soils, whereas for rainy days, the rainfall pattern controls the diurnal variations. For the soils deeper than 10 cm, the RH_{pa} and SM have a negligible diurnal variation on both non-rainy and rainy days, indicating that the upper boundary conditions mainly affect the dynamics of soil water and vapor within the top 10 cm layer. For a given non-rainy day, the RH_{pa} and SM in the top 5 cm soils tended to decrease with increase of T_a, become minimal around 2:00 pm when T_a was probably warmest, and then increase with decrease of T_a. This is consistent with the finding of Goss and Madliger (2007). The possible reason is that the top 5 cm soils are mostly dry and have a temperature either close to or higher than T_a at any time within the day. As a result, the soil water and pore vapor in the top 5 cm soils can be more quickly moved up into the ambient atmosphere around 2:00 pm, when the solar radiation is maximal, than at other times, when the solar radiation becomes relatively smaller.

In contrast, in the 5–10 cm soils, the RH_{pa} reached a maximum around 2:00 pm and then reduced to a smaller value, whereas the SM was almost invariant. One possible reason is that around 2:00 pm the temperature of the 5 cm soils was higher than that of the 5–10 cm soils, resulting in a downward positive temperature gradient, retarding the upward movement of vapor. Another possible reason is that around 2:00 pm, more solar heat was transferred down into the soils, vaporizing more soil water. At other times, because T_a was lower than the temperature of the 5 and 10 cm soils, the downward positive temperature gradient might be reversed to an upward positive gradient and thus the cumulated vapor could move up through the 5 cm soils into the ambient atmosphere, causing the decrease of RH_{pa}. The study results indicate that on dry days, the RH_{pa} was minimal at midnight, when the total heat flux out of the soils was greatest, while it reached a maximum around 2:00 pm, when the total heat flux into the soils was largest. This is a typical feature of phase III evaporation, which has a very low but near-constant evaporation rate, as discussed in section 4.4.4. Note that for phase III, some vapor near the soil surface could be condensed into liquid water, replenishing the top 5 cm soils (Wang 2015) if T_a is lower than the dew point temperature and/or the actual vapor pressure is larger than the saturated vapor pressure (Viessman and Lewis 2003).

For a rainy day the RH_{pa} and SM in the top 5 cm soils reached a peak during the rainfall period. As a result of the rain, the soil evaporation could switch from phase III either to phase II or phase I (Wang 2015). At these two latter phases, vaporization mainly occurs near the soil surface. However, with the progress of vaporization a DSL may be formed again, switching soil evaporation back to phase III. In comparison with that for a non-rainy day, the RH_{pa} in the 5 and 10 cm soils had a much smaller variation. One possible reason is that on a rainy day, T_a was cooler, leading to a smaller downward temperature gradient and thus less retardation of the upward movement of vapor. Another possible reason is that less heat might be transferred into the soils to vaporize the soil water. Moreover, the higher atmospheric humidity could increase the chance for vapor in the ambient atmosphere to be condensed to replenish the soil water of the top 5 cm soils and also retard the movement of pore vapor up into the ambient atmosphere.

At the annual time scale, for non-rainy days, the average RH_{pa} in the surface soil layer was slightly higher in 2014 (48.88%) than 2013 (45.57%); in a deeper layer, however, the RH_{pa} was the same. For rainy days, regardless of the soil layer, the average RH_{pa} was the same for both years. This indicates that the increased precipitation in 2014 had a minimal influence on the pore vapor within the soils. The higher RH_{pa} in the surface soil layer for the non-rainy days probably resulted from the condensation of vapor from the ambient atmosphere (McHugh *et al* 2015) because the year 2014 was more humid and thus had a larger air relative humidity than the year 2013. In contrast, at the monthly time scale, for the months of July to November, regardless of non-rainy or rainy days, the increased precipitation in 2014 did result in a higher RH_{pa} within any soil layer. The total precipitation of these five months in 2014 was 96.5 mm more than that in 2013, most of which might be converted into pore vapor. For the other months, the increased precipitation in 2014 caused the RH_{pa} to increase in a month but decrease in another month, regardless of the non-rainy or rainy days. One possible reason is that the latent heat for vaporization was less in 2014 because of more rainy days. Another possible reason is that the cooler air temperature could prompt the emission of pore vapor into the ambient atmosphere, leading to the decrease of RH_{pa} within the soils. On the other hand, as expected and overall, the SM was higher in 2014 than 2013 regardless of the time scales, soil layers, and non-rainy or rainy days. This indicates that more precipitation is always beneficial to infiltration and replenishment of soil moisture. For a given month and a soil layer, the SM increase from 2013 to 2014 was mainly dependent on the increase of precipitation intensity rather than that of precipitation amount. Herein, precipitation intensity is defined as the ratio of total precipitation to the number of rainy days. For instance, while the increase of monthly precipitation in August was larger than that in July (44.3 versus 24.3 mm), the SM increase in August was smaller. The increase of precipitation intensity in August was almost half smaller than that in July (13.3 versus 26.2 mm d^{-1}). The reason is that because of the very high evaporation rate at the study site, a larger proportion of the precipitation with a smaller intensity might be evaporated before reaching the ground, leading to less water being available for infiltration into soils to replenish moisture.

The study results are consistent with those reported by others (Kobayashi *et al* 1989, Goss and Madliger 2007, Han and Zhou 2013). Using Rotronic Hygroclip (RH) SC04 probes, Goss and Madliger (2007) measured T_a and RH_{pa} in the pore air of dry sandy soils at a Tanzanian site (6°51′ S, 37°38′ E) for an observation period of consecutive 30 days. The measured data indicate that for a typical day, during phases I and II, RH_{pa} increases with increase of T_a (reaching a maximum value around noon) regardless of soil depth, during phase III, however, RH_{pa} decreases with increase of T_a for the top DSL (reaching a minimum value around noon) but increases with increase of T_a for the evaporation zone (reaching a maximum value around noon). The data also indicate that continuous evaporation tends to move the evaporation zone down into a deeper soil layer (i.e. increase the thickness of top DSL) but sporadic rain can bring the evaporation zone up back to a shallower soil layer (i.e. reduce the thickness of top DSL). The simulated values of SM are comparable with those in other desert lands (e.g. the Namib Desert and Negev Desert). In an observation study conducted in a 30 km^2 dune field located within the Negev Desert, where the average annual precipitation ranges from 70 to 150 mm, Zaady *et al* (2014) found that the soil moisture varied from 0.025 to 0.08, depending on the soil infiltration rate and rainfall intensity. Using observed data at three sites within the Namib Desert, where the average annual precipitation ranges from 5 to 100 mm, Li *et al* (2016b) developed and used a stochastic model to examine the effect of rainfall on soil moisture. They found that the soil moisture varied from 0.0045 to 0.037 and was mainly controlled by the rainfall pattern as well as fog and dew. Although the precipitation at the HSL site is slightly larger than that in the Namib Desert and Negev Desert, the soil moisture levels at all three locations are not incomparable. One possible explanation is that the repellent impact of DSL can prevent the moist soils beneath from being dried up further (Wang 2015). Another possible explanation is that vapor in ambient atmosphere can be condensed into liquid water to replenish the moisture of the (in particular top 5 cm) soils (Zaady *et al* 2014, McHugh *et al* 2015).

This study focused on the temporal variations of SM and RH_{pa} at various depths of interest and examined the vertical thermal fluxes of vapor. The results have practical implications for restoring and/or protecting the fragile ecosystems of desert lands across the world, which are characterized by high solar radiation, prevalent and high-speed wind, low atmospheric humidity, and sporadic and low-magnitude precipitation. In a desert environment, because the climatic evaporation demand is usually much greater than the available surface water, soil water is evaporated to meet the evaporation demand deficit. However, the evaporation process of soil water can be effectively retarded by the top DSL, which has an albedo of 0.4 or larger and can reflect much solar radiation back into the atmosphere, reducing the heat to be transferred into the moist soils beneath. As a result, the moist soils can be protected by the top DSL from being further dried up. Hitherto, because of the incomplete understanding of the retardation role of such a DSL, trees have been planted as a malpractice to prevent dust winds and improve desert ecosystems (Wang *et al* 2016b). The roots of trees can penetrate through the top DSL into the moist soils beneath to transpire a large amount of water, drying up the moist soils. Thus by growing trees, the roots will penetrate and dry up deeper moist soils, greatly increasing the thickness of the dry soils. At some point, because the moist soils would be buried by dry soils to a depth that is too deep to be reached by the roots, the trees will start to die off, leading to a worse eco-environment. In terms of the results of this study, it is suggested that native sand-fixing plants (e.g. *Salsolar tragus*, *Salix gordejevii*, and *Caragana microphylla*) be planted to restore and protect desert ecosystems. Such plants have a root depth of around 10 cm, do not penetrate into the deep moist soils, do not transpire much soil water (<2.8 mm d^{-1}), but can stabilize the top DSL. They have needle-shaped leaves with a very small leaf area index and can close their leaf stomata when water is insufficient for transpiration. Their roots can capture, store, and utilize soil water from sources of infiltration, capillary rise, and vapor condensation from ambient air (McHugh *et al* 2015). The vegetation species of *Salsolar tragus*, *Salix gordejevii*, and *Caragana microphylla* are native to and currently exist in the HSL (Duan *et al* 2015).

At the HSL site, the actual evaporation (AE) accounted for less than 15% of the potential evapotranspiration (PET). One reason is that the small (<520 mm) annual precipitation led to limited amount of water available for evaporation. For a given year, there were fewer than 60 rainy days, most of which had a daily precipitation of less than 20 mm. Such a small amount of precipitation might just temporarily increase the soil moisture of the zero to 10 cm deep soils (Yao *et al* 2013). Once the precipitation event ends, the replenished soil water could be evaporated quickly because of the very high PET (i.e. atmospheric evaporation demand). In addition, although the few (<15) days with a daily precipitation larger than 20 mm could have a soil moisture larger than the wilting point $\theta_r = 0.02$ in the entire vertical soil profile, the soil moisture in the zero to 10 cm deep soils could decrease to less than θ_r within an hour or two if a non-rainy day followed the rainy day. Thus, a top dry soil layer (DSL) could be formed very quickly and exist on most days (Han and Zhou 2013).

Another reason is that the top DSL can retard the soil water beneath from being further vaporized to satisfy the PET (Wang 2015). For a given soil, its thermal conductivity decreases with a decrease in soil moisture (Andersland and Anderson 1978, Farouki 1981, Tarnawski and Leong 2000, Ghuman and Jalota 2006). The DSL had a soil moisture of less than 0.02, at which the thermal conductivity became very small (<0.1 W m^{-1} K^{-1}), leading to much less solar radiation heat being transferred into the moist soils beneath. As a result, the heat available for soil water vaporization was reduced and thus less soil water might be evaporated. In addition, the soil albedo, defined as the fraction of the incident radiation that is reflected from the surface (Dobos 2003), increases exponentially as soil moisture decreases (Sugathan *et al* 2014). Because the DSL had a very low soil moisture and a very large albedo (>0.4), most of the incident radiation might be reflected back to the atmosphere, leaving much less radiation on the surface to be transferred downward into the soils. Further, while the thermal diffusivity is almost independent of soil moisture (van Lier and Durigon 2012), the vapor in the soils could not be transported upward through the DSL into the ambient atmosphere when the air temperature was higher than the soil temperature (i.e. there was a downward thermal gradient) (Bittelli *et al* 2008). Because the thermal gradient is smaller for a thicker DSL, the retardation effect increases with an increase of DSL thickness.

The HSL site had a top DSL (thickness $\delta = 0$–30 cm) most of the days and thus the hydrological cycle was dominated by phase III evaporation. The DSL could have prevented the 90 cm deep moist soils beneath being further dried out, which was very beneficial for the fragile ecosystem. For a non-rainy day, the evaporation rate increased from early to late morning and then started to decrease until midnight, which is mainly because the relative changes of air temperature and soil temperature led to thermal gradients with different magnitudes and directions. In the afternoon, the soil moisture of the zero to 10 cm soils tended to increase, probably resulting from vapor condensation and the upward movement of liquid water from the deeper soils. In contrast, for a rainy day, the evaporation rate was lower during raining than non-raining time periods, which can be attributed to the reduced atmospheric evaporation demand during raining time periods, when the air humidity increases and the air temperature decreases. Once the rain event ended, the evaporation rate started to increase because of an increased atmospheric evaporation demand. For most of the rainy days, the rainfall was small (<20 mm) and could only replenish the soil moisture of the zero to 10 cm deep soils, whereas for a few rain days, the rainfall was relatively large (>20 mm) and might replenish the entire vertical soil profile and recharge the shallow aquifer.

During the growing season (May to September), the peak evaporation rate of soil water was determined to be 0.5–2.8 mm d^{-1}, depending on the month of interest. The evaporation in May was lowest, whereas the evaporation rate in July or August was highest. Given that the HSL site was bare, the evaporation was non-beneficial (Ursino *et al* 2014). To protect the bare soils from being eroded by wind and sustain the fragile ecosystem, an appropriate vegetation coverage needs to be established.

Herein, the basic requirement for the vegetation species is that they can survive in semi-arid environment, have a root depth of less than 10–20 cm, and have a peak transpiration rate of less than 2.8 mm d^{-1}. The limitation of root depth ensures that the DSL thickness should not be increased after the vegetation coverage, while the limitation of maximum transpiration rate ensures that such vegetation species provide a very good protection of the bare soils while should not consume more water than the existing non-beneficial evaporation. The promising vegetation species are *Piptochaetium napostaense*, *Stipa tenuis*, *Pappophorum subbulbosum*, and *Artemisia* (Distel and Fernández 1987, Wu and Yang 2013).

5.5.5 Summary and conclusions

This study examined the dynamics of soil water and pore vapor in the dry sandy soils at a typical site within the Horqin Sandy Land (HSL). The examination was done using a HYDRUS-1D model that was calibrated and validated using the measured data on climate and soil water in 2013 and 2014. The results revealed a threshold depth of 10 cm, below which the soils had almost constant soil moisture (SM) and pore air relative humidity (RH_{pa}). For a given non-rainy day, the RH_{pa} and SM of the top 5 cm soils tended to decrease with increase of atmosphere temperature (T_a), attaining a minimum around 2:00 pm. In the 5–10 cm soils, the SM had a negligible diurnal variation, whereas the RH_{pa} tended to increase with increase of T_a. In contrast, for a rainy day the diurnal variations of RH_{pa} and SM of the top 10 cm soils were mainly controlled by the rainfall pattern and amount, with maximums at a time during the rain event. Moreover, the top dry soil layer (DSL) can effectively retard the moist soils beneath from being further dried out. Planting native vegetation that has a shallow root depth (–10 cm) may be preferred to enhance role of the DSL in protecting the fragile desert ecosystem.

For the HSL site the actual evaporation (AE) only accounted for about 15% of the potential evapotranspiration (PET), while AE exponentially increases with an increase of PET ($R^2 = 0.95$). Such a low AE-PET ratio can be attributed to limited water available for evaporation as well as retardation (i.e. repellent) effects of top DSL on heat and vapor transfer. For a given study year, there were less than 60 rainy days, with a total precipitation of less than 520 mm and many dry days between any two rainy days. More than 80% of the rainy days had a daily precipitation of less than 20 mm, which could just temporally replenish the soil moisture of the zero to 10 cm deep soils and provide a small amount of water for evaporation. In addition, a top DSL with a 10–30 cm thickness existed most of the days. The DSL reduced the available heat for vaporization of water in the moist soils beneath and retarded the upward movement of vapor into the ambient atmosphere. Because of the DSL, the 90 cm soils beneath might be protected from being further dried, which was very beneficial for the fragile ecosystem. Further, the 5–10 cm soils functioned as the evaporation zone most of the time, whereas the top 5 cm soils might be an evaporation zone in the morning but become a condensation zone from late afternoon onwards. Moreover, vegetation coverage needs to be established to protect the bare soils from being eroded by wind. The vegetation species should be able to survive in semi-arid environment, have a root depth of less than the median depth of the DSL (i.e. 10–20 cm), and have a peak transpiration rate of less than the peak non-beneficial evaporation rate from the bare soils (i.e. 2.8 mm d^{-1}).

5.6 Grass productivity as influenced by climate and livestock grazing

Net primary production (NPP) depends on cycles of carbon, water, and nutrients. Its accurate prediction is needed but incomplete in the existing literature for understanding the effects of climate change on ecohydrological processes. Existing NPP models rarely have site-specific parameters for characterizing dryland ecosystems, including the Balagaer River watershed in northeast China. For instance, although evapotranspiration (ET) is a crucial variable for biosphere–climate interactions, it is oversimplified, likely biasing NPP predictions. For this reason, some of the previous studies overpredicted NPP and erroneously predicted the observed temporal trend. This case study aims to present a new NPP model by overcoming the above problems and then applies this model to examine NPP spatiotemporal variations in the Balagaer River watershed. The model estimates potential ET using the Penman–Monteith formula and actual ET using a theoretical solution to the Budyko's hypothesis. Also, the model incorporates CLIGEN as a stand-alone module, which can perturb the monthly statistics of daily observations to generate hypothetical climates. For the studies watershed, the model predicted the observed NPP magnitude and spatiotemporal variability well. Across the 56 historical years, the model showed a weak declining trend in NPP and a spatial pattern that are consistent with the intrinsic site properties. However, climate change will either slightly increase or not affect grass productivity in the foreseeable future.

5.6.1 Background

Net primary production (NPP) is the total biomass (i.e. net organic matter) of plant stems, leaves, and roots at a time of interest (Chapin *et al* 2002). NPP consists of two components, namely aboveground net primary production (ANPP) and belowground net primary production (BNPP). ANPP is the organic matter of stems and leaves, whereas BNPP is the organic matter of roots. For a given elapsed time, the NPP increment is the difference of the carbon assimilated by photosynthesis from the carbon consumed by autotrophic respiration. In practice, ANPP is commonly used as a surrogate for NPP because ANPP can be measured by nondestructive cost-effective techniques (e.g. remote sensing). NPP bridges the biosphere and climate system (Melillo *et al* 1993, Zhou *et al* 1998, Cramer and Field 1999, Scurlock *et al* 1999, Ni 2004). Given the importance of understanding global cycles of carbon, water, and nutrients, various studies have been carried out to determine NPP (Day and Monk 1974, Lauenroth *et al* 1986, Sala *et al* 1988, Paruelo *et al* 1997, Asner *et al* 2003, Bai *et al* 2004, Ni 2004, Wang *et al* 2011, Irisarri *et al* 2012), develop NPP prediction models (Lieth 1977, Uchijima and Seino 1985, Zhou and Zhang 1995, Thornley 1998), and scrutinize the responses of NPP to climate change (Xiao *et al* 1995, Thornley and Cannell 1997, Paruelo *et al* 1999, Yu *et al* 2004, Yang *et al* 2008, Svoray and Karnieli 2011, Zhang *et al* 2011).

Collected using various techniques and at different spatial scales from the 1980s to 1990s (Ni 2004), the data on ANPP and NPP for temperate northern China reveal that ANPP tended to decrease with an increase of air temperature but to increase with an increase of precipitation. In addition, using the inventory ANPP data of Japan's forests from 1980 to 2005, Wang *et al* (2011) regressed NPP and its components (i.e. either stem or leaf but not both) over ANPP. The regressions showed a positive relationship between ANPP and NPP as well as an upward trend of ANPP across the 25 recorded years. Further, using global data on NPP and climate, researchers (Lieth 1977, Uchijima and Seino 1985) developed three NPP models, namely Miami, Thornthwaite Memorial, and Chikugo. However, these models had large prediction errors for dryland ecosystems because the evapotranspiration (ET) algorithms were not developed for water-limited environments. To address the prediction errors, Zhou and Zhang (1995) formulated the Synthetic model and showed that this model performed better than the previous three models in reproducing measured NPPs in Chinese drylands. Nonetheless, because the Synthetic model oversimplifies ET processes, it needs to be further refined and improved.

Moreover, Xiao *et al* (1995) examined temporal variations of ANPP from 1980 to 1989 as influenced by air temperature and precipitation in the Xilin River Basin (115°32′ to 117°12′ E, 43°26′ to 44°39′ N) in the northeast Inner Mongolia Autonomous Region (IMAR) of China. The examination revealed that ANPP tended to increase with an increase of precipitation but decrease with an increase of air temperature and that ANPP exhibited no significant (p-value > 0.05) trend over the record period. Zhang *et al* (2011) analyzed climate data at 47 gauges and ANPP data at 7 ecological research sites in IMAR from 1980 to 2005. The analysis showed a significant (p-value $= 0.003$) downward trend in ANPP, which was also predicted by the Synthetic model. Given that the Xilin River Basin is part of the study area of Zhang *et al* (2011), the inconsistent ANPP trends from the above two studies indicate the importance of considering the interactions of hydroclimate and site-specific environmental properties. Such interactions are currently not included in the Synthetic model.

This case study intends to (i) propose a new NPP prediction model with a widely accepted ET formula and site-specific environment properties incorporated and (ii) use this model to predict the spatiotemporal variations of NPP for historical and hypothetical climates. The testbed was the Balagaer River watershed (section 5.1.1), which is about 180 km northeast of the Xilin River Basin. Because of climate change, IMAR has been drying out since 1980 and will continue to be drier in the future (Angerer *et al* 2008, Yi *et al* 2011, Wang *et al* 2014a, Gao *et al* 2017a, Wang *et al* 2021b), making this case study vital both theoretically and practically.

5.6.2 Model development

5.6.2.1 Inputs and operation

Figure 5.35 shows the general framework of the NPP model. The model requires inputs of soil–water properties, topographical feature, vegetation characteristic, and

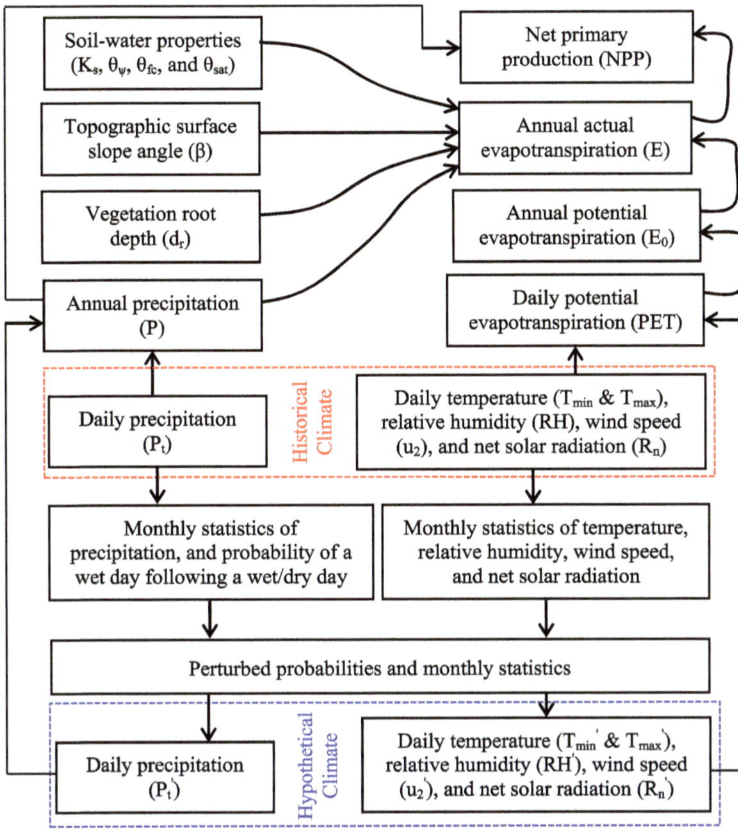

Figure 5.35. General framework of the new net primary production (NPP) model. Flowchart showing model inputs, how to process historical data, and how to generate hypothetical climates. K_s: saturated hydraulic conductivity; θ_ψ: permanent wilting point; θ_{fc}: field capacity; and θ_{sat}: saturated soil moisture. (Reproduced with permission from figure 1 in Wang *et al* (2014b). Copyright 2014 Elsevier.)

climatic variables. The soil–water properties include saturated hydraulic conductivity (K_s), permanent wilting point (θ_ψ), field capacity (θ_{fc}), and saturated soil moisture (θ_{sat}). The topographic feature is landscape slope angle (β), whereas the vegetation characteristic is root depth (d_r). The climatic variables include precipitation, minimum and maximum temperatures, relative humidity or dew point temperature, wind speed, and net solar radiation.

For a given historical year, daily precipitations (P_t) are cumulated to obtain the annual precipitation (P). For a given day of the year, the daily potential evapotranspiration (PET) is computed using the daily values of the other six climatic variables. The computed PET values are cumulated to obtain the annual potential evapotranspiration (E_0). The annual actual evapotranspiration (E) is estimated using β, d_r, E_0, and soil–water properties. The NPP of this year is computed in terms of E and P.

Across the entire record period, for a given climatic variable, the daily values are used to compute three monthly statistics of mean, standard deviation, and skew coefficient (Johnson *et al* 1996). For the month of interest, the daily values of P_t are processed to compute the probabilities of a wet day following a dry day (pr_wd) and a wet day following a wet day (pr_ww) (Yu 2005). Herein, pr_wd is the ratio of the number of wet days following dry days to total number of dry days, whereas pr_ww is the ratio of the number of wet days following wet days to the total number of wet days.

Previous studies (Angerer *et al* 2008, Zhang *et al* 2011) have shown that climate change tends to alter the above three statistics, pr_wd, and/or pr_ww. Hence, to generate daily data on hypothetical climates, the computed monthly statistics, pr_wd, and pr_ww are increased or decreased (i.e. perturbed) by empirical deltas, and the perturbed statistics and probabilities are input into CLIGEN (Kinnell and Yu 2020) to generate future values of the climatic variables required by the new NPP model. This approach is an alternative to conventional approaches for downscaling climate change predictions by Global Circulation Models (GCMs) or Regional Circulation Models (RCMs) (Zhang 2007, IPCC 2011). This alternative approach uses empirically perturbed statistics and probabilities to represent the impacts of climate change: a larger delta signifies a larger impact and vice versa. In contrast, the GCM–RCM predictions correspond to prior scenarios of socioeconomic development, greenhouse-gas emission, technological advancement, and population. The conventional downscaling approaches assume that the prior GCM–RCM scenarios at a larger spatial scale (e.g. region) also hold at a smaller spatial scale (e.g. watershed), to which the predictions are transformed. That is, this alternative approach generates hypothetical climates that are not directly associated with any prior scenarios. However, the hypothetical climates are hypothesized to be more appropriate for examining NPP responses to climate change than the downscaled data because of unknown uncertainties and/or errors in the prior scenarios, GCM–RCM predictions, and downscaling approaches (Zhang 2007). For a given hypothetical climate, the corresponding annual precipitation and evapotranspiration are computed as for the historical climate described above. By assuming that the NPP function as E and P is stationary and independent of climates, NPP is computed using the function with the annual precipitation and evapotranspiration of this hypothetical climate as inputs.

5.6.2.2 Computation of annual P and E

The model computes annual P as the summation of daily precipitations and annual E_0 as the summation of daily PETs calculated using the Penman–Monteith method (equations (3.39)–(3.50)). The counts of the non-rainy days are taken as the number of wet days (D_w). The model estimates annual E using a theoretical (i.e. Fu's) solution to the Budyko's hypothesis (Budyko 1978, Fu 1981, Fu 1996, Zhang *et al* 2004). Fu's solution assumes that E is a nonlinear function of P and E_0. It has been well tested across the world (Yang *et al* 2007, Yang and Zhou 2011, Istanbulluoglu *et al* 2012). To consider the soil–water and vegetation enhancing effects on E, this case study revised the original Fu's solution as

$$\frac{E}{\alpha_{sm} P} = 1 + \frac{E_0}{\alpha_{sm} P} - \left[1 + \left(\frac{E_0}{\alpha_{sm} P}\right)^{\omega}\right]^{\frac{1}{\omega}} \tag{5.6}$$

annual actual evapotranspiration [mm]
annual potential evapotranspiration [mm]
shape factor of integrated effects of vegetation, soil, and topography on water balance (Eq. 5.7) (> 1.0) [–]
soil-water enhancement factor (> 1.0) [–]
annual precipitation [mm]

$$\omega = \alpha_{veg}\left[1 + 8.652\left(\frac{K_s}{P / D_w}\right)^{-0.368}\left(\frac{S_{max}}{E_0}\right)^{0.463} e^{-4.464\tan(\beta)}\right] \tag{5.7}$$

shape factor of integrated effects of vegetation, soil, and topography on water balance (> 1.0) [–]
saturated hydraulic conductivity [mm d⁻¹]
available water in rootzone soils (Eq. 5.8) [mm]
multiplication factor of influences of leaf area and root structure on water balance (> 1.0) [–]
annual precipitation [mm]
number of wet days [d]
topographic surface slope angle [°]
annual potential evapotranspiration [mm]

$$S_{max} = \left(\theta_{fc} - \theta_{\psi}\right) d_r \tag{5.8}$$

available water in rootzone soils [mm]
field capacity [–]
root depth [mm]
permanent wilting point [–]

5.6.2.3 Formulization of NPP

The model computes NPP, in g DM m^{-2}, as (Zhou and Zhang 1995, Zhou *et al* 1998)

$$NPP = 100(RDI)(E)\,e^{-[9.87+6.25(RDI)]^{0.5}} \tag{5.9}$$

climatic dryness index (Eq. 5.10) [–]
net primary production [g DM m⁻²]
unit conversion factor [–]
annual actual evapotranspiration [mm]

$$RDI = 0.629 + 0.237\left(\frac{E_0}{P}\right) - 0.00313\left(\frac{E_0}{P}\right)^2 \tag{5.10}$$

climatic dryness index [–]
annual potential evapotranspiration [mm]
annual precipitation [mm]

5.6.2.4 Generation of hypothetical climates

A Microsoft Excel® macro called WGNmaker4.xls or WGNmaker4.xlsm (Boisramé 2011) is embedded into the model to calculate the monthly statistics, pr_wd, and pr_ww using historical data. It executes a set of algorithms and can deal with missing values and unequal record periods (e.g. 50 years of precipitation but 20 years of wind speed). The days with missing values are simply excluded from calculations. The calculated monthly statistics, pr_wd, and pr_ww are increased or decreased by empirical deltas to obtain the corresponding perturbed values (figure 5.35).

Using the perturbed values, CLIGEN (Kinnell and Yu 2020) generates daily time series of precipitation, temperature, dew point, wind speed, and solar radiation. CLIGEN is a stochastic weather generator for a single geographic point, with inputs in English Customary units but outputs in International System (SI) units (Flanagan *et al* 2001). For instance, the monthly precipitation statistics should be in inches, while the CLIGEN-generated daily time series of precipitation will be in millimeters. Previous studies (Elliot and Arnold 2001, Yu *et al* 2004, Yu 2005, Zhang *et al* 2008) have shown that CLIGEN is applicable all over the world, including the semi-arid region where the Balagaer River watershed is located.

5.6.3 Materials and methods

This case study was performed on the Balagaer River watershed (figure 5.1). The soils, with more than 70% sand and silt particles (0.002–2.0 mm) and less than 30% clay particles (<0.002 mm), have a loamy texture (figure 5.2(a)) and can be classified into four types of I-Mo-2c, Kh1–2b, I-K-2c, and Zm2–2/3a (FAO 2008), with upper (0–30 cm) and lower (30–100 cm) layers (figure 5.2(c)). The first three types of soils have a similar permeability, while the last type of soil is the least permeable. Across the watershed, K_s varies from 1.5 to 107.0 mm h^{-1} (figure 5.2(b)), with a mean of 28.5 mm h^{-1}.

5.6.3.1 Data

This case study used daily data on precipitation, air temperature, relative humidity, wind speed, solar radiation, and evapotranspiration from West Ujimqin Climate Station 54012 (figure 5.1) from 1 January 1955 to 31 December 2010. The data were downloaded from http://www.cma.gov.cn/en/aboutcma/Organizational/InstitutionsUnderCMA/201203/t20120319_166513.html. In addition, this case study used a soil map in a geographic information system (GIS) format. The map was acquired from the Food and Agriculture Organization of the United Nations (http://www.fao.org/home/en). The map consists of 22 polygons (i.e. soil map units), each of which represents one type of soil in the study watershed. The associated attributes of soil texture and organic matter content were used to determine the corresponding soil–water properties (e.g. θ_ψ and θ_{sat}). Further, this case study used a 30 m digital elevation model (DEM) to delineate the boundary of the study watershed and its subbasins. The DEM was downloaded from the International Scientific Data Service Platform website http://www.scidatas.com.

5.6.3.2 Model parameterization

For each soil map unit, the soil–water properties of θ_ψ, θ_{fc}, θ_{sat}, and K_s were determined using the soil–plant–air–water (SPAW) model (Saxton and Rawls 2006), with percentages of sand, clay, and organic matter as the model inputs. That is, SPAW estimates those soil–water properties as empirical functions of soil texture and organic matter content. The functions, whose details can be found in Saxton *et al* (1986), are valid for a wide range of textures with two reasonable assumptions. The first assumption is that soil moisture is equal to θ_{sat} if the suction head is smaller than the air-entry pressure head, decreases linearly from the air-entry pressure head to 10 kPa, and decreases nonlinearly with increase of suction head from 10 to 1500 kPa. The second assumption is that hydraulic conductivity nonlinearly decreases with a decrease of soil moisture from θ_{sat} to θ_ψ. In addition, the DEM was processed in ArcGIS® to delineate the boundary of the study watershed and its subbasins, resulting in 27 subbasins. A uniform gradient (i.e. single value for β) was assumed for each subbasin and β was automatically computed by ArcGIS® using the DEM elevations. Further, the areal proportions of the soil map units that are included in each of the subbasins were determined by overlaying the subbasin map and the soil map in ArcGIS®. For a given soil–water property (i.e. θ_ψ, θ_{fc}, θ_{sat}, or K_s) and a given subbasin, the value of this property for this subbasin was computed as the area-weighted average of the corresponding values across the soil map units within this subbasin. Moreover, a constant plant root depth of $d_r = 60$ cm and a constant atmosphere pressure of $p_a = 101.3$ kPa were used in this case study.

5.6.3.3 Hypothetical climates

The historical observations from 1955 to 2010 revealed that the air temperature of the studied watershed had been increasing at about 0.04 °C per year (figure 5.15(b)). A similar upward trend was true for the Xilin River Basin (Xiao *et al* 1995, Yi *et al* 2011), which is about 180 km away from the studies watershed, and for the IMAR as a whole (Angerer *et al* 2008). A step change in air temperature occurred in 1980. The daily air temperatures before 1980 fluctuated but did not exhibit an obvious upward trend on an annual basis, whereas the daily air temperatures after 1980 fluctuated and gradually shifted up toward a higher value from year to year. Thus, the 'reference period' was from 1955 to 1979, during which the mean daily average temperature was 1.04 °C and the mean annual precipitation was 297.9 mm, while the 'changed period' was from 1980 to 2010, during which the mean daily average temperature was 2.08 °C and the mean annual precipitation was 322.8 mm.

For each monthly statistics, pr_wd, or pr_ww, its perturbed values were generated by linearly interpolating between the corresponding lower bound for the 'reference period' (table 5.17) and upper bound for the 'changed period' (table 5.18). If a perturbed value is smaller than the lower limit of the corresponding range in tables 5.17 and 5.18, let $X_{i,j}^m$ be the lower limit, whereas if a perturbed value is

Table 5.17. The monthly probabilities and statistics calculated using the 1955 to 1979 (i.e. 'reference period') daily climate data[1].

Statistics or probability[2]	Definition	Range	Jan	Feb	Mar	Apr	May	Jun	Jul	Aug	Sep	Oct	Nov	Dec
$\overline{T}_{max}$ (°C)	Average daily maximum temperature	[−40, 40]	−13.31	−9.93	−1.05	10.58	18.72	23.18	25.63	24.20	18.37	10.18	−1.35	−10.60
$\overline{T}_{min}$ (°C)	Average daily minimum temperature	[−40, 40]	−24.82	−23.44	−14.99	−3.80	3.11	9.08	13.43	11.35	3.86	−4.39	−14.03	−21.79
$s_{T\,max}$ (°C)	Standard deviation of daily maximum temperature	⩾ 0	5.62	6.51	7.01	6.60	5.95	4.52	3.57	4.05	4.87	6.54	7.00	6.58
$s_{T\,min}$ (°C)	Standard deviation of daily minimum temperature	⩾ 0	5.35	6.45	7.14	5.29	5.17	3.96	3.10	3.79	4.84	5.21	6.42	5.76
$\overline{P}$ (mm d^{-1})	Ratio of average total monthly precipitation to number of wet days	⩾ 0	0.00	0.00	1.52	1.48	3.08	4.75	6.70	6.13	3.28	3.08	2.20	0.00
s_P (mm d^{-1})	Standard deviation of daily precipitation	⩾ 0	0.00	0.00	0.17	0.60	2.53	4.58	7.28	6.73	2.92	1.76	0.13	0.00
sk_P (mm d^{-1})	Skew coefficient of daily precipitation	(−∞, +∞)	0.00	0.00	17.33	10.23	7.78	3.67	4.03	4.47	4.62	16.81	25.79	0.00
pr_wd	Probability of a wet day following a dry day	[0, 1]	0.00	0.00	0.01	0.05	0.16	0.35	0.40	0.33	0.22	0.05	0.00	0.00
pr_ww	Probability of a wet day following a wet day	[0, 1]	0.00	0.00	0.17	0.26	0.46	0.52	0.55	0.55	0.45	0.16	0.00	0.00
N_w (d)	Average number of days of precipitation (i.e. number of wet days)	⩽ 31	0.00	0.00	0.24	1.88	6.52	12.84	14.56	13.16	8.80	2.04	0.08	0.00
$\overline{R}_n$ (MJ m^{-2} d^{-1})	Average daily net solar radiation	⩾ 0	7.53	10.94	15.41	19.22	22.17	22.61	21.51	19.59	16.24	11.75	8.11	6.59
s_{Rn} (MJ m^{-2} d^{-1})	Standard deviation of daily net solar radiation	⩾ 0	1.61	2.25	3.39	4.44	4.88	5.17	5.51	5.13	4.08	3.11	1.99	1.27
$\overline{T}_d$ (°C)	Average daily dew point temperature	⩾ 0	0.71	0.69	0.59	0.44	0.43	0.59	0.68	0.70	0.61	0.54	0.61	0.68
$\overline{u}_2$ (m s^{-1})	Average daily wind speed	⩾ 0	4.81	4.48	4.68	5.42	5.03	3.78	3.45	3.22	3.78	4.14	4.60	4.91
s_{u2} (m s^{-1})	Standard deviation of daily wind speed	⩾ 0	2.48	2.35	2.38	2.50	2.32	1.83	1.70	1.70	2.00	2.21	2.23	2.34
sk_{u2} (m s^{-1})	Skew coefficient of daily wind speed	(−∞, +∞)	0.18	0.33	0.70	0.42	0.38	0.72	0.65	0.71	0.82	0.91	0.29	0.17

[1] Reproduced with permission from table 1 in Wang *et al* (2014b).Copyright 2014 Elsevier.)

[2] s_{Rn}, s_{u2}, and sk_{u2} were manually computed in Microsoft Excel®, while the others were computed using the WGNmaker4.xls macro (Boisramé 2011).

Table 5.18. The monthly probabilities and statistics calculated using the 1980 to 2010 (i.e. 'changed period') daily climate data[1].

Statistics or probability[2]	Definition	Range	Jan	Feb	Mar	Apr	May	Jun	Jul	Aug	Sep	Oct	Nov	Dec
$\overline{T}_{max}$ (°C)	Average daily maximum temperature	[−40, 40]	−12.84	−7.86	0.44	11.59	19.17	24.30	26.26	25.01	19.34	10.44	−1.19	−10.00
$\overline{T}_{min}$ (°C)	Average daily minimum temperature	[−40, 40]	−23.97	−20.91	−12.74	−2.67	4.23	10.52	14.22	12.15	5.04	−3.23	−12.94	−20.58
$s_{T\,max}$ (°C)	Standard deviation of daily maximum temperature	⩾ 0	5.48	6.51	6.67	6.77	5.69	4.72	4.03	3.87	5.07	6.73	7.15	5.99
$s_{T\,min}$ (°C)	Standard deviation of daily minimum temperature	⩾ 0	5.34	6.27	6.30	5.30	4.84	4.10	3.15	3.67	4.89	5.28	6.25	5.49
$\overline{P}$ (mm d^{-1})	Ratio of average total monthly precipitation to number of wet days	⩾ 0	0.40	0.54	0.91	1.76	4.18	4.59	6.70	5.67	3.94	2.30	1.08	0.50
s_P (mm d^{-1})	Standard deviation of daily precipitation	⩾ 0	0.31	0.43	0.75	1.14	3.48	5.57	7.58	5.19	3.29	1.61	0.89	0.35
sk_P (mm d^{-1})	Skew coefficient of daily precipitation	(−∞, +∞)	9.44	12.30	6.78	5.33	5.60	7.82	4.71	3.77	4.78	6.07	8.84	4.55
pr_wd	Probability of a wet day following a dry day	[0, 1]	0.15	0.16	0.17	0.16	0.17	0.32	0.36	0.28	0.21	0.13	0.16	0.19
pr_ww	Probability of a wet day following a wet day	[0, 1]	0.35	0.38	0.37	0.33	0.46	0.57	0.55	0.51	0.43	0.39	0.37	0.36
N_w (d)	Average number of days of precipitation (i.e. number of wet days)	⩽ 31	6.06	5.47	6.75	5.45	7.52	12.35	14.16	11.23	8.10	5.71	6.00	7.19
$\overline{R}_n$ (MJ m^{-2} d^{-1})	Average daily net solar radiation	⩾ 0	7.46	10.75	14.98	18.79	21.46	22.56	21.46	19.86	16.30	11.65	7.85	6.24
s_{Rn} (MJ m^{-2} d^{-1})	Standard deviation of daily net solar radiation	⩾ 0	1.62	2.28	3.45	4.39	5.17	5.35	5.32	4.76	3.87	3.03	1.99	1.40
$\overline{T}_d$ (°C)	Average daily dew point temperature	⩾ 0	0.69	0.66	0.58	0.42	0.42	0.55	0.66	0.65	0.57	0.53	0.61	0.68
$\overline{u}_2$ (m s^{-1})	Average daily wind speed	⩾ 0	3.57	3.31	3.53	4.05	3.79	3.00	2.65	2.53	2.84	3.22	3.57	3.69
s_{u2} (m s^{-1})	Standard deviation of daily wind speed	⩾ 0	2.15	1.98	1.93	2.10	2.06	1.64	1.41	1.39	1.73	1.81	1.99	2.09
sk_{u2} (m s^{-1})	Skew coefficient of daily wind speed	(−∞, +∞)	0.71	1.05	0.79	1.02	1.09	0.97	0.95	0.93	1.23	0.80	0.81	0.57

[1] Reproduced with permission from table 2 in Wang *et al* (2014b). Copyright 2014 Elsevier.

[2] s_{Rn}, s_{u2}, and sk_{u2} were manually computed in Microsoft Excel®, while the others were computed using the WGNmaker4.xls macro (Boisramé 2011).

greater than the upper limit of the corresponding range in those two tables, let $X^{m}_{i,j}$ be the upper limit. Such perturbed values were input into CLIGEN to generate hypothetical climates. For a hypothetical climate of interest, the perturbed values were generated as

$$X^{m}_{i,j} = X^{b_m}_{i,j} + C^{m}\left(X^{cp}_{i,j} - X^{rp}_{i,j}\right) \tag{5.11}$$

indicator of base value of X [mm or %]
indicator of hypothetical climate [–]
indicator of "changed period" [–]
statistics [mm] or probability [%] of interest
indicator of "reference period" [–]
precipitation or temperature [–]
month of interest [–]
perturbation seed (≥ 0) [–]

This case study generated and analyzed nine hypothetical climates (table 5.19), each of which consisted of 30 years daily time series of precipitation, temperature, dew point, wind speed, and solar radiation. The daily time series of relative humidity was derived using equation (3.45) from the daily time series of dew point.

5.6.3.4 NPP calculation

For each of the 27 subbasins, the values of NPP were calculated using equations (5.9) and (5.10) on the annual basis for 30 years, yielding thirty sets of subbasin-level NPPs. In addition, for each of the soil–water properties or β, the watershed value was computed as the area-weighted average of the corresponding values across the subbasins. The watershed properties and β were used in equations (5.6), (5.7), (5.9), and (5.10) to compute the values of NPP on the annual basis for 30 years, yielding thirty watershed-level NPPs. Further, the subbasin- and watershed-level NPPs were plotted and mapped to visually scrutinize the spatiotemporal variations and influencing variables.

Table 5.19. The hypothetical climates analyzed in this study[1].

Climate	Period to compute base $X^{b_m}_{i,j}$ for perturbation	Perturbation seed C^m
C1	'reference period' (i.e. 1 January 1955 to 31 December 1979)	0
C2	'changed period' (i.e. 1 January 1980 to 31 December 2010)	0
C3	'record period' (i.e. 1 January 1955 to 31 December 2010)	0
C4-1	'changed period' (i.e. 1 January 1980 to 31 December 2010)	1
C4-2	'changed period' (i.e. 1 January 1980 to 31 December 2010)	2
C4-3	'changed period' (i.e. 1 January 1980 to 31 December 2010)	3
C5-1	'record period' (i.e. 1 January 1955 to 31 December 2010)	1
C5-2	'record period' (i.e. 1 January 1955 to 31 December 2010)	2
C5-3	'record period' (i.e. 1 January 1955 to 31 December 2010)	3

[1] The symbols are defined in equation (5.11). Reproduced with permission from table 3 in Wang *et al* (2014b). Copyright 2014 Elsevier.

5.6.4 Results and discussion

This section presents the results and discusses the findings.

5.6.4.1 The model parameters

The 27 subbasins were delineated to have drainage areas from 3 to 458 km^2, with an average size of 198 km^2 (table 5.20). As expected, a subbasin at a lower location tended to have a gentler slope than that at an upper location: the slope angles varied from 1.29° to 8.67°, with an overall average of 4.17°. In addition, the soil–water properties varied spatially: subbasin 20 had the largest θ_ψ, θ_{fc}, and θ_{sat}, subbasin 9

Table 5.20. The adopted model parameters in this study[1].

Subbasin[2]	Area (km^2)	β (°)	θ_ψ	θ_{fc}	θ_{sat}	K_s (mm h^{-1})
1 (outlet)	140	3.36	0.155	0.323	0.481	11.33
2	222	2.87	0.155	0.323	0.481	11.33
3	3	1.29	0.137	0.317	0.480	12.59
4	163	3.45	0.237	0.392	0.514	4.64
5	195	2.65	0.164	0.320	0.473	9.75
6	40	1.54	0.156	0.311	0.471	11.31
7	172	2.70	0.103	0.232	0.458	34.18
8	453	2.58	0.103	0.232	0.458	34.18
9	67	1.89	0.050	0.104	0.462	107.48
10	66	2.68	0.051	0.120	0.460	96.74
11	458	2.30	0.131	0.286	0.465	15.75
12	351	4.04	0.050	0.107	0.461	105.30
13	455	3.39	0.126	0.270	0.460	18.24
14	211	3.39	0.140	0.295	0.467	13.86
15	23	2.64	0.140	0.295	0.467	13.86
16	169	6.46	0.237	0.382	0.504	4.64
17	23	2.89	0.113	0.283	0.469	18.80
18	127	3.27	0.113	0.283	0.469	18.80
19	91	4.68	0.123	0.280	0.465	17.70
20	323	6.98	0.315	0.433	0.514	1.27
21	419	4.42	0.159	0.304	0.464	11.16
22	148	6.55	0.159	0.304	0.464	11.16
23	243	5.18	0.159	0.304	0.464	11.16
24	174	5.22	0.159	0.304	0.464	11.16
25	201	4.54	0.159	0.304	0.464	11.16
26	141	8.67	0.159	0.304	0.464	11.16
27 (upmost)	270	7.19	0.159	0.304	0.464	11.16
Watershed	5350	4.17	0.149	0.287	0.470	22.69

[1] β: landscape slope angle; θ_ψ: permanent wilting point; θ_{fc}: field capacity; θ_{sat}: saturated soil moisture; and K_s: saturated hydraulic conductivity. Reproduced with permission from table 4 in Wang *et al* (2014b). Copyright 2014 Elsevier.

[2] A subbasin with a smaller number is located at a lower location than a subbasin with a larger number, and vice versa.

the smallest θ_ψ and θ_{fc}, and subbasins 7 and 8 the smallest θ_{sat}. Averaged across the watershed, the soil retention and detention storage values were $\theta_{fc} - \theta_\psi = 0.138$ and $\theta_{sat} - \theta_{fc} = 0.183$, respectively. Further, the subbasins in the middle and upper parts of the watershed were determined to be more permeable than those in the lower part, as indicated by an upstream–downstream decrease pattern of K_s. The saturated hydraulic conductivity for the watershed was determined to be 22.69 mm h^{-1}.

5.6.4.2 Calculated NPP for historical climate

For the historical climate from 1955 to 2010, the watershed was predicted to have an annual mean NPP = 284.6 g DM m^{-2}. However, the annual biomass productivity could fluctuate largely (figure 5.36), with a minimum NPP = 164.6 g DM m^{-2} in 2007, and a maximum NPP = 391.8 g DM m^{-2} in 1998, and an insignificant downward trend at a decreasing rate of 0.034 g DM m^{-2} per year (Hamed and Rao 1998). The NPP in a wetter year tended to be larger than that in a drier year. Across the record years, E was always controlled by precipitation because $E_0 > P$, leading to synchronic fluctuations of NPP with P and E and strong positive correlations thereof (coefficient of determination $R^2 > 0.96$) (figure 5.37). The spatial distribution of NPP barely changed from one year to another and was apparently independent of climate (figure 5.38). For any two locations of interest, if the NPP at one location increased, the NPP at another location would also increase; and vice versa. That is, NPP was found to have positive spatial autocorrelations (Sokal and Oden 1978).

Across the recorded years, subbasins 9, 10, and 12 had the lowest NPP, subbasins 4 and 20 the highest NPP, and the other subbasins a comparable NPP (figure 5.39(a)). Such a spatial distribution of NPP was also true for each recorded year. On the other hand, across the study watershed, NPP and E fluctuated

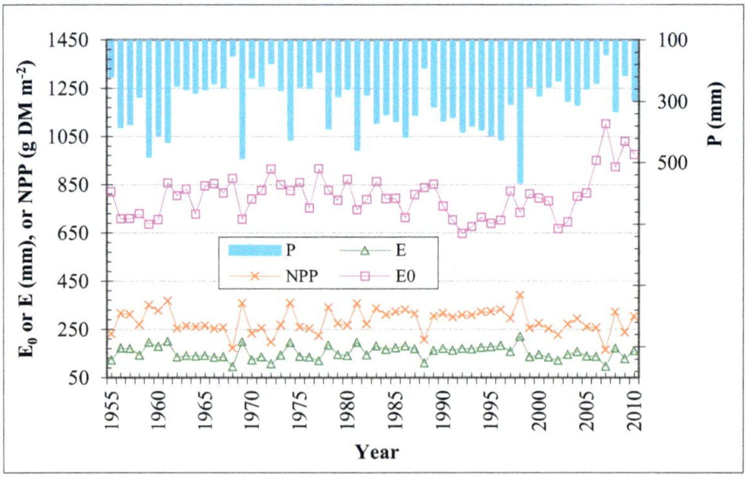

Figure 5.36. The modeled biomass productivity under historical climate. The results are for the Balagaer River watershed from 1955 to 2010. Plots show fluctuations of net primary production (NPP), potential evapotranspiration (E_0), actual evapotranspiration (E), and precipitation (P) on the annual basis. (Adopted with permission from figure 5 in Wang *et al* (2014b). Copyright 2014 Elsevier.)

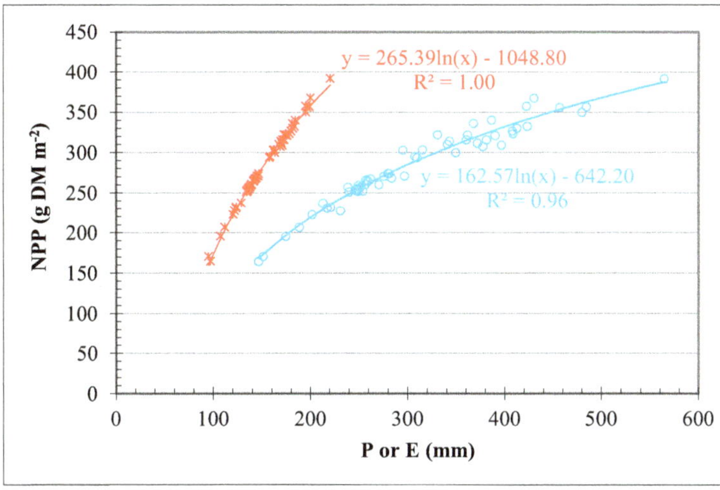

Figure 5.37. The modeled influence of historical climate on biomass productivity. The results are for the Balagaer River watershed from 1955 to 2010. Plots showing regressions of NPP over precipitation (P) (in light blue color) or actual evapotranspiration (E) (in red color) on the annual basis. (Reproduced with permission from figure 6 in Wang *et al* (2014b). Copyright 2014 Elsevier.)

synchronically (figure 5.36) and matched spatially (figure 5.39(a) versus figure 5.39(b)). That is, regardless of the years, a subbasin with a higher E had a larger NPP, whereas a subbasin with a lower E had a smaller NPP.

Driven by the historical climate of 1956 to 2006, the Synthetic model developed by Zhou *et al* (2002) showed that NPP in the Xilin River Basin would vary from 325 to 525 g DM m^{-2} (Zhang *et al* 2011). For any historical year, the NPP of the above study was larger than that of this case study, whereas spanning the record years, the annual NPPs were predicted by the two studies to have a similar interannual fluctuation pattern. One explanation is that the Synthetic model used overestimated R_n values, but this case study used reliable R_n observations. Another explanation is that the Synthetic model used underestimated PET values from an empirical formula (Yang and Zhou 2011), but this case study used reasonable PET estimates from the Penman–Monteith equation (Monteith 1965). The Penman–Monteith equation is physically based and superior to the empirical formula. Ni (2004) found that in arid and semi-arid ecosystems, NPP is a negative function of PET, which explains the overestimation of NPP by the Synthetic model, probably because of the underestimated PET.

Wang *et al* (2011) developed a regression relationship between ANPP and NPP (ANPP = 0.88NPP − 100.11, R^2 = 0.98) for the study watershed. To assess the results of this case study against literature values of ANPP, the regression formula was used to compute the ANPPs from the predicted NPPs. From 1980 to 2003, as shown in figure 5.40, the computed ANPPs of this case study were comparable to the measured ANPPs at the ecosystem research stations in northeast IMAR (Bai *et al* 2004, Zhang *et al* 2011); the computed and measured ANPPs had a similar interannual variability. Thus, the model proposed by this case study had a good

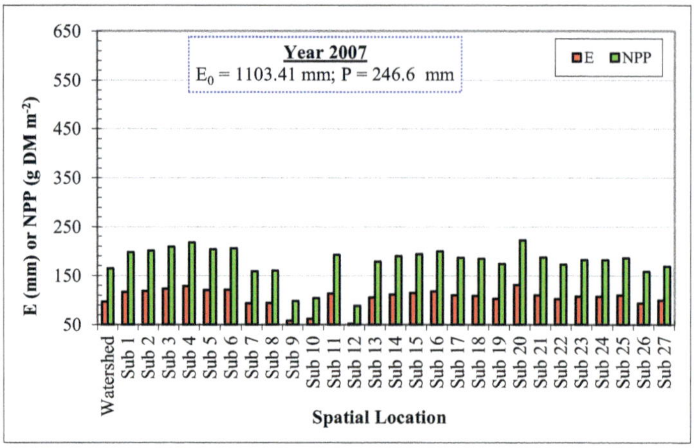

(a)

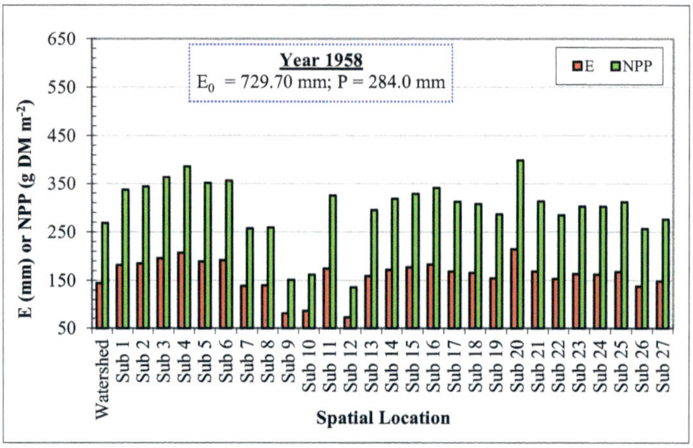

(b)

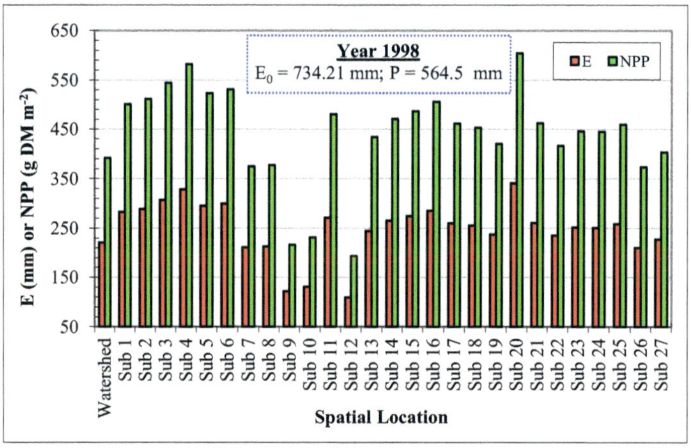

(c)

Figure 5.38. The modeled spatial distributions of biomass productivity for selected years. Plots showing annual actual evapotranspiration (E) and net primary production (NPP) in (a) 1968 with the lowest NPP, (b) 1958 with median NPP, and (c) 1998 with highest NPP. P is the annual precipitation and E_0 is the annual potential evapotranspiration. (Adopted with permission from figure 7 in Wang *et al* (2014b). Copyright 2014 Elsevier.)

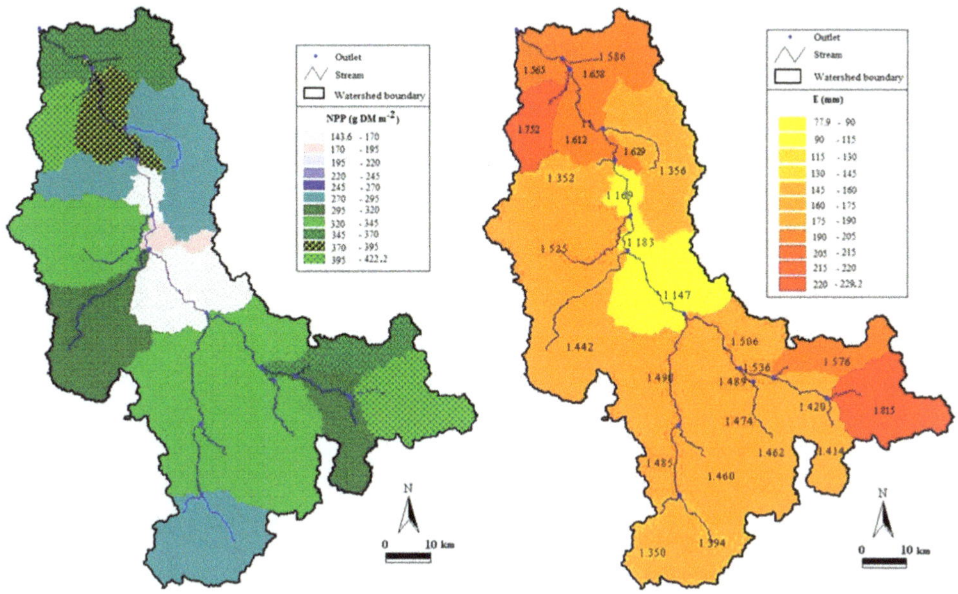

(a) (b)

Figure 5.39. Spatial patterns under historical climate from 1955 to 2010. Map showing (a) annual average net primary production (NPP) and (b) mean annual actual evapotranspiration (E), superimposed with the annual average ω in equations (5.6) and (5.7). (Adopted with permission from figures 8 and 9 in Wang *et al* (2014b). Copyright 2014 Elsevier.)

performance in predicting NPP and the regression formula was appropriate for computing the ANPPs from the predicted NPPs. For 1988 the computed ANPP was over 100 g DM m^{-2} lower than the corresponding measured value because those ecosystem research stations received more precipitation than the Balagaer River watershed (250 mm versus 189 mm), leading to distinctly different ANPPs (Paruelo *et al* 1997, Yang *et al* 2008, Svoray and Karnieli 2011).

The interannual fluctuations of NPP did not lead to a statistically significant declining trend (p-value $\gg$ 0.05). The non-trending variation can be attributed to the cancellation effects of increasing precipitation and rising PET (figures 5.15 and 5.36). The NPP spatial pattern was apparently independent of climate and mainly controlled by site-specific physiographic characteristics that are represented by parameter ω in this case study (figure 5.39(a) versus figure 5.39(b)). The subbasins with a larger ω tended to have a higher NPP than those with a smaller ω. A larger ω signifies larger soil retention storage (equation (5.7)), gentler topographic gradients (i.e. smaller β), less permeable soils (i.e. smaller K_s), and thus more soil water for plant growth.

5.6.4.3 Predicted NPP for hypothetical climates

As shown in table 5.19, this case study generated nine hypothetical climates. Three of the climates, namely C1, C2, and C3, had a mean smaller than the historical mean

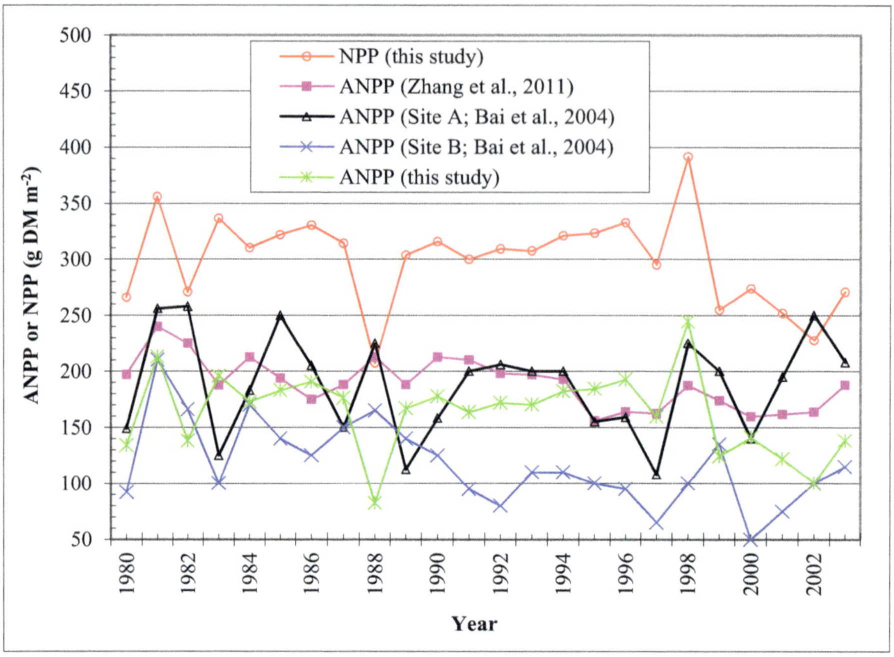

Figure 5.40. Performance of the proposed model in predicting biomass productivity for the Balagaer River watershed. The biomass productivity was represented by net primary production (NPP) and aboveground net primary production (ANPP). Plots showing the modeled values for the study watershed versus the literature values for the Inner Mongolia Autonomous Region grasslands. (Adopted with permission from figure 10 in Wang *et al* (2014b). Copyright 2014 Elsevier.)

annual precipitation of 312 mm (figure 5.41), whereas the other six, namely C4-1, C4-2, C4-3, C5-1, C5-2, and C5-3, had a mean greater than the historical mean. C4-3 and C5-3 had the largest mean of 532 mm, which was 210 mm more than the historical mean. The increases in mean annual daily air temperature were assumed to be 0.1 °C to 5.3 °C.

Across IMAR, annual precipitation was predicted to decrease or have no statistically significant trend in some areas (Yi *et al* 2011, Zhang *et al* 2011) and to increase in amount and seasonal variability in other areas (Gao *et al* 2017a). Regardless of the areas, air temperature has increased and will continue to increase in the future. The nine hypothetical climates were generated to represent possible future climates in IMAR.

Except for C1 and C3, the other seven hypothetical climates had smaller net solar radiation and wind speeds than the historical climate, leading to a smaller E_0. C4-3 and C5-3 had a largest annual precipitation but a much smaller E_0 and the lowest E. For a hypothetical climate of interest, the E_0, P, and E were dependent on site-specific physiographic properties, which were represented by parameter ω defined in equation (5.7). ω was determined to be from 1.37 to 1.54 at the watershed scale (figure 5.41(a)), whereas ω varied by up to 0.95 interannually at the subbasin scale (figure 5.41(b)).

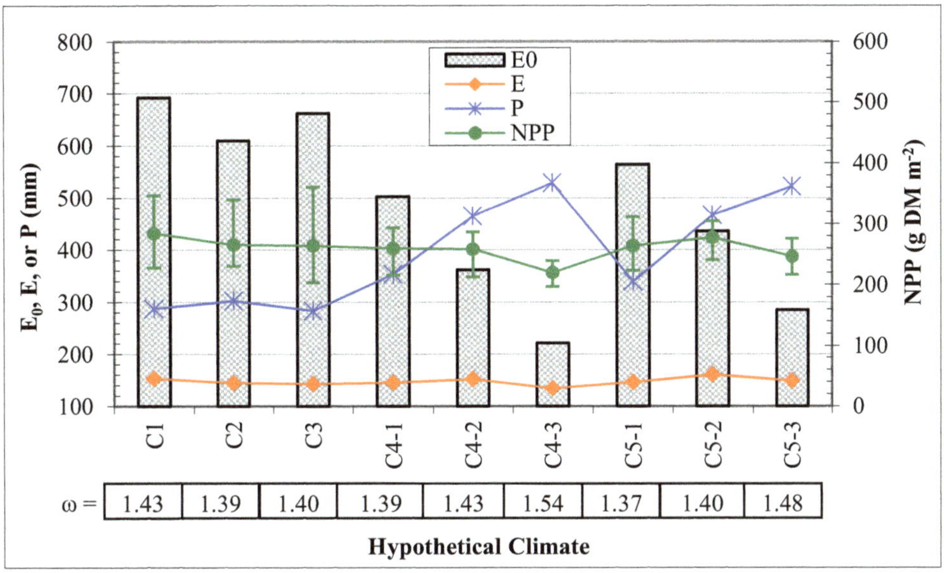

(a)

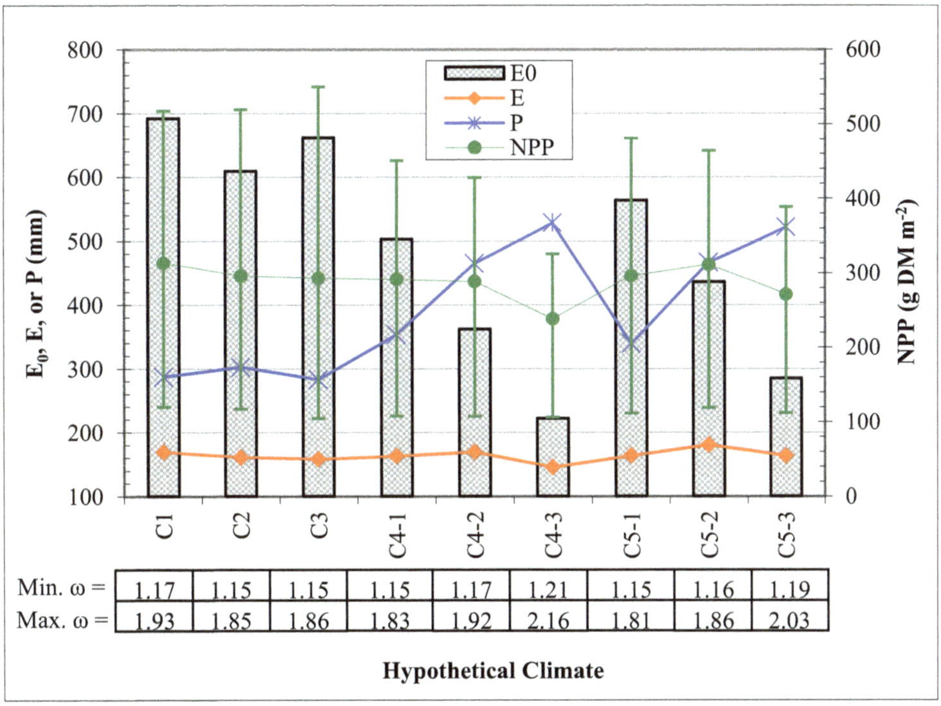

(b)

Figure 5.41. The predicted net primary production (NPP) for the hypothetical climates defined in table 5.19. The results are for the (a) watershed-level parameters and (b) subbasin-level parameters. E_0: mean annual potential evapotranspiration; E: actual evapotranspiration; and P: precipitation. For NPP, the means and one standard deviation error bars spanning 30 years are shown. ω is a shape factor defined in equation (5.7). (Reproduced with permission from figures 11 and 12 in Wang *et al* (2014b). Copyright 2014 Elsevier.)

At the watershed scale, the annual average NPP was predicted to be smaller and have a lower interannual variability for the nine hypothetical climates than for the historical climate (figure 5.41). As a result of climate change, the biomass productivity would be reduced from 284.6 g DM m^{-2} (historical) to 219.8 to 284.3 g DM m^{-2} (future). However, when the subbasin-scale NPPs were predicted and then averaged across the 27 subbasins, the annual average NPP was slightly larger for the nine climate scenarios except for C4-3 and C5-3 than for the historical climate; but the interannual variability was comparable for hypothetical and historical climates (figure 5.41(b) versus figure 5.36). C4-3 and C5-3 were predicted to have the lowest E, resulting in the lowest NPP (figure 5.37).

The prediction discrepancies from using watershed-scale versus subbasin-scale parameters indicate the influences of site-specific physiographic properties. The watershed-scale parameters, which smooth out spatial heterogeneities, tend to cause overpredictions for some areas and underestimations for others. The underestimations exceed the overestimations, leading to an overall underestimated NPP. On the other hand, the subbasin-scale parameters, which account for spatial heterogeneities, can somewhat reduce erroneous predictions, yielding a relatively reasonable NPP. Using the subbasin-scale predictions, the overall NPP in the study watershed will slightly increase or have no detectable change, depending on the future climates. NPP has been steadily increasing in the middle and high northern latitudes ($\geqslant 42°$ N) since the 1980s (Myneni *et al* 1997, Myneni *et al* 2001, Hicke *et al* 2002, Fang *et al* 2003, Nemani *et al* 2003, Piao *et al* 2005) and in Japan's forests Wang *et al* (2011). However, the NPP in the Balagaer River watershed would decrease drastically if future climates were like C4-3 or C5-3.

5.6.5 Conclusions

This case study presented a model for predicting effects of climate change on vegetation biomass productivity. The model estimates PET and E using widely accepted methods, considers site-specific physiographic properties, and generates hypothetical climates using CLIGEN with perturbed values of historical monthly statistics. Such perturbed statistics can characterize possible future climates resulting from climate change. Using the Balagaer River watershed as a testbed, the model was parameterized and used to predict NPPs for the historical climate and nine future climates. The results showed a positive relationship of NPP with precipitation (P) and actual evapotranspiration (E). The NPP in a wetter was higher than that in a drier year. At the watershed scale, the historical NPPs fluctuated interannually but did not exhibit any statistically significant trend over the 56 years. However, regardless of the historical years, NPP had a consistent spatial pattern: the NPP of a subbasin with a higher E tended to be larger than that of a subbasin with a lower E, and vice versa. As a result of climate change, NPP would slightly increase or statistically remain unchanged, depending on the concerned spatial scale (e.g. watershed or subbasin) and the future climate of interest. The above takeaways are also supported by previous studies (Xiao *et al* 1995, Piao *et al* 2005, Zhang *et al* 2005) which were conducted in the region encompassing the Balagaer River watershed.

5.7 Interactive effects of land use and soil on water and sediment yields

Land cover change and soil erosion have been extensively studied separately, but their interactive effects are poorly understood. This case study aims to investigate such interactive effects on changes to water and sediment yields. The investigation was conducted in the 117 845 ha upper Cowhouse Creek watershed in north central Texas using the Soil and Water Assessment Tool (SWAT), a distributed hydrological model for tackling problems of nonpoint source pollution. Calibrated and validated using observed daily streamflow at the watershed outlet, a SWAT model was used to predict how conversions of range brush to range grasses on an individual soil basis would impact water and sediment yields. Overall, the removal of range brush was predicted to increase the annual water yield by 24 mm per hectare treated area, whereas the annual water yield would be increased by 80 mm per hectare treated area if range brush were removed from an upland soil with a moderately high permeability, with a small increase of annual sediment loading by 4.2 tones per hectare treated area and a minimal alteration to the existing spatial sources of sediment. A larger increase in water yield would be achieved by removing range brush from a soil in the vicinity of floodplains. For a treated soil, the increase in water yield was predicted to be larger in a wetter than a drier hydrological condition. With a reasonable generalization, land-use–soil interactive effects need to be considered in designing and implementing the best management practices for watershed health and sustainability.

5.7.1 Background

Water is vital for socioeconomics and eco-environments, but it can be in short supply for various reasons (TSSWCB 2003). For most places, the paramount reason is nonpoint source pollution caused by sedimentation (Bosch *et al* 2004, Chu *et al* 2004). Sediments can reduce the capacity of storage structures (e.g. reservoirs), decrease the conveyance of waterways, lower the usability of waterbodies, and transport toxic contaminants (e.g. pesticides) and harmful diseases (e.g. pathogens) into waterbodies (Ashmore 1993, Harden 1993, Conroy *et al* 2006, Grismer 2007, Juracek and Ziegler 2007). To reduce sediments, best management practices (BMPs) have been developed and implemented (Vaché *et al* 2002, Bracmort *et al* 2006). However, in arid and semi-arid regions, including north central Texas in the United States of America (USA), BMPs are expected to play dual roles—reducing sediments while increasing water yields. Brush control is hypothesized as such a BMP (Hibbert *et al* 1974, Richardson *et al* 1979, Hibbert 1983, Dugas and Mayeux 1991, Bednarz *et al* 2001, Greer 2005). In this regard, the rationale is to increase runoff by reducing evapotranspiration (Bosch and Hewlett 1982). Previous studies (Wilcox *et al* 2005) revealed that in the arid and semi-arid regions of the USA, water yields were increased by controlling dense stands of saltcedar in riparian zones, Ashe juniper in areas with rapid subsurface flows (e.g. springs), and mesquite on dry soils with deep macropores. Nonetheless, it is not conclusive of how much more water can be salvaged and utilized and few studies have addressed the sediment issues associated with brush control.

The chaparral conversion in Arizona was found to increase water yield by about 1 mm for every 4 mm increase in precipitation for years with an annual precipitation of above 400 mm (Hibbert *et al* 1974, Hibbert 1983). For instance, in a year with an annual precipitation of 350 mm, the water yield was not affected by the conversion, whereas in a year with an annual precipitation of 410 mm, the water yield was increased by (410–400) × ¼ mm = 2.5 mm. In contrast, the treatment of brushes on Texas rangelands was found to decrease water yield at the beginning due to the increased surface roughness and depression storage, but to increase water yield sometime later because of decreasing depression storage with elapsing time (Richardson *et al* 1979). The removal of 8700 ha saltcedar along the 132 km Acme–Artesia reach of the Pecos River in New Mexico did not cause any detectable change in streamflow along this study reach (Welder 1988), although saltcedar had an annual water usage of about 305 mm more than the replacement plant species (Weeks *et al* 1987).

Mature Ashe junipers transpire 305–430 mm water per year (Owens and Ansley 1997), which is about 38 mm more than the annual transpiration of herbaceous plants (Dugas *et al* 1998). In the Rolling Plains of Texas, the removal of mesquite was found to result in a detectable but small increase of water yield (Dugas and Mayeux 1991), whereas in the Leon River watershed of Texas, the infiltration rate was found to increase following the removal of Ashe junipers but started to decrease eleven months later (Greer 2005). These field observational studies indicate that the effect of brush removal on water yield depends on the spatial scale concerned, such as small plot, hillslope, or small catchment. Given the great interest in using brush control to increase water yield at a watershed scale, such small-scale results need to be scaled up (Wilcox *et al* 2005). In this regard, spatially distributed hydrological models, such as SWAT, can be calibrated using the small-scale results and then used to evaluate the watershed-scale effects of vegetation removal on water and sediment yields (Bednarz *et al* 2001, Wu *et al* 2001, TAES 2002, TSSWCB 2003, Afinowicz *et al* 2005).

For comparison purposes, changes of water and sediment yields were standardized by the treated rather than total drainage area. Herein, the treated area was the area within which range brushes were replaced by range grasses. Previous modeling studies showed a wide range of increases in water yield, depending on geographic locations. For instance, the annual water yield was predicted to increase by 127 mm per hectare treated area from removing the 80 970 ha shrubs in the 323 885 ha Pedernales River watershed of Texas (Bednarz *et al* 2001), by 75 mm from removing the Ashe junipers in the Cusenberry Draw in western Edwards Plateau (Wu *et al* 2001), by 127 mm from removing the mesquites in the Lake Arrowhead watershed in north central Texas (TAES 2002), by 165 mm from removing the range brushes in the Palo Pinto Reservoir watershed (TSSWCB 2003), and by only 38 mm from removing range brushes in the North Fork of the Upper Guadalupe River watershed (Afinowicz *et al* 2005). These modeled increases in water yield were much larger than the field observations presented by Wilcox *et al* (2005) and discussed above.

Among others, the above-cited field observational and modeling studies did not examine land-use–soil interactive effects on water yield or address the impacts of brush control on sediment yield. Such interactive effects play regulatory roles in the hydrological cycle (Wilcox *et al* 2005) and thus their neglect might be the primary reason for the overpredicted increases in water yield by the modeling studies. Greer (2005) showed that brush removal would measurably increase sediment yield. To fill these two information gaps, this case study intends to examine the land-use–soil interactive effects on water and sediment yields in the 117 845 ha Cowhouse Creek watershed. As described in section 5.1.3, the watershed is in north central Texas and drains the area upstream of the US Geological Survey (USGS) flow gauge 08101000 (figures 5.5–5.7).

5.7.2 Materials and methods

5.7.2.1 Data and model set-up

This case study used four physiographic datasets, namely the 30 m National Elevation Dataset (NED), the 30 m 2011 National Land Cover Data (NLCD), the State Soil Geographic database (STATSGO), and the National Hydrography Dataset (NHD). These datasets were downloaded from the websites of US Geological Survey (USGS 2008) and US Department of Agriculture Natural Resources Conservation Service (USDA-NRCS 2008) in the formats required by AvSWAT-X, an upgraded ArcView GIS interface for SWAT (Di Luzio *et al* 2002, Neitsch *et al* 2002, Di Luzio *et al* 2004). NHD is a set of digital spatial information on the surface water features of lakes, ponds, streams, rivers, springs, and wells. STATSGO, which was collected in the USGS 1:250 000 scale, 1 by 2 degree topographic quadrangle units and merged and distributed as statewide coverage, defines seven types of soils for the study watershed (figure 5.6 and table 5.1). It has a county-level resolution and is generally used for regional, multistate, river-basin, state, and multicounty resource planning, management, and monitoring. NLCD distinguishes 16 different land uses (Homer *et al* 2004). It defines 14 types of land uses for the study watershed; figure 5.7 shows the spatial pattern of the three major land uses, namely range brush, range grasses, and forest. NED presents high-quality topographical elevations across the United States in a seamless raster format.

For modeling purposes, this case study used the NED in AvSWAT-X to delineate subbasins by referencing the surface water drainage network presented in the NHD, yielding boundaries of the watershed and its subbasins and associated drainage channels. The subbasins were delineated by trial and error to closely match the delineated drainage channels with the NHD-presented drainage network, resulting in 130 subbasins of 5.4–4372 ha in size (figure 5.5). In addition, for each subbasin, its hydrological response units (HRUs) were defined by overlaying the NLCD and STATSGO with threshold levels of 20% for land use and 10% for soil (Neitsch *et al* 2002, Wang and Melesse 2005). As a basic unit for water balance analysis, a HRU is a unique combination of land use, soil, topography, and climate (i.e. spatially homogeneous). These two thresholds were applied to eliminate minor land uses in the subbasin and minor soils within a land-use area (Di Luzio *et al* 2002,

Neitsch *et al* 2002). For a given subbasin, this case study eliminated land uses that cover less than 20% of the subbasin area and re-apportioned the remaining land uses to make 100% of the subbasin area modeled; for a remaining land use of interest, soils that covered less than 10% of the land-use area were eliminated and the remaining soils were re-apportioned to make 100% of the land-use area modeled. As a result, 546 HRUs were defined for the watershed, with 1–8 HRUs for a subbasin.

The daily data on precipitation and minimum and maximum temperatures at the National Weather Service gauge PR413485 (figure 5.5) from 1 January 1980 to 31 December 2006 were reformatted into SWAT input files. The few missing precipitations were automatically filled by the SWAT's weather generator (Neitsch *et al* 2002). The warm-up period was from 1 January 1980 to 30 November 1984 to stabilize the initial values of the model parameters, while the assessment period was from 1 December 1984 to 30 November 2006 to examine the land-use–soil interactive effects on water and sediment yields.

5.7.2.2 Model evaluation

The model was calibrated against the observations of daily streamflow at the watershed outlet from 1 December 1984 to 30 November 1996 and validated using the observations of daily streamflow from 1 December 1996 to 30 November 2006. The calibration was performed by manually adjusting selected model parameters (table 5.21) to make the predictions match closely the observations of (i) daily streamflow, (ii) monthly runoff, and (iii) the exceedance times for various flow thresholds (Wang and Melesse 2005, 2006).

In addition to visualization plots and duration curves showing predictions versus observations of streamflow at the watershed outlet, the coefficient of determination (R^2) (equation (4.56)) and Nash–Sutcliffe coefficient (NSE) (equation (4.58)) were used to measure the model performance. As stated in section 4.5.3, R^2 measures the model's capability to explain the variability presented in the streamflow observations; it can range from 0 to 1, with a higher value indicating a better model performance. NSE measures the goodness-of-fit of the model to the observations; it can range from $-\infty$ to 1, with a higher value indicating a better overall fit and 1 indicating a perfect fit. The predictions are less reliable than using the average of the streamflow observations if the NSE is negative, whereas the predictions are more reliable than using this average otherwise.

The sediment component of the model could not be calibrated and validated because no sediment data were available. Instead, six sediment-related parameters, namely ADJ_PKR, PRF, SPCON, SPEXP, CH_EROD, CH_COV (table 5.21), were empirically adjusted to make the predicted sediment yield at the watershed outlet closely match the literature value (Narasimhan *et al* 2007). In the Cedar Creek watershed, Narasimhan *et al* (2007) surveyed 56 sites and estimated the average annual total sediment yield to be 1.72 tones ha^{-1}, of which 0.59 tones ha^{-1} was from erosions of the stream bank and bed and 1.13 tones ha^{-1} from overland erosion. The estimation was performed using the Rapid Assessment Point Method (RAP-M) developed by Windhorn (2001) and applied by NRCS. Given that the Cowhouse Creek watershed is just 150 km south of the Cedar Creek watershed, this case study

Table 5.21. List of adjusted parameters for calibrating the SWAT model[1].

Parameter	Definition	Range[2]	Value
SURLAG (day)	Surface runoff lag coefficient	1.0–12.0	2.0
MSK_CO1	Muskingum translation coefficient for normal flow	0.0–10.0	2.5
MSK_CO2	Muskingum translation coefficient for low flow	0.0–10.0	2.5
REVAPMN (mm H_2O)	Threshold depth of water in the shallow aquifer for 'revap' or percolation to the deep aquifer to occur	0.0–500.0	1.0
GW_REVAP	Groundwater 'revap' coefficient	0.0–0.2	0.15
GWQMN (mm H_2O)	Threshold depth of water in the shallow aquifer required for return flow to occur	0.0–5000.0	500
ESCO	Soil evaporation compensation factor	0.01–1.0	0.27
EPCO	Plant water uptake compensation factor	0.01–1.0	0.85
CN2	SCS curve number for soil moisture condition II	60.0–95.0	Default[3]
CH_COV	Channel cover factor	0.0–1.0	0.022
CH_EROD (cm h^{-1} pa^{-1})	Channel erodibility factor	0.0–1.0	0.090
ADJ_PKR	Peak rate adjustment factor for sediment routing in subbasin	1.0–3.0	1.52
PRF	Peak rate adjustment factor for sediment routing in reach	1.0–3.0	1.00
SPCON	Coefficient in sediment transport equation	0.0001–0.01	0.05
SPEXP	Exponent in sediment transport equation	1.0–2.0	1.50

[1] Reproduced with permission from table 2 in Wang *et al* (2010). Copyright 2010 Elsevier.
[2] The ranges for the parameters were based on Neitsch *et al* (2002) and Wang and Melesse (2005).
[3] The values for this parameter were derived by the AVSWAT-X for SWAT 2005 and varied by hydrological response unit (HRU).

assumed that these two watersheds had similar sediment yields and source proportions: 34.3% of the sediment from erosions of the stream bank and bed and 65.7% from overland erosion.

5.7.2.3 Assessment of land-use–soil interactive effects

For a given soil, the range brush that covers the soil was replaced by range grasses, while the land covers of other soils remained unchanged. In this regard the HRUs of the soil and range brush were changed to those of the soil and range grasses, substituting the physiological parameters (e.g. maximum leaf area index) and overland hydrological parameters (e.g. Manning's n) of range grasses for the corresponding values of range brush (Neitsch *et al* 2002). Also, for the redefined HRUs, the USLE (Universal Soil Loss Equation) (Williams 1975, 1982, 1995) C-factor of range brush was replaced by that of range grasses. Range grasses are less resistant and thus have a smaller C-factor than range brush (Hedrick *et al* 1966). In this case study, the C-factors of range brush and range grasses were automatically extracted by AvSWAT-X from the Land Cover and Plant Growth database embedded in the SWAT software package.

Table 5.22. The simulated options[1].

	Range brush			Option							
Soil code	Area (ha)	% of the soil area	Base condition	1	2	3	4	5	6	7	8
TX176	1361.2	18.2	NC	B→G	NC	NC	NC	NC	NC	NC	B→G
TX244	2653.7	53.1	NC	NC	B→G	NC	NC	NC	NC	NC	B→G
TX251	9676.3	33.0	NC	NC	NC	B→G	NC	NC	NC	NC	B→G
TX260	148.0	19.7	NC	NC	NC	NC	B→G	NC	NC	NC	B→G
TX463	5447.4	41.6	NC	NC	NC	NC	NC	B→G	NC	NC	B→G
TX609	16 471.2	45.7	NC	NC	NC	NC	NC	NC	B→G	NC	B→G
TX629	14 105.5	58.9	NC	NC	NC	NC	NC	NC	NC	B→G	B→G

[1] Reproduced with permission from table 3 in Wang *et al* (2010). Copyright 2010 Elsevier. NC: no change; B → G: range brush converted to range grasses.

This case study simulated eight scenarios or options. The first seven options (table 5.22) corresponded to the seven types of soils in the study watershed, while the eighth option was to convert all range brush into range grasses. These scenarios were assessed against the baseline condition with the NLCD land uses in terms of the SWAT-predicted water and sediment yields at the watershed outlet, revealing how the conversion effect of range brush into range grasses would be different soil by soil.

For a subbasin of interest, the mean daily sediment yield was computed as the arithmetic average of the predicted daily values across the entire assessment period (1 December 1984 to 30 November 2006) and the sediment occurrence or frequency was computed as the ratio of the number of days with a non-zero sediment yield to the total number of assessment days (i.e. 8358). These computations were performed for the baseline condition and the eight options, yielding nine means and nine frequencies for each subbasin. The means or frequencies were broken into and mapped as low, medium, and high categories, which were grouped by the first and third quartiles of the corresponding pooled dataset (with $130 \times 9 = 1170$ values) across the 130 subbasins for the baseline condition and the eight options. As a result, two quartiles were derived for mean daily sediment yield and another two quartiles were obtained for sediment frequency. Each quartile pair was taken as the broken points of the three categories: the low category included values of less than the first quartile, the medium category values between these two quartiles, and the high category values of greater than the third quartile. The created maps were compared visually to scrutinize the land-use–soil interactive effects on sediment contributions from the subbasins of the watershed.

5.7.3 Results and discussion

5.7.3.1 Calibrated model

The SWAT model adopted the CN2 values estimated by AvSWAT-X and the adjusted values shown in table 5.21 for the other eight calibration parameters, namely SURLAG, MSK_CO1, MSK_CO2, REVAPMN, GW_REVAP, GWQMN, ESCO, and EPCO. The six sediment-related parameters were determined to be CH_COV = 0.022, CH_EROD = 0.090 cm h^{-1} pa^{-1}, ADJ_PKR = 1.52, PRF = 1.0, SPCON = 0.05, and SPEXP = 1.5. The calibrated model gave an annual average sediment yield of 1.76 tones ha^{-1} at the watershed outlet, with 0.66 tones ha^{-1} from erosions of the stream bank and bed and 1.10 tones ha^{-1} from overland erosion. The modeled sediment yield and source contributions were comparable to those presented by Narasimhan *et al* (2007).

The calibrated model performed well (table 5.23) in reproducing the observed monthly and seasonal discharges at the watershed outlet, as indicated by NSE > 0.63, and the observed streamflow volume, as indicated by a prediction error of <1% (Moriasi *et al* 2007). The model's good performance was also verified by capturing the streamflow fluctuations, as shown by the visualization plots showing the observed versus predicted streamflow at annual (figure 5.42(a)) and monthly (figure 5.42(b)) time scales.

The annual average discharge was overestimated for one year but underpredicted for another year (figure 5.42(a)). The model mostly underestimated the record-breaking peak discharge of 20 December 1991 (USGS 1998), when an extreme storm occurred. The storm generated more than 102 mm daily rainfall in many locations of the watershed. At the daily time scale, the discharges of larger than 50 m^3 s^{-1} tended to be underestimated, whereas the discharges smaller than this tended to be overestimated (figure 5.43). Overall, the simulated daily discharges were less reliable than using the average of the observed values, as indicated by daily NSE < 0. The poor model performance at the daily time scale can be attributed to SWAT's limitation in modeling long-lasting dry conditions and/or short-term heavy thunderstorms (Feyereisen *et al* 2007). For this reason, the assessment was performed using the model predictions at the monthly and annual time scales, without further analyzing the daily predictions.

The record-breaking peak discharge in 1991 was poorly reproduced, indicating SWAT's limitation in simulating runoff from an extreme heavy storm (Wang *et al* 2006), which usually results in large and quick fluctuations of soil moisture within a period much shorter than a day (Brutsaert 2005). However, the SWAT model was

Table 5.23. Statistics of discharges at the outlet of the study watershed[1].

	Calibration[2]				Validation[2]			
Statistics	Obs. (m^3 s^{-1})	Pred. (m^3 s^{-1})	R^2	NSE	Obs. (m^3 s^{-1})	Pred. (m^3 s^{-1})	R^2	NSE
Daily		3.04	0.05	−0.11	3.65	3.96	0.03	−0.47
Monthly	3.02	3.04	0.64	0.63	3.65	3.96	0.22	−0.07
Seasonal	3.02	3.04	0.79	0.76	3.65	3.96	0.04	−0.37

[1] Reproduced with permission from table 5 in Wang *et al* (2010). Copyright 2010 Elsevier. Obs.: observed; Pred.: predicted; R^2: coefficient of determination; and NSE: Nash–Sutcliffe coefficient.

[2] Calibration: 1 December 1984 to 30 November 1996; and Validation: 1 December 1996 to 30 November 2006.

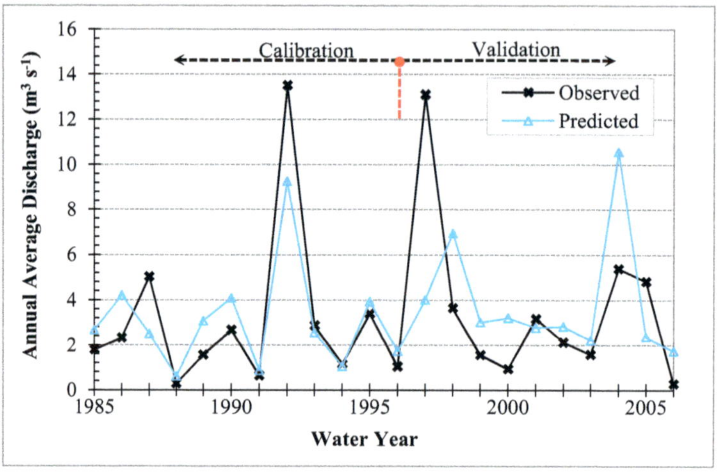

(a)

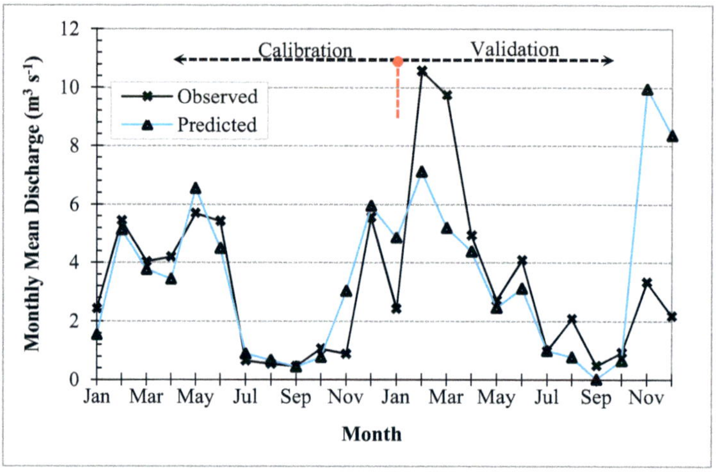

(b)

Figure 5.42. The predicted versus observed discharges at the watershed outlet. Plots showing the (a) annual discharges and (b) monthly discharges. Calibration: 1 December 1984–30 November 1996; and Validation: 1 December 1996–30 November 2006. (Reproduced with permission from figure 4 in Wang *et al* (2010). Copyright 2010 Elsevier.)

run at a daily time step with daily average soil moisture, which is lower than the soil moisture at peaking, causing the underprediction of the peak discharge (Wang *et al* 2008a). Also, the SWAT model used the SCS curve number method, which did not consider the influence of rainfall intensity on runoff generation, although such an influence might be crucial for an extreme storm.

For the validation, the model failed to reproduce the discharges from two extreme large storms of 1997 and 2004 (figure 5.42), as indicated by negative NSEs (table 5.23). This can again be attributed to the SWAT's limitations discussed in the last paragraph. Similarly, the SWAT model underestimated daily discharges

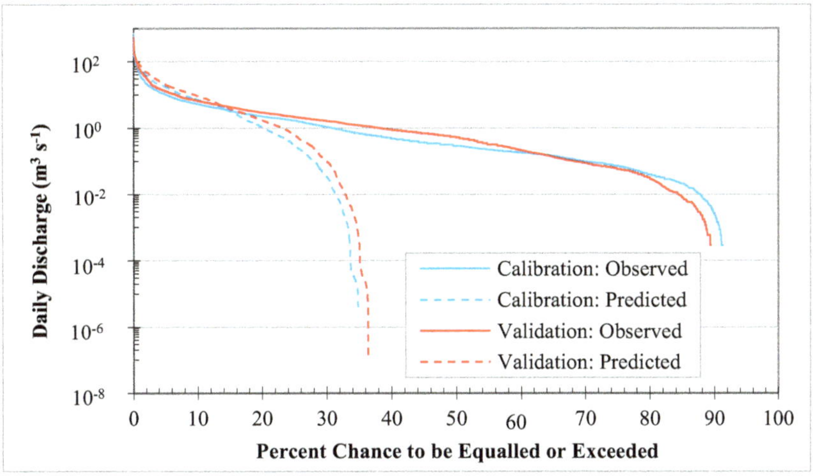

Figure 5.43. Duration curves of the predicted and observed daily discharges at the watershed outlet. Calibration: 1 December 1984–30 November 1996; and Validation: 1 December 1996–30 November 2006. (Reproduced with permission from figure 5 in Wang *et al* (2010). Copyright 2010 Elsevier.)

larger than 50 $m^3\ s^{-1}$, whereas the model overestimated smaller daily discharges (figure 5.43). The model performed very well in reproducing the observed runoff volume, as indicated by a small prediction error of <8% (Moriasi *et al* 2007). Overall, the SWAT model of this case study had a performance comparable to or better than the SWAT models of previous studies (Vazquez-Amábile and Engel 2005, Du *et al* 2006, van Liew *et al* 2007). Thus, this calibrated SWAT model was good enough for examining the land-use–soil interactive effects on water and sediment yields in the study watershed.

5.7.3.2 Effects on water and sediment yields at the watershed outlet

The water yield at the watershed outlet was predicted to have a detectable increase if a sufficiently large area of range brush were replaced by range grasses, which agrees with the conclusion drawn from 94 field experiments (Bosch and Hewlett 1982). Options 1, 2, 4, and 5 (table 5.22) were predicted not to affect the water yield (table 5.24) because the conversion areas of 148.0–5447.4 ha might not be sufficiently large. For the studies watershed, the threshold conversion (i.e. just sufficiently large) area was determined to be between 5500 and 10 000 ha to result in a detectable increase of the water yield. Following the principles of the hydrological cycle, for the threshold conversion area the evapotranspiration decrease from the removal of range brush is just equal to the evapotranspiration increase from growing range grasses, while for a conversion area larger than the threshold, the evapotranspiration decrease from the removal of range brush will be greater than the evapotranspiration increase from growing range grasses. Welder (1988) did not detect any change in streamflow from removal of the 8700 ha saltcedar, probably because the removal area was smaller than the threshold for a detectable effect. In contrast, Dugas and Mayeux (1991) detected a measurable but small increase in

Table 5.24. The modeled annual average discharge and sediment loading at the watershed outlet[1].

Period	Base Condition	Option 1	2	3	4	5	6	7	8[4]
Calibration[2]									
Discharge (m^3 s^{-1})	3.04	3.05	3.04	3.08	3.04	3.04	3.06	3.38	3.38
Sediment (tonnes $year^{-1}$)	203 602	204 557	203 715	203 230	203 981	196 660	201 715	211 590	205 265
Validation[2]									
Discharge (m^3 s^{-1})	3.96	3.96	3.96	4.07	3.96	3.96	4.07	4.34	4.37
Sediment (tonnes $year^{-1}$)	226 108	339 136	337 850	340 971	338 229	325 342	338 113	346 947	343 163
Evaluation[3]									
Discharge (m^3 s^{-1})	3.46	3.46	3.46	3.53	3.46	3.46	3.52	3.82	3.84
Sediment (tonnes $year^{-1}$)	213 832	265 730	325 656	328 449	265 003	255 152	263 714	273 116	267 946

[1] Reproduced with permission from table 5 in Wang *et al* (2010). Copyright 2010 Elsevier.
[2] Calibration: 1 December 1984 to 30 November 1996; and Validation: 1 December 1996 to 30 November 2006.
[3] The evaluation period includes both the calibration and validation periods.
[4] The predicted sediment loadings for option 8 were lower than those for option 7 because the conversion of all range brush into range grasses would retard sediment from the uplands being transported downstream to the outlet of the study area.

water yield, indicating that the removal area of mesquite might be just larger than the required threshold.

Options 3, 6, 7, and 8 were predicted to increase the water yield, as indicated by the predicted discharges for these four options being greater than the predicted discharges for the baseline condition (table 5.24). The increases in water yield were predicted to be larger for the validation than the calibration period because the validation period was wetter than the calibration period in terms of the observed stream flows (3.65 versus 3.02 m^3 s^{-1}) (table 5.23). A wetter hydrological condition is more favorable for increasing water yield from brush removal than a drier condition (Wilcox *et al* 2005). Option 8 was predicted to increase the annual water yield by 24 mm from the 49 863 ha treated area, within which range brush was replaced by range grasses. This predicted increase was judged to be realistic because it was around half of what was reported by Afinowicz *et al* (2005). The increase is likely caused by the reductions in interception and transpiration due to the replacement of range brush by range grasses (TSSWCB 2003). For options 3, 6, and 7, the increases in annual water yield were predicted to be 11–80 mm in the corresponding treated areas. Regardless of the options, the conversion on the TX629 soil was predicted to be more effective than the conversions on the TX251 and TX609 soils, because TX629 soil has a moderately high permeability of about 10 mm h^{-1} (table 5.1) and is in the upland area (figure 5.6), the rapid movement of water through the soil tends to increase runoff coefficient once range brush was replaced by range grasses (Wilcox *et al* 2005).

Although the TX609 soil was more permeable than the TX251 soil, the increase in annual water yield was predicted to be larger by converting range brush into range grasses on the TX251 than TX609 (23 mm versus 11 mm from the corresponding treated areas). One reason is that the TX251 soil is closer to the Cowhouse Creek and its major tributaries, leading to less evapotranspiration and infiltration losses of the salvaged water by the conversion of range brush into range grasses before the water discharges into the streams (figure 5.6). Another reason is that less soil water would be consumed by range grasses than range brush (Wilcox *et al* 2005). Rationally, the runoff potential varies with hydrological soil group (HSG) (USDA-NRCS 2004). HSG D has the highest runoff potential, whereas HSG A soils have the lowest runoff potential. HSG B and HSG C soils have a moderate runoff potential. For this reason, the conversion of range brush into range grasses on HSG D soils was more effective in increasing water yield than on the other three HSGs soils. This explains the inconclusive increases in water yield reported by previous studies, such as Bednarz *et al* (2001), Wu *et al* (2001), TAES (2002), TSSWCB (2003), and Afinowicz *et al* (2005).

While likely increasing water yield, the conversion of range brush to range grasses may induce more soil erosion and increase sediment yield (table 5.24) because the erosion resistance of range grasses is smaller than that of range brush as indicated by the USLE C-factors (Hedrick *et al* 1966) and because the increased overland runoff and discharge have more erosive energy to detach and transport topsoil particles (Duan 2004). On the one hand, options 1, 2, 4, and 5 were predicted not to increase water yield, but they would increase the annual sediment yield by 7.7–346.0 tones ha^{-1} treated area. On the other hand, options 3, 6, 7, and 8 were predicted to increase the water yield, option 8 would increase the sediment yield by about 1.0 tons $year^{-1}$ ha^{-1} treated area, and the other three options would increase the sediment yield by 4.2–11.9 tones ha^{-1} treated area.

As expected, the increase in sediment yield was predicted to be soil-specific. The conversion of range brush into range grasses tended to induce a greater increase in sediment yield for a soil with higher USLE erodibility factor and runoff potential than a soil with lower ones. For instance, because of the conversion of range brush into range grasses, the increase in annual water yield was predicted to be almost seven times larger for the TX629 than TX609 soil, but the increase in sediment yield was predicted to be comparable for these two types of soils (4.2 versus 3.0 tones ha^{-1} treated area for the TX609 and TX629 soil, respectively). The TX629 soil is much more resistant than the TX609 soil, as indicated by the USLE erodibility factor of the TX629 soil being just 37% of that of the TX609 soil (table 5.1). For the TX629 soil, the increase in erosive stress from the larger increase in water yield can be partially offset by the greater resistance of the soil, leading to a moderate increase in sediment yield, whereas for the TX609 soil, the increase in erosive stress from the smaller increase in water yield may be exacerbated by the smaller resistance of the soil, leading to a slightly greater increase in sediment yield. For the TX251 soil, the predicted large increase in sediment yield can be attributed to this soil being close to the Cowhouse Creek and its tributaries (figure 5.6), providing a favorable natural condition for eroded soils to be transported downward into the streams. Depending on the soil of interest, the streamflow sediment concentration would be increased by 300–1000 mg L^{-1} from the conversion of range brush into range grasses. In terms of balancing increases in water and sediment yields, for the studies watershed, the

TX629 soil was predicted to be most appropriate for substituting range grasses for range brush.

5.7.3.3 The effects on sediment yields at the subbasin outlets

For the baseline condition, more subbasins in the upper part of the watershed were predicted to have a medium (5–10 tones ha^{-1}) or high (>10 tones ha^{-1}) sediment yield, while more subbasins in the lower part were predicted to have a low (<5 tones ha^{-1}) or medium sediment yield (figure 5.44(a)). Such a spatial pattern can be attributed to the upper subbasins having a greater USLE-length-slope factor than the lower subbasins (Moore and Burch 1986) and the lower part of the watershed having a higher forest coverage (figure 5.7), protecting topsoil from erosion and retarding the transport of eroded soils downward into the streams. However, the spatial pattern of sediment occurrence (i.e. percent chance of having a non-zero daily sediment yield) was inconsistent with the above-discussed spatial pattern of sediment yield (figure 5.44(b) versus figure 5.44(a)). A subbasin with a low sediment yield could contribute sediment at a high frequency (>15%), whereas a subbasin with a higher sediment yield might contribute sediment at a low frequency (<10%), and vice versa. The subbasins with a medium sediment yield could contribute sediment at a low frequency, a high frequency, or a medium frequency (i.e. 10%–15%).

The conversion of range brush into range grasses would or would not change the baseline spatial patterns of sediment yield and frequency, depending on the option of interest. Options 1, 2, 4, and 5 were predicted to barely change the spatial patterns and thus are not shown. Option 1 was predicted to change the sediment yield from the medium to high category for one upper subbasin and that from the low to medium category for one lower subbasin, and the sediment occurrence from the medium to high frequency for one lower subbasin and from the high to medium frequency for another lower subbasin. Options 2, 4, and 5 would change the sediment yield categories and sediment occurrences for only one or two subbasins. The reasons for so few subbasins having such changes from the baseline spatial patterns can be attributed to the spatial topology and physical properties of the treated soils. The TX176 soil stretches along the floodplain of Cowhouse Creek and the TX244 and TX260 soils are clustered with small sizes in the upper and lower areas (figure 5.6). The TX463 soil was predicted to have the smallest increase in annual sediment yield of 7.7 tones $year^{-1}$ ha^{-1} treated area (table 5.24), indicating the least sensitivity of this soil to the conversion of range brush into range grasses.

In contrast, options 3, 6, 7, and 8 were predicted to change the baseline spatial patterns (figures 5.45 to 5.48 versus figure 5.44). As expected, for an option of interest, changes would occur in the subbasins that have the soil treated by this option. For instance, because option 3 was to convert range brush into range grasses on TX176 soil, this option was predicted to change the sediment yield and sediment occurrence in a subbasin in the vicinity of Cowhouse Creek (figure 5.45(a),(b)) if this subbasin has TX176 soil (figure 5.6). The land-use–soil interactive effects on sediment yields and sediment occurrences were reflected by the predicted spatial patterns for option 8 (figures 5.48(a) and (b)). In terms of minimizing changes to the baseline spatial patterns of sediment yield and sediment occurrence, option 7, which

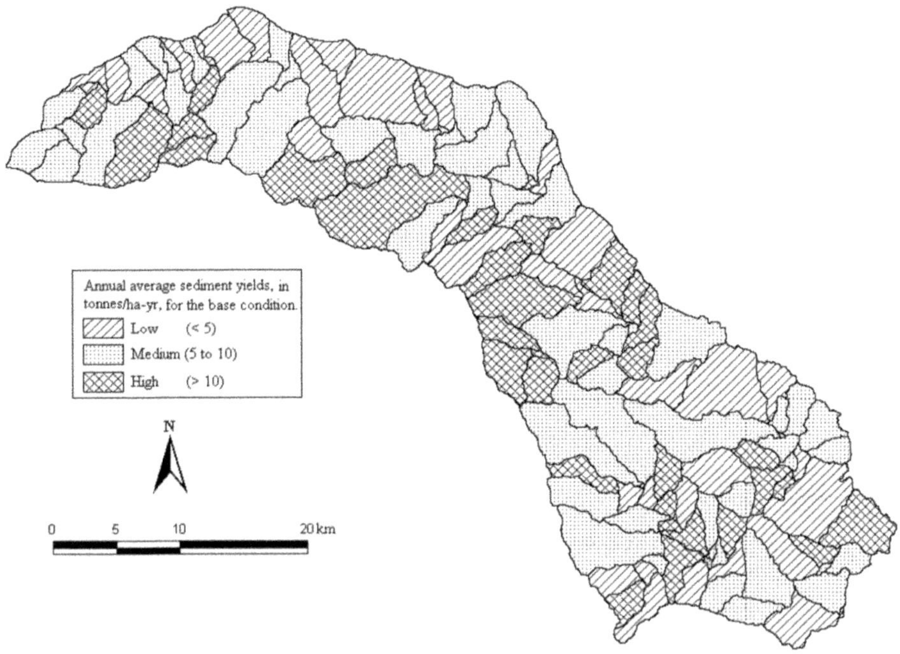

(a)

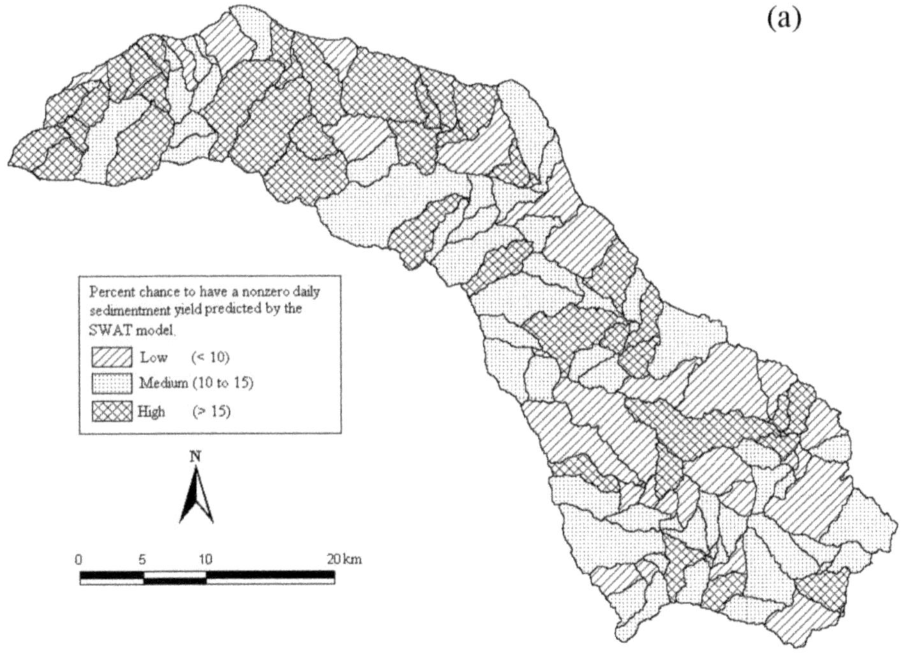

(b)

Figure 5.44. The predicted subbasin-scale sediments under the baseline condition. Map showing the sediment (a) yields and (b) occurrences. (Reproduced with permission from figure 6 in Wang *et al* (2010). Copyright 2010 Elsevier.)

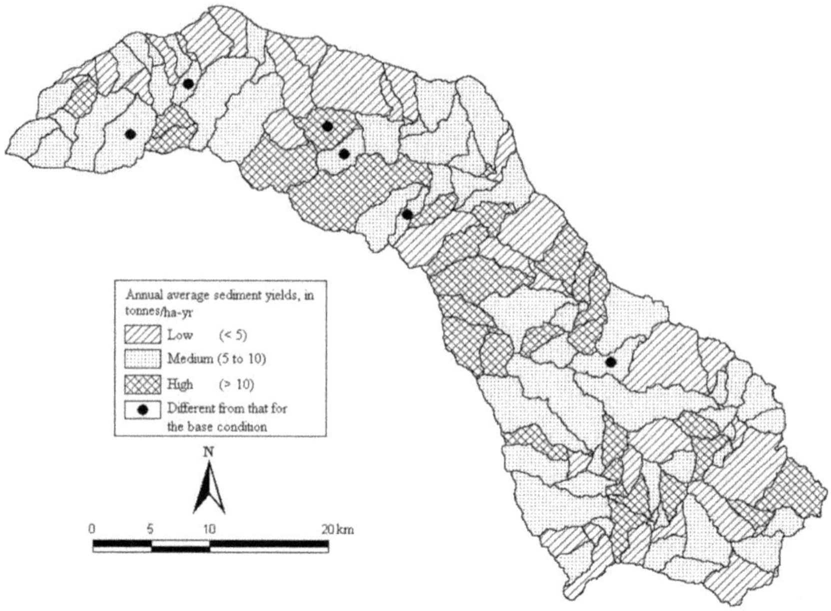

(a)

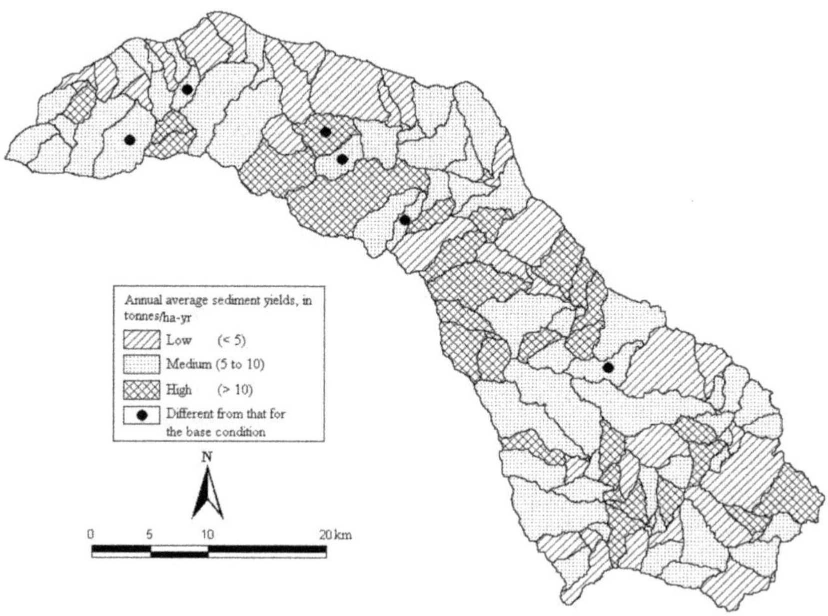

(b)

Figure 5.45. The predicted subbasin-scale sediments for option 3. Map showing the sediment (a) yields and (b) occurrences. (Reproduced with permission from figure 7 in Wang *et al* (2010). Copyright 2010 Elsevier.)

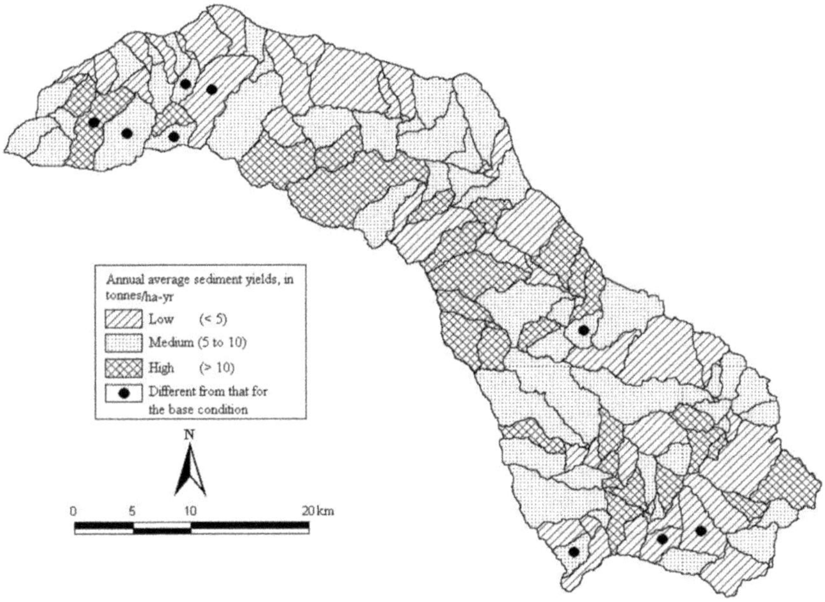

(a)

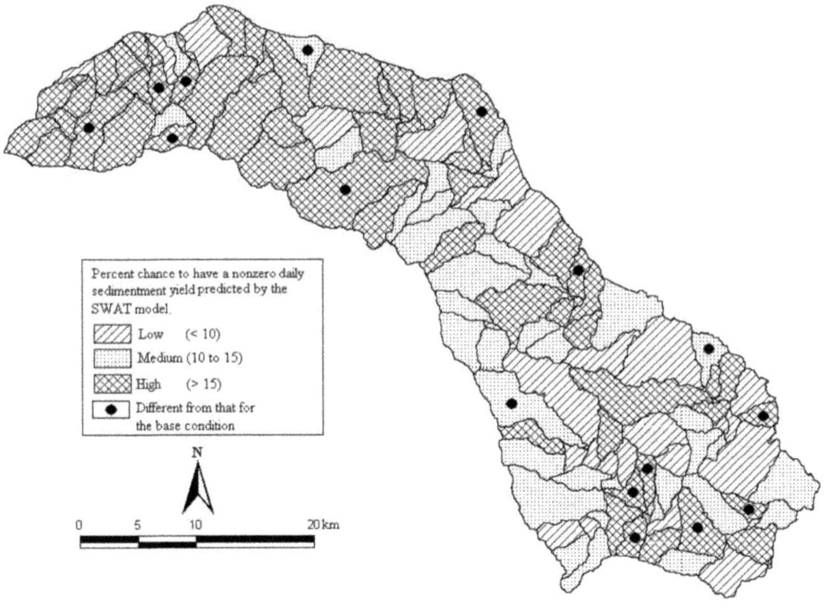

(b)

Figure 5.46. The predicted subbasin-scale sediments for option 6. Map showing the sediment (a) yields and (b) occurrences. (Reproduced with permission from figure 8 in Wang *et al* (2010). Copyright 2010 Elsevier.)

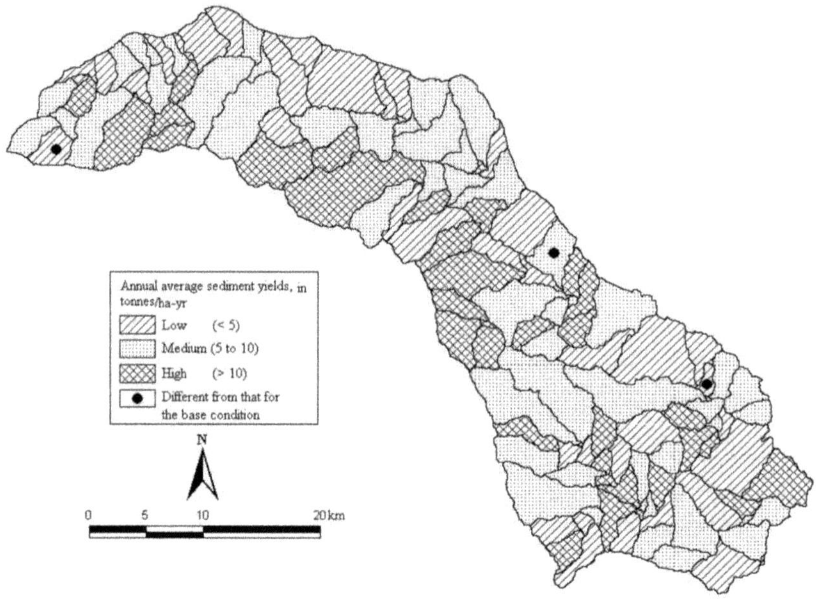

(a)

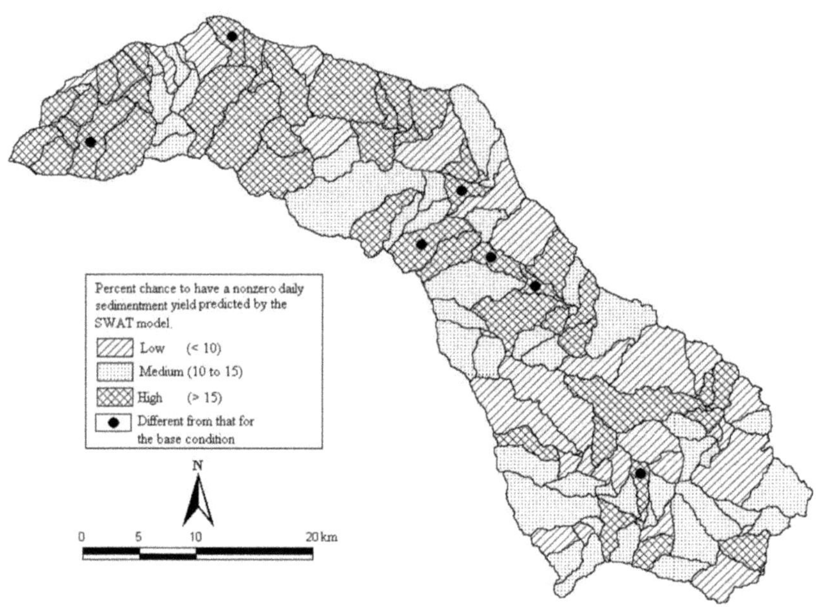

(b)

Figure 5.47. The predicted subbasin-scale sediments for option 7. Map showing the sediment (a) yields and (b) occurrences. (Reproduced with permission from figure 9 in Wang *et al* (2010). Copyright 2010 Elsevier.)

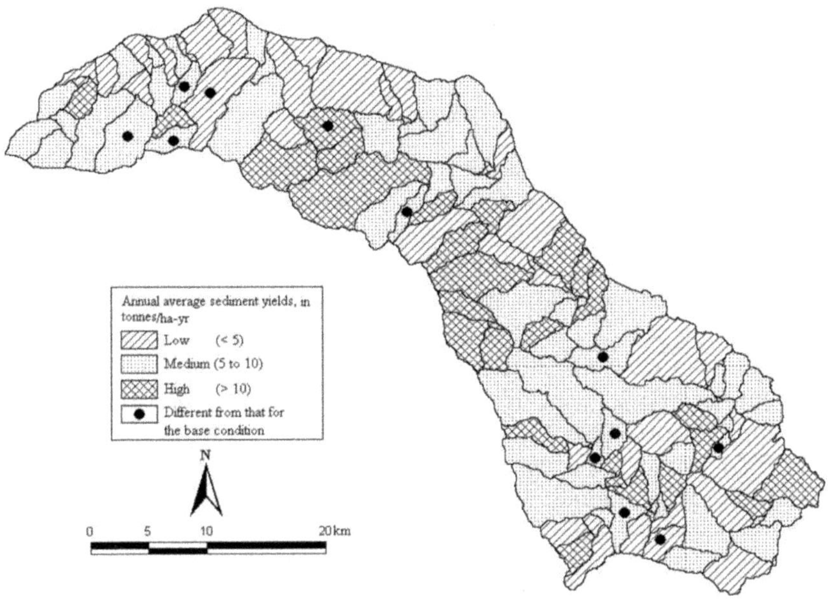

(a)

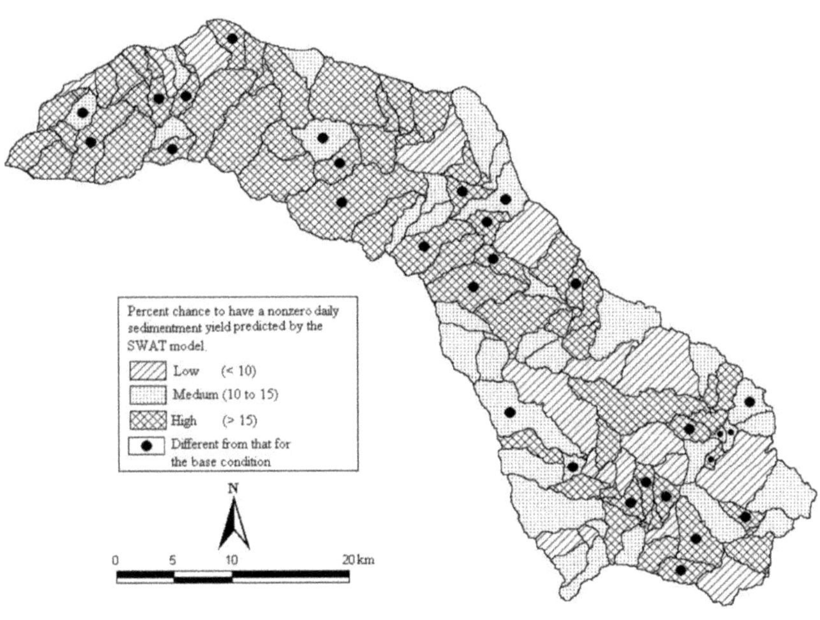

(b)

Figure 5.48. The predicted subbasin-scale sediments for option 8. Map showing the sediment (a) yields and (b) occurrences. (Reproduced with permission from figure 10 in Wang *et al* (2010). Copyright 2010 Elsevier.)

was to convert range brush into range grasses on the TX629 soil, was judged to be the most favorable (figure 5.47 versus figure 5.44).

5.7.4 Summary and conclusions

This case study calibrated and used a SWAT model to predict the land-use–soil interactive effects on water and sediment yields from the 117 845 ha Cowhouse Creek watershed in north central Texas. The model was run on a daily time step from 1 December 1984–30 November 2006. This calibrated model, which had a simulation accuracy comparable with or better than most SWAT models of previous studies, was used to predict changes to the baseline water and sediment yields from the conversion of range brush into range grasses soil by soil.

At the watershed scale, the average annual water yield was predicted to increase by 24 mm ha^{-1} treated area if all 49 863 ha range brush was replaced by range grasses. At the subbasin scale, conversions on the TX176, TX244, TX260, or TX463 soil would barely result in any change, whereas conversions on the TX609, TX251, and TX629 soils would increase the annual water yield by 11, 23, and 80 mm, respectively. Conversion on the TX629 soil would result in the largest increase in water yield because this HSG D soil has a moderately high permeability and is in the upland area. For a soil of interest, the calibration period was drier and thus had a smaller increase in water yield than the validation period. The TX251 soil was predicted to have a larger increase in water yield than the TX609 soil because the former soil was geographically closer to the streams than the latter soil.

On the other hand, the baseline sediment yield would or would not be increased, depending on the treated soils. For some soils, the conversion of range brush into range grasses would have no detectable effect on water yield but was predicted to increase sediment yield, whereas for other soils, the conversion was predicted to increase both water and sediment yields. Relative to the other six soils, the TX629 soil would be most favored because the relevant conversion was predicted to result in the largest increase of 80 mm ha^{-1} treated area in annual water yield, the smallest increase of 4.2 tones ha^{-1} treated area in annual sediment yield, and a minimal alteration to the baseline spatial patterns of sediment yields and occurrences. Thus, to improve the watershed health and sustainability, such land-use–soil interactive effects should be considered in developing brush control measures. This case study augmented previous field experiments (Bosch and Hewlett 1982, TAES 2002) and modeling studies (TSSWCB 2003). It generated useful information for assessing relative changes in water and sediment yields as a function of brush conversion options on an individual soil basis. However, caution should be taken in interpreting the absolute changes in sediment yield because the sediment module could not be calibrated due to lack of data.

5.8 The effects of biocrusts on water balance and soil erosion

Biocrusts are intimate associations of soil particles with cyanobacteria, algae, fungi, lichens and bryophytes, uncountable bacteria, and mosses. They are omnipresent as living cover in the open spaces surrounding vascular plants in arid and semi-arid steppe ecosystems and can play active roles in improving the physicochemical properties (e.g. water holding capacity) of top materials, increasing surface resistance to erosion and water availability at the soil–vegetation–atmosphere interface. However, few studies have been conducted to model such effects for a long-term period, partially because existing ecohydrological models do not have explicit algorithms for biocrusts. The objectives of this section are to (i) adapt SWAP to assess the long-term effects of biocrusts on major water balance components, such as evaporation, transpiration, and percolation, and (ii) take SWAP's outputs as inputs into IFAP to evaluate the long-term effects of biocrusts on aeolian and fluvial erosions. The testbed was the Balager River watershed described in section 5.1.1. In this regard, the SWAP and IFAP models in the absence (designated SWAP-soil and IFAP-soil for description purposes) and presence (designated SWAP-moss and IFAP-moss for description purposes) of biocrusts were created. The SWAP-soil was calibrated and validated in section 5.2, while the IFAP-soil was parameterized in section 5.4. The topmost 1 cm sublayer in the SWAP-moss model is parameterized using the literature properties of the moss biocrusts and the parameters for the sublayers beneath were same as those in the SWAP-soil model. The IFAP-moss model inherited the parameters from the IFAP-soil model, with the erosivity thresholds adjusted by empirical crust resistance factors. The four models were run from 1955 to 2017 at a daily time step—the initial three years were used as the warm-up period and the remaining 60 years for assessments. The effects of the moss biocrusts were assessed by comparing the predictions of the SWAP-moss model with those of the SWAP-soil model as well as the predictions of the IFAP-moss model with those of the IFAP model. The results indicated that the presence of the biocrusts will reduce the wind erosion occurrence by 21.9% and the wind erosion rate by 1.4 mm year^{-1}. A maximal reduction was predicted to occur in April, when wind speeds were relatively high while soil moisture and grass coverage were very low. Biocrusts will increase the average annual evaporation by 3.7 cm and decrease the average annual transpiration and percolation by 1.6 and 2.2 cm, respectively. A reasonable generalization of these results is that biocrusts are very beneficial for sustaining dryland ecosystems that are susceptible to soil erosion and degradation.

5.8.1 Introduction

For a dryland ecosystem, biocrusts play profound roles in improving the physicochemical properties (e.g. water and nutrient holding capacity) of soils in the rootzone, increasing topsoil resistance to erosion, fixing carbon dioxide and nitrogen, and regulating cycles of energy, carbon, and water at the soil–vegetation–atmosphere interface (Mager and Thomas 2011, Reed *et al* 2019). However, the science of biocrusts did not emerge until the 1970s and did not evolve into a real discipline until 2003 (Sancho *et al* 2014, Bowker *et al* 2018), when an edited volume by Belnap and Lange (2003) was first published. From then on this discipline has been advanced greatly by presentations at international conferences held in 2010 (BioCrust1), 2013 (BioCrust2), 2016 (BioCrust3), 2019 (BioCrust4), and 2021 (BioCrust5), and by various journal papers (Belnap 2006). More international conferences have been planned in 2023 (BioCrust6) and 2025 (BioCrust7), where new ideas are expected to be shared, evaluated, and synthesized. Given biocrust science's four evolving research focuses, namely scaling, modeling, improved understanding of *in situ* processes, and restoration (Rozenstein and Adamowski 2017, Reed *et al* 2019), a variety of knowledge gaps need to be filled. In the existing literature (Bowker *et al* 2018), some studies devoted to measure the structure of biocrusts, while others evaluated the ecological and hydrological functions of biocrusts.

The structure of a biocrust can be characterized by micro-organism composition, abundance of dominant phototropic species, chlorophyll *a* content, color, successional stage, community mottled and patchy pattern, and spatial cover and distribution. The biocrust can have a dominant phototropic species of cyanobacteria, lichens, or mosses, which corresponds to one of the three successional stages (Belnap 2006). As the crust matures (i.e. grows through the successional stages), the biomass of cyanobacteria, lichens, and mosses increases, which in turn increases the aggregate stability, shear strength, and roughness of the ground surface. During the first stage, filamentous cyanobacteria that live just below the soil surface aggregate soil particles by releasing extracellular polymeric substances (EPSs) to form sheaths as well as by excreting polysaccharides (Adessi *et al* 2018), while during the second or third stage, lichens or mosses live atop and colonize the soil surface (Belnap 2006). Cyanobacteria are less likely to clog soil pores than lichens and mosses. Cyanobacteria biocrusts may be further classified into one of the six levels of development (LODs) in terms of the visual classification system presented by Belnap *et al* (2008). A lower LOD biocrust tends to have a lighter color, fewer lichens and/or mosses, and a smoother surface than a higher LOD biocrust; and vice versa. Li *et al* (2016c) and Becerra-Absalón *et al* (2019) used DNA extraction techniques to examine the microbial species in biocrusts, and found that the cyanobacteria biocrusts had more bacteria and phototrophs but fewer fungi than the lichen and moss biocrusts and that the species compositions of lichen and moss biocrusts were statistically same. The species composition of cyanobacteria biocrusts seemed to be sensitive to site-specific environments. Román *et al* (2019) used a nondestructive methodology to quantify chlorophyll *a* content, which is a well-proven proxy of biocrust development and photosynthetic species status. As expected, those authors found that the moss biocrust had a much higher chlorophyll *a* content than the lichen biocrust, which in turn had a marginally higher chlorophyll *a* content than the cyanobacteria biocrust. Li *et al* (2016c) found that at the community level, the micro-organisms within a cyanobacteria biocrust tended to be homogenously distributed, but those within either a lichen or moss biocrust tended to have a mottled and patchy distribution pattern. For the biocrusts of second and/or third stages, the well-developed (i.e. moss and lichen) patches are usually separated by less-developed (i.e. cyanobacteria) patches, with a well-developed patch surrounded by less-developed patches. Navas Romero *et al* (2019) analyzed variations of the biocrust coverage of central-west Argentina. They found that the (i) arid system had a higher biocrust coverage than the hyper-arid system, which in turn had a higher biocrust coverage than the semi-arid system and (ii) the hyper-arid system had the highest coverage of cyanobacteria biocrusts, while the arid system had the highest coverage of lichen and moss biocrusts. The biocrust coverage of an ecosystem depends on the ecosystem's characteristics and environments regardless of its spatial extent. Gao *et al* (2017b) found that the biocrust and vegetation coverages of an ecosystem were complementary; that is, the biocrust coverage tended to increase with a decrease of the vegetation coverage, and vice versa.

Biocrusts can enhance the sustainability and resilience of fragile ecosystems in arid and semi-arid regions by increasing the beneficial micro-organisms and water–carbon–nutrient holding capacity of rootzone soils. Steven *et al* (2014) employed an rRNA gene and shotgun metagenomics sequencing technique presented by Janda and Abbott (2007) to compare the microbial communities inhabiting the rootzones of the dominant shrub and the interspace biocrusts in a Mojave desert shrubland in Nevada. They found that most of the numerically abundant bacteria and fungi were present in both habitats (i.e. the biocrusts and rootzones) though the proportional abundance of those members differed significantly between the habitats. Adessi *et al* (2018) measured the soil retention curves (matric potential versus relative water content) for intact biocrusts versus EPS-extracted biocrusts (i.e. same biocrust soils without living micro-organisms) and showed that the intact biocrusts had a much larger water holding capacity than the EPS-extracted biocrusts. The presence of EPS made the matric potential remain unchanged up to 80% of the saturated water content, whereas without EPS, the matric potential starts to decrease from 20% of the saturated water content. This indicates that biocrusts can delay the onset of restrictive matric potentials by increasing the water holding capacity.

In addition, Belnap *et al* (2012) did *in situ* artificial rainfall simulations at 83 experiment plots (covered by cyanobacteria biocrusts) within a cool desert region of the Colorado Plateau. They found that the eroded soil per unit plot area consistently decreased with increasing LOD (i.e. a plot covered by the biocrusts at an earlier LOD was more susceptible to water erosion than a plot covered by the biocrusts at a later LOD). Gao *et al* (2017b) performed artificial rainfall simulations in the laboratory using undisturbed biocrusts box-sampled at 59 sites (covered by biocrusts ranging from cyanobacteria to moss) across the Loess Plateau of China. They found that (i) the soil erodibility was almost halved with the presence of a biocrust (regardless of its successional stage) relative to without the biocrust, (ii) the organic matter contents (OMCs) were doubled or more in the biocrusts than in the bare soils, and (iii) OMCs were higher in the moss biocrusts than in the cyanobacteria biocrusts. In comparison to that in the bare soils, the OMC in the soils underneath the cyanobacteria biocrusts was increased by one and half times while the OMC in the top 10 cm soils underneath the moss biocrusts was increased by four times. Such OMC increases can be attributed to the biocrusts' stage-dependent roles in carbon sequestration. Cantón *et al* (2014) performed *in situ* artificial rainfall simulations on four 1 m square plots (covered by a physical crust, an incipient cyanobacteria crust, a well-developed cyanobacteria crust, and a lichen crust, respectively) in the Tabernas Desert of Spain. They found that (i) the OMC was higher in a more-developed than a less-matured biocrust (e.g. lichen versus well-developed cyanobacteria) and (ii) the OMCs in the three biocrusts were higher than the OMC in the physical crust. The OMC in the top 1 cm soils underneath the well-developed cyanobacteria crust was the highest, while the OMCs in the soils underneath the other three types of crusts were barely different. The carbon loss was highest from the physical crust and higher from a more-developed biocrust than a less-mature biocrust. The removal of the lichen biocrust induced a highest carbon loss, followed by the removal of the physical and cyanobacteria crusts. Regardless, most of the carbon losses were associated with topsoil erosion and transport. Chamizo *et al* (2017) observed organic carbon losses under natural rainfall in the Tabernas Desert of Spain. Their findings were consistent with those of Gao *et al* (2017ba) and Cantón *et al* (2014). A few studies (Chamizo *et al* 2018) showed that as with the aforementioned effects on organic carbon, biocrusts can also increase nutrients (e.g. nitrogen and phosphorus) in underlying soils and for nearby plants.

Further, Belnap (2006) reviewed the pre-2004 studies on the hydrological effects of biocrusts, leading to two major conclusions: (i) the presence of biocrusts can increase infiltration in comparison to the absence of biocrusts or the presence of physical crusts, and (ii) a biocrust at a later successional stage (or higher LOD) tends to have a rougher surface, is more favorable for infiltration, and has a larger soil water retention than a biocrust at an earlier stage (or lower LOD). The first conclusion is consistently (Bu *et al* 2014, Bu *et al* 2015, Chamizo *et al* 2018), but the second conclusion is not always (Li *et al* 2016c), supported by studies published since Belnap's review. Based on *in situ* artificial rainfall simulations in the Loess Plateau of China, Zhao and Xu (2013) found that infiltration was decreasing with increasing successional stages, which is contradictory to the relationship between infiltration and biocrust stage that was shown by Chamizo *et al* (2017) using the observations of natural rainfalls in the Tabernas desert of Spain. The presence of a biocrust was found to either increase (Zhao and Xu 2013, Cantón *et al* 2014, Fick *et al* 2019) or decrease (Bu *et al* 2015, Xiao *et al* 2015, Chamizo *et al* 2017) runoff relative to the absence of the biocrust. Such inconsistent effects on runoff have yet to be explained (Belnap 2006, Kuske *et al* 2012, Faist *et al* 2017) and may be caused by the varying hydroclimatic, topographical, physiographical, and geomorphological features.

Most existing studies, including those cited above, were conducted in deserts with sparse vegetation coverage. Because of the importance of semi-arid steppe grasslands in sustaining the local-to-global climate, socioeconomics, and eco-environment, it is imperative to fill such a knowledge gap and understand the ecohydrological roles of biocrusts in steppe ecosystems. Steppe grasslands across the world are suffering from moderate to severe degradation (Bai *et al* 2004), which has become a key constraint to the sustainability of grassland agriculture. Peart (2008) examined policy and economic reasons for grassland degradation, while Gőri *et al* (2005) assessed the impacts of grassland degradation on wildlife species. Christensen *et al* (2004) and Endo *et al* (2006) analyzed climate change and its impact on the vulnerability of steppe grasslands. The climate in the Inner Mongolian steppe has been shown to follow a significant spatiotemporal trend, with increasing air temperatures during winter and more frequent droughts in spring (Shao *et al* 2008, Wang *et al* 2014a). Other researchers (Christensen *et al* 2003, Christensen *et al* 2005, Li 2007, Blank *et al* 2009) studied the ecological functions of steppe grasslands and how such functions are affected by grazing and overgrazing, demonstrating that a grazing intensity (i.e. the ratio of the biomass in a non-grazed area to the biomass in the same area if grazed) of 1.5 or above is unlikely to pose a sustainability problem. However, only a few studies (Zhou *et al* 2002, Zhang *et al* 2005, Li *et al* 2007, Hao *et al* 2008) have assessed the hydrological processes in steppe grasslands; these few studies, which generally focused on how evapotranspiration (ET) varies with precipitation and/or soil water content, revealed that ET is limited by the availability of soil water (i.e. atmospheric heat stress is weaker than soil water stress and soil evaporation is the primary contributor to ET) and that ET has maximum daily values ranging from 3.0 to 6.2 mm d^{-1} in such grasslands. Daily ET rates tend to increase with increasing soil water content, which in turn is highly dependent on the drying–wetting cycles. Overall, soil water content is higher when grass coverage is denser (Liu *et al* 2006). These few studies also indicated that the aboveground biomass of grass species has a positive linear relationship with precipitation and that precipitation in the first half of the year (January to July) is the primary climatic factor causing fluctuations in community biomass production (Bai *et al* 2004, Luo *et al* 2014, Yang 2014, Wang *et al* 2014c). To date, there is a rudimentary understanding of steppe hydrological processes (Sprenger *et al* 2016, Goldsmith *et al* 2018) or precisely how erosion of the 10–20 cm topsoil affects the degradation of covering grasses (Wang *et al* 2014b).

The vulnerability and resilience of steppe grasslands depend on the phenomena linking the hydrological cycle with the properties of biocrusts and topsoil and the interactions thereof. Understanding the perturbations in the cycle that trigger unwanted changes in the spatiotemporal characteristics of grasslands remains a challenge because such an effort generally transcends the disciplinary and geographical boundaries; it is, however, key to developing adaptive measures for mitigating the degradation of grasslands. Most previous studies were designed to address the trilogy of grassland degradation, topsoil erosion, and hydrology alteration individually (Sugita *et al* 2015, Sun *et al* 2017) and were thus unable to examine the important nexuses among them. However, in spite of being a key subecosystem, biocrusts in steppe grasslands have barely been studied (Faist *et al* 2017, Fick *et al* 2019). In consequence, the results from previous studies are of only limited utility for those seeking to develop practical solutions for sustaining grassland agriculture. The objective of this study was to investigate the effects of biocrusts on water balance and topsoil erosion, using the Balagaer River watershed (see section 5.1.1) as the testbed. Herein, the new foci stem from a desire to contextualize biocrusts more robustly within the larger framework of ecosystem function and to compare their roles more quantitatively to those of other biotic and abiotic elements.

5.8.2 Materials and methods

5.8.2.1 Adapted models

This study adapted two models, namely the SWAP model calibrated in section 5.2 and the IAFP model presented in section 5.4, to evaluate the effects of moss biocrusts. In the SWAP model, the optimal grazing intensity of GI = 3.0 cattle ha^{-1} was used.

To model the situation with biocrusts, the 200 cm soil profile was re-discretized into six sublayers (figure 5.49). Two versions of the model were created in terms of

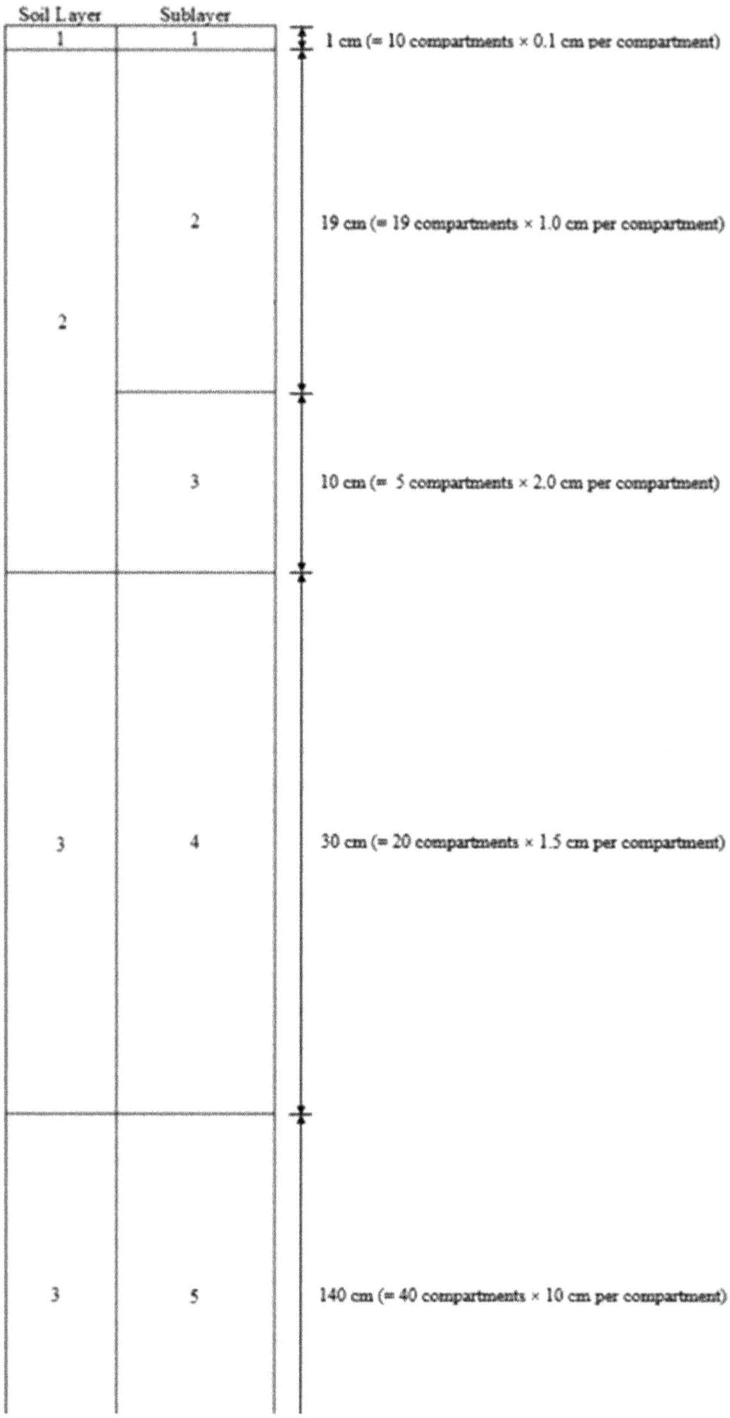

Figure 5.49. The modeled 200 cm soil profile. For modeling purposes, the second soil layer was discretized into two sublayers and each of the other four soil layers was not further treated as one sublayer. Each sublayer was subdivided into several compartments with a constant height.

the properties of the first-layer soils (table 5.24): one version (designated SWAP-soil for description purposes) used the properties of the first-layer soils, whereas another version (designated SWAP-moss for description purposes) changed the properties of the first-layer soils into those of the moss biocrusts. Both versions used the properties of the layered soils beneath. Herein, based on the data presented by Sun *et al* (2021), the thickness of the moss biocrusts was assumed to be a constant of 1.0 cm, although it may vary slightly by year and/or season (Bu *et al* 2018, Xiao *et al* 2019). The properties of the non-crusted soils were determined as stated in section 5.1.1, while the properties of the moss biocrusts were determined by inversely fitting the Fayer–Simmons soil water retention curve (equation (4.23)) to the aeolian-sand data presented by Sun *et al* (2021). Because those authors' study site is only 800 km away from the Balagaer River watershed with a variety of common features in hydro-climate, topography, physiography, and phenology, the determined properties of the moss biocrusts were reasonably assumed to also be applicable to the Balagaer River watershed. The soils at those authors' sites consist of 90.74% sand particles (20–2000 μm), 8.89% silt particles (2–20 μm), and 0.37% clay particles (<2 μm), which are similar to those of the Balagaer River watershed (figure 5.2(a)). The Fayer–Simmons retention curve fits the data very well (figure 5.50), as indicated by a small sum squared error (SSE) of 0.047, a small root mean squared error (RMSE) of 0.031, and a large coefficient of determination (R^2) of 0.96.

Similarly, two versions of the IAFP models, designated IAFP-soil and IAFP-moss for description purposes, were parameterized to predict changes of soil erosion because of the presence of biocrusts. The IAFP-soil model was parameterized using

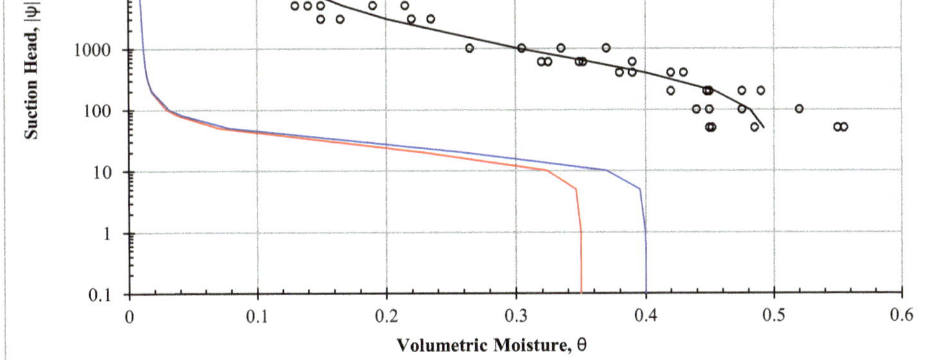

Figure 5.50. The water retention curves used in this study. The moss biocrusts increase the water holding capacity of the upper soil, which is consistent with Adessi *et al* (2018).

the properties of the first-layer soils, whereas the IAFP-moss model was parameterized using the properties of the moss biocrusts. The parameters of these two models (table 5.26) were determined from table 5.25 and section 5.4. The effects of biocrusts on erosion can be considered by introducing the resistance factors in equations (3.106) and (3.107), which may be estimated by empirical formulas in terms of the sizes of soil particles and biocrust aggregates.

Previous studies (Fick *et al* 2020) showed that biocrusts reduce the erodibility of ground surfaces by (i) binding soil particles together through growth of cellular filaments and exudation of extracellular polysaccharides (EPS), (ii) shielding surfaces from abrasion by sand, micro aggregates, and organic matter with lichen thalli and other cryptogrammic structures, and (iii) reducing wind energy at the soil surface via generation of surface microtopography (i.e. enhancing effective aerodynamic roughness length). That is, for a given soil, its erodibility is reduced partially because biocrust aggregates have a larger diameter than the soil particles. The combined effects of biocrusts on wind erodibility can be reflected by the resistance factor in equation (3.106), which can be hypothesized as

$$f_{crust,a} = \sqrt{\frac{d_{s,biocrust}}{d_s}} \tag{5.12}$$

crust resistance factor to wind erosion [> 1.0] → $f_{crust,a}$; median diameter of biocrust aggregates [L] → $d_{s,biocrust}$; median diameter of surface soil particles [L] → d_s

Based on the data presented by Gao *et al* (2017b) and Hashim *et al* (2020), the combined effects of biocrusts on water erodibility can be reflected by the resistance factor in equation (3.107), which can be hypothesized as

$$f_{crust,w} = \left(\frac{d_s}{d_{s,biocrust}}\right)^{\frac{1}{3}} \tag{5.13}$$

median diameter of surface soil particles [L] → d_s; crust resistance factor to water erosion [< 1.0] → $f_{crust,w}$; median diameter of biocrust aggregates [L] → $d_{s,biocrust}$

Table 5.25. Properties of the soil layers.

Layer	Material	Residual soil moisture θ_r [–]	Saturated soil moisture θ_{sat} [–]	Soil hysteresis factor α [cm^{-1}]	Exponent in equation (4.24) n [–]	Saturated hydraulic conductivity K_s [cm d^{-1}]	Air-entry pressure head ψ_b [cm]
1	Soils	0.02	0.35	0.05	2.5	286.98	−2.0
	Moss biocrusts	0.17	0.50	0.0027	1.435	216.00	−36.5
2	Soils	0.02	0.35	0.05	2.5	286.98	−2.0
3	Soils	0.02	0.40	0.05	2.5	312.78	−2.0

Table 5.26. Parameters of the IAFP models.

Parameter	Definition	Value
Wind		
z (m)	Aboveground height at which wind speeds are measured	2.0
Topography		
α (°)	Slope angle of the ground surface	6.3
Soils[1]		
n_s (–)	Porosity of soils	0.35
d_s (mm)	Mean diameter of concerned soil particles	0.15
ρ_s (kg m^{-3})	Particle density of the soils	2650
P_{clay} (%)	Mass content of clay particles (<0.002 mm)	1.5
Grasses		
h_{max} (m)	Maximum height of an individual grass	0.2
D_{max} (mm)	Maximum diameter of an individual grass	5.0
$PWST_{max}$ (kg ha^{-1})	Maximum dry mass of grass stems per unit area	2693
D_v (m^{-2})	Population density (i.e. number of individual grasses per unit area)	5.0
LAI_{worst} (–)	Smallest leaf area index	0.01
LAI_{best} (–)	Largest leaf area index	10
Moss biocrusts[1]		
$n_{s,biocrus}$ (–)	Porosity of moss biocrusts	0.50
$d_{s,biocrus}$ (mm)	Median diameter of biocrust aggregates	0.5
Erosion		
C_a (g m^{-5} s^2)	Wind erosion modulus (i.e. wind erodibility of bare and dry soils)	4.6×10^{-4}
C_f (g m^{-2} s^{-1})	Water erosion modulus (i.e. water erodibility of bare and saturated soils with a flat surface)	4.4×10^{-5}
b_1 (–)	Shape factor of runoff in the water erosion formula (equation (3.103))	0.7
b_2 (–)	Shape factor of topographic gradient in the water erosion formula (equation (3.103))	0.8
b_3 (–)	Shape factor of vegetation in the water erosion formula (equation (3.103))	0.03

[1] The soils and moss biocrusts are assumed to have the same particle density of 265° g m^{-3}.

5.8.2.2 Evaluation methods

On the one hand, the SWAP-soil and SWAP-moss models were run separately from 1 January 1955 to 31 December 2017. The beginning three years were used as a warm-up period to stabilize the initial values of the model variables, whereas the years from 1958 to 2017 were used to evaluate the influences of the moss biocrusts. First, changes in water balance components, namely storage, runoff, infiltration, evaporation, transpiration, and percolation, were evaluated using a variety of comparison tables and visualization plots. Second, the dynamics and vertical profiles of soil moisture were evaluated using virtualization plots and statistical tests. In this regard, two-way analyzes of variance (ANOVA) were conducted to examine any influences of soil depth and year (as a proxy for climate) on changes of soil moisture in the absence and in the presence of moss biocrusts. Finally, regression analyzes were carried out to elucidate any relationships between the moss biocrusts' effects and rainfall. All tabulations, plots, and statistical analyzes were performed in Microsoft Excel® and its Analysis ToolPak add-in.

On the other hand, the daily soil moisture of the first 1 cm sublayer and the daily LAIs simulated by the SWAP-soil or SWAP-moss model were, respectively, input into the IAFP-soil or SWAP-moss model to predict soil erosions by wind, water, and both. The differences of erosion in the absence of and in the presence of moss biocrusts were evaluated by comparison tables, visualization plots, and ANOVA tests.

5.8.3 Results

5.8.3.1 The effects on water balance

Averaged from 1958 to 2017, the annual runoff, storage, and storage change in the presence of moss biocrusts were predicted to be slightly smaller than those in the absence of the biocrusts, while the interceptions for these two types of ground cover were predicted to be the same (table 5.27).

At a significance level of $\alpha = 0.05$ the effects of the moss biocrusts on these four water balance components (i.e. runoff, storage, storage change, and interception) were statistically insignificant (p-value >0.1) and negligible. In contrast, the presence of the biocrusts was predicted to significantly increase the evaporation but reduce the transpiration and percolation, indicating that more water might penetrate into the ground surface but be held by the biocrusts without being able to further move downward. Such effects of the biocrusts were also true for each of the assessment years (figure 5.51).

The biocrusts effects on evaporation and percolation tended to increase, while those on transpiration tended to decrease, with an increase of precipitation (figure 5.52), although the relationships between these three water balance components and precipitation were very weak ($R^2 < 0.27$). That is, the moss biocrusts might enhance evaporation but diminish percolation more in a wetter year than in a drier one, whereas the diminishing effects of the biocrusts on transpiration would be lessened by increasing precipitation.

At a monthly time scale, the predicted effects of the moss biocrusts on runoff, storage, storage change, and interception were also insignificant, whereas those on transpiration, evaporation, and percolation were significant (figure 5.53). As expected, in cold months (i.e. October to March), the effects on transpiration were predicted to be near nil as the grasses did not rejuvenate and/or grow until late April. In the month of April, May, July, August, or September, the predicted transpirations were smaller in the presence than in the absence of the moss biocrusts, while the interannual variability of the transpiration reductions was opposite to that of the precipitations. In June, the effects on transpiration were predicted to be zero, indicating that from late May, livestock might fully graze the rejuvenating grasses. In addition, the predicted increasing effects of the moss biocrusts on evaporation

Table 5.27. The simulated annual average water balance components with and without moss biocrusts[1].

Component	Mean (cm)	Standard deviation (cm)	Skew coefficient	p-value
Precipitation	33.123	8.876	0.586	–
Interception				
Moss biocrusts	0.175	0.040	−0.281	
No biocrusts	0.175	0.040	−0.281	1.000
Runoff				
Moss biocrusts	0.006	0.028	5.580	
No biocrusts	0.007	0.029	5.559	0.159
Transpiration				
Moss biocrusts	9.688	2.195	0.230	
No biocrusts	11.209	2.266	0.414	0.000*
Evaporation				
Moss biocrusts	13.731	2.521	0.550	
No biocrusts	10.000	1.972	0.835	0.000*
Percolation				
Moss biocrusts	−9.483	6.885	−1.245	
No biocrusts	−11.673	7.270	−1.147	0.000*
Storage change				
Moss biocrusts	0.040	2.304	0.831	
No biocrusts	0.059	2.141	0.921	0.753
Initial storage				
Moss biocrusts	18.071	1.521	−0.560	
No biocrusts	18.083	1.419	−0.584	0.803
End storage				
Moss biocrusts	18.111	1.411	−0.073	
No biocrusts	18.143	1.260	0.267	0.545

[1] Averaged across the assessment years of 1958 to 2017.
* Signifies that the effects of the moss biocrusts on this component are statistically significant.

tended to be larger in a wetter than a drier month. For a given month, the interannual variability of the evaporation increases was consistent with that of the precipitations. Further, the decreasing effects of the biocrusts on percolation were predicted to be significant in July to November and larger in a wetter than a drier month. Although June was not dry, the percolation was barely different in the presence of or in the absence of biocrusts because of livestock grazing. In each of the five months with significant effects, the interannual variability of the percolation decreases was consistent with that of the precipitation. Moreover, at a daily time scale, similar biocrust effects on transpiration, evaporation, and percolation were found to be true as well.

5.8.3.2 The effects on soil moisture

The moss biocrusts were predicted to significantly increase the moisture levels of the top 1 cm soils, and slightly decrease the moisture levels of the soils beneath

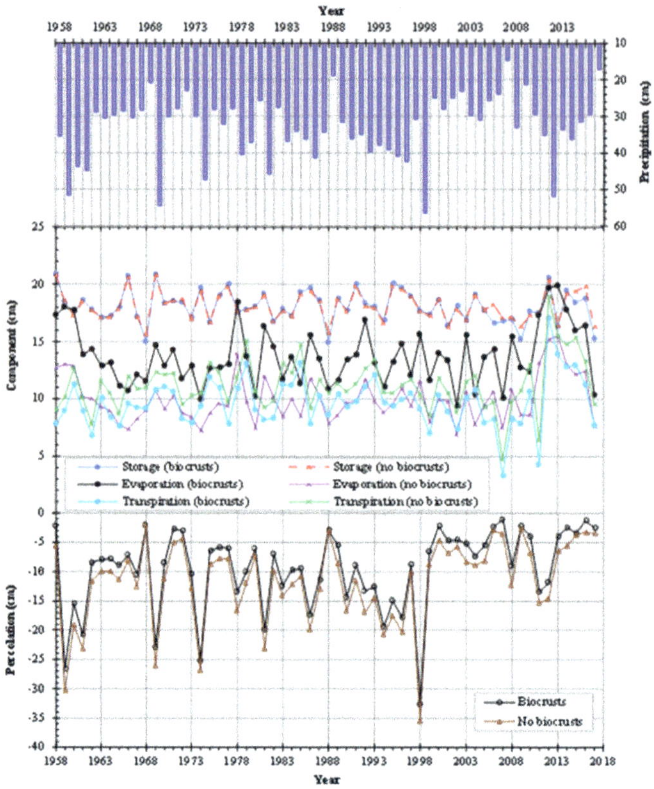

Figure 5.51. The simulated effects of the moss biocrusts on water balance components. Plots show the annual values of precipitation, storage, runoff, evaporation, transpiration, and percolation.

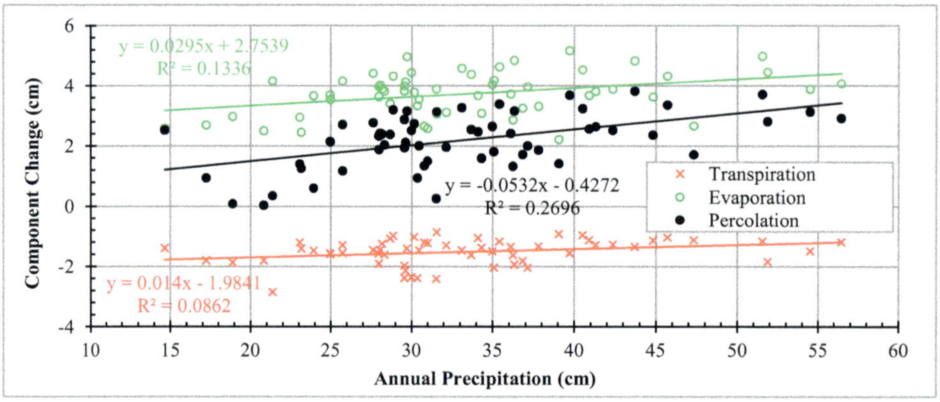

Figure 5.52. The influences of precipitation on the simulated effects of the moss biocrusts on water balance components. The plots show the annual values of transpiration, evaporation, and percolation versus annual precipitation. For a given component, its change is the subtraction of the value in the absence of the biocrusts from that in the presence of the biocrusts.

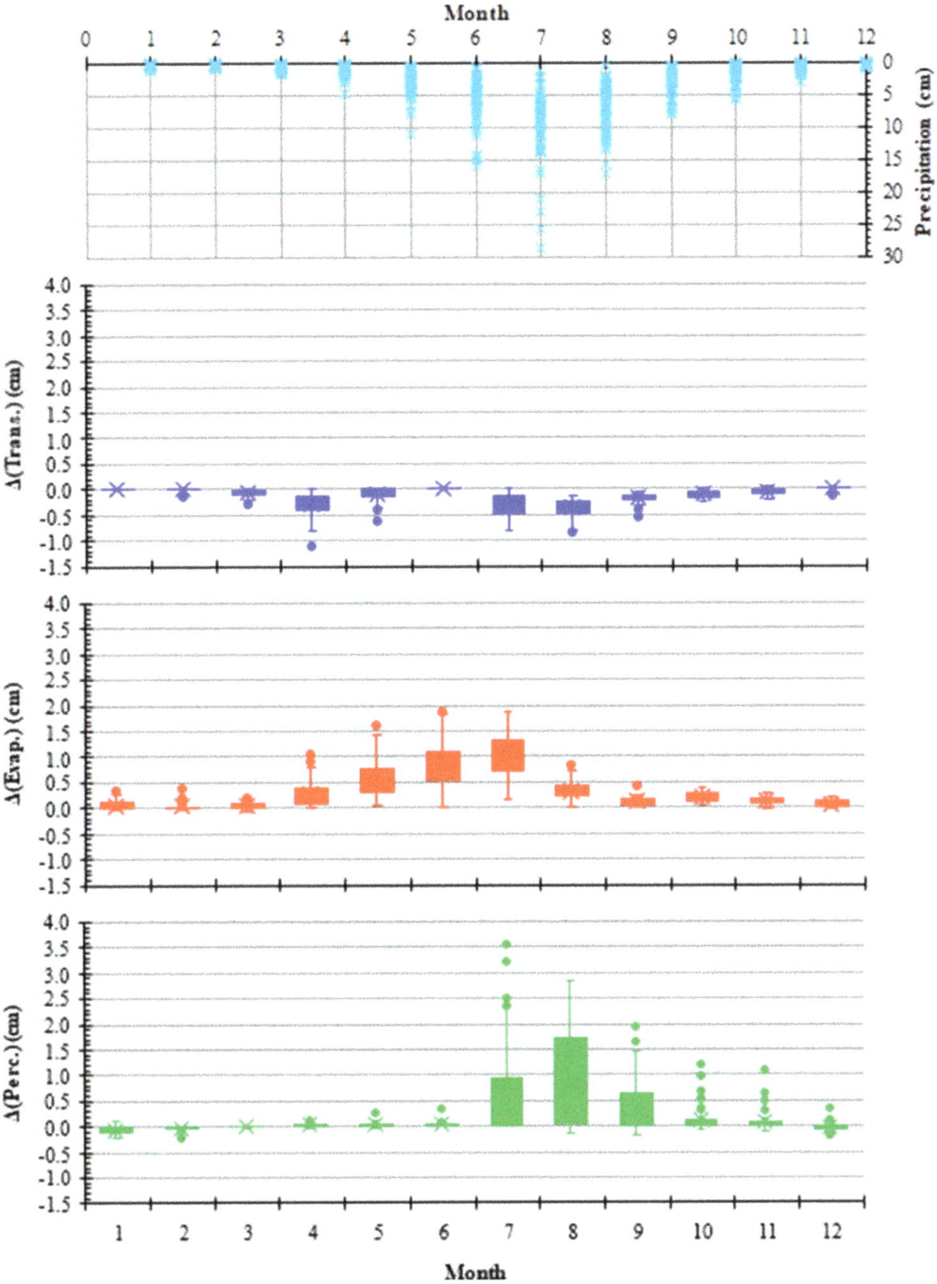

Figure 5.53. The simulated effects of the moss biocrusts on monthly water balance components. Box–whisker plots of the monthly values of transpiration, evaporation, and percolation from 1958 to 2017. The change of a component is the difference between the value in the absence of biocrusts and that in the presence of the biocrusts. Δ(Trans.): change of monthly transpiration; Δ(Evap.): change of monthly evaporation; and Δ (Perc.): change of monthly percolation.

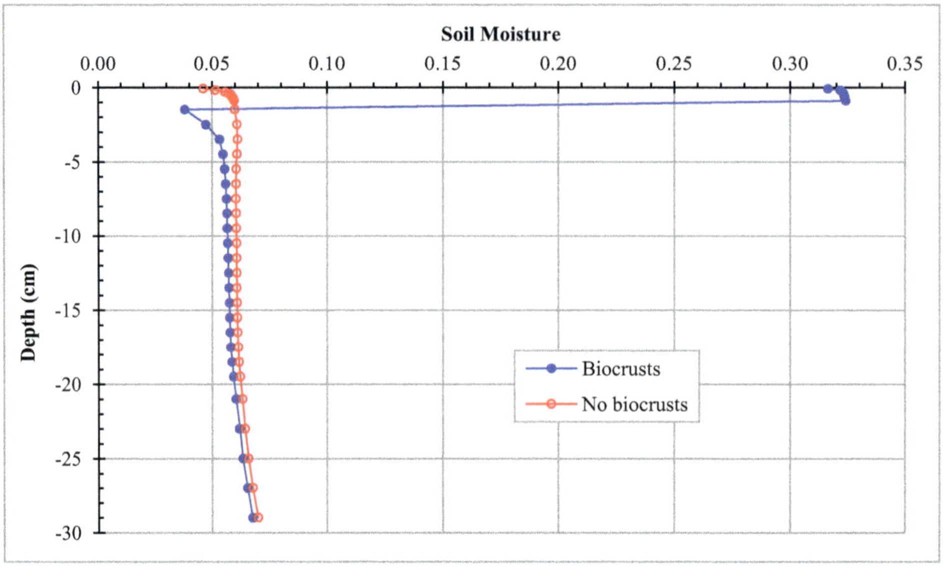

Figure 5.54. The vertical distributions of the simulated annual average soil moisture. The predicted effects of the moss biocrusts on moisture levels at different soil depths.

(figure 5.54). As expected, such predicted effects were smaller at a deeper than a shallower soil depth, with an asymptote of zero.

For a given assessment year, the moisture could be increased by up to 0.27 or decreased by up to 0.022 (figure 5.55), depending on the soil depth of interest; the depth-averaged moisture for the entire soil profile, however, was barely affected by the presence or absence of the biocrusts. In addition, despite being very weak ($R^2 \leqslant 0.04$), for a given soil depth, the biocrust effects tended to decrease with increasing precipitation (figure 5.56), whereas for the entire soil profile, the biocrust effects might be negligible.

From 1958 to 2017, the presence of the moss biocrusts would significantly reduce the number of days with soil moisture of less than a threshold of interest (figure 5.57). For instance, for the top 1 cm soils, the daily moisture was predicted to always be larger than 0.15 if the biocrusts were present, whereas more than 95% of the assessment days (i.e. 20 862 out of 21 960 days) might have moisture of less than 0.15 if the biocrusts were absent. Again, regardless of the thresholds, the differences of the nonexceedance probabilities in the presence of the biocrusts from those in the absence of the biocrusts would monotonically become smaller with increasing soil depths, indicating smaller biocrust effects on the moisture of shallower than deeper soils.

5.8.3.3 The effects on soil erosion

Regardless of the absence or presence of the moss biocrusts, for the 21 915 assessment days, the study watershed was predicted to incur wind erosion more frequently and more severely than water erosion (table 5.28). Nonetheless, averaged

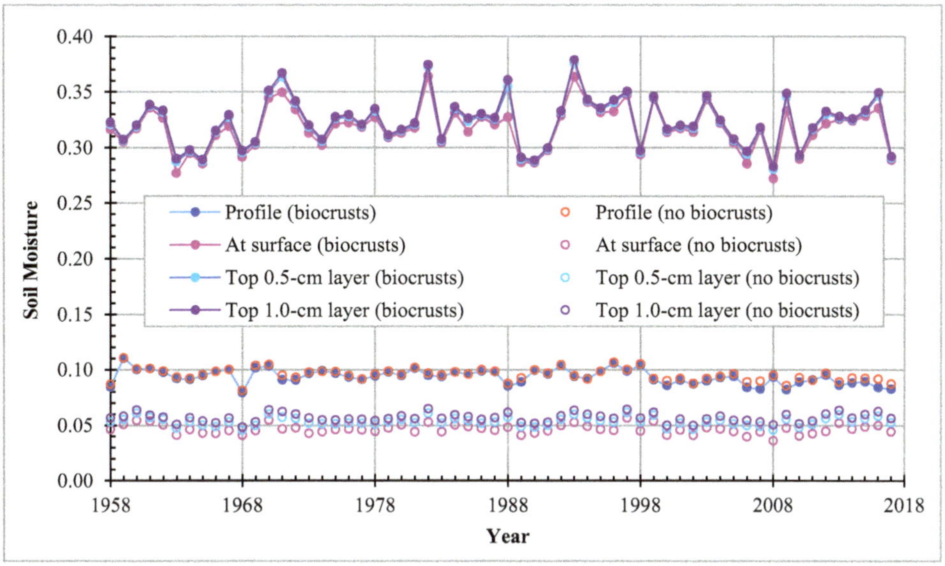

Figure 5.55. The temporal variations of the simulated annual average soil moisture. Plots show the interannual fluctuations of the depth-averaged moisture for the entire 200 cm profile, the soil surface (i.e. top 0.1 cm soil layer), the top 0.5 cm soil layer, and the top 1.0 cm soil layer. For the profile or a soil layer, one plot (solid line with filled circles) shows the moisture in the presence of the biocrusts, while another plot (hallow circles only) shows the moisture in the absence of biocrusts.

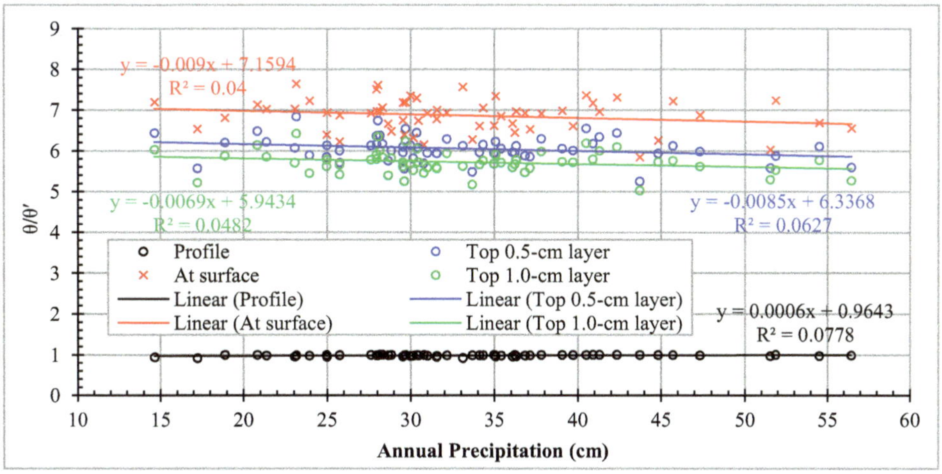

Figure 5.56. The simulated effects of the moss biocrusts as affected by precipitation. The effects are expressed as the ratios of the annual average soil moisture in the presence of the biocrusts, θ, to the corresponding values in the absence of the biocrusts, θ′. Plots show θ/θ′ versus annual precipitation for the entire 200 cm profile, the soil surface (i.e. top 0.1 cm soil layer), the top 0.5 cm soil layer, and the top 1.0 cm soil layer.

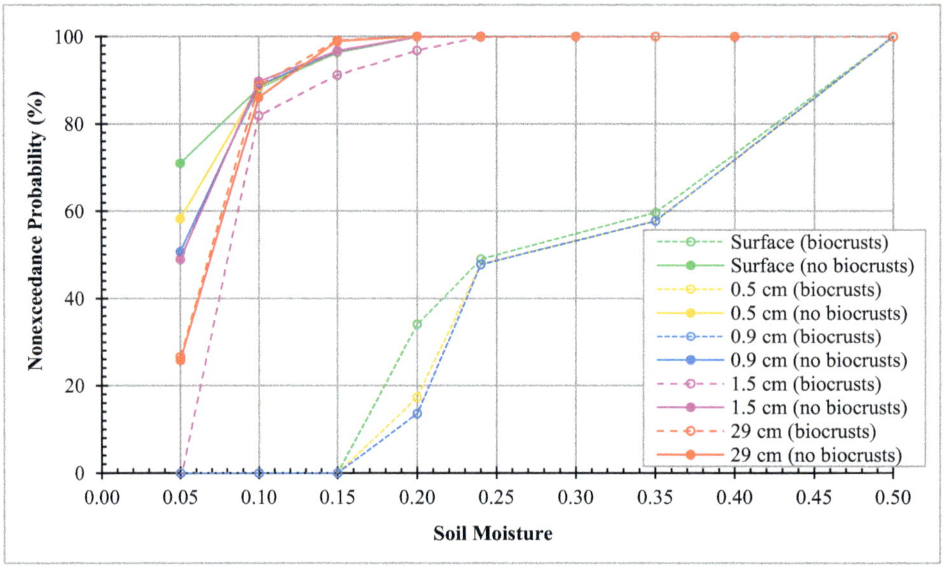

Figure 5.57. The simulated effects of the moss biocrusts on daily soil moisture. The nonexceedance probabilities of the daily soil moisture in the presence versus in the absence of the moss biocrusts for selected five soil layers.

across the erosion days, the wind and water erosion rates would be reduced from 4.2 to 2.8 mm year^{-1} and from 0.034 to 0.024 mm year^{-1}, respectively, due to the presence of the moss biocrusts.

In the absence of the biocrusts, wind and water erosion would occur on 8949 and 4 days, respectively. Across the erosion days, the average wind erosion rate (2.2 mm year^{-1} or 2.1×10^7 t year^{-1}) was predicted to be about 65 times larger than the average water erosion rate (0.034 mm year^{-1} or 3.2×10^5 t year^{-1}). The maximum daily wind erosion rate would be over 1333 times greater than the maximum daily water erosion rate, whereas the minimum daily wind erosion rate was predicted to be over 4570 times smaller than the minimum daily water erosion rate. Among the 8949 wind erosion days, the daily soil moisture (θ) was smaller than 0.2 for 8943 days and greater than or equal to 0.2 for the other 6 days. The four days with water erosion had $\theta < 0.2$. As expected, the maximum daily wind erosion rate for the days with $\theta < 0.2$ was over five times larger than that for the days with $\theta \geqslant 0.2$, but the minimum daily wind erosion rate for the days with $\theta < 0.2$ was about 11 667 times smaller than that for the days with $\theta \geqslant 0.2$.

In the presence of the biocrusts, wind and water erosion would occur on 4142 and 4 days, respectively. Across the erosion days, the average wind erosion rate (2.8 mm year^{-1} or 2.3×10^7 t year^{-1}) was predicted to be over 115 times larger than the average water erosion rate (0.024 mm year^{-1} or 2.1×10^5 t year^{-1}). The maximum daily wind erosion rate would be over 1333 times greater than the maximum daily water erosion rate, whereas the minimum daily wind erosion rate was predicted to be over 9310 times smaller than the minimum daily water erosion rate. Among the

Table 5.28. Daily erosions in the absence (in black) versus presence (in red) of moss biocrusts[1].

Criterion statistics	Aeolian Erosion[2] Mass per unit area ($g\ m^{-2}\ d^{-1}$)	Aeolian Erosion[2] Mass for the watershed ($t\ yr^{-1}$)	Aeolian Erosion[2] Depth for the watershed ($mm\ yr^{-1}$)	Fluvial Erosion[3] Mass per unit area ($g\ m^{-2}\ d^{-1}$)	Fluvial Erosion[3] Mass for the watershed ($t\ yr^{-1}$)	Fluvial Erosion[3] Depth for the watershed ($mm\ yr^{-1}$)
Erosion days for either[4]		8949 days	4142 days		4 days	4 days
minimum	1.8×10^{-5}	3.5×10^{1}	3.8×10^{-6}	8.1×10^{-2}	1.6×10^{5}	1.7×10^{-2}
	5.8×10^{-6}	1.1×10^{1}	1.4×10^{-6}	5.4×10^{-2}	1.1×10^{5}	1.1×10^{-2}
maximum	3.1×10^{2}	6.1×10^{8}	6.1×10^{1}	3.3×10^{-1}	6.4×10^{5}	6.8×10^{-2}
	2.8×10^{2}	5.5×10^{8}	5.6×10^{1}	2.1×10^{-1}	4.1×10^{5}	5.0×10^{-2}
average	1.1×10^{1}	2.1×10^{7}	2.2×10^{0}	1.6×10^{-1}	3.2×10^{5}	3.4×10^{-2}
	1.2×10^{1}	2.3×10^{7}	2.8×10^{0}	1.1×10^{-1}	2.1×10^{5}	2.4×10^{-2}
Erosion days for both[5]		4142 days	4142 days		4 days	4 days
average	2.0×10^{1}	3.9×10^{7}	4.2×10^{0}	1.6×10^{-1}	3.2×10^{5}	3.4×10^{-2}
	1.2×10^{1}	2.3×10^{7}	2.8×10^{0}	1.1×10^{-1}	2.1×10^{5}	2.4×10^{-2}
Erosion days with $\theta < 0.2$		8943 days	2567 days		4 days	0 days
median	4.5×10^{0}	8.7×10^{6}	9.3×10^{-1}	1.2×10^{-1}	2.4×10^{5}	2.6×10^{-2}
	5.4×10^{0}	1.1×10^{7}	1.3×10^{0}	0.0×10^{0}	0.0×10^{0}	0.0×10^{0}
mean	1.1×10^{1}	2.1×10^{7}	2.2×10^{0}	1.6×10^{-1}	3.2×10^{5}	3.4×10^{-2}
	1.2×10^{1}	2.3×10^{7}	2.8×10^{0}	0.0×10^{0}	0.0×10^{0}	0.0×10^{0}
minimum	1.8×10^{-5}	3.5×10^{1}	3.8×10^{-6}	8.1×10^{-2}	1.6×10^{5}	1.7×10^{-2}
	5.8×10^{-6}	1.1×10^{1}	1.4×10^{-6}	0.0×10^{0}	0.0×10^{0}	0.0×10^{0}
maximum	3.1×10^{2}	6.1×10^{8}	6.1×10^{1}	3.3×10^{-1}	6.4×10^{5}	6.8×10^{-2}
	2.0×10^{2}	3.9×10^{8}	4.8×10^{1}	0.0×10^{0}	0.0×10^{0}	0.0×10^{0}
standard deviation	1.8×10^{1}	3.5×10^{7}	3.7×10^{0}	1.1×10^{-1}	2.2×10^{5}	2.3×10^{-2}
	1.9×10^{1}	3.7×10^{7}	4.6×10^{0}	0.0×10^{0}	0.0×10^{0}	0.0×10^{0}
Erosion days with $\theta \geq 0.2$		6 days	1575 days		0 days	4 days
median	4.1×10^{0}	8.1×10^{6}	7.8×10^{-1}	0.0×10^{0}	0.0×10^{0}	0.0×10^{0}
	5.3×10^{0}	1.0×10^{7}	1.2×10^{0}	7.9×10^{-2}	1.6×10^{5}	1.8×10^{-2}
mean	1.7×10^{1}	3.4×10^{7}	3.3×10^{0}	0.0×10^{0}	0.0×10^{0}	0.0×10^{0}
	1.2×10^{1}	2.2×10^{7}	2.7×10^{0}	1.1×10^{-1}	2.1×10^{5}	2.4×10^{-2}
minimum	2.1×10^{-1}	4.1×10^{5}	3.9×10^{-2}	0.0×10^{0}	0.0×10^{0}	0.0×10^{0}
	4.1×10^{-4}	8.0×10^{2}	9.1×10^{-5}	5.4×10^{-2}	1.1×10^{5}	1.1×10^{-2}
maximum	6.0×10^{1}	1.2×10^{8}	1.1×10^{1}	0.0×10^{0}	0.0×10^{0}	0.0×10^{0}
	2.8×10^{2}	5.5×10^{8}	5.6×10^{1}	2.1×10^{-1}	4.1×10^{5}	5.0×10^{-2}
standard deviation	2.5×10^{1}	4.8×10^{7}	4.7×10^{0}	0.0×10^{0}	0.0×10^{0}	0.0×10^{0}
	2.0×10^{1}	3.9×10^{7}	4.6×10^{0}	7.1×10^{-2}	1.4×10^{5}	1.7×10^{-2}

[1] Assessment days: 21,915; and θ: daily soil moisture.
[2] Days with a non-zero wind erosion: 4142 for the presence of the biocrusts and 8949 for the absence of the biocrusts, leading to the occurrences of 18.9 and 40.8%, respectively.
[3] Days with a non-zero water erosion: 4 both for the presence and absence of the biocrusts.
[4] Computed separately using the days (i.e., 4142 for wind and 4 for water) with a non-zero erosion in the presence of the biocrusts and the days (i.e., 8949 for wind and 4 for water) with a non-zero erosion in the absence of the biocrusts.
[5] Computed using the days (i.e., 4142 for wind and 4 for water) with non-zero erosions for both the presence and absence of the biocrusts. Note that a day with a non-zero erosion in the presence of the biocrusts was predicted to consistently have a non-zero erosion in the absence of the biocrusts.

wind 4142 erosion days, the daily θ values were smaller than 0.2 for 2567 days and greater than or equal to 0.2 for the other 1575 days. The four days with water erosion had $\theta \geqslant 0.2$. Unlike that in the absence of biocrusts, the maximum daily wind erosion rate for the days with $\theta < 0.2$ was about 1.4 times smaller than that for the days with $\theta \geqslant 0.2$, and the minimum daily wind erosion for the days with $\theta < 0.2$

was about 116 times smaller than that for the days with $\theta \geqslant 0.2$. The median wind erosion rate for the days with $\theta < 0.2$ was almost same as that for the days with $\theta \geqslant 0.2$.

Regardless of the absence or presence of moss biocrusts, the wind erosion rate in a month with a larger wind speed tended to be larger than that in a month with a smaller wind speed (figure 5.58). Wind speed was the primary driver for wind

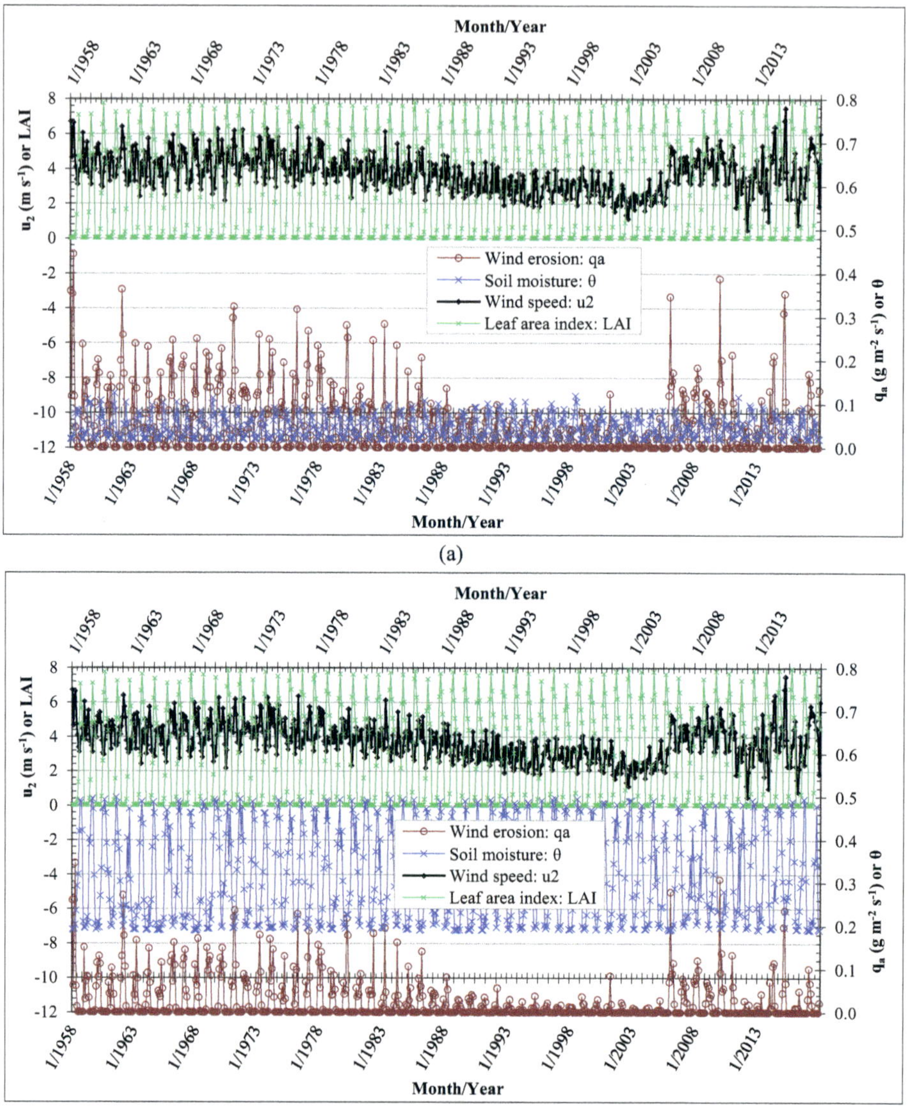

Figure 5.58. The monthly mean wind erosion rate and its influencing variables. The simulations for the moss biocrusts to be (a) absent and (b) present.

erosion. Across the 60 assessment years, the years of 1988–2005 had smaller wind speeds and overall lower wind erosion rates that other years. For a given month of a given year, the wind erosion rate was also dependent on the soil moisture and leaf area index (LAI). The highest wind erosion rate occurred in April 1958, when both the soil moisture and LAI were the smallest. The wind erosion would be greatly alleviated by maintaining $\theta \geqslant 0.1$ and LAI $\geqslant 3.5$. A month with larger θ and/or LAI tended to have a lower wind erosion rate than a month with smaller θ and/or LAI. Nonetheless, the presence of the moss biocrusts was predicted to consistently reduce wind erosion: the highest erosion rate in April 1958 would be reduced from 0.444 to 0.347 g m^{-2} s^{-1} (or 7.5×10^{10} to 5.9×10^{10} t $year^{-1}$).

On the other hand, overland runoff events were generated by snowmelt because of the sporadically rising air temperatures in the Januarys of eight years (figure 5.59). In the absence of biocrusts, these events occurred in 1964, 1991, 2005, and 2013, whereas in the presence of the biocrusts, they occurred in 1961, 1988, 2002, and 2010, indicating that the moss biocrusts might interact with ambient atmosphere to affect snowpack melting processes. The eight runoff events caused varying degrees of water erosion, but the event-averaged water erosion rate was reduced from 0.16 g m^{-2} s^{-1} (or 3.2×10^5 t $year^{-1}$) in the absence of the biocrusts to 0.11 g m^{-2} s^{-1} (or 2.1×10^5 t $year^{-1}$) in the presence of the biocrusts, showing the protective effects of the biocrusts on bare soils from water erosive forces.

In the absence of biocrusts, the annual mean soil moisture is predicted to be 0.02 to 0.06, whereas in the presence of biocrusts, the annual mean soil moisture would vary from 0.26 to 0.37 (figure 5.60), indicating that the moss biocrusts could retain more water than the parent soils and that the moisture of the biocrusts was much less influenced by climate than that of the parent soils. As a result, the annual mean wind erosion rates in the presence of biocrusts were consistently lower than the corresponding values in the absence of biocrusts. In contrast, the presence of the biocrusts was predicted to have no influence on the LAIs (figure 5.59), which varied but were not correlated with precipitation (figure 5.61). Again, the wind erosion rates were primarily controlled by wind speeds (figure 5.62), as indicated by the nearly synchronic fluctuations of these two variables, and secondarily rectified by soil moisture and grass coverage. The water erosion was predicted to occur in January and thus the erosion rates would be solely dependent on the overland runoff amounts (figure 5.63) because the land was apparently bare (LAI = 0.03).

Across the 60 assessment years, wind erosion was predicted to mainly occur in January to June, when the wind speeds were relatively large, the soil moisture very small, and the LAIs almost zero (figure 5.64). In the absence of the biocrusts, the average monthly wind erosion rates varied from 2.1 g m^{-2} s^{-1} in June to 12.0 g m^{-2} s^{-1} in April, whereas in the presence of the biocrusts, the average monthly wind erosion rates varied from 0.5 g m^{-2} s^{-1} in June to 7.5 g m^{-2} s^{-1} in April. The biocrusts would reduce the peak erosion rate in April by 37.5%. In contrast, wind erosion was predicted to be minimal in July to December, when the soil moisture and LAIs were relatively high. This indicates that moist soils and/or soils covered by dense grasses could resist erosive forces induced by wind.

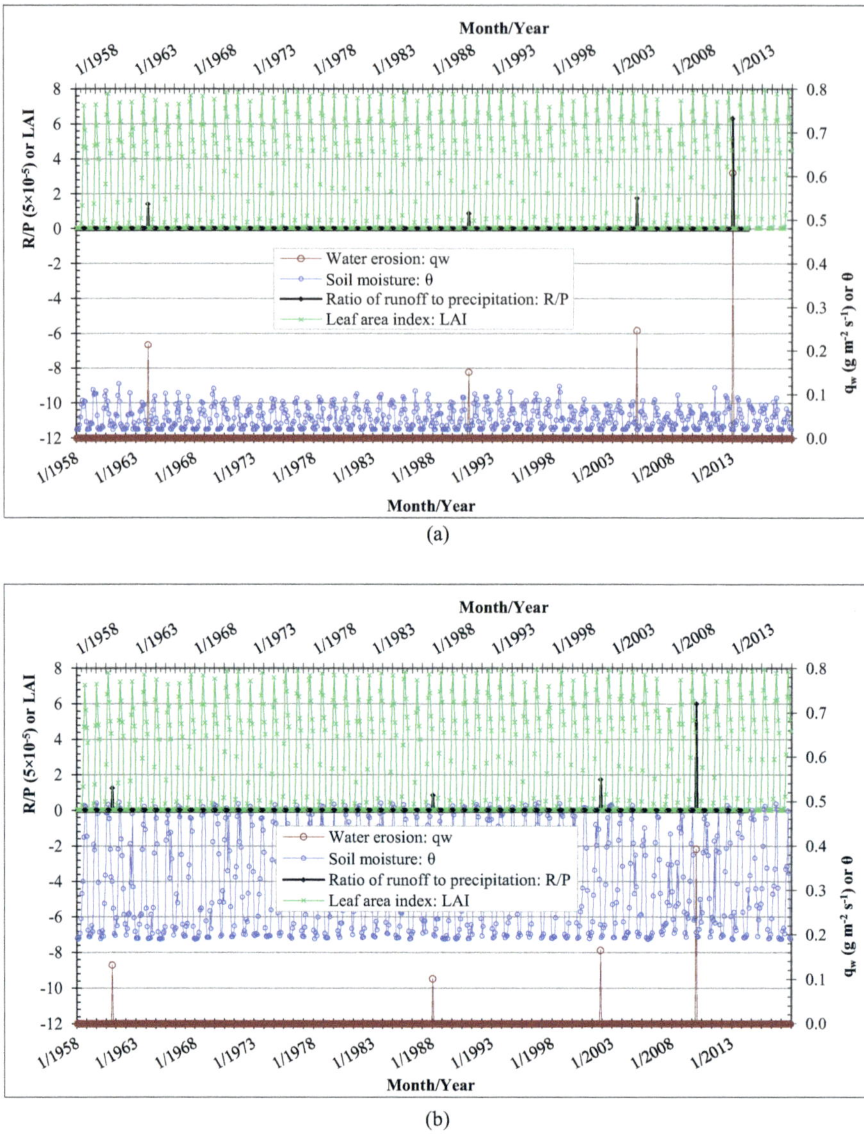

Figure 5.59. The monthly mean water erosion rate and its influencing variables. The simulations for the moss biocrusts to be (a) absent and (b) present.

5.8.4 Discussion

Moss biocrusts can hold more water than the parent soils, leading to more water being available for evaporation while less water is available for transpiration and percolation (table 5.27). The biocrust organisms excrete exopolysaccharides, releasing extracellular polymetric substances (EPSs) (Adessi *et al* 2018). The EPSs glue fine soil particles into larger-sized aggregates or bundles, which are assembled with coarse soil particles to form the skeletons of biocrusts. As a result, the porosity or

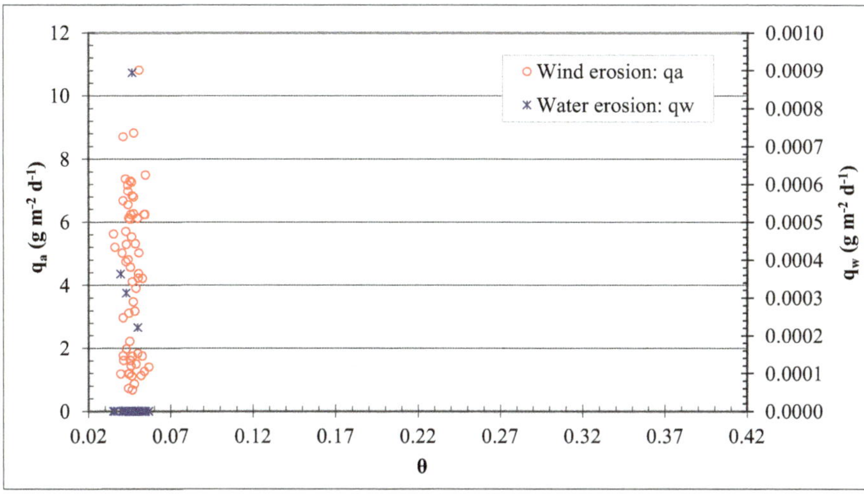

(a)

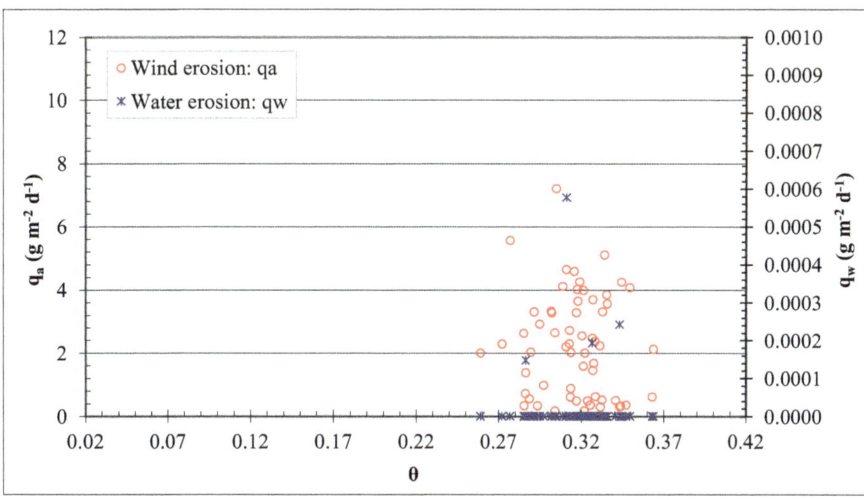

(b)

Figure 5.60. The annual mean wind and water erosion rates versus soil moisture. The simulations for the moss biocrusts to be (a) absent and (b) present.

saturated soil moisture of the biocrusts is larger than that of the parent soils (Chamizo *et al* 2016), providing more pores of large sizes for water storage. Because the biofilms established by the EPSs protect pore water from being lost to desiccations, the biocrusts have a higher residual soil moisture than the parent soils, and for a given suction head, the soil moisture of the biocrusts is higher than that of the parent soils, likely supplying more water for evaporation. However, the EPSs increase water repellency, retarding the movement of water through pores, and cause pore clogging, reducing the connectivity of pores and sinuosity of pore

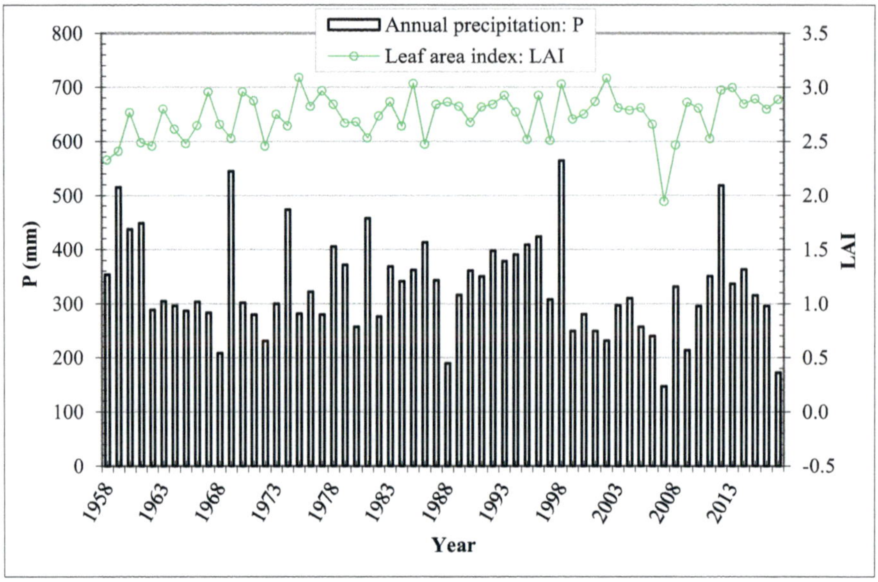

(a)

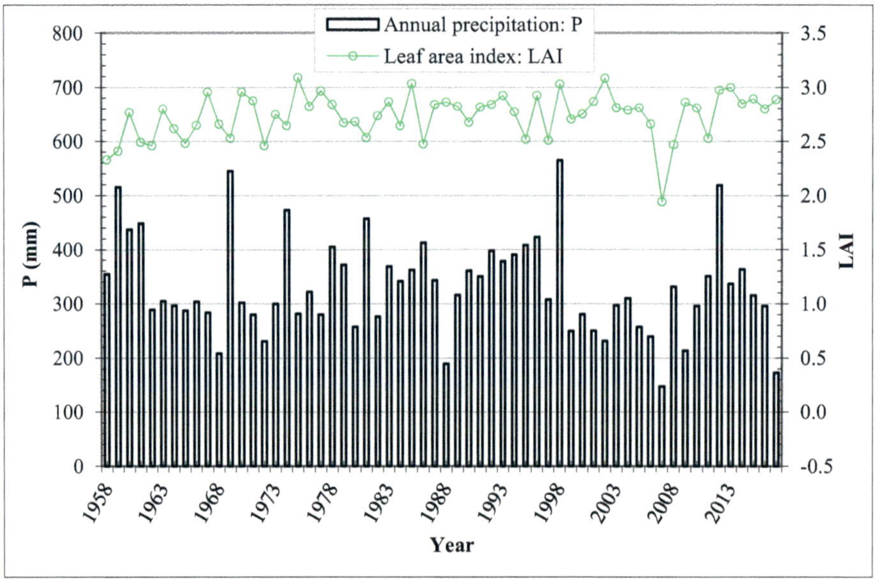

(b)

Figure 5.61. The annual mean leaf area index. The simulations for the moss biocrusts to be (a) absent and (b) present.

channels (Xiao *et al* 2019). As a result, the biocrusts have a smaller saturated hydraulic conductivity and allow less water to infiltrate into the rootzone for plant uptake and percolation than the parent soils (Adessi *et al* 2018, Guan and Cao 2021, Sun *et al* 2021).

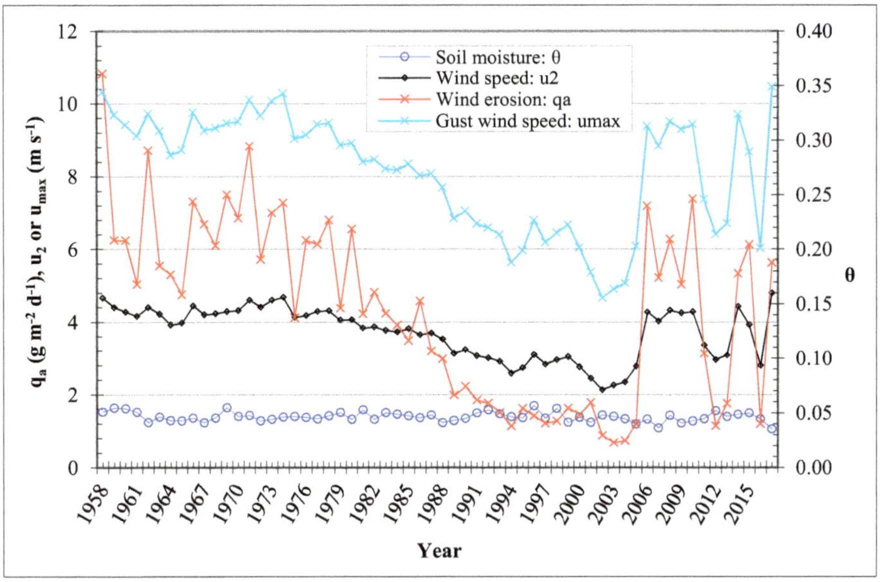

(a)

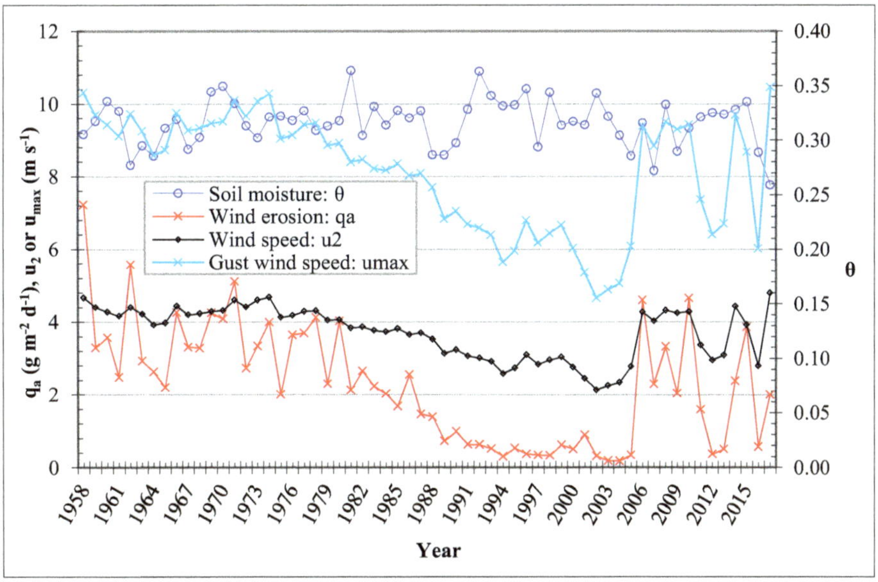

(b)

Figure 5.62. The annual mean wind erosion rate and its influencing variables. The simulations for the moss biocrusts to be (a) absent and (b) present.

The moisture of the biocrusts varies gently from time to time and is much higher than that of the soils beneath (figures 5.54, 5.55, and 5.57). The biocrust moisture replenished by rainwater is protected by the biofilms from being lost quickly to evaporation and infiltration, leading to a relatively high moisture during the

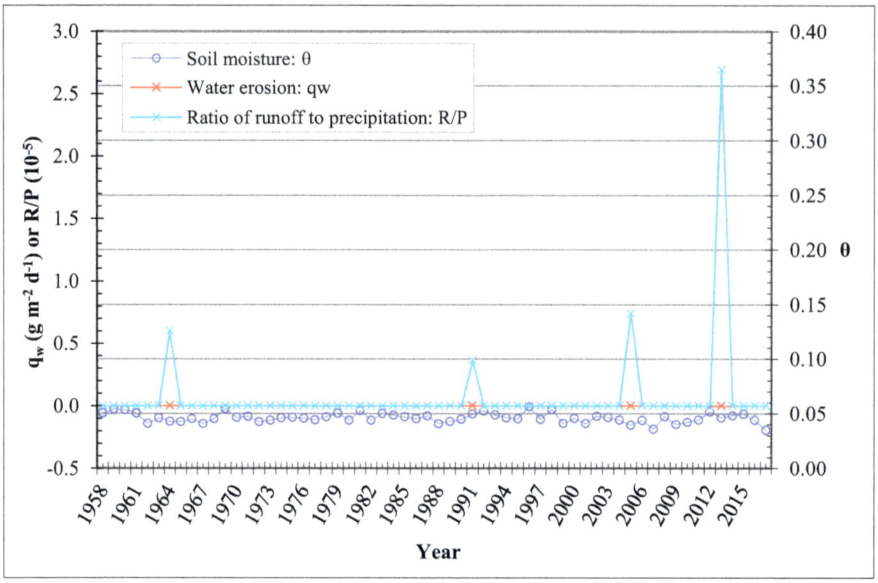

(a)

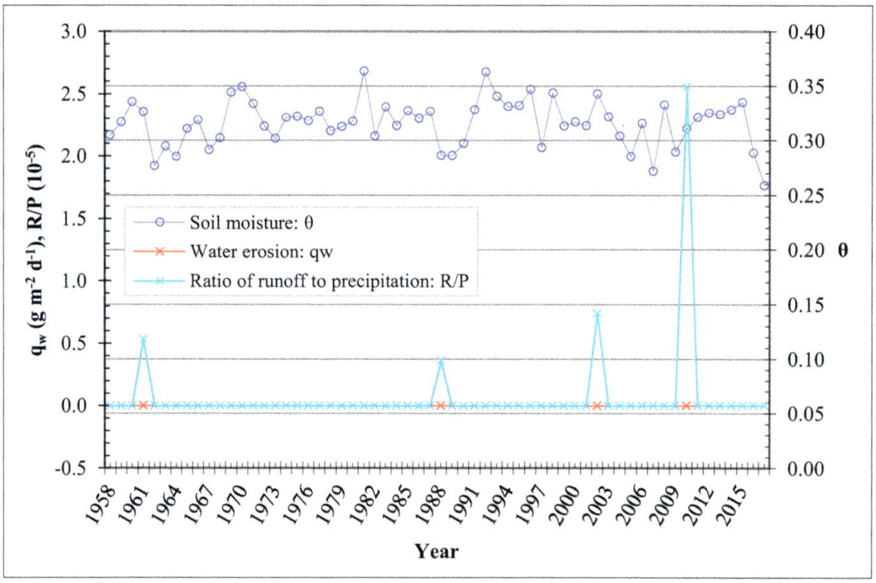

(b)

Figure 5.63. The annual mean water erosion rate and its influencing variables. The simulations for the moss biocrusts to be (a) absent and (b) present.

subsequent dry days. The large volume of pores functions as a 'reservoir' that can regulate the fluctuations of biocrust moisture. Because the penetration water is mostly stored in the biocrust pores and may slowly move downward, the moisture of the soils beneath can rarely be replenished by normal storms. The rainwater of a

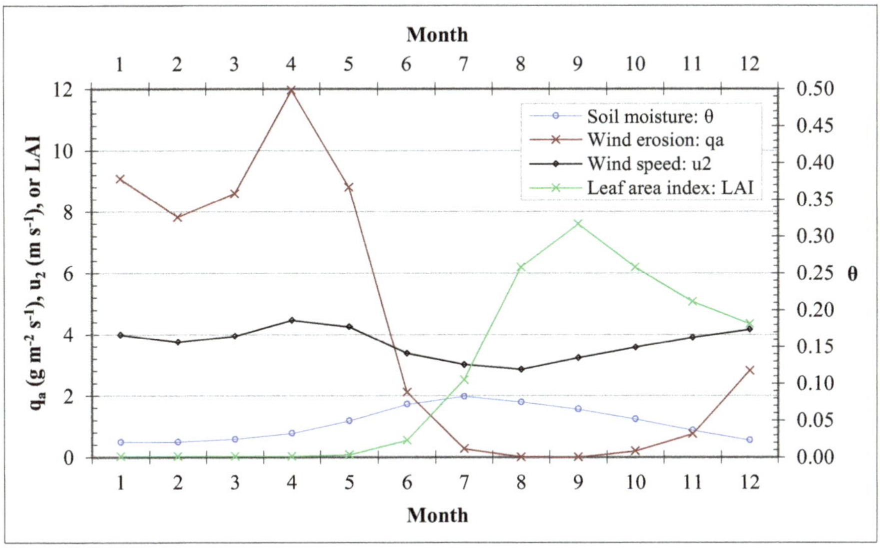

(a)

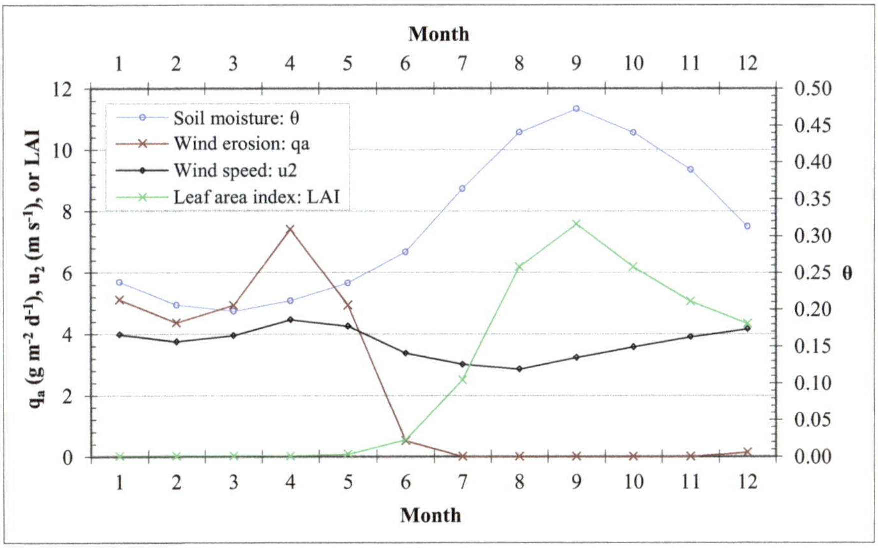

(b)

Figure 5.64. The average monthly wind erosion and its influencing variables. The simulations for the moss biocrusts to be (a) absent and (b) present.

sporadic extreme storm may be sufficient to saturate the biocrusts and then penetrate through the biocrust layer to wet the soils beneath. In contrast, in the absence of biocrusts, the topsoil moisture is very low and the moisture at a deeper soil depth tends to be higher than that at a shallower soil depth (figures 5.54, 5.55, and 5.57). Although the topsoil can be somewhat wetted immediately from a storm, the replenished moisture is lost to evaporation within a very short period (e.g. minutes) after the storm ends. For an extreme storm, the topsoil (<1.0 cm) may be intermittently saturated, and the soil water above the field capacity can move down to replenish the moisture of the soils beneath or deep soils; once the storm stops, the wetted topsoil will dry out quickly because of a very high evaporation rate. Thus, the moisture of the parent topsoil has a larger temporal variability than that of the biocrusts, whereas for a given depth larger than 1.0 cm, the soil moisture tends to be larger in the absence than in the presence of biocrusts (figure 5.54); this is consistent with the field measurements of Nevins *et al* (2020).

The moss biocrusts can effectively protect the topsoil from aeolian erosion (table 5.28 and figures 5.58–5.60). The presence of the biocrusts can reduce the occurrence of wind erosion by 21.9% and the average erosion depth by 1.4 mm year^{-1}. One reason is that the biocrust aggregates have a larger size than the parent soil particles (Herrick *et al* 2001). It is common sense that the larger-sized biocrust aggregates are more resistant to be mobilized by wind than the smaller-sized parent soil particles. The second reason is that the moisture of the biocrusts is relatively high for most days, increasing the threshold surface shear velocity (equations (3.93) and (3.94)) and thus the threshold shear stress (equations (3.90) and (3.91)). That is, the minimum wind velocity above which the biocrust aggregates can be mobilized is greater than that above which the parent soil particles can be mobilized. As a result, the winds on some days can be strong enough to mobilize the parent soil particles but may be too weak to mobilize the biocrust aggregates. The third reason is that the EPSs can shield the biocrust aggregates from direct wind abrasion, decreasing the chance and severity of the aggregates being mobilized. In contrast, the parent fine soil particles do not have such a protection shield and are easily mobilized by wind-induced shear forces. The last reason is that the biocrusts have a higher organic matter content than the parent topsoil (Nevins *et al* 2020). This factor is not explicitly reflected in the IFAP model, but it is implicitly considered in the higher biocrust moisture. Similarly, the biocrusts can better protect the topsoil from sporadic fluvial erosions (table 5.27 and figures 5.58–5.60). Such protection effects can be attributed to the biocrust aggregates with a larger size and the shielding feature of the EPSs. Because the parent topsoil is very loose, the fine particles can be easily mobilized once runoff is generated.

The complements between biocrusts and grasses resulting from climate (e.g. rainfall and air temperature) variations have been reported by Gao *et al* (2017b), but they could not be modeled in this study because relevant algorithms are lacking in the existing literature and need to be developed in future studies. Thus, the LAIs are same in the absence and in the presence of the biocrusts, while they are mainly dependent on livestock grazing and precipitation (figure 5.61). Regardless of whether the biocrusts are presented, wind and water erosions are primarily controlled by wind speed and overland runoff, respectively, while both types of erosion are secondarily rectified by soil moisture and LAI (figures 5.62 and 5.63). One reason is that wind- and runoff-induced shear stresses are erosive forces. Another reason is that soil moisture and LAI have positive influences on resistant strengths. A higher soil moisture can enhance the resistance of surface materials (i.e. aggregates or particles) to wind-induced erosive forces while tends to increase the erodibility of surface materials from water-induced abrasions. A larger LAI can enhance the resistance and reduce the erodibility and thus plays a vital role in stabilizing surface materials. For wind erosion, the worst condition is the combination of a high wind speed, a low soil moisture, and a small LAI, which are true for the Balagaer River watershed in April (figure 5.64). Given that wind speed and LAI in this spring month cannot be managed, a well-developed biocrust should be established and maintained to stabilize the fertile topsoil.

5.8.5 Conclusions

This study assessed the effects of moss biocrusts on water balance, soil moisture, and soil erosion using the SWAP and IFAP models set up for the Balagaer River watershed. The results indicated that the presence of biocrusts will increase the average annual evaporation by 3.7 cm but decrease the average annual transpiration and percolation by 1.6 and 2.2 cm, respectively. The biocrusts can hold more water with higher moisture most of the time than the parent topsoil. The presence of the biocrusts was predicted to effectively reduce the wind erosion occurrence by 21.9% and the wind erosion rate by 1.4 mm year^{-1}, with a maximal effect in April when wind speeds were relatively high while soil moisture and vegetation coverage were very low. New algorithms need to be developed to model complements between biocrusts and grasses resulting from rainfall and air temperature variations.

5.9 Resilience and degradation of semi-arid ecosystems

Resilience has been widely used to characterize whether an ecosystem of interest subject to a pressure would be able to stay at its original state or shift to an alternative state. After being disturbed by the pressure, the ecosystem may shift to a new state. If the pressure is lower than a threshold, the disturbed ecosystem can bounce back to its original state upon removal of the pressure, implying that the ecosystem is fully resilient. In contrast, if the pressure is higher than the threshold, the disturbed ecosystem may recover to an intermediate state between the new and original states, implying that the ecosystem is partially resilient, or will permanently stay at the new state, implying that the ecosystem is non-resilient. Thus, resilience can be considered as a measure of the threshold pressure that the ecosystem can experience before it shifts to an alternative state or starts to degrade irreversibly to a catastrophic condition (e.g. semidesert or desert). In the existing literature, resilience is defined and computed differently, making it infeasible to compare one study to another. The objectives of this study are to (i) establish a new theory of resilience, hereinafter designated the Wang's ecosystem resilience theory (WERT), (ii) apply WERT to elucidate the effects of livestock grazing on resilience, and (iii) determine the equilibria and shifts of ecosystem states for varying livestock carrying capacities. The testbed is the Balagaer River watershed, described in section 5.1.1. The WERT provided a handy way of visualizing and determining resilience. For the studied watershed, the threshold carrying capacity was determined to be 0.8 cattle ha^{-1} with a rotational grazing practice that has a grazing period of 15 years once every

3–5 years. A logical generalization is that the WERT will advance the understanding and application of resilience in managing ecosystems.

5.9.1 Introduction

For a dryland ecosystem of interest, the resilience, which measures the threshold pressure that the ecosystem can experience before it shifts to an alternative state, needs to be determined to prevent land degradation and desertification resulting from climatic variations and inappropriate land management practices. However, the definitions and quantifications of resilience are inconsistent and conclusive in the existing literature because the ecosystem is usually subject to various stressors induced by climate and livestock grazing (Ford 2001, Christensen *et al* 2003, Wang *et al* 2012a, Ganjurjav *et al* 2019, Xu *et al* 2021). For instance, historical data reveal statistically significant downward trends in annual precipitation, summer rainfall, and the number of wet days (Wang *et al* 2014a, Gao *et al* 2017a, Wang *et al* 2021b), inevitably causing water shortages for vegetation growth and biomass production. On top of declining precipitation, livestock grazing removes the leaves and stems of vegetation, lowers soil moisture, and reduces the protective coverage of the land surface and thus the critical shear strength of the topsoil from aeolian and fluvial erosion (Wang *et al* 2014a, Wang *et al* 2021a). In practice, to quantify the resilience, a state (i.e. status) of the ecosystem can be indicated by the aboveground biomass of grass. The pristine state can be assumed to be the one when there is no livestock grazing, whereas the worst state would occur if the livestock grazing intensity (i.e. removal percent of the pristine-state biomass) is very high (e.g. >0.49). The aboveground biomass is maximal at the pristine state and tends to decrease with an increase of grazing intensity. At the worst state, the aboveground biomass will be minimal or even near-zero and the land is almost bare. In principle, for a below-threshold pressure the ecosystem is resilient (i.e. temporally shifts to a new state that would be able to bounce back to the previous state once the pressure is alleviated), whereas for an above-threshold pressure, the ecosystem would become non-resilient (i.e. permanently shift to an alternative state). If the ecosystem keeps on shifting to alternative states because of a series of above-threshold pressures, it will be irreversibly degrading toward desertification.

Christensen *et al* (2003) examined the resilience of the 3680 km^2 Baiinxile livestock farm (43.5°N, 116.5°E), on a typical steppe in Inner Mongolia, China. In this regard, a savanna model (Coughenour 1994) was parameterized using field data and literature values and validated using the biomass data extracted from the Landsat TM images from 1980 to 1989. Driven by a 50 year long weather time series, the model was run to simulate the aboveground biomasses (i.e. primary production), vegetation compositions, and root biomasses at 14 different livestock carrying capacities ranging from 5.5 to 59.8 AUY km^{-2}. Herein, AUY stands for animal unit per year, meaning the number of animals that continuously graze throughout a year at a dry matter intake rate of 8 kg d^{-1} (Hinnant 1994). The simulation results indicated a threshold livestock carrying capacity of 40 AUY km^{-2} (i.e. a grazing intensity of 0.49), above which the farm system would shift from an herbaceous-vegetation state to a stable shrub-dominated state. The herbaceous-vegetation state is superior to the shrub-dominated state in the long run.

Ganjurjav *et al* (2019) conducted a five-year grazing experiment with Tibetan sheep (*Ovis aries*, live weight ~40 kg) in a 60.1 ha semi-arid alpine steppe (32°10′N, 92°33′E, 4600 m a.s.l.) on the Qinghai-Tibetan Plateau, China. The authors collected data on the aboveground primary productions of legumes, forbs, and graminoids in response to five sheep carrying capacities, namely no grazing, light grazing (2.4 sheep ha^{-1}), moderate grazing (3.6 sheep ha^{-1}), and heavy grazing (6.0 sheep ha^{-1}). The experimental site was subdivided into 36 plots, of which the non-zero grazing plots were grazed once every ten days from June to September. Those authors computed the short-term resilience as the remanent aboveground biomass immediately after grazing. The results indicated that regardless of the plant species, the aboveground biomass of a greater carrying capacity was significantly less than that of a smaller carrying capacity. For a given plant species, the resilience of a greater carrying capacity could be either higher or lower than that of a smaller carrying capacity, while for a given carrying capacity, the resilience of one plant species might be either higher or lower than that of another plant species. Such grazing-species complementary performances in resilience can facilitate the overall tolerance of the alpine steppe to grazing. However, because of the reduction in gross biomass, the steppe would risk degradation with the combined effects of large carrying capacities and low precipitations over a long period of time, which needs to be carefully considered in grassland management for the future climate.

Xu *et al* (2021) examined resilience by imposing an extreme drought for two years followed by two years of recovery in a semi-arid grassland (50°10′46″N, 119°22′56″E) of Inner Mongolia, China. Those authors defined resilience as the ratio of the aboveground net primary production (ANPP) after a perturbation to that in the control. The results indicated that the ANPP decreased by up to 60% as the drought progressed. After the first year following the drought, the average ANPP across the drought plots recovered to 83% of that across the ambient control plots, while by the second year, the average ANPP recovered fully. Those authors attributed the high resilience of the ecosystem to the compensating effects of grasses for forb productivity reduction as well as to the short-term drought stress because the impact of a drought progressively tends to amplify as the drought persists.

In the existing literature, including that cited above, the definitions and formulizations of resilience are inconsistent and conclusive, which makes it impossible to compare one study to another. Also, few studies on dryland ecosystems have examined resilience in the long run. However, given the changing climate, a future drought is expected to have a longer duration and a greater severity (Wang *et al* 2014a). Thus, the resilience of an ecosystem should be examined in terms of long-term ANPPs. The objectives of this study were to (i) put forward a new theory of resilience, hereinafter designated the Wang's ecosystem resilience theory (WERT), (ii) apply the theory to elucidate the effects of livestock grazing on resilience, and (iii) determine the equilibria and shifts of ecosystem states for varying livestock carrying capacities. The testbed was the semi-arid Balagaer River watershed described in section 5.1.1.

5.9.2 The WERT

5.9.2.1 Concepts

For a given ecosystem (e.g. steppe grassland), the stability indicates the closeness of its annual statuses to the long-term mean status. Generally, the aboveground biomass of steppe grasses can be used to measure the grassland status. The scatteredness of the annual statuses around the mean can be described by the coefficient of variation (C_V), which is the ratio of the standard deviation to the mean. A larger C_V indicates a greater interannual variability and thus a smaller stability of the ecosystem, and vice versa. The ecosystem is unstable when $C_V \geqslant 1.0$. In contrast, the reliability indicates the chance for the ecosystem to have statuses same as or better than an upper threshold in a specified number of consecutive years, whereas the vulnerability indicates the chance for the ecosystem to have statuses the same as or worse than a lower threshold in specified number of consecutive years. Herein, the specified number of consecutive years indicates the concerned time period in terms of managing the ecosystem. For this case study, the retrospective biomasses of the steppe grassland were simulated by assuming no grazing. The lower and upper thresholds were taken as the first and third quantiles, respectively, of the retrospective biomasses. The typical planning interval of pasturing practices was empirically specified to be three consecutive years.

The stability of an ecosystem can be defined as

standard deviation

stability [0 ~ 1] →

$$S_{ta} = \max\left\{0, 1-\frac{s}{\bar{z}}\right\} \tag{5.14}$$

mean

The reliability of the ecosystem can be calculated as

number of years with a biomass of larger than or equal to the third quantile

reliability [0 ~ 1] →

$$R_{el} = 1-\left(1-\frac{N_{\geq Z75}}{N}\right)^{m} \tag{5.15}$$

total number of years

specified number of consecutive years

The vulnerability of the ecosystem can be calculated as

number of years with a biomass of less than or equal to the first quantile

vulnerability [0 ~ 1] →

$$V_{ul} = 1-\left(1-\frac{N_{\leq Z25}}{N}\right)^{m} \tag{5.16}$$

total number of years

specified number of consecutive years

S_{ta}, R_{el}, or V_{ul} varies from zero to one. A larger S_{ta} or R_{el} indicates a more stable or reliable status, whereas a larger V_{ul} indicates a more vulnerable status, and vice versa. That is, a status with larger values of S_{ta} and R_{el} and a smaller value of V_{ul} tends to be better than that with smaller values of S_{ta} and R_{el} and a larger value of V_{ul}. When S_{ta}, R_{el}, and V_{ul} are plotted in a triangular diagram as shown in figure 5.65, the vertices of the equilateral triangle represent three extreme statuses. Vertex A_e, at which $S_{ta} = 0.5$, $R_{el} = 0.0$, and $V_{ul} = 0.0$, signifies the 'acceptable' status, while vertex B_e, at which $S_{ta} = 1.0$, $R_{el} = 1.0$, and $V_{ul} = 0.0$, signifies the 'best' status. Vertex C_e, at which $S_{ta} = 0.0$, $R_{el} = 0.0$, and $V_{ul} = 1.0$, signifies the 'crucial' status.

For an ecosystem of interest, S_{ta}, R_{el}, and V_{ul} can be computed using equations (5.14) to (5.16). The vertical line of the S_{ta} and the slopped lines of the R_{el}, and V_{ul}

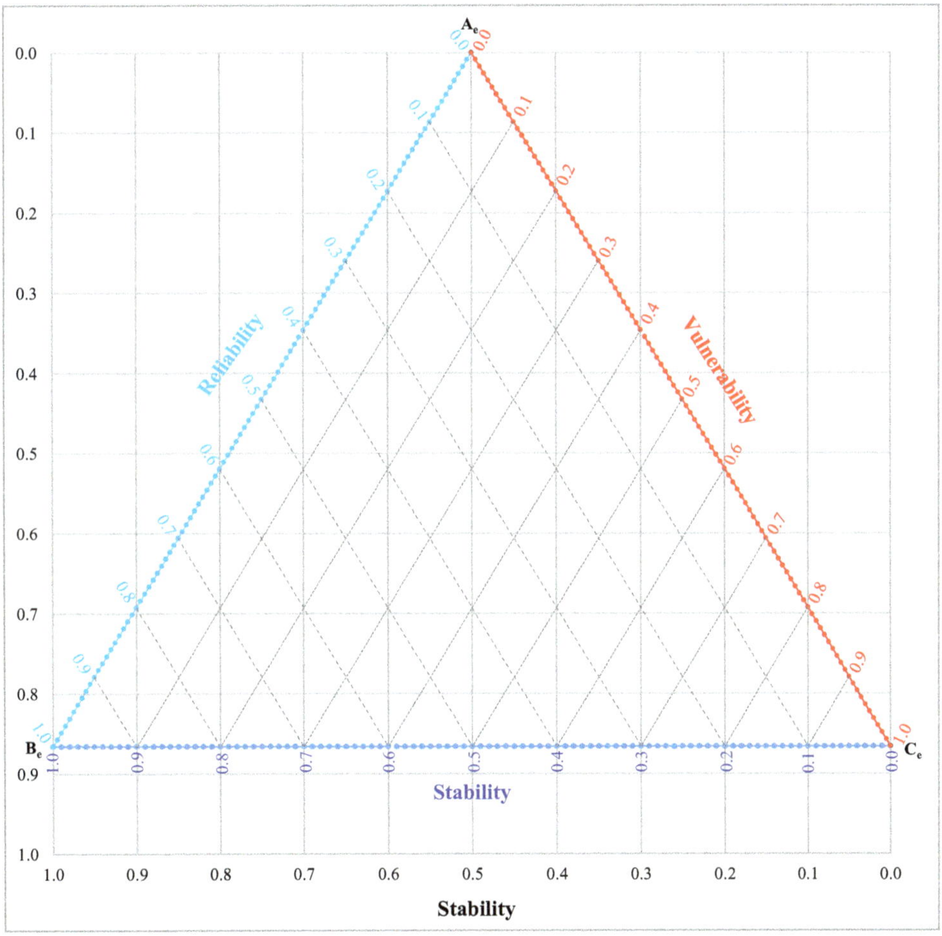

Figure 5.65. Wang's ecosystem resilience theory (WERT). The triangular diagram of stability, reliability, and vulnerability.

form a scalene triangle. The overlap between the inscribed circle of the scalene triangle and the equilateral triangle defines the resilience of the ecosystem and is termed the resilience circle hereafter for description purposes. A status of the ecosystem corresponds to a resilience circle. If the resilience circles intersect, the corresponding statuses can be switched from one to another depending on the stress levels, implying that the ecosystem is resilient; otherwise, the corresponding statuses will not be switchable regardless of the stress levels, implying that the ecosystem is non-resilient. The ecosystem may be changed from being resilient to non-resilient by a very high-level persistent stress (e.g. overgrazing for years). For instance, because of a stress, the ecosystem will be changed from its existing status to a new status. If the resilience circle for the new status intersects the resilience circle for the existing status, the ecosystem is resilient because it will bounce back to the existing status once the stress is removed; otherwise, the ecosystem will be non-resilient because it will stay at the new status even though the stress is removed.

For a given status, the resilience can be measured using the distance from the center of the resilience circle to vertex C_e in figure 5.65. If the status is better than the 'crucial' status, the resilience is positive, whereas if the status is worse than the 'crucial' status, the resilience is negative. Numerically, the resilience for the status can be computed as

x-coordinate of vertex C_e in Figure 5.65 (= 0.0)
y-coordinate of vertex C_e in Figure 5.65 (= $3^{1/2}/2$)

resilience [≤ 1] →
$$R_{es} = \pm\sqrt{(x_0 - x_{Ce})^2 + (y_0 - y_{Ce})^2} \tag{5.17}$$

x-coordinate of the center of the resilience circle
y-coordinate of the center of the resilience circle

If $R_{el} = 0.0$ and $V_{ul} = 1.0$, R_{es} is negative; otherwise, R_{es} is positive. In figure 5.65 $x_{Ce} = 0.0$ and $y_{Ce} = \frac{\sqrt{3}}{2}$. The coordinates of vertex A_e are $x_{Ae} = 0.5$ and $y_{Ae} = 0.0$, and the coordinates of vertex B_e are $x_{Be} = 1.0$ and $y_{Be} = \frac{\sqrt{3}}{2}$. The resiliencies for the statuses represented by vertices A_e and B_e are computed as $\sqrt{(x_{Ae} - x_{Ce})^2 + (y_{Ae} - y_{Ce})^2} = \sqrt{(0.5-0.0)^2 + \left(0.0 - \frac{\sqrt{3}}{2}\right)^2} = 1.0$ and $\sqrt{(x_{Be} - x_{Ce})^2 + (y_{Be} - y_{Ce})^2} = \sqrt{(1.0-0.0)^2 + \left(\frac{\sqrt{3}}{2} - \frac{\sqrt{3}}{2}\right)^2} = 1.0$, respectively, while the resilience for the status represented by vertex C_e is computed as $\sqrt{(x_{Ce} - x_{Ce})^2 + (y_{Ce} - y_{Ce})^2} = 0.0$. This indicates that the statuses represented by vertices A_e and B_e are fully resilient, whereas the status represented by vertex C_e is not resilient at all.

5.9.2.2 Mathematical description of the equilateral triangle

The stability line is the bottom edge (i.e. B_eC_e) of the equilateral triangle shown in figure 5.65. The coordinates of a point on B_eC_e can be determined as x-coordinate of the point on B_eC_e in Figure 5.65 [0 ~ 1] → $x^*_{sta} = S_{ta}$ ← stability [0 ~ 1] (5.18)

y-coordinate of the point on B_eC_e in Figure 5.65 → $y^*_{sta} = \frac{\sqrt{3}}{2}$ (5.19)

The stability gridlines are vertical. A gridline passing through a point on B_eC_e can be defined as

stability [0 ~ 1]

$$0y = x - S_{ta} \quad (5.20)$$

y-coordinate of a point on the gridline passing through the point on B_eC_e in Figure 5.65

x-coordinate of the point on the gridline passing through the point on B_eC_e in Figure 5.65

The reliability line is the left edge (i.e. A_eB_e) of the equilateral triangle shown in figure 5.65. The coordinates of a point on A_eB_e can be determined as

x-coordinate of the point on A_eB_e in Figure 5.65 [0.5 ~ 1] → $x^*_{rel} = 0.5R_{el} + 0.5$ (5.21)

reliability [0 ~ 1]

y-coordinate of the point on A_eB_e in Figure 5.65 [0 ~ $3^{1/2}/2$] → $y^*_{rel} = \frac{\sqrt{3}}{2}R_{el}$ (5.22)

reliability [0 ~ 1]

Derivation:

$x^*_{rel} = 1 - (1 - R_{el})\cos 60^\circ = 1 - (1 - R_{el})\left(\frac{1}{2}\right) = 0.5R_{el} + 0.5 \rightarrow$ equation (5.21)

$y^*_{rel} = R_{el}\sin 60^\circ = \frac{\sqrt{3}}{2}R_{el} \rightarrow$ equation (5.22)

The reliability gridlines are parallel to the vulnerability line, which is the right edge (i.e. A_eC_e) of the equilateral triangle. A gridline passing through a point on A_eB_e can be defined as reliability [0 ~ 1]

$$y = -\sqrt{3}x - \sqrt{3}(R_{el} + 0.5) \tag{5.23}$$

y-coordinate of a point on the gridline passing through the point on A_eB_e in Figure 5.65

x-coordinate of a point on the gridline passing through the point on A_eB_e in Figure 5.65

Derivation:
Let $y = ax + b$, which passes through the point: (x^*_{rel}, y^*_{rel}).
Parallel to the right edge → $a = -2\sin 60° = -\sqrt{3}$ → $y = -\sqrt{3}x + b$.
Substituting → $\frac{\sqrt{3}}{2}R_{el} = -\sqrt{3}(0.5R_{el} + 0.5) + b$ → $b = \sqrt{3}(R_{el} + 0.5)$ → equation (5.23).

The vulnerability line is the right edge (i.e. A_eC_e) of the equilateral triangle shown in figure 5.65. The coordinates of a point on A_eC_e can be determined as

x-coordinate of the point on A_eC_e in Figure 5.65 [0 ~ 0.5] →

$$x^*_{vul} = -0.5V_{ul} + 0.5 \tag{5.24}$$

vulnerability [0 ~ 1]

y-coordinate of the point on A_eC_e in Figure 5.65 [0 ~ $3^{½}/2$] →

$$y^*_{vul} = \frac{\sqrt{3}}{2}V_{ul} \tag{5.25}$$

vulnerability [0 ~ 1]

Derivation:

$x^*_{vul} = (1 - V_{ul})\cos 60° = (1 - V_{ul})\left(\frac{1}{2}\right) = -0.5V_{ul} + 0.5$ → equation (5.24)

$y^*_{vul} = V_{ul}\sin 60° = \frac{\sqrt{3}}{2}V_{ul}$ → equation (5.25)

The vulnerability gridlines are parallel to the reliability line, which is the left edge (i.e. A_eB_e) of the equilateral triangle. A gridline passing through a point on A_eC_e can be defined as vulnerability [0 ~ 1]

$$y = \sqrt{3}x + \sqrt{3}\left(V_{ul} - 0.5\right) \tag{5.26}$$

y-coordinate of a point on the gridline passing through the point on A_eC_e in Figure 5.65

x-coordinate of a point on the gridline passing through the point on A_eC_e in Figure 5.65

Derivation:
Let $y = ax + b$, which passes through the point: (x^*_{vul}, y^*_{vul}).
parallel to the left edge → $a = 2\sin 60° = \sqrt{3}$ → $y = \sqrt{3}x + b$.
Substituting → $\frac{\sqrt{3}}{2}V_{ul} = \sqrt{3}(-0.5V_{ul} + 0.5) + b$ → $b = \sqrt{3}(V_{ul} - 0.5)$ → equation (5.26).

5.9.2.3 Formulizations of resilience

In geometry, the incircle or inscribed circle of a triangle is the largest circle contained in the triangle; it touches (i.e. is tangent to) the three sides. The center of the incircle is the intersection of the three angle bisectors (figure 5.66).

For vertex A, its coordinates are computed as

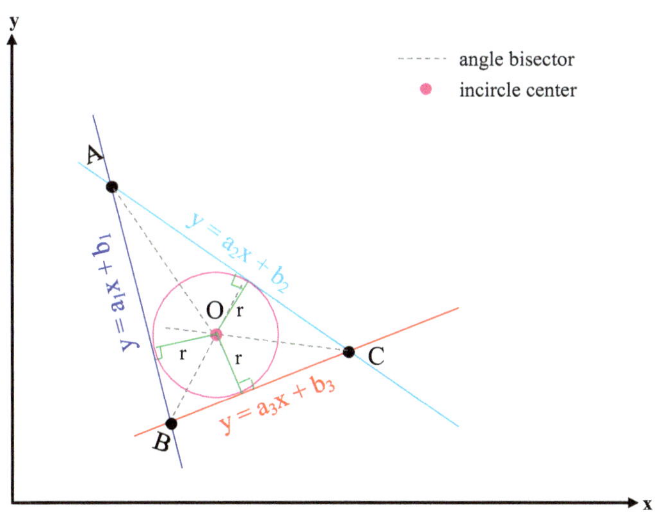

Figure 5.66. The incircle of a triangle. The center is the intersection of the three angle bisectors.

intercept of line AB in Figure 5.66 — intercept of line AC in Figure 5.66

x-coordinate of vertex A of the triangle shown in Figure 5.66 →

$$x_A = -\frac{b_1 - b_2}{a_1 - a_2} \tag{5.27}$$

slope of line AB in Figure 5.66 — slope of line AC in Figure 5.66

intercept of line AC in Figure 5.66 — intercept of line AB in Figure 5.66

y-coordinate of vertex A of the triangle shown in Figure 5.66 →

$$y_A = \frac{a_1 b_2 - a_2 b_1}{a_1 - a_2} \tag{5.28}$$

slope of line AB in Figure 5.66 — slope of line AC in Figure 5.66

> **Derivation:**
> Line BA passes through A → $y_A = a_1 x_A + b_1$.
> Line CA passes through A → $y_A = a_2 x_A + b_2$.
> Solving these two equations simultaneously → equations (5.27) and (5.28).

For vertex B, its coordinates are computed as

intercept of line AB in Figure 5.66 — intercept of line BC in Figure 5.66

x-coordinate of vertex B of the triangle shown in Figure 5.66 →

$$x_B = -\frac{b_1 - b_3}{a_1 - a_3} \tag{5.29}$$

slope of line AB in Figure 5.66 — slope of line BC in Figure 5.66

intercept of line BC in Figure 5.66 — intercept of line AB in Figure 5.66

y-coordinate of vertex B of the triangle shown in Figure 5.66 →

$$y_B = \frac{a_1 b_3 - a_3 b_1}{a_1 - a_3} \tag{5.30}$$

slope of line AB in Figure 5.66 — slope of line BC in Figure 5.66

> **Derivation:**
> Line AB passes through B → $y_B = a_1 x_B + b_1$.
> Line CB passes through B → $y_B = a_2 x_B + b_2$.
> Solving these two equations simultaneously → equations (5.29) and (5.30).

For vertex C, its coordinates are computed as

intercept of line AC in Figure 5.66; intercept of line BC in Figure 5.66

x-coordinate of vertex C of the triangle shown in Figure 5.66 →

$$x_C = -\frac{b_2 - b_3}{a_2 - a_3} \tag{5.31}$$

slope of line AC in Figure 5.66; slope of line BC in Figure 5.66

intercept of line BC in Figure 5.66; intercept of line AC in Figure 5.66

y-coordinate of vertex C of the triangle shown in Figure 5.66 →

$$y_C = \frac{a_2 b_3 - a_3 b_2}{a_2 - a_3} \tag{5.32}$$

slope of line AC in Figure 5.66; slope of line BC in Figure 5.66

Derivation:
Line AC passes through C → $y_C = a_2 x_C + b_2$.
Line BC passes through C → $y_C = a_3 x_C + b_3$.
Solving these two equations simultaneously → equations (5.31) and (5.32).

The lengths of the triangle sides are computed as

x-coordinate of vertex B; y-coordinate of vertex B

length of side AB of the triangle shown in Figure 5.66 →

$$L_{AB} = \sqrt{(x_A - x_B)^2 + (y_A - y_B)^2} \tag{5.33}$$

x-coordinate of vertex A; y-coordinate of vertex A

x-coordinate of vertex C; y-coordinate of vertex C

length of side AC of the triangle shown in Figure 5.66 →

$$L_{AC} = \sqrt{(x_A - x_C)^2 + (y_A - y_C)^2} \tag{5.34}$$

x-coordinate of vertex A; y-coordinate of vertex A

x-coordinate of vertex C; y-coordinate of vertex C

length of side BC of the triangle shown in Figure 5.66 →

$$L_{BC} = \sqrt{(x_B - x_C)^2 + (y_B - y_C)^2} \tag{5.35}$$

x-coordinate of vertex B; y-coordinate of vertex B

The area of the triangle is computed as

$$A_{ABC} = \frac{1}{2} L_{BC} \frac{|a_3 x_A - y_A + b_3|}{\sqrt{a_3^2 + 1}} \tag{5.36}$$

x-coordinate of vertex A; y-coordinate of vertex A; area of the triangle shown in Figure 5.66; intercept of line BC in Figure 5.66; length of side BC of the triangle; slope of line BC in Figure 5.66

Derivation:
As shown in the following context, the perpendicular distance from A to line BC is

$$d = \frac{| a_3 x_A - y_A + b_3 |}{\sqrt{a_3^2 + 1}}$$

$$A_{ABC} = \frac{1}{2} L_{BC} d = \frac{1}{2} L_{BC} \frac{| a_3 x_A - y_A + b_3 |}{\sqrt{a_3^2 + 1}}.$$

The radius of the incircle is computed as

$$r = \frac{2A_{ABC}}{L_{AB} + L_{AC} + L_{BC}} \tag{5.37}$$

incircle radius; area of the triangle; length of side AB; length of side AC; length of side BC

Derivation:
$A_{ABC} = 1/2(L_{AB})(r) + 1/2(L_{AC})(r) + 1/2(L_{BC})(r) = 1/2(L_{AB} + L_{AC} + L_{BC})(r)$.
Dividing $1/2(L_{AB} + L_{AC} + L_{BC})$ by the two sides → equation (5.37).

The coordinates of the incircle center are computed as

$$y_0 = \frac{[(a_2 - a_3)x_0 + (b_2 - b_3)][(a_2 + a_3)x_0 + (b_2 + b_3)] - (a_2^2 - a_3^2)r^2}{2[(a_2 - a_3)x_0 + (b_2 - b_3)]} \tag{5.38}$$

x-coordinate of incircle center; incircle radius; y-coordinate of incircle center; slope of line AC in Figure 5.66; slope of line BC in Figure 5.66; intercept of line BC in Figure 5.66; intercept of line BC in Figure 5.66

y-coordinate of incircle center

x-coordinate of incircle center

intercept of line BC in Figure 5.66

incircle radius

$$r = \frac{|a_2 x_0 - y_0 + b_2|}{\sqrt{a_2^2 + 1}} \tag{5.39}$$

slope of line AC in Figure 5.66

Derivation:

The perpendicular distance from O to line AC $\rightarrow r = \frac{|a_2 x_0 - y_0 + b_2|}{\sqrt{a_2^2 + 1}}$

$$\rightarrow (a_2^2 + 1)r^2 = y_0^2 - 2(a_2 x_0 + b_2)y_0 + (a_2 x_0 + b_2)^2.$$

The perpendicular distance from O to line BC $\rightarrow r = \frac{|a_3 x_0 - y_0 + b_3|}{\sqrt{a_3^2 + 1}}$

$$\rightarrow (a_3^2 + 1)r^2 = y_0^2 - 2(a_3 x_0 + b_3)y_0 + (a_3 x_0 + b_3)^2.$$

Subtracting the second equation from the first equation

$$\rightarrow (a_2^2 - a_3^2)r^2 = -2[(a_2 - a_3)x_0 + (b_2 - b_3)]y_0 + [(a_2 - a_3)x_0 + (b_2 - b_3)][(a_2 + a_3)x_0 + (b_2 + b_3)]$$

$$\rightarrow y_0 = \frac{[(a_2 - a_3)x_0 + (b_2 - b_3)][(a_2 + a_3)x_0 + (b_2 + b_3)] - (a_2^2 - a_3^2)r^2}{2[(a_2 - a_3)x_0 + (b_2 - b_3)]} \rightarrow \text{equation (5.38).}$$

Rearranging the first equation $\rightarrow (a_2 x_0 + b_2)^2 - 2y_0(a_2 x_0 + b_2) + y_0^2 - (a_2^2 + 1)r^2 = 0.$

Using quadratic formula $\rightarrow a_2 x_0 + b_2 = \frac{-(-2y_0) \pm \sqrt{(-2y_0)^2 - 4*1*[y_0^2 - (a_2^2 + 1)r^2]}}{2*1} = y_0 \pm r\sqrt{a_2^2 + 1}$

$$\rightarrow x_0 = \frac{\left(y_0 \pm r\sqrt{a_2^2 + 1}\right) - b_2}{a_2} \rightarrow \text{equation (5.39).}$$

Upon solving equations (5.38) and (5.39) for x_0 and y_0, the incircle can be described as

x-coordinate of a point on incircle

y-coordinate of a point on incircle

$$(x - x_0)^2 + (y - y_0)^2 = r^2 \tag{5.40}$$

incircle radius

x-coordinate of incircle center

y-coordinate of incircle center

Derivation: perpendicular distance formula
Point (x_0, y_0).
Line: $ax + by + c = 0 \rightarrow y = -(a/b)x - (c/b)$.
Distance squared:

$$d^2 = (x_0 - x)^2 + (y_0 - y)^2 = (x_0 - x)^2 + \left(y_0 + \frac{a}{b}x + \frac{c}{b}\right)^2 = (x - x_0)^2 + \left(\frac{a}{b}x + y_0 + \frac{c}{b}\right)^2.$$

Differentiating with respect to x: $\rightarrow 2d\frac{dd}{dx} = 2(x - x_0) + 2\left(\frac{a}{b}x + y_0 + \frac{c}{b}\right)\frac{a}{b}$

$$\rightarrow \frac{dd}{dx} = \frac{(x - x_0) + \left(\frac{a}{b}x + y_0 + \frac{c}{b}\right)\frac{a}{b}}{d} = \frac{\left[1 + \left(\frac{a}{b}\right)^2\right]x - x_0 + \left(y_0 + \frac{c}{b}\right)\frac{a}{b}}{d}.$$

Let $\frac{dd}{dx} = \frac{\left[1 + \left(\frac{a}{b}\right)^2\right]x - x_0 + \left(y_0 + \frac{c}{b}\right)\frac{a}{b}}{d} = 0 \rightarrow \left[1 + \left(\frac{a}{b}\right)^2\right]x - x_0 + \left(y_0 + \frac{c}{b}\right)\frac{a}{b} = 0$

$$\rightarrow x = \frac{x_0 - \left(y_0 + \frac{c}{b}\right)\frac{a}{b}}{1 + \left(\frac{a}{b}\right)^2}.$$

The perpendicular distance: $d_{min}^2 = \left(\frac{x_0 - \left(y_0 + \frac{c}{b}\right)\frac{a}{b}}{1 + \left(\frac{a}{b}\right)^2} - x_0\right)^2 + \left(\frac{a}{b}\frac{x_0 - \left(y_0 + \frac{c}{b}\right)\frac{a}{b}}{1 + \left(\frac{a}{b}\right)^2} + y_0 + \frac{c}{b}\right)^2$

$$\rightarrow d_{min}^2 = \left[\frac{1}{1 + \left(\frac{a}{b}\right)^2}\right]^2 \left[\left(\frac{a}{b}\right)^2\left(\frac{a}{b}x_0 + y_0 + \frac{c}{b}\right)^2 + \left(\frac{a}{b}x_0 + y_0 + \frac{c}{b}\right)^2\right]$$

$$\rightarrow d_{min}^2 = \left[\frac{1}{1 + \left(\frac{a}{b}\right)^2}\right]^2 \left[1 + \left(\frac{a}{b}\right)^2\right]\frac{1}{b^2}(ax_0 + by_0 + c)^2$$

$\rightarrow d_{min}^2 = \frac{1}{a^2 + b^2}(ax_0 + by_0 + c)^2 \rightarrow d_{min} = \frac{|ax_0 + by_0 + c|}{\sqrt{a^2 + b^2}}$.

The coordinates of the tangential point between line AB and the incircle are computed as

$$x_{AB} = \frac{x_0 - a_1(b_1 - y_0) \pm \sqrt{(1 + a_1^2)r^2 - [a_1x_0 + (b_1 - y_0)]^2}}{1 + a_1^2} \tag{5.41}$$

y-coordinate of a point on incircle
incircle radius
x-coordinate of incircle center
x-coordinate of the tangential point
intercept of line AB in Figure 5.66
slope of line AB in Figure 5.66

x-coordinate of the tangential point

y-coordinate of the tangential point

$$y_{AB} = a_1 x_{AB} + b_1 \quad (5.42)$$

slope of line AB in Figure 5.66

intercept of line AB in Figure 5.66

The coordinates of the tangential point between line AC and the incircle are computed as

y-coordinate of a point on incircle

incircle radius

x-coordinate of incircle center

x-coordinate of the tangential point

$$x_{AC} = \frac{x_0 - a_2(b_2 - y_0) \pm \sqrt{(1 + a_2^2) r^2 - [a_2 x_0 + (b_2 - y_0)]^2}}{1 + a_2^2} \quad (5.43)$$

intercept of line AC in Figure 5.66

slope of line AC in Figure 5.66

x-coordinate of the tangential point

y-coordinate of the tangential point

$$y_{AC} = a_2 x_{AC} + b_2 \quad (5.44)$$

slope of line AC in Figure 5.66

intercept of line AC in Figure 5.66

The coordinates of the tangential point between line BC and the incircle are computed as

y-coordinate of a point on incircle

incircle radius

x-coordinate of incircle center

x-coordinate of the tangential point

$$x_{BC} = \frac{x_0 - a_3(b_3 - y_0) \pm \sqrt{(1 + a_3^2) r^2 - [a_3 x_0 + (b_3 - y_0)]^2}}{1 + a_3^2} \quad (5.45)$$

intercept of line BC in Figure 5.66

slope of line BC in Figure 5.66

x-coordinate of the tangential point

y-coordinate of the tangential point

$$y_{BC} = a_3 x_{BC} + b_3 \quad (5.46)$$

slope of line BC in Figure 5.66

intercept of line BC in Figure 5.66

Derivation: Take x_{AB} and y_{AB} as an example because the derivations for x_{AC}, y_{AC}, x_{BC}, and y_{BC} are similar.

At tangential point (x_{AB}, y_{AB}):

$(x_{AB} - x_0)^2 + (y_{AB} - y_0)^2 = r^2$

$y_{AB} = a_1 x_{AB} + b_1$ → equation (5.42).

Substitute the second equation into the first equation, yield:

$(x_{AB} - x_0)^2 + (a_1 x_{AB} + b_1 - y_0)^2 = r^2$

$\rightarrow (x_{AB})^2 - 2x_0\, x_{AB} + (x_0)^2 + (a_1 x_{AB})^2 + 2a_1(b_1 - y_0)x_{AB} + (b_1 - y_0)^2 = r^2$

$\rightarrow (1 + a_1{}^2)(x_{AB})^2 - 2[x_0 - a_1(b_1 - y_0)]x_{AB} + x_0{}^2 + (b_1 - y_0)^2 - r^2 = 0.$

Using the quadratic formula, this leads to

$$x_{AB} = \frac{2[x_0 - a_1(b_1 - y_0)] \pm \sqrt{(2[x_0 - a_1(b_1 - y_0)])^2 - 4[(1 + a_1^2)[x_0^2 + (b_1 - y_0)^2 - r^2]]}}{2(1 + a_1^2)}$$

$$\rightarrow x_{AB} = \frac{2[x_0 - a_1(b_1 - y_0)] \pm 2\sqrt{(1 + a_1^2)r^2 - [a_1 x_0 + (b_1 - y_0)]^2}}{2(1 + a_1^2)}$$

$$\rightarrow x_{AB} = \frac{x_0 - a_1(b_1 - y_0) \pm \sqrt{(1 + a_1^2)r^2 - [a_1 x_0 + (b_1 - y_0)]^2}}{1 + a_1^2} \rightarrow \text{equation (5.41).}$$

5.9.2.4 Resilience circle

As shown in figure 5.65, a triad of stability (S_{ta}), reliability (R_{el}), and vulnerability (V_{ul}) indicate a state of the system of interest. The corresponding gridlines of the triad form a triangle, whose incircle (hereafter termed the resilience circle) defines the system's resilience (R_{es}) for a state. For a given disturbing stress, the system can be shifted to a new state, forming a new resilience circle. If the new resilience circle intersects the current resilience circle, the system will be able to bounce from the new state back to the current state once the disturbing stress is removed; otherwise, the system will stay at the new state. Referring to figure 5.66, let AB, AC, and BC be the gridlines for S_{ta}, R_{el}, and V_{ul}, respectively, forming a triangle ABC.

Rearranging equation (5.20) yields the function for side AB expressed as

stability [0 ~ 1]

x-coordinate of a point on side AB

$$x + 0y - S_{ta} = 0 \tag{5.47}$$

y-coordinate of the point on side AB

Rearranging equation (5.23) yields the function for side AC expressed as

y-coordinate of the point on side AC

$$\sqrt{3}x + y - \sqrt{3}(R_{el} + 0.5) = 0 \tag{5.48}$$

x-coordinate of a point on side AC   reliability [0 ~ 1]

Rearranging equation (5.26) yields the function for side BC expressed as

y-coordinate of the point on side BC

$$\sqrt{3}x - y + \sqrt{3}(V_{ul} - 0.5) = 0 \tag{5.49}$$

x-coordinate of a point on side BC   vulnerability [0 ~ 1]

The coordinates of the triangle vertices are computed as

x-coordinate of vertex A → $x_A = S_{ta}$ ← stability [0 ~ 1]

$$x_A = S_{ta} \tag{5.50}$$

reliability [0 ~ 1] stability [0 ~ 1]

y-coordinate of vertex A →

$$y_A = 0.5\sqrt{3} + \sqrt{3}\left(R_{el} - S_{ta}\right) \tag{5.51}$$

x-coordinate of vertex B → $x_B = S_{ta}$ ← stability [0 ~ 1]

$$x_B = S_{ta} \tag{5.52}$$

vulnerability [0 ~ 1] stability [0 ~ 1]

y-coordinate of vertex B →

$$y_B = -0.5\sqrt{3} + \sqrt{3}\left(V_{ul} + S_{ta}\right) \tag{5.53}$$

reliability [0 ~ 1] vulnerability [0 ~ 1]

x-coordinate of vertex C →

$$x_C = 0.5 + 0.5\left(R_{el} - V_{ul}\right) \tag{5.54}$$

reliability [0 ~ 1] vulnerability [0 ~ 1]

y-coordinate of vertex C →

$$y_C = 0.5\sqrt{3}\left(R_{el} + V_{ul}\right) \tag{5.55}$$

Substituting equations (5.43) and (5.44) into equation (5.48) as well as equations (5.45) and (5.46) into equation (5.49), rearranging and simplifying, yield $a_2 = -\sqrt{3}$, $b_2 = \sqrt{3}(R_{el} + 0.5)$, $a_3 = \sqrt{3}$, and $b_3 = \sqrt{3}(V_{ul} - 0.5)$. Given that side AB is vertical, equations (5.41) and (5.42) are reduced as

x-coordinate of the tangential point between side AB and incircle → $x_{AB} = S_{ta}$ ← stability [0 ~ 1]

$$x_{AB} = S_{ta} \tag{5.56}$$

y-coordinate of the tangential point between side AB and incircle → $y_{AB} = y_0$ ← y-coordinate of incircle center

$$y_{AB} = y_0 \tag{5.57}$$

5.9.3 Methods

The calibrated soil–water–atmosphere–plant (SWAP) model, which is detailed in section 5.2, was run from 1955 to 2017 for livestock carrying capacities of 0.0–5.0 cattle ha^{-1} (i.e. 0.0–5.0 × 4305/2928 = 7.35 AUY ha^{-1}). Herein, the dry mass intake rate per cattle is 4305 kg $year^{-1}$ (Nelle and Reinke 2020), while the intake rate per animal unit per year (AUY) is 2928 kg $year^{-1}$ (Hinnant 1994). Unlike the non-stopping year-round grazing practice used by Christensen *et al* (2003), our study assumed a restrictive grazing practice: a grazing cycle starts once the aboveground dry matter is more than 350 kg ha^{-1} and continues until the dry matter is reduced to 150 kg ha^{-1}, with a dropping and treading loss of 0.43–0.5 kg ha^{-1}. This restrictive grazing practice has been implemented by the ranchers of the study watershed. The livestock grazing pressures in these two studies should not be directly compared because for a given year, the number of grazing days in our study depends on the available grass aboveground biomass and would be much shorter. In addition, for a given livestock carrying capacity, the simulated annual aboveground biomasses from 1958 to 2017 were used to compute the stability, reliability, and vulnerability, which in turn were used to construct the resilience circle and determine the resilience. The situation with no grazing was considered as the baseline, against which the situations with non-zero livestock grazing pressures were assessed. The first and third quartiles of the simulated aboveground biomasses for zero grazing were computed and used as the thresholds (i.e. $N_{\leqslant Z25}$ and $N_{\geqslant Z75}$) in equations (5.14) to (5.16) to compute the stabilities, reliabilities, and vulnerabilities in response to both zero and non-zero livestock carrying capacities. Further, to retrospect the history, the data on ANPP from 2000 to 2009 were extracted in ENVI 4.7 from 1 km MOD17A3 images (Luo *et al* 2014) and used to scrutinize the historical resilience in relative to the baseline. In practice, ANPP has been widely used as a surrogate of aboveground biomass of vegetation including steppe grasses. Moreover, equations (3.1) to (3.6) were used to determine the equilibria and shifts of the states because of varying livestock carrying capacities.

5.9.4 Results

5.9.4.1 The baseline resilience

For a zero grazing intensity, the simulated ANPPs varied from 12 904 to 18 011 kg DM ha^{-1} (figure 5.67), with a mean of 15 008 kg DM ha^{-1} and a standard deviation of 1005 kg DM ha^{-1}. The first quantile was 14 222.5 kg DM ha^{-1}, while the third quantile was 15 676.0 kg DM ha^{-1}. The stability and reliability were computed to be 0.933 and 0.606, respectively, whereas the vulnerability was determined as 0.578. The baseline resilience was 0.756, with the resilience circle (figure 5.68) close to the best-status vertex (i.e. B_e).

5.9.4.2 The impacts of grazing on resilience

With an increase of livestock carrying capacity, the reliability was predicted to decrease while the vulnerability was predicted to increase, both monotonically (figure 5.69). On the other hand, the stability was predicted to decrease gradually until a carrying capacity of 3.0 cattle ha^{-1} and to plunge to near zero at a carrying capacity of 3.5 cattle ha^{-1}. For a carrying capacity of larger than 3.5 cattle ha^{-1}, the ANPPs were predicted to be very low with a small variability (figure 5.67), leading to an increase in stability.

In correspondence, the resilience was predicted to decrease gradually to a carrying capacity of 0.8 cattle ha^{-1} and to plunge to a negative value (figure 5.69). From this threshold carrying capacity onwards, the absolute resilience was predicted to

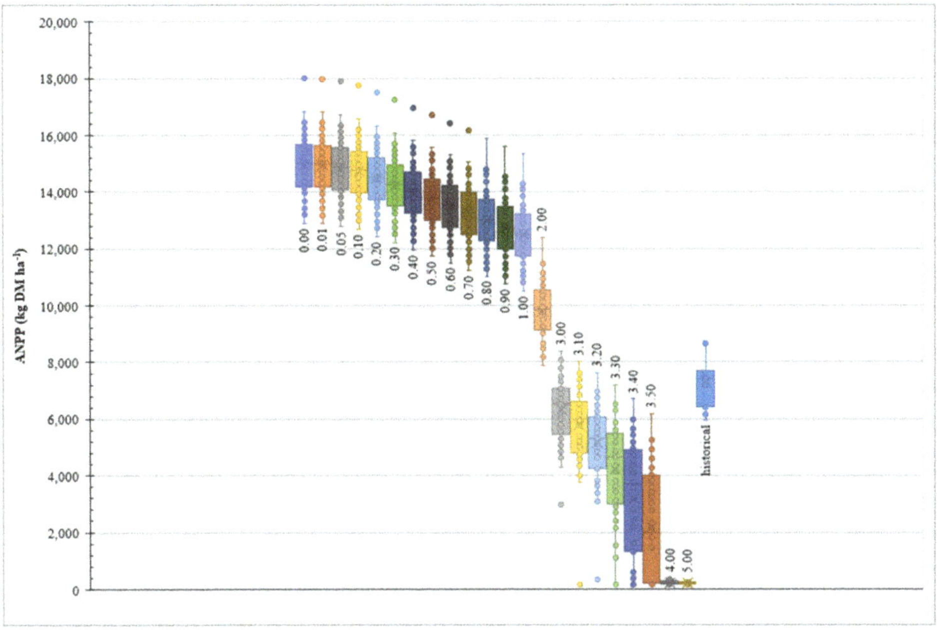

Figure 5.67. Box–whisker plots of the aboveground net primary productions (ANPPs). The ANPPs were simulated for the labeled livestock carrying capacities, in cattle ha^{-1}, or extracted from the 1 km MOD17A3 images for historical grazing practices. The results are for the Balagaer River watershed.

decrease steadily up to a carrying capacity of 3.5 cattle ha^{-1}, followed by an increasing trend. For a given carrying capacity, the negative resilience indicates that the corresponding resilience circle does not completely enclose the baseline resilience circle; these two circles either intersect or are apart (figure 5.68). For the Balagaer River watershed, for a carrying capacity of $\leqslant$ 3.2 cattle ha^{-1}, the steppe ecosystem would be able to bounce back to its baseline condition if grazing is ceased; otherwise, the steppe ecosystem will permanently be shifted to a worse condition. For a carrying capacity of $\geqslant$ 3.5 cattle ha^{-1}, the steppe ecosystem would shift to the crucial or an irreversible degradation status. The historical data indicated that the steppe ecosystem is currently very vulnerable and unreliable, with a stability of 0.878 and a resilience of −0.47 (figure 5.69) but it can fully recover if the historical grazing (–2.5 cattle ha^{-1} in average) were reduced or stopped (figure 5.68).

5.9.4.3 The equilibria and shifts of states

For a livestock carrying capacity of $\leqslant$ 0.8 cattle ha^{-1}, the steppe ecosystem was predicted to be able to bounce back to its initial equilibrium state provided that the fertile-soil depth was thicker than 0.05 m (figure 5.70(a)). The ecosystem's biomass would decrease to zero in a year but recover in another year, indicating a high resilience (figure 5.68). However, once the fertile-soil depth was reduced to be thinner than 0.05 m, the steppe ecosystem would shift to its catastrophic equilibrium state, irreversibly degrading to a semidesert with sparse vegetation and various sand dunes.

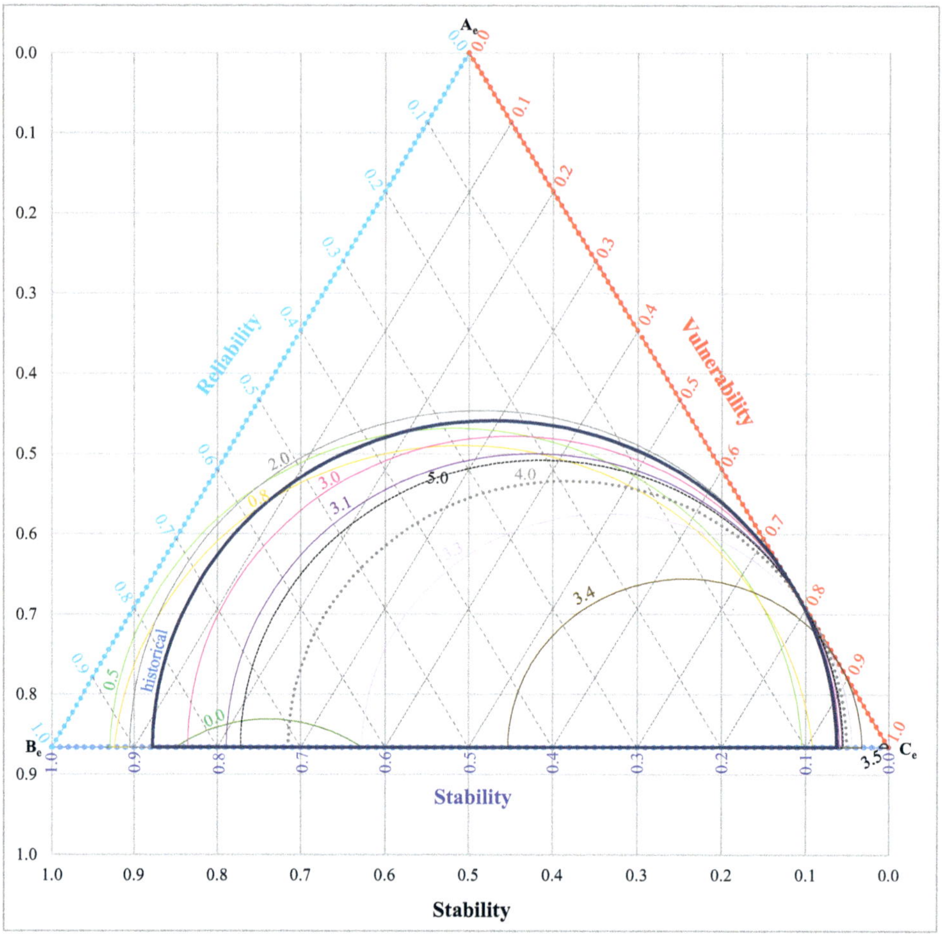

Figure 5.68. Resilience circles. Effects of livestock carrying capacity on the resilience of the Balagaer River watershed. The livestock carrying capacity, in cattle ha^{-1}, is labeled in the same color as the corresponding resilience circle.

With an increase of livestock carrying capacity, the thresholds of biomass and fertile-soil depth below which the steppe ecosystem would start to degrade irreversibly were predicted to increase as well. For a carrying capacity of 2.5 cattle ha^{-1}, the thresholds were predicted to be 0.15 kg m^{-2} and 0.08 m (figure 5.70(b)), while for a carrying capacity of 4.3 cattle ha^{-1}, the thresholds were predicted to be 0.85 kg m^{-2} and 0.16 m (figure 5.70(f)). For a carrying capacity of between 2.5 and 4.3 cattle ha^{-1}, the thresholds were predicted to be 0.15–0.85 kg m^{-2} and 0.08–0.16 m (e.g. figures 5.70(c)–(e)). That is, the steppe ecosystem would more likely shift to its catastrophic equilibrium state for a higher livestock carrying capacity than for a lower one. Moreover, for a carrying capacity of $\leqslant$ 3.0 cattle ha^{-1}, the steppe ecosystem was predicted to shift from the initial to catastrophic state during 13–17 years in terms of biomass and 20–24 years in terms of fertile-soil depth (figure 5.71),

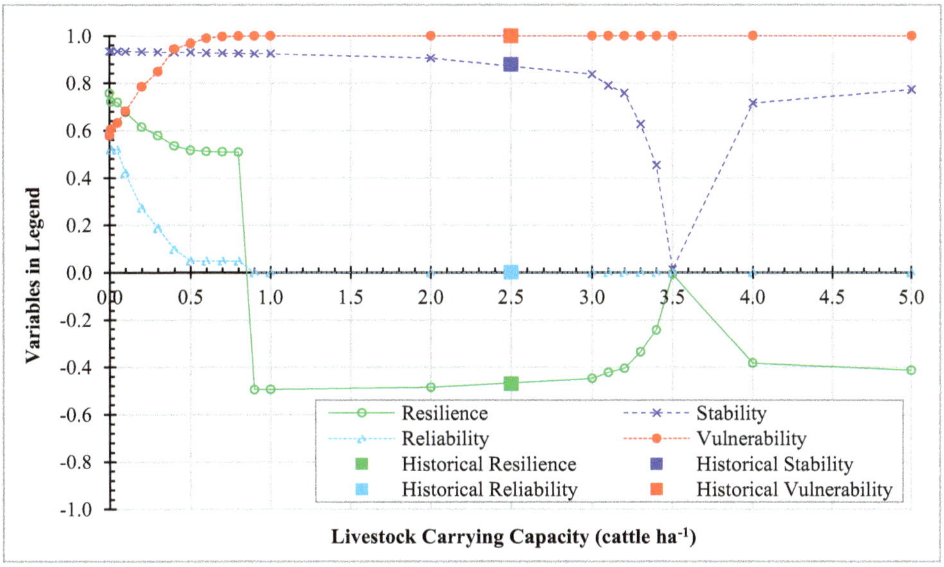

Figure 5.69. Impacts of livestock carrying capacity. The simulated and historical values of resilience, stability, reliability, and vulnerability of the Balagaer River watershed.

whereas for a carrying capacity of larger than 4.3 cattle ha^{-1}, the steppe ecosystem was predicted to shift to its catastrophic equilibrium state right away.

5.9.5 Discussion

Wang's ecosystem resilience theory (WERT) provides a straightforward means for visualizing and computing ecosystem resilience as affected by climate and/or livestock grazing. It has resolved the subjectivity and inconsistency of how resilience is defined and determined in the existing literature. WERT computes resilience as a function of reliability, stability, and vulnerability, which in turn are determined from simulated or measured biomasses (e.g. aboveground net primary productions or ANPPs). The incircle of the triangle formed by reliability, stability, and vulnerability, which are standardized to have a value from zero to one, defines the corresponding resilience circle; a resilience circle signifies a state of the ecosystem of interest. If the resilience circle of a post-shifted state encloses or intersects the resilience circle of the pre-shifted state, the ecosystem will be able to bounce back to the pre-shifted state if the shift-inducing pressure (e.g. grazing) is removed; otherwise, the ecosystem will stay at the post-shifted state even though the pressure is removed. The resilience of a state is computed as the Euclidean distance from the center of the resilience circle to the point with a reliability of zero, a stability of zero, and a vulnerability of one. The resilience can range from negative one to positive one: a positive value indicates that the ecosystem can fully recover from the post-shifted state to the pre-shifted state (i.e. is fully resilient), while a negative value shows that the ecosystem can partially recover to an intermediate state between the post-shifted state and the pre-shifted state (i.e. is partially resilient) or will

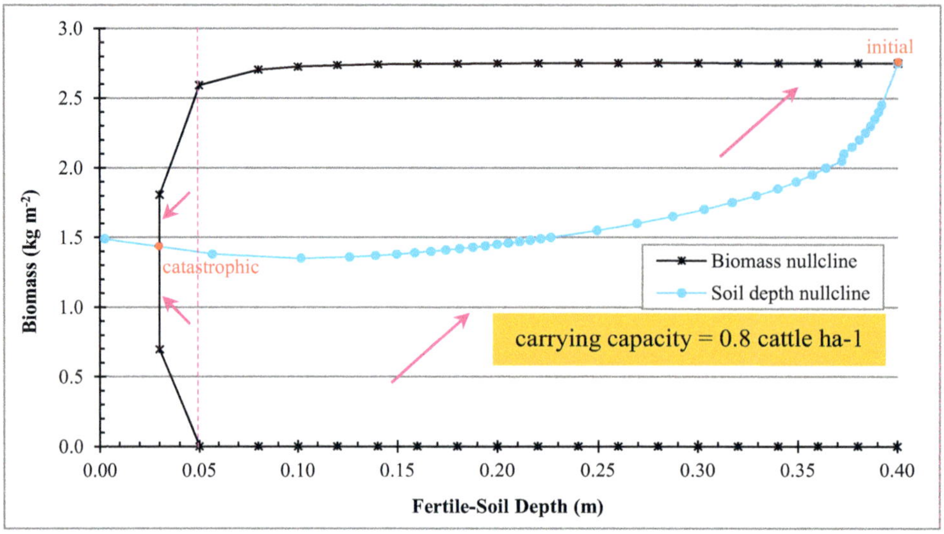

(a)

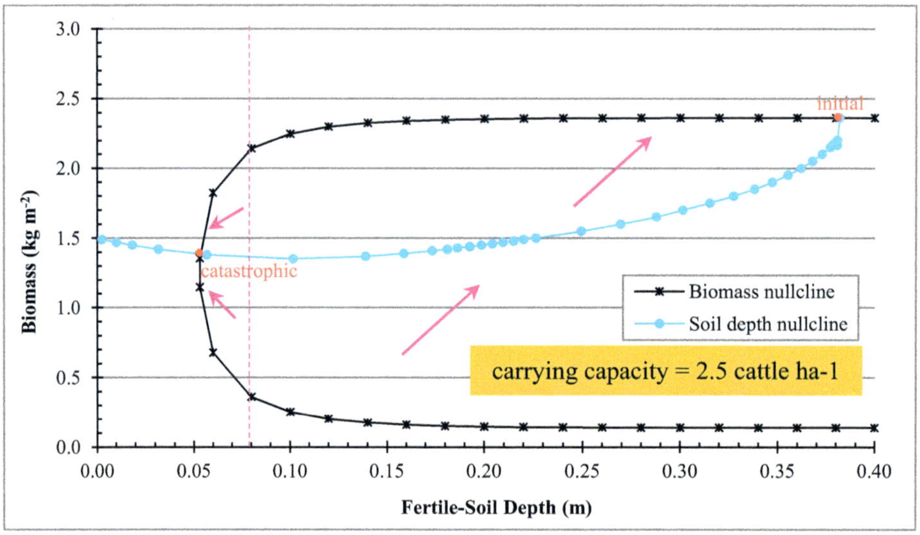

(b)

Figure 5.70. Nullclines of fertile-soil depth and biomass. Plots for livestock carrying capacities, in cattle ha^{-1}, of (a) 0.8, (b) 2.5, (c) 3.0, (d) 3.5, (e) 4.0, and (f) 4.3. On each plot, a pink arrow signifies the state shifting direction, while the pink dashed line is the border between the zone towards the initial state and the zone towards the catastrophic state.

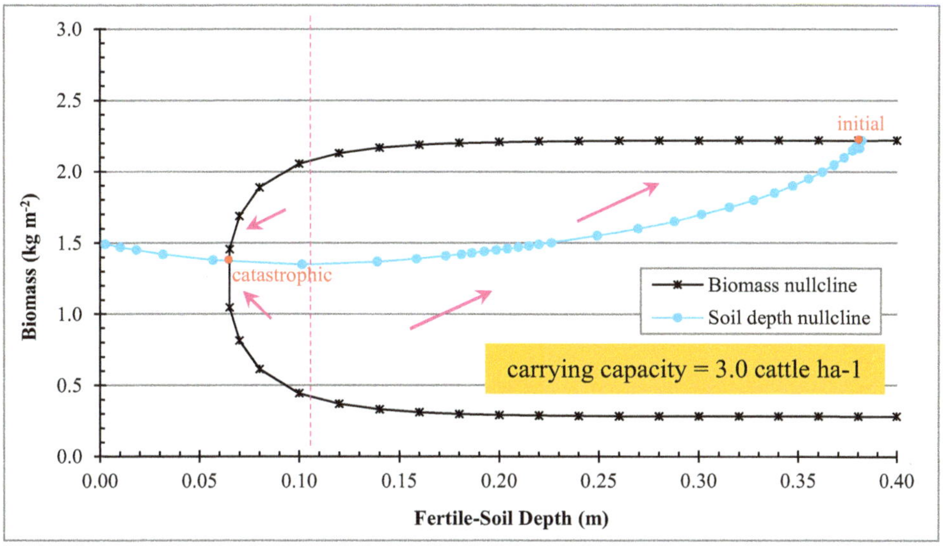

(c)

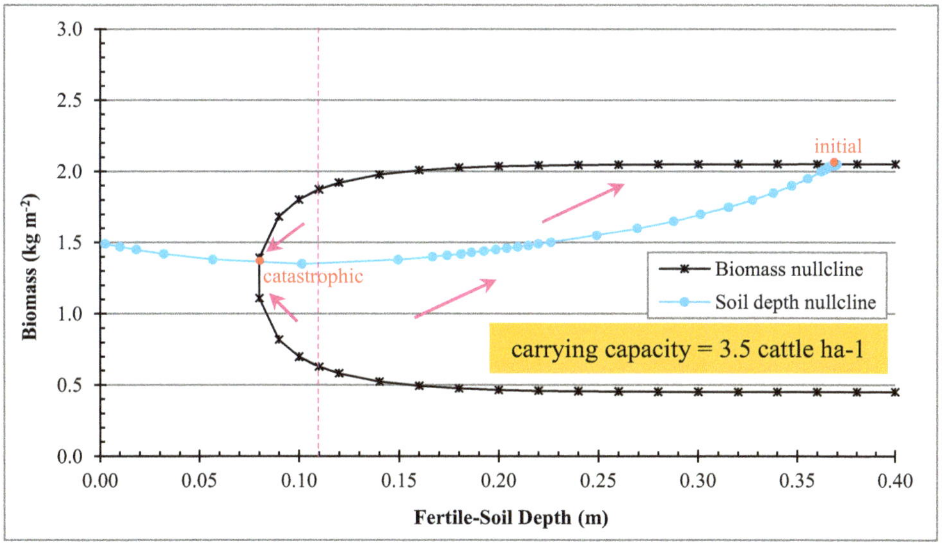

(d)

Figure 5.70. (Continued.)

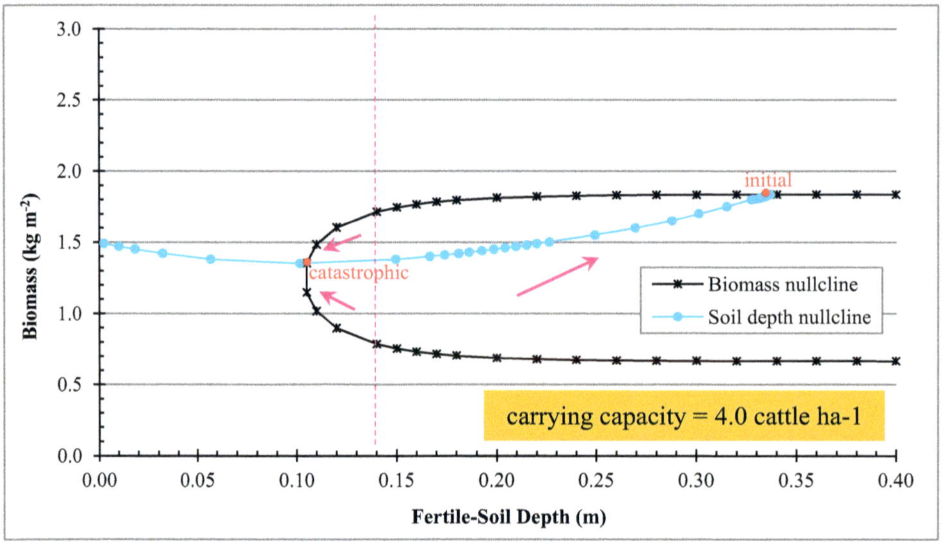

(e)

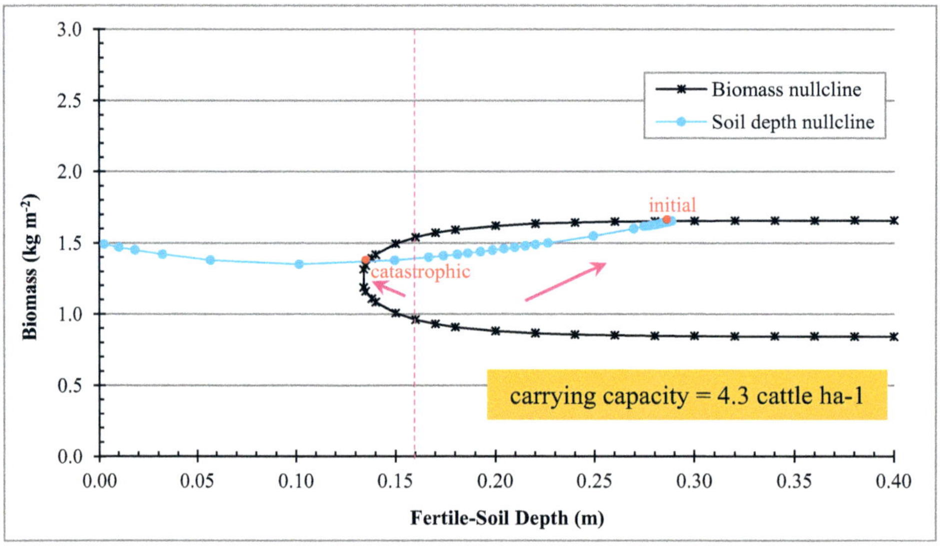

(f)

Figure 5.70. (Continued.)

permanently stay at the post-shifted state (i.e. is non-resilient). If the ecosystem is fully resilient, the resilience circle of the post-shifted state encloses the resilience circle of the pre-shifted state, whereas if the ecosystem is non-resilient, the resilience circle of the post-shifted state does not overlap the resilience circle of the pre-shifted state. If the ecosystem is partially resilient, the resilience circle of the post-shifted state intersects the resilience circle of the pre-shifted state.

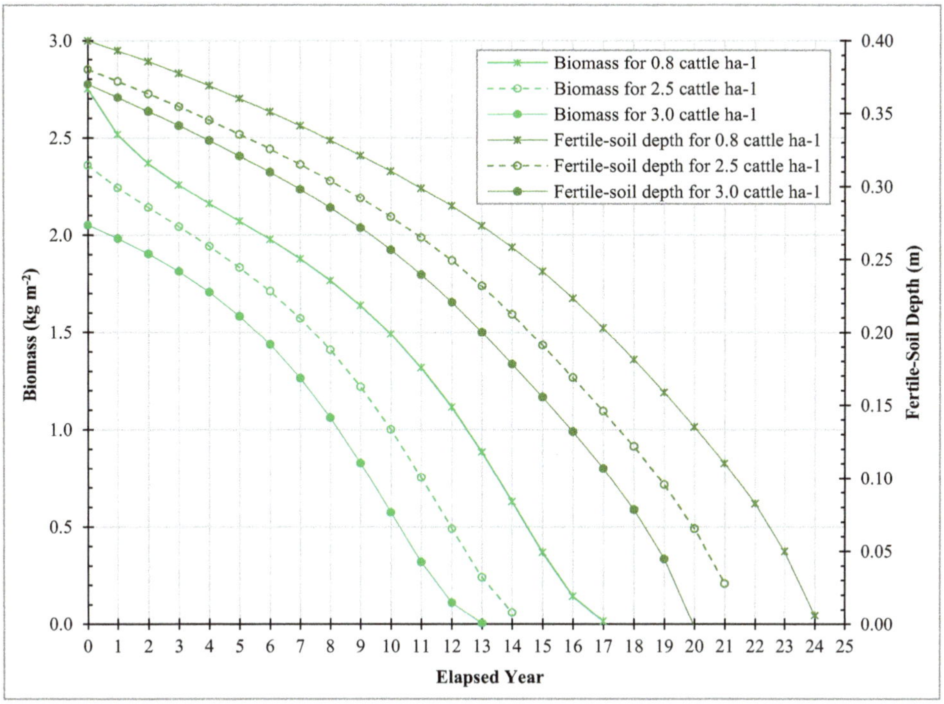

Figure 5.71. Years from an initial to a catastrophic state. The predictions for three selected livestock carrying capacities of the Balagaer River watershed.

For the Balagaer River watershed, grass production will steadily decrease with an increase of livestock carrying capacity, with plunges for carrying capacities of larger than 1.0, 2.0, and 3.5 cattle ha^{-1}. This steppe ecosystem will be fully resilient if the carrying capacity is below 0.8 cattle ha^{-1}, whereas it will be non-resilient if the carrying capacity is above 3.3 cattle ha^{-1} (figure 5.68). Between these two thresholds of carrying capacity, the steppe ecosystem will be partially resilient, which is the current situation. Christensen *et al* (2003) reported a threshold carrying capacity of 0.27 cattle ha^{-1} for a steppe ecosystem that shares common characteristics of hydroclimate and physiography with our study watershed. This threshold is smaller than 0.8 cattle ha^{-1} because of the different grazing practices discussed in section 5.9.3. Moreover, for a carrying capacity of $\leqslant$ 3.0 cattle ha^{-1}, the steppe ecosystem will shift from the initial to catastrophic state during 13–17 years in terms of biomass and 20–24 years in terms of fertile-soil depth (figure 5.71). To avoid an irreversible degradation, the carrying capacity should be lower than 0.8 cattle ha^{-1} and the grazing period of 15 years can be rotated once every 3–5 years, depending on the annual precipitation.

5.9.6 Conclusions

This study put forward a new resilience theory, called Wang's ecosystem resilience theory (WERT). This theory eliminated the subjectivity and inconstancy of how to define and determine resilience. The Balagaer River watershed was predicted to be fully resilient for a carrying capacity of lower than 0.8 cattle ha^{-1} with a rotational grazing practice that has a grazing period of 15 years once every 3–5 years.

5.10 Summary and general conclusions

Using the three semi-arid ecosystems described in section 5.1, namely the Balagawer River watershed in north China, the Horqin Sandy Land (HSL) site in north China, and the Upper Cowhouse Creek watershed in south USA, this chapter demonstrates how to model water–soil–vegetation dynamic interrelations as well as how to define and predict the resilience of such an ecosystem subject to a pressure.

A calibrated SWAP model is used to simulate the long-term responses of soil moisture to climate variability and livestock grazing in the Balagaer River watershed. Livestock grazing will decrease the top 1.5 cm soil moisture but increase the moisture of the soils beneath. The optimal grazing intensity is around 3.0 cattle ha^{-1}, above which the number of days with a topsoil moisture of greater than 0.15 will reduce phenomenally. The grazing impacts on soil moisture will monophonically decrease with an increase of reference evapotranspiration and annual precipitation when it is larger than 220 mm. For the years with an annual precipitation of less than 220 mm, the grazing impacts could either increase or decrease with an increase of precipitation, depending on the relative magnitude of daily reference evapotranspiration. As a generalization, across the semi-arid Eurasian ecoregion, it is imperative to practice pasturing below a threshold grazing intensity to addressland degradation being intensified by climate change.

The SWAP model is also used to evaluate the long-term effects of climate variability and livestock grazing intensity (GI) on non-beneficial (NBPC) versus beneficial (BPC) precipitation consumption. The NBPC–BPC ratios for GI $\geqslant$ 5.0 cattle ha^{-1} will be significantly larger than those for GI $\leqslant$ 3.0 cattle ha^{-1}. A grazing intensity of around 3.0 cattle ha^{-1} can be considered as the threshold, above which livestock grazing will likely trigger irreversible degradation or even desertification of the steppe grassland. For a given grazing intensity, the NBPC–BPC ratio tends to be higher in a drier than a wetter year. Such a negative correlation is stronger for 'critical' years that have an annual precipitation of less than 60 cm or an aridity index larger than 1.5. The NBPC–BPC ratios will increase with an increase of grazing intensity regardless of the amount of annual precipitation. The ratio could become larger than one in a 'critical' year with GI $\geqslant$ 5.0 cattle ha^{-1}, implying that more percent precipitation will be reallocated to be non-beneficial to hydro-ecological sustainability and resilience. It is recommended that policies on pasturing practices be made to ensure grazing intensities less than an appropriate threshold (e.g. 3.0 cattle ha^{-1}) to sustain the hydro-ecosystems of the study watershed and the Great Steppe.

The parameterized IAFP model for the Balagaer River watershed predicts an aeolian erosion occurrence frequency of 0.9%, but it predicts a higher fluvial erosion occurrence frequency of 22.0%. The mean aeolian erosion rate or depth (3.8×10^7 t $year^{-1}$ or 4.1 mm $year^{-1}$) across the 179 occurrence days is predicted to be much greater than the mean fluvial erosion rate or depth (1.5×10^6 t $year^{-1}$ or 0.2 mm $year^{-1}$)

across the 4532 occurrence days. However, across the evaluation (i.e. entire simulation) period, the predicted fluvial erosion rate (3.3×10^5 t year^{-1}) is equivalent to the aeolian erosion rate (3.3×10^5 t year^{-1}), implying that aeolian and fluvial processes probably have a 'maximum interaction' at the 'existing' annual mean soil moisture of 0.24. That is, aeolian and fluvial processes have almost equivalent contributions to the predicted annual mean total soil erosion of 660 375 t year^{-1} (or 0.07 mm year^{-1}). For years with a soil moisture of less than 0.25, aeolian erosion is predicted to dominate fluvial erosion; otherwise, fluvial erosion might dominate. Further, the effect of the reduction of vegetation cover is predicted to be larger on soil erosion for a drier than a wetter year as well as on aeolian than fluvial erosion for any year. Aeolian erosion is predicted to be more sensitive to vegetation disturbance than fluvial erosion, both erosion processes, however, are predicted to be almost equally sensitive to the variation of soil moisture. For the studied watershed, the 'existing' vegetation coverage provides good protection of the soils from wind and water erosive forces.

For the Balagaer River watershed, as expected, the NPP is a positive function of annual precipitation and annual actual evapotranspiration and tends to be higher in a wetter than a drier year. The NPP neither increases nor declines with a statistically significant trend during the 56 historical years, and it goes up in one year and goes down in another year. In contrast, the NPP exhibits a clear spatial pattern: the subbasins with a higher evapotranspiration tend to have a larger NPP and vice versa. Moreover, the NPP is most likely to either increase slightly or have no noticeable change because of climate change. The predicted NPP variations depend on the spatial (e.g. watershed versus subbasin) scale of interest as well as on how the climate would change.

A calibrated HYDRUS-1D model for the HSL site reveals a threshold depth of 10 cm, below which the soils have almost constant soil moisture (SM) and pore air relative humidity (RH_{pa}). For a given non-rainy day, the RH_{pa} and SM of the top 5 cm soils tend to decrease with increase of atmosphere temperature (T_a), attaining a minimum around 2:00 pm. In the 5–10 cm soils, the SM has a negligible diurnal variation, whereas the RH_{pa} tends to increase with increase of T_a. In contrast, for a rainy day, the diurnal variations of RH_{pa} and SM of the top 10 cm soils are mainly controlled by the rainfall pattern and amount, with maximums at a time during the rain event. Moreover, the top dry soil layer (DSL) can effectively retard the moist soils beneath from being further dried out. Planting native vegetation that has a shallow root depth (–10 cm) may be preferred to enhance the DSL's role in protecting the fragile desert ecosystem. The vegetation species should survive in a semi-arid environment, have a root depth of less than the median depth of the DSL (i.e. 10–20 cm), and have a peak transpiration rate of less than the peak non-beneficial evaporation rate from the bare soils (i.e. 2.8 mm d^{-1}).

For the upper Cowhouse Creek watershed, the removal of all 49 863 ha range brush will increase the annual water yield by 24 mm ha^{-1} treated area. The linkage between brush removal and increase in water yield is stronger in areas adjacent to the stream channels. On the other hand, the removal of range brush increases sediment loading, although there are no detectable effects on water yield. The removal of range brush on the TX629 soil is most beneficial because it results in the largest increase of annual water yield (80 mm ha^{-1} treated area) but in only a small increase in annual sediment loading (4.2 tonnes ha^{-1} treated area) and a minimal alteration to the existing spatial patterns of sediment contributions within the study area. The land-use–soil interactive effects need to be considered to develop best management practices (e.g. brush control programs) for improving watershed health and sustainability.

The SWAP and IFAP models for the Balagaer River watershed show that the presence of a moss biocrust will increase the average annual evaporation by 3.7 cm but decrease the average annual transpiration and percolation by 1.6 and 2.2 cm, respectively. The biocrusts can hold more water with higher moisture most of the time than the parent topsoil. The presence of the biocrusts is predicted to effectively reduce the wind erosion occurrence by 21.9% and the wind erosion rate by 1.4 mm $year^{-1}$, with a maximal effect in April when wind speeds are relatively high while soil moisture and vegetation coverage are very low.

The Wang's ecosystem resilience theory (WERT) advances the understanding and application of resilience for protecting, utilizing, and restoring semi-arid ecosystems in a changing climate. It eliminates the subjectivity and inconstancy in the existing literature. The Balagaer River watershed is predicted to be fully resilient for a carrying capacity (i.e. GI) of lower than 0.8 cattle ha^{-1} with a rotational grazing practice that has a grazing period of 15 years once every 3–5 years.

The thresholds of livestock carrying capacity are 3.0 and 0.8 cattle ha^{-1} in accordance with the sustaining hydrology and biomass productivity, respectively. For a given grazing intensity, the alteration of natural hydrology occurs more slowly than the reduction of biomass productivity. Livestock grazing can quickly and directly impact grass growth and remove aboveground leaves and stems, which in turn influence hydrological processes such as evapotranspiration, infiltration, and runoff.

Declaration

Parts of this chapter have been reprinted from Wang *et al* (2010, 2014a, 2014b, 2021) and Pedram *et al* (2018). Copyright 2010, 2014, 2018, 2021 Elsevier. Copyright 2014 Wiley.

References

Adessi A, de Carvalho C, de Philippis R, Branquinho C and da Silva J M 2018 Microbial extracellular polymeric substances improve water retention in dryland biological soil crusts *Soil Biol. Biochem.* **116** 67–9

Afinowicz J D, Munster C L and Wilcox B P 2005 Modeling effects of brush management on the rangeland water budget: Edwards Plateau, Texas *J. Am. Water Resour. Assoc.* **41** 181–93

AHDB 2020 Calculating dry matter intakes for rotational grazing of cattle https://ahdb.org.uk/knowledge-library/calculating-dry-matter-intakes-for-rotational-grazing-of-cattle

Allen R G, Pereira L S, Raes D and Smith M 1998 Crop evapotranspiration: guidelines for computing crop water requirements *FAO Irrigation and drainage paper* 56 Food and Agriculture Organization of the United Nations

Andersland O B and Anderson D M 1978 *Geotechnical Engineering for Cold Regions* (New York: McGraw-Hill)

Angerer J, Han G, Fujisaki I and Havstad K 2008 Climate change and ecosystems of Asia with emphasis on Inner Mongolia and Mongolia *Rangelands* **30** 46–51

Ashmore P 1993 Contemporary erosion of the Canadian landscape *Prog. Phys. Geog.* **17** 190–204

Asner G P, Scurlock J M O and Hicke J A 2003 Global synthesis of leaf area index observations: implications for ecological and remote sensing studies *Global Ecol. Biogeogr.* **12** 191–205

Austin A T, Yahdjian L, Stark J M, Belnap J, Porporato A, Norton U, Ravetta D A and Schaeffer S M 2004 Water pulses and biogeochemical cycles in arid and semiarid ecosystems *Oecologia* **141** 221–35

Bai Y, Han X, Wu J, Chen Z and Li L 2004 Ecosystem stability and compensatory effects in the Inner Mongolia grassland *Nature* **431** 181–4

Barnes R F, Nelson C J, Collins M and Moore K J 2003 *Forages: An Introduction to Grassland Agriculture* (Ames, IA: Iowa State Press)

Becerra-Absalón I, Muñoz-Martín M Á, Montejano G and Mateo P 2019 Differences in the cyanobacterial community composition of biocrusts from the drylands of central Mexico: are there endemic species? *Front. Microbiol.* **10** 1–21

Bednarz S T, Dybala T, Muttiah R S, Rosenthal W and Dugas W A 2001 *Brush/Water Yield Feasibility Studies* (Temple, TX: Blackland Research Center)

Belnap J 2006 The potential roles of biological soil crusts in dryland hydrologic cycles *Hydrol. Process.* **20** 3159–78

Belnap J and Lange O L (ed) 2003 *Biological Soil Crusts: Structure, Function and Management Ecological Studies* vol 150 (Berlin: Springer)

Belnap J, Munson S M and Field J P 2011 Aeolian and fluvial processes in dryland regions: the need for integrated studies *Ecohydrology* **4** 615–22

Belnap J, Phillips S L, Witwicki D L and Miller M E 2008 Visually assessing the level of development and soil surface stability of cyanobacterially dominated biological soil crusts *J. Arid. Environ.* **72** 1257–64

Belnap J, Wilcox B P, van Scoyoc V and Philips S L 2012 Successional stage of biological soil crusts: an accurate indicator of ecohydrological condition *Ecohydrology* **6** 474–82

Bittelli M, Ventura F, Campbell G S, Snyder R L, Gallegati F and Pisa P R 2008 Coupling of heat, water vapor, and liquid water fluxes to compute evaporation in bare soils *J. Hydrol.* **362** 191–205

Blank B, Breuer L, Butterbach-Bahl K and Frede H G 2009 Indicators of grazing impact in Inner Mongolian steppe ecosystems *Geophysical Research Abstracts* **11** EGU2009–8683

Boisramé G 2011 An Excel Macro to calculate weather station statistics: WGNmaker4.xls or WGNmaker4.xlsm https://swat.tamu.edu/media/41586/wgen-excel.pdf

Bosch D D, Sheridan J M, Batten H L and Arnold J G 2004 Evaluation of the SWAT model on a coastal plain agricultural watershed *Trans. ASAE* **47** 1493–506

Bosch J M and Hewlett J D 1982 A review of catchment experiments to determine the effect of vegetation changes on water yield and evapotranspiration *J. Hydrol.* **55** 3–23

Bowker M A, Reed S C, Maestre F T and Eldridge D J 2018 Biocrusts: the living skin of the earth *Plant Soil* **429** 1–7

Bracmort K S, Arabi M, Frankenberger J R, Engel B A and Arnold J G 2006 Modeling long-term water quality impact of structural BMPs *Trans. ASABE* **49** 367–74

Breshears D D, Whicker J J, Johansen M P and Pinder J E 2003 Wind and water erosion and transport in semi-arid shrubland, grassland and forest ecosystems: quantifying dominance of horizontal wind-driven transport *Earth Surf. Processes Landforms* **28** 1189–209

Bristow K L, Campbell G S and Calissendorff K 1993 Test of a heat-pulse probe for measuring changes in soil water content *Soil Sci. Am. J.* **57** 930–4

Bristow K L and Horton R 1996 Modeling the impact of partial surface mulch on soil heat and water flow *Theor. Appl. Climatol.* **54** 85–98

Brutsaert W 2005 *Hydrology: An Introduction* (Cambridge: Cambridge University Press)

Bu C, Li R, Wang C and Bowker M A 2018 Successful field cultivation of moss biocrusts on disturbed soil surfaces in the short term *Plant Soil* **429** 227–40

Bu C, Wu S, Han F, Yang Y and Meng J 2015 The combined effects of moss-dominated biocrusts and vegetation on erosion and soil moisture and implications for disturbance on the Loess Plateau, China *PLosS ONE* **10** e0127394

Bu C, Wu S and Yang K 2014 Effects of physical soil crusts on infiltration and splash erosion in three typical Chinese soils *Int. J. Sediment Res.* **29** 491–501

Budyko M I 1978 *The Heat Balance of the Earth's Surface* (Washington, DC: National Technical Information Service at US Department of Commerce) p 259

Bullard J E and Livingstone I 2002 Interactions between aeolian and fluvial systems in dryland environments *Area* **34** 8–16

Cantón Y, Román J R, Chamizo S, Rodríguez-Caballero E and Moro M J 2014 Dynamics of organic carbon losses by water erosion after biocrust removal *J. Hydrol. Hydromech.* **62** 258–68

Chamizo S, Cantón Y, Rodríguez-Caballero E and Domingo F 2016 Biocrusts positively affect the soil water balance in semiarid ecosystems *Ecohydrology* **9** 1208–21

Chamizo S, Mugnai G, Rossi F, Certini G and de Philippis R 2018 Cyanobacteria inoculation improves soil stability and fertility on different textured soils: gaining insights for applicability in soil restoration *Front. Environ. Sci.* **6** 49

Chamizo S, Rodríguez-Caballero E, Román J R and Cantón Y 2017 Effects of biocrust on soil erosion and organic carbon losses under natural rainfall *Catena* **148** 117–25

Chapin F S III, Matson P A and Mooney H A 2002 *Principles of Terrestrial Ecosystem Ecology* (New York: Springer)

Charbeneau R J 2006 *Groundwater Hydraulics and Pollutant Transport* (Long Grove, IL: Waveland Press)

Chen S, Bai Y, Zhang L and Han X 2005 Comparing physiological responses of two dominant grass species to nitrogen addition in Xilin River Basin of China *Environ. Exp. Bot.* **53** 65–75

Chen W, Dong Z, Li Z and Yang Z 1996 Wind tunnel test of the influence of moisture on the erodibility of loessial sandy loam soils by wind *J. Arid. Environ.* **34** 391–402

Chepil W S 1946 Dynamics of wind erosion: V. cumulative intensity of soil drifting across eroding fields *Soil Sci.* **61** 257

Chepil W S 1956 Influence of moisture on erodibility of soil by wind *Soil Sci. Soc. Am. Proc.* **20** 288–92

Christensen L, Coughenour M B, Ellis J E and Chen Z 2003 Sustainability of Inner Mongolian grasslands: application of the Savanna model *J. Range Manag.* **56** 319–27

Christensen L, Coughenour M B, Ellis J E and Chen Z 2004 Vulnerability of the Asian typical steppe to grazing and climate change *Clim. Change* **63** 351–68

Christensen L, Shauna B and Coughenour M B 2005 Integrated assessment of the dynamics, stability and resilience of the Inner Mongolian grazing ecosystems *Nomadic Peoples* **9** 131–45

Chu T W, Shirmohammadi A, Montas H and Sadeghi A 2004 Evaluation of the SWAT model's sediment and nutrient components in the Piedmont physiographic region of Maryland *Trans. ASAE* **47** 1523–38

CCMIC 2020 National meteorological data http://nmic.gov.cn/web/index.htm

Conroy W J, Hotchkiss R H and Elliot W J 2006 A coupled upland-erosion and instream hydrodynamic-sediment transport model for evaluating sediment transport in forested watersheds *Trans. ASABE* **49** 1713–22

Coughenour M B 1994 *Savanna—Landscape and Regional Ecosystem Model* (Ft Collins, CO: Natural Resource Ecology Laboratory, Colorado State University) pp 1–62

Cramer W and Field C B 1999 Comparing global models of terrestrial net primary productivity (NPP): introduction *Glob. Change Biol.* **5** iii–iv

D'Odorico P, Bhattachan A, Davis K F, Ravi S and Runyan C W 2013 Global desertification: drivers and feedbacks *Adv. Water Res.* **51** 326–44

Daamen C C and Simmonds L P 1996 Measurement of evaporation from bare soil and its estimation using surface resistance *Water Resour. Res.* **32** 1393–402

Davies J, Barchiesi S, Ogali C J, Welling R, Dalton J and Laban P 2016 *Water in Drylands: Adapting to Scarcity Through Integrated Management* (Gland: International Union for Conservation of Nature (IUCN))

Day F P and Monk C D 1974 Vegetation patterns on a southern Appalachian watershed *Ecology* **55** 1064–74

Demidenko E 2019 *Advanced Statistics with Applications in R* (Hoboken, NJ: Wiley-Blackwell)

Dey S, Tripathi S N, Singh R P and Holben B N 2004 Influence of dust storms on the aerosol optical properties over the Indo-Gangetic basin *J. Geophys. Res. Atmos.* **109** D20211

Di Luzio M, Srinivasan R and Arnold J G 2004 A GIS-coupled hydrological model system for the watershed assessment of agricultural nonpoint and point sources of pollution *Trans. GIS* **8** 113–36

Di Luzio M, Srinivasan R, Arnold J G and Neitsch S L 2002 *ArcView Interface for SWAT2000—User's Guide* (Temple, TX: Grassland, Soil and Water Research Laboratory, Agricultural Research Service; Blackland Research Center, Texas Agricultural Experiment Station)

Distel R A and Fernández O A 1987 Leaf water potential trends in three grasses native to semiarid Argentina *J. Range Manag.* **40** 203–7

Dobos E 2003 Albedo *Encyclopedia of Soil Science* (Marcel Dekker: New York) pp 1–3

Doherty J 2004 *PEST: Model-Independent Parameter Estimation User Manual* 5th edn (Brisbane: Watermark Numerical Computing)

Dong W, Yu Z and Weber D 2003 Simulations on soil water variation in arid regions *J. Hydrol.* **275** 162–81

Dong Z, Liu X and Wang X 2002 Wind initiation thresholds of the moistened sands *Geophys. Res. Lett.* **29** 1585

Du B, Saleh A, Jaynes D B and Arnold J G 2006 Evaluation of SWAT in simulating nitrate nitrogen and atrazine fates in a watershed with tiles and potholes *Trans. ASABE* **49** 949–59

Duan J G 2004 Simulation of flow and mass dispersion in meandering channels *J. Hydraul. Eng.* **130** 566–76

Duan L, Liu T, Wang X and Luo Y 2015 Spatio-temporal patterns of water table and vegetation status of a deserted area *Water* **7** 5788–805

Duan L, Liu T, Wang X, Wang G, Ma L and Luo Y 2011 Spatio-temporal variations in soil moisture and physicochemical properties of a typical semiarid sand-meadow-desert landscape as influenced by land use *Hydrol. Earth Syst. Sci.* **15** 1865–77

Dugas W A, Hicks R A and Wright P 1998 Effect of removal of *Juniperus ashei* on evapotranspiration and runoff in the Seco Creek watershed *Water Resour. Res.* **34** 1499–506

Dugas W A and Mayeux H S 1991 Evaporation from rangeland with and without honey mesquite *J. Range Manag.* **44** 161–70

Dunn O J 1961 Multiple comparisons among means *J. Am. Stat. Assoc.* **56** 52–64

Eichelberger S, McCaa J, Nijssen B and Wood A 2008 Climate change effects on wind speed *North-Am. Windpower* **7** 68–72

Elliot W J and Arnold C D 2001 Validation of the weather generator CLIGEN with precipitation data from Uganda *Trans ASAE* **44** 53–8

Endo N, Kadota T, Matsumoto J, Ailikun B and Yasunari T 2006 Climatology and trends in summer precipitation characteristics in Mongolia for the period 1960–98 *J. Meteorol. Soc. Japan* **84** 543–51

Faist A M, Herrick J E, Belnap J, van Zee J W and Barger N N 2017 Biological soil crust and disturbance controls on surface hydrology in a semi-arid ecosystem *Ecosphere* **8** 1–24

Fang J, Piao S, Field C B, Pan Y, Guo Q, Zhou L, Peng C and Tao S 2003 Increasing net primary production in China from 1982 to 1999 *Front. Ecol. Environ.* **1** 293–7

Farouki O T 1981 *CRREL Monograph 81–1: Thermal Properties of Soils* (Hanover, NH: United States Army Corps of Engineers)

Fécan F, Marticorena B and Bergametti G 1999 Parameterization of the increase of the Aeolian erosion threshold wind friction velocity due to soil moisture for arid and semi-arid areas *Ann. Geophys.* **17** 149–57

Feddes R A, Kowalik P J and Zaradny H 1978 Simulation of field water use and crop yield *Simulation Monographs* (Wageningen: Centre for Agricultural Publishing and Documentation) (New York: Wiley) p 189

Feyereisen G W, Strickland T C, Bosch D D and Sullivan D G 2007 Evaluation of SWAT manual calibration and input parameter sensitivity in the Little River watershed *Trans. ASABE* **50** 843–55

FFAO 2021 Action against desertification http://fao.org/in-action/action-against-desertification/overview/en/

Fick S E, Barbier N, Tatarko J and Dunniway M 2020 Induced biological soil crust controls on wind erodibility and dust (PM10) emissions *Earth Surf. Process. Landf.* **45** 224–36

Fick S E, Barger N N and Duniway M C 2019 Hydrological function of rapidly induced biocrusts *Ecohydrology* **12** 1–38

Field J P, Breshears D D and Whicker J J 2009 Toward a more holistic perspective of soil erosion: why aeolian research needs to explicitly consider fluvial processes and interactions *Aeolian Res.* **1** 9–17

Field J P, Breshears D D, Whicker J J and Zou C B 2011 On the ratio of wind- to water-driven sediment transport: conserving soil under global-change-type extreme events *J. Soil Water Conserv.* **66** 51A–56A

Flanagan D C, Ascough J C, Wagner L E and Geter W F 2006 Development of an integrated water and wind erosion model *AGU Fall Meeting (San Francisco, CA)*

Flanagan D C, Meyer C R, Yu B, Scheele D L, Ascough J C and Flanagan D C 2001 Evaluation and enhancement of CLIGEN weather generator *Soil Erosion Research for the 21st Century* (St Joseph, MI: American Society of Agricultural Engineering) pp 107–10

Ford P 2001 Scale, ecosystem resilience, and fire in shortgrass steppe *Trans. Ecol. Environ.* **46** 447–56

Fu B P 1981 On the calculation of the evaporation from land surface *Sci. Atmos. Sin.* **5** 23–31

Fu B P 1996 On the calculation of evaporation from land surface in mountainous areas *Sci. Atmos. Sin.* **16** 328–35

Fullen M A 1991 A comparison of runoff and erosion rates on bare and grassed loamy sand soils *Soil Use Manag.* **7** 136–9

Fullen M A and Reed A H 1986 Rainfall, runoff and erosion on bare arable soils in east Shropshire, England *Earth Surf. Processes Landforms* **11** 413–25

Ganjurjav H, Zhang Y, Gornish E S, Hu G, Li Y, Wan Y and Gao Q 2019 Differential resistance and resilience of functional groups to livestock grazing maintain ecosystem stability in an alpine steppe on the Qinghai-Tibetan Plateau *J. Environ. Manage.* **251** 109579

Gao L, Bowker M A, Xu M, Sun H, Tuo D and Zhao Y 2017b Biological soil crusts decrease erodibility by modifying inherent soil properties on the Loess Plateau, China *Soil Biol. Biochem.* **105** 49–58

Gao Q, Ci L and Yu M 2002 Modeling wind and water erosion in northern China under climate and land use changes *J. Soil Water Conserv.* **57** 46–55

Gao R, Li F, Wang X, Liu T, Du D and Bai Y 2017a Spatiotemporal variations of precipitation across the Mongolian Plateau of China over the past half century *Atmos. Res.* **193** 204–15

Gao T, Xu Y, Bo Y and Yu X 2006 Synoptic characteristics of dust storms observed in Inner Mongolia and their influence on the downwind area (the Beijing–Tianjin Region) *Meteorol. Appl.* **13** 393–403

Ghanbarian-Alavijeh B, Liaghat A, Huang G and van Genuchten M T 2010 Estimation of the van Genuchten soil water retention properties from soil textural data *Pedosphere* **20** 456–65

Ghuman B S and Jalota S K 2006 Heat capacity of soil *Encyclopedia of Soil Science* (New York: Springer) pp 811–13

Goldsmith G R, Allen S T, Braun S, Engbersen N, González-Quijano C R, Kirchner J W and Siegwolf R T W 2018 Spatial variation in throughfall, soil, and plant water isotopes in a temperate forest *Ecohydrology* **12** e2059

Gőri S, Kapocsi I, Szilágyi A, Zsoldos I, Molnár A, Nagy G and Gábor T 2005 *Restoration of Pannonic Steppes and Marshes: LIFE-Nature Project in the Hortobágy National Park* (Debrecen, Hungary: Hortobá National Park Directorate)

Goss K and Madliger M 2007 Estimation of water transport based on *in situ* measurements of relative humidity and temperature in a dry Tanzanian soil *Water Resour. Res.* **43** W05433

Gowing J W, Konukcu F and Rose D A 2006 Evaporative flux from a shallow water table: the influence of a vapour–liquid phase transition *J. Hydrol.* **321** 77–89

Greer C H 2005 *Hydrologic Impacts of Mechanical Shearing of Ashe Juniper in Coryell County* (College Station, TX: Texas A&M University)

Greve P, Roderick M L, Ukkola A M and Wada Y 2019 The aridity index under global warming *Environ. Res. Lett.* **14** 124006

Grismer M E 2007 Soil restoration and erosion control: quantitative assessment and direction *Trans. ASABE* **50** 1619–26

Guan H and Cao R 2021 Biocrusts and its disturbance controls on soil infiltrability in a semi-arid ecosystem *Ecohydrology* **14** e2324

Hamed K H and Rao A R 1998 A modified Mann–Kendall trend test for autocorrelated data *J. Hydrol.* **204** 182–96

Han J and Zhou Z 2013 Dynamics of soil water evaporation during soil drying: laboratory experiment and numerical analysis *Sci. World J.* **2013** 240280

Hao Y, Wang Y, Mei X, Huang X, Cui X, Zhou X and Niu H 2008 CO_2, H_2O and energy exchange of an Inner Mongolia steppe ecosystem during a dry and wet year *Acta Oecol.* **33** 133–43

Harden G P 1993 Land use, soil erosion, and reservoir sedimentation in an Andean drainage basin in Ecuador *Mountain Res. Develop.* **13** 177–84

Hashim Z E, Al-Madhhachi A T and Alzubaidi L A 2020 Behavior of soil erodibility parameters due to biological soil crusts using jet erosion tests *Ecol. Eng.* **153** 105903

He S, Qiu L, Jiang D, Lamusa A, Liu Z and Luo Y 2008 Sandfixing effects of *Caragana microphylla* shrub in Horqin Sandy *Front. For. China* **3** 31–5

Hedrick D W, Hyder D N, Sneva F A and Poulton C E 1966 Ecological response of sagebrush–grass range in central Oregon to mechanical and chemical removal of *Artemisia Ecology* **47** 432–9

Heitman J L, Xiao X, Horton R and Sauer T J 2008 Sensible heat measurements indicating depth and magnitude of subsurface soil water evaporation *Water Resour. Res.* **44** W00D05

Henwood W 1998 The world's temperate grasslands: a beleaguered biome *Parks* **8** 1–2

Herrick J E, Whitford W G, de Soyza A G, va Zee J W, Havstad K M, Seybold C A and Walton M 2001 Field soil aggregate stability kit for soil quality and rangeland health evaluations *Catena* **44** 27–35

Hibbert A R 1983 Water yield improvement potential by vegetation management on western rangelands *Water Resour. Bul.* **19** 375–81

Hibbert A R, Davis E A and Scholl D G 1974 Chaparral conversion potential in Arizona. Part I: water yield response and effect on other resources *USDA Forest Service Research Paper* 36 Rocky Mountain Forest and Range Experiment Station, Fort Collins, CO

Hicke J A, Asner G P, Randerson J T, Tucker C, Los S, Birdsey R A, Jenkins J C and Field C 2002 Trends in North American net primary productivity derived from satellite observations, 1982–1998 *Global Biogeochem. Cycles* **16** 1–15

Hinnant R T 1994 What is an animal-unit? A time to conform *Rangelands* **16** 33–5

Hjelmfelt A T 1991 Investigation of curve number procedure *J. Hydraul. Eng.* **117** 725–37

Homer C, Huang C, Yang L, Wylie B and Coan M 2004 Development of a 2001 national land-cover database for the United States *Photogram. Eng. Remote Sens.* **70** 829–40

Huxman T E, Snyder K A, Tissue D, Leffler A J, Ogle K, Pockman W T, Sandquist D R, Potts D L and Schwinning S 2004 Precipitation pulses and carbon fluxes in semiarid and arid ecosystems *Oecologia* **141** 254–68

IIPCC 2001 Climate change 2001: the scientific basis *Working Group 1 to Third Assessment Report of the International Panel on Climate Change (IPCC)* ed J Houghton and Y Ding (Cambridge: Cambridge University Press) p 881

IIPCC 2011 Climate model output http://ipcc-data.org/sim/gcm_global/index.html

Imster E 2021 China's largest dust storm in a decade hits Beijing *EarthSky*

Irisarri G J N, Oesterheld M, Paruelo J M and Texeira M A 2012 Patterns and controls of above-ground net primary production in meadows of Patagonia: a remote sensing approach *J. Veg. Sci.* **23** 114–26

Istanbulluoglu E, Wang T, Wright O M and Lenters J D 2012 Interpretation of hydrologic trends from a water balance perspective: the role of groundwater storage in the Budyko hypothesis *Water Resour. Res.* **48** W00H16

Janda J M and Abbott S L 2007 16 rRNA gene sequencing for bacterial identification in the diagnostic laboratory: pluses, perils, and pitfalls *J. Clin. Microbiol.* **45** 2761–4

Johnson G L, Hanson C L, Hardegree S P and Ballard E B 1996 Stochastic weather simulation: overview and analysis of two commonly used models *J. Appl. Meteorol.* **35** 1878–96

Juracek K E and Ziegler A C 2007 *Estimation of Sediment Sources using Selected Chemical Tracers in the Perry Lake and Lake Wabaunsee Basins, Northeast Kansas. Scientific Investigations Report 2007–5020* (Washington, DC: US Department of the Interior and US Geological Survey)

Kamai T 2013 Development of heat pulse sensors to measure vadose zone thermal properties, water content, and water flux density *PhD thesis* University of California-Davis

Kassas M 1995 Desertification: a general review *J. Arid. Environ.* **30** 115–28

Kinnell P I A and Yu B 2020 CLIGEN as a weather generator for predicting rainfall erosion using USLE based modelling systems *Catena* **194** 104745

Kirkby M J 1978 The stream head as a significant geomorphic threshold *Working Paper* 216 Department of Geography, University of Leeds

Kobayashi T, Matsuda A and Kamichika M 1989 A simple method for estimating the rate of evaporation from a dry sand surface *J. Agri. Meteorol.* **44** 269–74

Kondo J, Saigusa N and Sato T 1992 A model and experimental study of evaporation from bare-soil surfaces *J. Appl. Meteorol.* **31** 304–12

Kroes J G, van Dam J C, Bartholomeus R P, Groenendijk P, Heinen M, Hendriks R F A, Mulder H M, Supit I and van Walsum P E V 2017 *SWAP version 4: Theory Description and User Manual* (Wageningen: Wageningen Environmental Research)

Kurc S A and Small E E 2007 Soil moisture variations and ecosystem-scale fluxes of water and carbon in semiarid grassland and shrubland *Water Resour. Res.* **43** W06416

Kuske C R, Yeager C M, Johnson S and Ticknor L O 2012 Response and resillence of soil biocrust bacterial communities to chronic physical disturbance in arid shrublands *ISME J.* **6** 886–97

Kwon H Y, Nkonya E, Johnson T, Graw V, Kato E and Kihiu E 2016 Global estimates of the impacts of grassland degradation on livestock productivity from 2001 to 2011 *Economics of Land Degradation and Improvement—A Global Assessment for Sustainable Development* ed E Nkonya, A Mirzabaev and J von Braun (New York: Springer)

Lal R 2001 Soil degradation by erosion *Land Degrad. Dev.* **12** 519–39

Lancaster N and Baas A 1998 Influence of vegetation cover on sand transport by wind: field studies at Owens Lake, California *Earth Surf. Proces. Landforms* **23** 69–82

Lauenroth W K, Hunt H W, Swift D M and Singh J S 1986 Estimating aboveground net primary production in grasslands: a simulation approach *Ecol. Model.* **33** 297–314

Li B, Shi P J and Lin X Q 1995 Development of a Chinese grassland–livestock balance dynamic monitoring system *Grassland Sci* **3** 95–102

Li B, Wang L, Kaseke K F, Li L and Seely M K 2016b The impact of rainfall on soil moisture dynamics in a foggy desert *PLoS One* **11** e0164982

Li H, Li R, Rossi F, Li D, de Philippis R, Hu C and Liu Y 2016c Differentiation of microbial activity and functional diversity between various biocrust elements in a heterogeneous community *Catena* **147** 138–45

Li J, Xu B, Yang X, Qin Z, Zhao L, Jin Y, Zhao F and Guo J 2017 Historical grassland desertification changes in the Horqin Sandy Land, Northern China (1985–2013) *Sci. Rep.* **7** 3009

Li Q, Zhang C, Shen Y, Jia W and Li J 2016a Quantitative assessment of the relative roles of climate change and human activities in desertification processes on the Qinghai-Tibet Plateau based on net primary productivity *Catena* **147** 789–96

Li S, Asanuma J, Kotani A, Davaa G and Oyunbaatar D 2007 Evapotranspiration from a Mongolian steppe under grazing and its environmental constraints *J. Hydrol.* **333** 133–43

Lieth H 1977 Modeling the primary productivity of the world *Primary Productivity of the Biosphere* ed H Leith and R H Whittaker (New York: Springer) pp 237–63

Liu G, Wen M, Chang X, Ren T and Horton R 2013 A self-calibrated dual probe heat pulse sensor for *in situ* calibrating the probe spacing *Soil Sci. Soc. Am. J.* **77** 417–21

Liu J, Qi Y, Shi H, Zhuang D and Hu Y 2008 Estimation of wind erosion rates by using 137Cs tracing technique: a case study in Tariat-Xilin Gol transect, Mongolian Plateau *Chin. Sci. Bull.* **53** 751–8

Liu M, Dries L, Huang J, Min S and T. J. 2019 The impacts of the eco-environmental policy on grassland degradation and livestock production in Inner Mongolia, China: an empirical analysis based on the simultaneous equation model *Land Use Policy* **88** 104167

Liu T, Liu X and Yu R 2006 Field measurement of the water retention curves and simple estimation of the parameters for soils in the central region of Horqin sandy land *2006 Western Pacific Geophysics Meeting (Beijing, China)*

Liu X 2011 Evaluation of the heat-pulse technique for measuring soil water content with thermo-TDR sensor *Proc. Environ. Sci.* **11** 1234–9

Loughran R J 1989 The measurement of soil erosion *Prog. Phys. Geog.* **13** 216–33

Luo Y, Liu T, Wang X and Duan L 2012 Influences of landform as a confounding variable on SOM-NDVI association in semiarid Ordos Plateau *J. Arid Land* **4** 450–6

Luo Y, Wang X, Li F, Gao R, Duan L and Liu T 2014 Responses of grass production to precipitation in a mid-latitude typical steppe watershed *Trans. ASABE* **57** 1595–610

Mace R E, Smyth R C, Xu L and Liang J 1999 *Transmissivity, Hydraulic Conductivity, and Storativity of the Carrizo-Wilcox Aquifer in Texas* (Austin, TX: Texas Water Development Board)

Mager D M and Thomas A D 2011 Extracellular polysaccharides from cyanobacterial soil crusts: a review of their role in dryland processes *J. Arid. Environ.* **75** 91–7

Mahfouf J F and Noilhan J 1991 Comparative study of various formulations of evaporation from bare soil using *in situ* data *J. Appl. Meteorol.* **30** 1354–65

Marshall J K 1971 Drag measurements in roughness arrays of varying density and distribution *Agric. Meteorol.* **8** 269–92

McCarthy D F *Essentials of Soil Mechanics and Foundations* 6th edn (Upper Saddle River, NJ: Prentice Hall)

McHugh T A, Morrissey E M, Reed S C, Hungate B A and Schwartz E 2015 Water from air: an overlooked source of moisture in arid and semiarid regions *Sci. Rep.* **5** 13767

Melillo J M, McGuire A D, Kicklighter D W, Moore I, Vorosmarty C J and Schloss A L 1993 Global climate change and terrestrial net primary production *Nature* **363** 234–40

METER Group, NI 2020 Measure the soil–plant–atmosphere continuum

Middleton N J, Middleton N and Thomas D S G 1997 *World Atlas of Desertification* (London: Hodder Arnold)

Monteith J L and Fogg G E 1965 Evaporation and the environment *The State and Movement of Water in Living Organisms* (New York: Cambridge University Press) pp 205–34

Moore I D and Burch G J 1986 Physical basis of the length-slope factor in the universal soil loss equation *Soil Sci. Soc. Am. J.* **50** 1294–8

Mori Y, Hopmans J W, Mortensen A P and Kluitenberg G J 2003 Multi-functional heat pulse probe for the simultaneous measurement of soil water content, solute concentration, and heat transport parameters *Vadose Zone J.* **2** 561–70

Moriasi D N, Arnold J G, van Liew M W, Bingner R L, Harmel R D and Veith T L 2007 Model evaluation guidelines for systematic quantification of accuracy in watershed simulations *Trans. ASABE* **50** 885–900

Motovilov Y G, Gottschalk L, England K and Rodhe A 1999 Validation of a distributed hydrological model against spatial observations *Agric. For. Meteorol.* **98–99** 257–77

Musick H B and Gillette D A 1990 Field evaluation of relationships between a vegetation structural parameter and sheltering against wind erosion *Land Degrad. Rehab.* **2** 87–94

Myneni R B, Dong J, Tucker C J, Kaufmann R K, Kauppi P E, Liski J, Zhou L, Alexeyev V and Hughes M K 2001 A large carbon sink in the woody biomass of northern forests *Proc. Natl Acad. Sci. USA* **98** 14784–9

Myneni R B, Keeling C D, Tucker C J, Asrar G and Nemani R R 1997 Increased plant growth in the northern high latitudes from 1981–1991 *Nature* **86** 241–51

Narasimhan B, Allen P M, Srinivasan R, Bednarz S T, Arnold J and Dunbar J A 2007 Streambank erosion and best management practice simulation using SWAT *ASABE Fourth Conf. on Watershed Management to meet Water Quality Standards and TMDLS (Total Maximum Daily Loads) San Antonio, TX*

Navas Romero A L, Herrera Moratta M A, Vento B, Rodriguez R A and Martinez Carretero E E 2019 Variations in the coverage of biological soil crusts along a gradient of aridity in the center-west of Argentina https://biorxiv.org/content/10.1101/725986v1

Nearing M A 1997 A single, continuous function for slope steepness influence on soil loss *Soil Sci. Am. J.* **61** 917–9

Neitsch S L, Arnold J G, Kiniry J R, Williams J R and King K W 2002 Soil and water assessment tool *Theoretical Documentation* version 2000 Grassland, Soil and Water Research Laboratory of Agricultural Research Service & Blackland Research Center of Texas Agricultural Experiment Station

Nelle S and Reinke S 2020 Animal unit equivalent chart—Texas

Nemani R R, Keeling C D, Hashimoto H, Jolly W M, Piper S C, Tucker C J, Myneni R B and Running S W 2003 Climate-driven increases in global terrestrial net primary production from 1982 to 1999 *Science* **300** 1560–3

Neumann I B and Wigen E 2013 The importance of the Eurasian steppe to the study of international relations *J. Int. Relat. Dev.* **16** 311–30

Nevins C J, Strauss S L and Inglett P W 2020 Biological soil crusts enhance moisture and nutrients in the upper rooting zone of sandy soil agroecosystems *J. Soil Sci. Plant Nutr.* **183** 615–26

Ni J 2004 Estimating net primary productivity of grasslands from field biomass measurements in temperate northern China *Plant Ecol.* **174** 217–34

Nordt L 2004 Late quaternary alluvial stratigraphy of a low-order tributary in central Texas, USA and its response to climate and sediment supply *Quat. Res.* **62** 289–300

Novak M D 2010 Dynamics of the near-surface evaporation zone and corresponding effects on the surface energy balance of a drying bare soil *Agric. For. Meteorol.* **150** 1358–65

Owens M K and Ansley J 1997 Ecophysiology and growth of Ashe and redberry juniper *Juniper Symp.* C A Taylor (San Angelo, TX: Agricultural Experiment Station) pp 19–32

Paruelo J M, Epstein H E, Lauenroth W K and Burke I C 1997 ANPP estimates from NDVI for the central grassland region of the United States *Ecology* **78** 953–8

Paruelo J M, Lauenroth W K, Burke I C and Sala O E 1999 Grassland precipitation-use efficiency varies across a resource gradient *Ecosystems* **2** 64–8

Peart B 2008 Lift in a working landscape: towards a conservation strategy for the world's temperate grasslands *A Record of the World Temperate Grasslands Conservation Initiative Workshop (Hohhot, China)*

Pedram S 2017 Simulated dynamics of soil water and pore vapor in a semiarid ecosystem *PhD thesis* Old Dominion University, Norfolk, VA

Pedram S, Wang X, Liu T and Duan L 2018 Simulated dynamics of soil water and pore vapor in a semiarid sandy ecosystem *J. Arid. Environ.* **151** 58–82

Penman H L 1948 Natural evaporation from open water, bare soil, and grass *Proc. R. Soc. Lond.* **193** 120–46

Pereyeara D A, Bucci S J, Arias N S, Ciano N, Cristiano P M, Goldstein G and Scholz F G 2017 Grazing increases evapotranspiration without the cost of lowering soil water storages in arid ecosystems *Ecohydrology* **10** e1850

Philip J R 1958 Evaporation, and moisture and heat fields in the soil *J. Meteorol.* **14** 354–66

Philip J R and de Vries D A 1957 Moisture movement in porous materials under temperature gradients *Trans. Am. Geophys. Union* **38** 222–32

Piao S, Fang J, Zhou L, Zhu B, Tan K and Tao S 2005 Changes in vegetation net primary productivity from 1982 to 1999 in China *Global Biogeochem. Cycles* **19** GB2027

Portnov B A and Safrielb U N 2004 Combating desertification in the Negev: dryland agriculture vs dryland urbanization *J. Arid. Environ.* **56** 659–80

Potter N M 2019 Measuring and modeling bare desert wind erosion from steppe grassland of northern China as affected by soil moisture and climate *MS thesis* Old Dominion University, Norfolk, VA

Qi Y, Shi L J H, Zhuang D and Hu Y 2010 Estimation of wind erosion rates of Mongolian Plateau by using 137Cs tracing technique *Ann. Warsaw Univ. Life Sci.—SGGW Land Reclam.* **42** 197–208

Ravi S R, Breshears D D, Huxman T E and D′Odorico P 2010 Land degradation in drylands: interactions among hydrologic–aeolian erosion and vegetation dynamics *Geomorphology* **116** 236–45

Reed S C, Delgado-Baquerizo M and Ferrenberg S 2019 Biocrust science and global change *New Phytol.* **223** 1047–51

Reiche M, Funk R, Zhang Z, Hoffmann C, Reche J, Wehrhan M, Li Y and Sommer M 2011 Application of satellite remote sensing for mapping wind erosion risk and dust emission-deposition in Inner Mongolia grassland, China *Grassland Sci.* **58** 8–19

Reid K D, Wilcox B P, Breshears D D and MacDonald L 1999 Runoff and erosion in a pinon-juniper woodland: influence of vegetation patches *Soil Sci. Soc. Am. J.* **63** 1869–79

Richardson C W, Burnett E and Bovey R W 1979 Hydrologic effects of brush control on Texas rangelands *Trans. ASAE* **22** 315–9

Rodriguez-Iturbe I 2000 Ecohydrology: a hydrologic perspective of climate–soil–vegetation dynamics *Water Resour. Res.* **36** 3–9

Román J R, Rodríguez-Caballero E, Rodríguez-Lozano B, Roncero-Ramos B, Chamizo S, Águila-Carricondo P and Cantón Y 2019 Spectral response analysis: an indirect and non-destructive methodology for the chlorophyll quantification of biocrusts *Remote Sens.* **11** 1350

Rooyen A F V 1998 Combating desertification in the southern Kalahari: connecting science with community action in South Africa *J. Arid. Environ.* **39** 285–97

Rozenstein O and Adamowski J 2017 A review of progress in identifying and characterizing biocrusts using proximal and remote sensing *Int. J. Appl. Earth Obs. Geoinf.* **57** 245–55

Saito H, Sakai M, Toride N and Šimůne K 2008 Advanced modeling of water flow and contaminant transport in porous media using the HYDRUS *Third HYDRUS Workshop (Tokyo, Japan)*

Sakai M, Jones S B and Tuller M 2011 Numerical evaluation of subsurface soil water evaporation derived from sensible heat balance *Water Resour. Res.* **47** W02547

Sala O E, Biondini M E and Lauenroth W K 1988 Bias in estimates of primary production: an analytical solution *Ecol. Modell.* **44** 43–55

Sancho L G, Maestre F T and Büdel B 2014 Biological soil crusts in a changing world: introduction to the special issue *Biodiver. Conser. J.* **23** 1611–7

Saravanapavan T and Salvucci G D 2000 Analysis of rate-limiting processes in soil evaporation with implications for soil resistance models *Adv. Water Res.* **23** 493–502

Saxton K E and Rawls W J 2006 Soil water characteristic estimates by texture and organic matter for hydrologic solutions *Soil Sci. Soc. Am. J.* **70** 1569–78

Saxton K E, Rawls W J, Romberger J S and Papendick R I 1986 Estimating generalized soil-water characteristics from texture *Trans. ASAE* **50** 1031–5

Scurlock J M O, Cramer W, Olson R J, Parton W J and Prince S D 1999 Terrestrial NPP: towards a consistent data set for global model evaluation *Ecol. Appl.* **9** 913–9

Šejna M, Šimůnek J and van Genuchten M Th 2022 *The HYDRUS Software Package for Simulating One-, Two-, and Three-Dimensional Movement of Water, Heat, and Multiple Solutes in Variably-Saturated Media. User Manual, Version 5.0* (Prague, Czech Republic: PC-Progress Engineering Software Developer) p 350

Shahraeeni E, Lehmann P and Or D 2012 Coupling of evaporative fluxes from drying porous surfaces with air boundary layer—characteristics of evaporation from discrete pores *Water Resour. Res.* **48** W09525

Shao C, Chen J, Li L, Xu W, Chen S, Gwen T, Xu J and Zhang W 2008 Spatial variability in soil heat flux at three Inner Mongolia steppe ecosystems *Agric. For. Meteorol.* **148** 1433–43

Shiyomi M, Akiyama T, Wang S, Yiruhan A, Hori Y, Chen Z, Yasuda T, Kawamura K and Yamamura Y 2011 A grassland ecosystem model of the Xilingol steppe, Inner Mongolia, China *Ecol. Model.* **222** 2073–83

Simanton J R, Weltz M A and Larsen H D 1991 Rangeland experiments to parameterize the water erosion prediction project model: vegetation canopy cover effects *J. Range Manag.* **44** 276–82

Šimůnek J, van Genuchten M Th and Šejna M 2022 *The HYDRUS Software Package for Simulating One-, Two-, and Three-Dimensional Movement of Water, Heat, and Multiple Solutes in Variably-Saturated Media. Technical Manual, Version 5.0* (Prague, Czech Republic: PC-Progress Engineering Software Developer) p 255

Singh J S, Milchunas D G and Lauenroth W K 1998 Soil water dynamics and vegetation patterns in a semiarid grassland *Plant Ecol.* **134** 77–89

Smits K M, Ngo V V, Cihan A, Sakaki T and Illangasekare T H 2012 An evaluation of models of bare soil evaporation formulated with different land surface boundary condition and assumptions *Water Resour. Res.* **48** W12526

Sokal R and Oden N L 1978 Spatial autocorrelation in biology: some biological implications and four applications of evolutionary and ecological interest *Biol. J. Linn. Soc.* **10** 229–49

Song L, Pan Y, Gong J, Li X, Liu M, Yang B, Zhang Z and Baoyin T 2020 Physiology of *Leymus chinensis* under seasonal grazing: implications for the development of sustainable grazing in a temperate grassland of Inner Mongolia *J. Environ. Manage.* **271** 110984

Sprenger M, Leistert H, Gimbel K and Weiler M 2016 Illuminating hydrological processes at the soil–vegetation–atmosphere interface with water stable isotopes *Rev. Geophys.* **54** 674–704

Steven B, Gallegos-Graves L V, Yeager C, Belnap J and Kuske C R 2014 Common and distinguishing features of the bacterial and fungal communities in biological soil crusts and shrub root zone soils *Soil Biol. Biochem.* **69** 302–12

Stockton P H and Gillette D A 1990 Field measurement of the sheltering effect of vegetation on erodible land surfaces *Land Degrad. Dev.* **2** 77–85

Stone R P and Hilborn D 2000 Universal soil loss equation (USLE) *Agricultural Engineering Factsheet* Ministry of Agriculture, Food and Rural Affairs, ON, Canada

Sugathan N, Biju V and Renuka G 2014 Influence of soil moisture content on surface albedo and soil thermal parameters at a tropical station *J. Earth Syst. Sci.* **123** 1115–28

Sugita M, Yoshizawa S and Byambakhuu I 2015 Limiting factors for nomadic pastoralism in Mongolian steppe: a hydrologic perspective *J. Hydrol.* **524** 455–67

Sun F, Xiao B, Li S and Kidron G J 2021 Towards moss biocrust effects on surface soil water holding capacity: soil water retention curve analysis and modeling *Geoderma* **399** 115120

Sun L, Yang L, Hao L, Fang D, Jin K and Huang X 2017 Hydrological effects of vegetation cover degradation and environmental implications in a semiarid temperate steppe, China *Sustainability* **9** 281

Svoray T and Karnieli A 2011 Rainfall, topography and primary production relationships in a semiarid ecosystem *Ecohydrology* **4** 56–66

TAES 2002 Brush management/water yield feasibility study for four watersheds in Texas *Texas Water Resources Institute Technical Report* No. 207 Texas Agricultural Experiment Station, College Station, TX

Takemi T 2005 Explicit simulations of convective-scale transport of mineral dust in severe convective weather *J. Meteorol. Soc. Japan (Ser. II)* **83** 187–203

Tarnawski V R and Leong W H 2000 Thermal conductivity of soils at very low moisture content and moderate temperatures *Transp. Porous Media* **41** 127–47

TCEQ 2007 *Updated Evaluation for the North-Central Texas—Trinity and Woodbine Aquifers—Priority Groundwater Management Study Area* (Austin, TX: Texas Commission on Environmental Quality)

Thornley J H M 1998 Dynamic model of leaf photosynthesis with acclimation to light and nitrogen *Ann. Bot.* **81** 421–30

Thornley J H M and Cannell M G R 1997 Temperate grassland responses to climate change: an analysis using the Hurley Pasture Model *Ann. Bot.* **80** 205–21

Tong C, Wu J, Yong S, Yang J and Yong W 2004 A landscape-scale assessment of steppe degradation in the Xilin River Basin, Inner Mongolia, China *J. Arid. Environ.* **59** 133–49

TSSWCB 2003 *Palo Pinto Reservoir Watershed: Brush Control Assessment and Feasibility Study* (Waco, TX: Texas State Soil and Water Conservation Board Brazos River Authority (TSSWCB))

Uchijima Z and Seino H 1985 Agroclimatic evaluation of net primary productivity of natural vegetation 1: Chikugo model for evaluating net primary productivity *J. Meteorol. Soc. Japan* **40** 343–52

UNEP 1991 *Status of Desertification and Implementation of the United Nations Plan of Action to Combat Desertification* United Nations Environment Programme, Nairobi, Kenya
UNEP 1992 *World Atlas of Desertification* (Nairobi: United Nations Environment Programme)
Ursino N, Cassiani G, Deiana R, Vignoli G and Boaga J 2014 Measuring and modeling water-related soil–vegetation feedbacks in a fallow plot *Hydrol. Earth Syst. Sci.* **18** 1105–18
USDA-NRCS 2004 *National Engineering Handbook* (Washington, DC: US Department of Agriculture Natural Resources Conservation Service)
UUSDA-NRCS 2008 *US General Soil Map (STATSGO2)*
UUSDA-NRCS 2020 *Animal Unit Equivalent—Texas: Domestic Livestock, Native Wildlife, and Exotic Wildlife*
USGS 1998 Extreme precipitation depths for Texas, excluding the Trans-Pecos region *Water-Resources Investigations Report* 98-4099 US Geological Survey (USGS), Washington, DC
USGS 2005 Ground Water Atlas of the United States: Oklahoma, Texas *Open-File Report* HA 730-E US Geological Survey (USGS), Washington, DC
USGS 2008 *USGS Geographic Data Download* US Geological Survey (USGS), Washington, DC
Vaché K B, Eilers J M and Santelman M V 2002 Water quality modeling of alternative agricultural scenarios in the US Corn Belt *J. Am. Water Resour. Assoc.* **38** 773–87
Valentin C, Poesen J and Li Y 2005 Gully erosion: impacts, factors and control *Catena* **63** 132–53
van Bavel C H M and Hillel D I 1976 Calculating potential and actual evaporation from a bare soil surface by simulation of concurrent flow of water and heat *Agri. Meteorol.* **17** 453–76
van Lier Q d J and Durigon A 2012 Soil thermal diffusivity estimated from data of soil temperature and single soil component properties *Rev. Bras. Ciênc. Solo* **37** 106–12
van Liew M W, Veith T L, Bosch D D and Arnold J G 2007 Suitability of SWAT for the conservation effects assessment project: a comparison on USDA-ARS experimental watersheds *J. Hydrol. Eng.* **12** 173–89
Vazquez-Amábile G G and Engel B A 2005 Use of SWAT to compute groundwater table depth and streamflow in the Muscatatuck River watershed *Trans. ASAE* **48** 991–1003
Viessman W and Lewis G L 2003 *Introduction to Hydrology* 5th edn (Upper Saddle River, NJ: Prentice Hall)
Wang L, Odorico P D, Evans J P, Eldridge D J, McCabe M F, Caylor K K and King E G 2012a Dryland ecohydrology and climate change: critical issues and technical advances *Hydrol. Earth Syst. Sci.* **16** 2585–603
Wang X 2015 Vapor flow resistance of dry soil layer to soil water evaporation in arid environment: an overview *Water* **7** 4552–74
Wang X, Chen F, Hasi E and Li J 2008c Desertification in China: an assessment *Earth Sci. Rev.* **88** 188–206
Wang X, Gao R and Yang X 2021a Responses of soil moisture to climate variability and livestock grazing in a semiarid Eurasian steppe *Sci. Total Environ.* **781** 146705
Wang X, Li F, Gao R, Luo Y and Liu T 2014b Predicted NPP spatiotemporal variations in a semiarid steppe watershed for historical and trending climates *J. Arid. Environ.* **104** 67–79
Wang X, Liu T, Li F, Gao R, Yang X, Duan L, Luo Y and Li R 2014b Simulated soil erosion from a semiarid typical steppe watershed using an integrated aeolian and fluvial prediction model *Hydrol. Process.* **28** 325–40
Wang X, Liu T and Yang W 2012b Development of a robust runoff-prediction model by fusing the Rational Equation and a modified SCS-CN method *Hydrol. Sci. J.* **57** 1118–40

Wang X, Luo Y, Liu T, Gao R, Li F and Duan L 2014c Estimated grazing removal rate of grasses in a semiarid Eurasian steppe watershed as influenced by climate *18th CIGR World Congress, Session I Land and Water Engineering, 2nd Inter-Regional Conf. on Land and water Challenges Ecohydrological Process and Pasture Management Issues (Hohhot, China)*

Wang X and Melesse A M 2005 Evaluation of the SWAT model's snowmelt hydrology in a northwestern Minnesota watershed *Trans. ASAE* **48** 1359–76

Wang X and Melesse A M 2006 Effects of STATSGO and SSURGO as inputs on SWAT models snowmelt simulation *J. Am. Water Resour. Assoc.* **42** 1217–36

Wang X, Melesse A M and Yang W 2006 Influences of potential evaporation estimation methods on SWAT's hydrologic simulation in a northwestern Minnesota watershed *Trans. ASABE* **49** 1755–71

Wang X, Pedram S, Liu T, Gao R, Li F and Luo Y 2016b Estimated grass grazing removal rate in a semiarid Eurasian steppe watershed as influenced by climate *Water* **8** 339

Wang X, Shang S, Yang W, Clary C R and Yang D 2010 Simulation of land use-soil interactive effects on water and sediment yields at watershed scale *Ecol. Eng.* **36** 328–44

Wang X, Shang S, Yang W and Melesse A M 2008a Simulation of an agricultural watershed using an improved curve number method in SWAT *Trans. ASABE* **51** 1323–39

Wang X, Yang W and Melesse A M 2008b Using hydrologic equivalent wetland concept within SWAT to estimate streamflow in watersheds with numerous wetlands *Trans. ASABE* **51** 55–72

Wang X, Yang X, Liu T, Li F, Gao R, Duan L and Luo Y 2014a Trend and extreme occurrence of precipitation in a midlatitude Eurasian steppe watershed at various time scales *Hydrol. Process.* **28** 5547–60

Wang Y, Fang J, Kato T, Guo Z, Zhu B, Mo W and Tang Y 2011 Inventory-based estimation of aboveground net primary production in Japan's forests from 1980 to 2005 *Biogeosciences* **8** 2009–106

Wang Y, Gao R, Wang X, Duan L, Liu T and Li D 2021b Long-term spatiotemporal variability in occurrences of wet and dry days across South Mongolian Plateau *Atmos. Res.* **262** 105795

Wang Y and Zhou G 2012 Light use efficiency over two temperate steppes in Inner Mongolia, China *PLoS One* **7** e43614

Wang Z, Ji L, Hou X and Schellenberg M P 2016a Soil respiration in semiarid temperate grasslands under various land management *PLoS One* **11** e0147987

Wang Z, Wan X, Tian M, Wang X, Chen J, Chen X, Chang S and Hou F 2020 Grazing season alters soil respiration in a semiarid grassland on the loess plateau *Nutr. Cycling Agroecosyst.* **118** 177–91

Weeks E P, Weaver H L, Campbell G S and Tanner B D 1987 Water use by saltcedar and by replacement vegetation in the Pecos River floodplain between Acme and Artesia *US Geological Survey Professional Paper* 491-G US Government Printing Office, Washington, DC

Welder G E 1988 Hydrologic effects of phreatophyte control, Acme-Artesia reach of the Pecos River, New Mexico, 1967–82 *US Geological Survey Water Resources Investigations* Report 87–4148 US Government Printing Office, Washington, DC

Wen G, Huang G, Hu K, Tao W and Gong H 2014 Changes in the characteristics of precipitation over northern Eurasia *Theor. Appl. Climatol.* **119** 653–65

White R P, Murray S and Rohweder M 2000 *Pilot Analysis of Global Ecosystems. Grassland Ecosystems* (Washington, DC: World Resource Institute)

Wickham J, Stehman S V, Sorenson D G, Gass L and Dewitz J A 2021 Thematic accuracy assessment of the NLCD 2016 land cover for the conterminous United States *Remote Sens. Environ.* **257** 112357

Wilcox B P, Dugas W A, Owens M K, Ueckert D N and Hart C R 2005 Shrub control and water yield on Texas rangelands: current state of knowledge *Texas Agricultural Experiment Station Research* Report 05-1 TexasA&M University, College Station, TX

Williams J R 1975 Sediment-yield prediction with Universal Equation using runoff energy factor *Present and Prospective Technology for Predicting Sediment Yield and Sources ARS-S-40* (Washington DC: US Department of Agriculture Research Service) pp 244–52

Williams J R 1982 Testing the modified universal soil loss equation *Proc. of the Workshop on Estimating Erosion and Sediment Yield on Rangelands ARM-W–26* (Washington DC: US Department of Agriculture) pp 157–61

Williams J R 1995 The EPIC model *Computer Models of Watershed Hydrology* ed V P Singh (Highlands Ranch, CO: Water Resources Publications) pp 909–1000

Windhorn R H 2001RAP-M: Rapid Assessment Point Method. Miscellaneous Pages (Champaign, IL: US Department of Agriculture (USDA) Natural Resources Conservation Service (NRCS))

Wischmeier W H and Smith D D 1978 *Predicting Rainfall Erosion Losses—A Guide to Conservation Planning, Agricultural Handbook No. 537* (Washington DC: US Department of Agriculture)

Wolfe S A and Nickling W G 1993 The protective role of sparse vegetation in wind erosion *Prog. Phys. Geog.* **17** 50–68

Wu B and Yang H 2013 Spatial patterns and natural recruitment of native shrubs in a semi-arid sandy land *PLoS One* **8** e58331

Wu X B, Redeker E J and Thurow T L 2001 Vegetation and water yield dynamics in an Edwards Plateau watershed *J. Range Manag.* **54** 98–105

Xiao B, Sun F, Yao X, Hu K and Kidron G J 2019 Seasonal variations in infiltrability of moss-dominated biocrusts on aeolian sand and loess soil in the Chinese Loess Plateau *Hydrol. Process.* **33** 2449–63

Xiao B, Zhao Y, Wang Q and Li C 2015 Development of artificial moss-dominated biological soil crusts and their effects on runoff and soil water content in a semi-arid environment *J. Arid. Environ.* **117** 75–83

Xiao X, Horton R, Sauer T J, Heitman J L and Ren T 2011 Cumulative soil water evaporation as a function of depth and time *Vadose Zone J.* **10** 1016–22

Xiao X, Wang Y, Jiang S, Ojima D S and Bonham C D 1995 Interannual variation in the climate and above-ground biomass of *Leymus* Chinense steppe and *Stipa grandis* steppe in the Xilin River basin, Inner Mongolia, China *J. Arid. Environ.* **31** 283–99

Xu C *et al* 2021 Resistance and resilience of a semi-arid grassland to multi-year extreme drought *Ecol. Indic.* **131** 108139

Yamanaka T, Takeda A and Shimada J 1998 Evaporation beneath the soil surface: some observational evidence and numerical experiments *Hydrol. Process.* **12** 2193–203

Yamanaka T and Yonetani T 1999 Dynamics of the evaporation zone in dry sandy soils *J. Hydrol.* **217** 135–48

Yan J and Li X 2002 The soil and water conservation and the utilization of rainfall resources in Inner Mongolia *12th Int. Soil Conservation (ISCO) Conf. (Beijing, China)*

Yang D, Sun F, Liu Z, Cong Z, Ni G and Lei Z 2007 Analyzing spatial and temporal variability of annual water–energy balance in nonhumid regions of China using the Budyko hypothesis *Water Resour. Res.* **43** W04426

Yang F and Zhou G 2011 Characteristics and modeling of evapotranspiration over a temperate desert steppe in Inner Mongolia, China *J. Hydrol.* **396** 139–47

Yang X 2014 Hydro-climate temporal variations and their influences on net primary production in a Eurasian Steppe watershed *MS thesis* Old Dominion University, Norfolk, VA

Yang Y, Fang J, Ma W and Wang W 2008 Relationship between variability in aboveground net primary production and precipitation in global grasslands *Geophys. Res. Lett.* **35** L23710

Yao S, Zhao C, Zhang T and Liu X 2013 Response of the soil water content of mobile dunes to precipitation patterns in Inner Mongolia, northern China *J. Arid. Environ.* **97** 92–8

Yi R, Ai L, Ma Z and Shiyomi M 2011 Forty-eight-year climatology of air temperature and precipitation changes in Xilinhot, Xilingol steppe (Inner Mongolia), China *Grassland Sci.* **57** 168–72

Yu B 2005 Adjustment of CLIGEN parameters to generate precipitation change scenarios in southeastern Australia *Catena* **61** 196–209

Yu M, Ellis J E and Epstein H E 2004 Regional analysis of climate, primary production, and livestock density in Inner Mongolia *J. Environ. Qual.* **33** 1675–81

Zaady E, Katra I, Yizhaq H, Kinast S and Ashkenazy Y 2014 Inferring the impact of rainfall gradient on biocrusts' developmental stage and thus on soil physical structures in sand dunes *Aeolian Res.* **13** 81–9

Zang C 2011 Quantification of the interrelation between hydrology alteration and degradation of typical steppe using remote sensing *2011 World Environmental & Water Resources Congress (Palm Springs, CA)*

Zeng Y, Su Z, Wan L and Wen J 2011a A simulation analysis of the advective effect on evaporation using a two-phase heat mass flow model *Water Resour. Res.* **47** 10W10529

Zeng Y, Su Z, Wan L and Wen J 2011b Numerical analysis of air–water–heat flow in unsaturated soil: is it necessary to consider airflow in land surface models? *J. Geophys. Res. Atmos.* **116** D20107

Zeng Y, Su Z, Wan L, Yang Z, Zhang T, Tian H, Shi X, Wang X and Cao W 2009 Diurnal pattern of the drying front in desert and its application for determining the effective infiltration *Hydrol. Earth Syst. Sci.* **13** 703–14

Zhai X, Zhang Y, Wang K, Chen Q, Li S and Huang D 2018 Grazing effects on the nutritive value of dominant species in steppe grasslands of northern China *BMC Ecol.* **18** 1–9

Zhang G, Kang Y, Han G and Sakurai K 2011 Effect of climate change over the past century on the distribution, extent and NPP of ecosystems of Inner Mongolia *Global Change Biol.* **17** 377–89

Zhang J, Zuo X, Zhou X, Lv P, Lian J and Yue X 2017 Long-term grazing effects on vegetation characteristics and soil properties in a semiarid grassland, northern China *Environ. Monit. Assess.* **189** 216

Zhang L, Hickel K and Dawes W R 2004 A rational function approach for estimating mean annual evaporation *Water Resour. Res.* **40** W02502

Zhang X 2007 A comparison of explicit and implicit spatial downscaling of GCM output for soil erosion and crop production assessments *Clim. Change* **84** 337–63

Zhang Y, Liu B, Wang Z and Zhu Q 2008 Evaluation of CLIGEN for storm generation on the semiarid Loess Plateau in China *Catena* **73** 1–9

Zhang Y, Munkhtsetseg E, Kadota T and Ohata T 2005 An observational study of ecohydrology of a sparse grassland at the edge of the Eurasian cryosphere in Mongolia *J. Geophys. Res.* **110** D14103

Zhao H, Li S, Zhang T, Ohkuro T and Zhou R 2004 Sheep gain and species diversity: in sandy grassland, Inner Mongolia *Rangeland Ecol. Manag.* **57** 187–90

Zhao M, Han G and Mei H 2006 Grassland resource and its situation in Inner Mongolia, China *Bull. Fac. Agric. Niigata Univ.* **58** 129–32

Zhao Y, Peth S, Horn R, Krümmelbein J, Ketzer B, Gao Y, Doerner J, Bernhofer C and Peng X 2010 Modeling grazing effects on coupled water and heat fluxes in Inner Mongolia grassland *Soil Tillage Res.* **109** 75–86

Zhao Y and Xu M 2013 Runoff and soil loss from revegetated grasslands in the hilly loess plateau region, China: influence of biocrust patches and plant canopies *J. Hydrol. Eng.* 18 387–93

Zhou G, Wang Y and Wang S 2002 Responses of grassland ecosystems to precipitation and land use along the northeast China transect *J. Veg. Sci.* **13** 361–8

Zhou G and Zhang X 1995 A natural vegetation NPP model *Acta Bot. Sin.* **19** 193–200

Zhou G, Zhen Y, Chen S and Luo T 1998 NPP model of natural vegetation and its application in China *Sci. Silvae Sin.* **34** 2–11

Zhu W, Gao Y, Zhang H and Liu L 2020 Optimization of the land use pattern in Horqin Sandy Land by using the CLUMondo model and Bayesian belief network *Sci. Total Environ.* **739** 139929

Zucca C, Julitta F and Previtali F 2011 Land restoration by fodder shrubs in a semi-arid agro-pastoral area of Morocco: effects on soils *Catena* **87** 306–12

Zuo X, Zhao X, Zhang T, Guo Y, Wang S and Drake S 2008 Spatial pattern and heterogeneity of soil properties in sand dune under grazing and restoration in Horqin Sandy Land, Northern China *Soil Tillage Res.* **99** 202–12

IOP Publishing

Xixi Wang

Chapter 6

Overall conclusions and discussion

This chapter summarizes the major findings from previous chapters, presents a broad discussion, and makes recommendations for additional research.

6.1 Major findings

First, water–soil–vegetation interactions determine semi-arid ecosystems to be fully resilient, partially resilient, or non-resilient. Such interactions can be altered, intensified, or weakened by climate change and/or inappropriate land practices (e.g. overgrazing and overexploitation) and need to be well understood and quantitatively predicted to utilize and at the same time to protect the ecosystem in the long run.

Second, water–soil–vegetation interactions are affected by a variety of abiotic and biotic processes, which need to be conceptualized well to develop usable algorithms and ecohydrological models. In this regard, a holistic approach, which consists of *in situ* monitoring and experiments, laboratory tests, and remote sensing analyses, should be carried out to advance our understanding of the processes and collect long-term data with high spatiotemporal resolution.

Third, the existing hydrological and agroecological models, such as HYDRUS-1D, SWAT, SWAP, and others, have been adapted to model semi-arid ecosystems. However, those models are not specifically designed for semi-arid ecosystems and thus may not be able to mimic some important processes, such as those relevant to biocrusts, dry soil layers, and aeolian–fluvial erosion. To overcome these issues, this book demonstrates how to modify soil–water retention curves in HYDRUS-1D and SWAP and integrate SWAP with IAFP.

Fourth, the presence of biocrusts will increase evaporation but decrease transpiration and percolation. Biocrusts can hold more water with higher moisture most of the time than the parent topsoil. Biocrusts can effectively reduce the occurrence and

doi:10.1088/978-0-7503-4020-5ch6

magnitude of wind erosion; this is particularly true for seasons with a relatively high wind speed, a low soil moisture, and a sparser vegetation coverage. A fragile desert ecosystem can be greatly protected by a top dry soil layer (DSL) because the DSL retards the moist soils beneath from being further dried up. To restore the ecosystem, selected vegetation species should survive in semi-arid environment, have a root depth of less than the median depth of DSL (i.e. 10–20 cm), and have a peak transpiration rate of less than the peak non-beneficial evaporation rate from the bare soils (i.e. 2.8 mm d^{-1}).

Fifth, the removal of vegetation by people and/or livestock adversely affect soil moisture, non-beneficial versus beneficial consumption of precipitation, and topsoil erosion. For the Balagaer River watershed, which is in north China and has a semi-arid steppe ecosystem, a livestock carrying capacity of larger than 3.0 cattle ha^{-1} will lower soil moisture and increase the non-beneficial consumption of precipitation. If the soil moisture is below 0.25 the steppe ecosystem will incur more frequent aeolian erosions with an elevated magnitude. For the Upper Cowhouse Creek watershed, which is in north central Texas of USA and has a semi-arid ecosystem of range brush and grasses, the artificial removal of range brush will either increase or has no effect on water yield, but it will increase soil erosion and sedimentation. The results from these two watersheds reveal the importance of adopting appropriate land management practices.

Finally, in terms of Wang's ecosystem resilience theory (WERT), the Balagaer River watershed would shift to a non-resilient catastrophic state for a livestock carrying capacity of larger than 3.0 cattle ha^{-1}. To make the ecosystem fully resilient, the livestock carrying capacity must be less than 0.8 cattle ha^{-1} with a rotational grazing practice that has a grazing period of 15 years once every 3 to 5 years; the current carrying capacity of 2.5 cattle ha^{-1} needs to be reduced. These may also be true for the Eurasian Steppe (i.e. Great Steppe), which is the vast steppe ecoregion of Eurasia stretching through Hungary, Bulgaria, Romania, Moldova, Ukraine, Western Russia, Siberia, Kazakhstan, Xinjiang, Mongolia, and Manchuria. Globally, because of climate change, semi-arid ecosystems need to be managed carefully for them to be resilient in biomass production and eco-environment protection.

6.2 Broad discussion

For dryland ecosystems, biocrusts play foundational roles in improving physico-chemical properties (e.g. water and nutrient holding capacities) of root-zone soils, increasing topsoil resistance to erosion, fixing carbon dioxide and nitrogen, and regulating cycles of energy, carbon, and water at the soil–vegetation–atmosphere interface (Mager and Thomas 2011, Reed *et al* 2019). However, biocrust science did not emerge until the 1970s and did not evolve into a real discipline until 2001 (Sancho *et al* 2014, Bowker *et al* 2018), when an edited volume by Belnap and Lange (2003) was first published. From then on, it has been greatly advanced by presentations at international conferences held in 2010 (BioCrust1), 2013 (BioCrust2), 2016 (BioCrust3), 2019 (BioCrust4), 2021 (BioCrust5) and by various journal papers (Belnap 2006). More international conferences will be held in 2023 (BioCrust6), and 2025 (BioCrust7), where new ideas are expected to be shared, evaluated, and synthesized. The biocrust science has four evolving research focuses, namely scaling, modeling, improved understanding of *in situ* processes, and restoration. In the existing literature (Bowker *et al* 2018), some studies are devoted to measuring biocrust structures, while others are aimed to evaluate biocrust functionalities. Given the importance of semi-arid ecosystems in sustaining climate, socioeconomics, and eco-environment at local to global scales, it is imperative to understand the ecohydrological dynamics of biocrusts.

Evaporation from bare sandy soils is the core component of the hydrological cycle (dominated by vertical water movement) in arid and semi-arid regions, in which any disturbance to the fragile hydrology is likely to trigger land desertification and deteriorate the very sparse vegetation ecosystem (Daamen and Simmonds 1996, Henwood 1998, Wang and Melesse 2006, Zhao *et al* 2006, Zhou *et al* 2008, Wang *et al* 2010, Duan *et al* 2011). Such a region is losing biome productivity (Christensen *et al* 2005) and incurring environmental problems, such as more frequent dust storms of greater magnitude and frequency (Rooyen 1998, Dey *et al* 2004, Portnov and Safrielb 2004, Takemi 2005). Similar desertification-related issues exist all over the world (Schwartz and Notini 1994, GEF and IFAD 2002, D'Odorico *et al* 2013). Addressing these bioenvironmental problems requires that the physical processes of soil evaporation (e.g. vapor and moisture advection and diffusion through varyingly saturated soils) are described accurately (Philip and de Vries 1957, Novak 2010, Shahraeeni *et al* 2012). Arid and semi-arid ecosystems are dominated by bare sandy soils that most of the time are covered by a thin (5 to 10 cm) top dry soil layer (DSL) (Duan *et al* 2011), within which soil moisture is dominantly in vapor phase and soil water potentials are usually below -1.5×10^4 cm (Goss and Madliger 2007). Such a DSL can have important vapor-flow-resistance (VFR) effects on the soil water evaporation rate and dynamic processes under natural conditions (Ahn *et al* 2012, Shahraeeni *et al* 2012, Wang 2015). That is, the DSL can retard the underlying soils from drying out further. Herein, the VFR effects are measured as the difference between the soil water evaporation rate with a DSL and that without the DSL. A good understanding of the VFR effects is obviously needed but currently lacking (Liu *et al* 2006).

Semi-arid steppe ecosystems across the world are suffering from moderate to severe degradation (Bai *et al* 2004), which has been a key constraint limiting the sustainability of grassland agriculture. Peart (2008) examined the policy and economic reasons for grassland degradation, while Gőri *et al* (2005) assessed the impacts of grassland degradation on wildlife species. Christensen *et al* (2004) and Endo *et al* (2006) analyzed climate change and its effect on the vulnerability of steppe grasslands. The climate has been shown to follow a significant spatiotemporal trend, with increasing air temperatures during winter and more frequent droughts in spring (Shao *et al* 2008, Wang *et al* 2013). Other researchers (Christensen *et al* 2003, Christensen *et al* 2005, Li 2007, Blank *et al* 2009) have studied the ecological functions of steppe grasslands and how such functions are affected by grazing and overgrazing, demonstrating that a grazing intensity (i.e. the ratio of the biomass in a grazed area to the biomass in the same area if not grazed) of 0.67 or above is unlikely to pose a sustainability problem. However, only a few studies (Zhou *et al* 2002, Zhang 2005, Li *et al* 2007, Hao *et al* 2008) have assessed the hydrological processes in steppe grasslands and the few studies that have generally focused on how evapotranspiration (ET) varies with precipitation and/or soil water content, revealing that ET is limited by the availability of soil water (i.e. atmospheric heat stress is weaker than soil water stress and soil evaporation is the primary contributor to ET) and that ET has maximum daily values ranging from 3.0 to 6.2 mm d^{-1} in such grasslands. Daily ET rates tend to increase with increasing soil water content, which in turn is highly dependent on the drying–wetting cycles. Overall, soil water content is higher when grass coverage is denser (Liu *et al* 2006). These studies also indicate that the above-ground biomass of grass species has a positive linear relationship with precipitation and that precipitation in the first half of the year (January to July) is the primary climatic factor causing fluctuations in community biomass production (Bai *et al* 2004, Luo *et al* 2014, Yang 2014, Wang *et al* 2014c). To date, there is only rudimentary understanding of the hydrological processes at the soil–vegetation–atmosphere interface (Sprenger 2016, Goldsmith *et al* 2018) and how erosion of the 10–20 cm topsoil affects vegetation degradation (Wang *et al* 2014b).

6.3 Further research

To accurately simulate semi-arid ecosystems, the existing mathematical models need to be improved by adding physically based algorithms of biocrusts, dry soils, and aeolian and fluvial erosions. Several such algorithms have been developed by Wang *et al* (2014b) and Wang (2015), while they need to be refined and validated using *in situ* observations across the world. In addition, more algorithms should be developed. Also, the WERT can be seamlessly coupled with the improved models to determine resilience from simulation results.

References

Ahn S, Im S and Doerr S 2012 Effect of thickness of a water repellent soil layer on soil evaporation rate *Geophys. Res. Abstr.* **14** EGU2012–14369

Bai Y, Han X, Wu J, Chen Z and Li L 2004 Ecosystem stability and compensatory effects in the Inner Mongolia grassland *Nature* **431** 181–4

Belnap J 2006 The potential roles of biological soil crusts in dryland hydrologic cycles *Hydrol. Processes* **20** 3159–78

Belnap J and Lange O L (ed) 2003 Biological soil crusts: structure, function and management *Ecological Studies* vol 150 (Berlin: Springer)

Blank B, Breuer L, Butterbach-Bahl K and Frede H G 2009 Indicators of grazing impact in Inner Mongolian steppe ecosystems *Geophys. Res. Abst.* **11** EGU2009–8683

Bowker M A, Reed S C, Maestre F T and Eldridge D J 2018 Biocrusts: the living skin of the Earth *Plant Soil* **429** 1–7

Christensen L, Coughenour M B, Ellis J E and Chen Z 2003 Sustainability of Inner Mongolian grasslands: application of the savanna model *J. Range Manag.* **56** 319–27

Christensen L, Coughenour M B, Ellis J E and Chen Z 2004 Vulnerability of the Asian typical steppe to grazing and climate change *Clim. Change* **63** 351–68

Christensen L, Shauna B and Coughenour M B 2005 Integrated assessment of the dynamics, stability and resilience of the Inner Mongolian grazing ecosystems *Nomad. Peoples* **9** 131–45

Daamen C C and Simmonds L P 1996 Measurement of evaporation from bare soil and its estimation using surface resistance *Water Resour. Res.* **32** 1393–402

Dey S, Tripathi S N, Singh R P and Holben B N 2004 Influence of dust storms on the aerosol optical properties over the Indo-Gangetic basin *J. Geophys. Res. Atmos.* **109** D20211

D'Odorico P, Bhattachan A, Davis K F, Ravi S and Runyan C W 2013 Global desertification: drivers and feedbacks *Adv. Water Res.* **51** 326–44

Duan L, Liu T, Wang X, Wang G, Ma L and Luo Y 2011 Spatio-temporal variations in soil moisture and physicochemical properties of a typical semiarid sand-meadow-desert landscape as influenced by land use *Hydrol. Earth Syst. Sci.* **15** 1865–77

Endo N, Kadota T, Matsumoto J, Ailikun B and Yasunari T 2006 Climatology and trends in summer precipitation characteristics in Mongolia for the period 1960–98 *J. Meteorol. Soc. Japan* **84** 543–51

GEF and IFAD 2002 Land Degradation and Desertification (http://ifad.org/events/wssd/gef/GEF_eng.pdf).

Goldsmith G R, Allen S T, Braun S, Engbersen N, González-Quijano C R, Kirchner J W and Siegwolf R T W 2018 Spatial variation in throughfall, soil, and plant water isotopes in a temperate forest *Ecohydrology* **12** e2059

Gőri S, Kapocsi I, Szilágyi A, Zsoldos I, Molnár A, Nagy G and Gábor T 2005 Restoration of pannonic steppes and marshes *LIFE-Nature Project in the Hortobágy National Park* HNP Directorate. Debrecen, Hungary

Goss K and Madliger M 2007 Estimation of water transport based on *in situ* measurements of relative humidity and temperature in a dry Tanzanian soil *Water Resour. Res.* **43** W05433

Hao Y, Wang Y, Mei X, Huang X, Cui X, Zhou X and Niu H 2008 CO_2, H_2O and energy exchange of an Inner Mongolia steppe ecosystem during a dry and wet year *Acta Oecol.* **33** 133–43

Henwood W 1998 The world's temperate grasslands: a beleaguered biome *Parks* **8** 1–2

Li S, Asanuma J, Kotani A, Davaa G and Oyunbaatar D 2007 Evapotranspiration from a Mongolian steppe under grazing and its environmental constraints *J. Hydrol.* **333** 133–43

Liu T, Liu X and Yu R 2006 Field measurement of the water retention curves and simple estimation of the parameters for soils in the central region of Horqin sandy land *2006 Western Pacific Geophysics Meeting* (Beijing, China)

Luo Y, Wang X, Li F, Gao R, Duan L and Liu T 2014 Responses of grass production to precipitation in a mid-latitude typical steppe watershed *Trans. ASABE* **57** 1595–610

Mager D M and Thomas A D 2011 Extracellular polysaccharides from cyanobacterial soil crusts: a review of their role in dryland processes *J. Arid. Environ.* **75** 91–7

Novak M D 2010 Dynamics of the near-surface evaporation zone and corresponding effects on the surface energy balance of a drying bare soil *Agric. For. Meteorol.* **150** 1358–65

Peart B 2008 Lift in a working landscape: towards a conservation strategy for the world's temperate grasslands *A Record of the World Temperate Grasslands Conservation Initiative Workshop (Hohhot, China)*

Philip J R and de Vries D A 1957 Moisture movement in porous materials under temperature gradients *Trans. Am. Geophys. Union* **38** 222–32

Portnov B A and Safrielb U N 2004 Combating desertification in the Negev: dryland agriculture vs dryland urbanization *J. Arid. Environ.* **56** 659–80

Reed S C, Delgado-Baquerizo M and Ferrenberg S 2019 Biocrust science and global change *New Phytol.* **223** 1047–51

Rooyen A F V 1998 Combating desertification in the southern Kalahari: connecting science with community action in South Africa *J. Arid. Environ.* **39** 285–97

Sancho L G, Maestre F T and Budel B 2014 Biological soil crusts in a changing world: introduction to the special issue *Biodivers. Conserv. J.* **23** 1611–7

Schwartz M L and Notini J 1994 *Desertification and Migration: Mexico and the United States* (San Francisco, CA: Natural Heritage Institute)

Shahraeeni E, Lehmann P and Or D 2012 Coupling of evaporative fluxes from drying porous surfaces with air boundary layer—characteristics of evaporation from discrete pores *Water Resour. Res.* **48** W09525

Shao C, Chen J, Li L, Xu W, Chen S, Gwen T, Xu J and Zhang W 2008 Spatial variability in soil heat flux at three Inner Mongolia steppe ecosystems *Agric. For. Meteorol.* **148** 1433–43

Sprenger M, Leistert H, Gimbel K and Weiler M 2016 Illuminating hydrological processes at the soil–vegetation–atmosphere interface with water stable isotopes *Rev. Geophys.* **54** 674–704

Takemi T 2005 Explicit simulations of convective-scale transport of mineral dust in severe convective weather *J. Meteorol. Soc. Japan (Ser. II)* **83** 187–203

Wang X 2015 Vapor flow resistance of dry soil layer to soil water evaporation in arid environment: an overview *Water* **7** 4552–74

Wang X, Liu T, Li F, Gao R, Yang X, Duan L, Luo Y and Li R 2014b Simulated soil erosion from a semiarid typical steppe watershed using an integrated aeolian and fluvial prediction model *Hydrol. Process.* **28** 325–40

Wang X, Luo Y, Liu T, Gao R, Li F and Duan L 2014c Estimated grazing removal rate of grasses in a semiarid Eurasian steppe watershed as influenced by climate *18th CIGR World Congress, Session I Land and Water Engineering, 2nd Inter-Regional Conf. on Land and Water Challenges Ecohydrological Process and Pasture Management Issues (Hohhot, China)*

Wang X and Melesse A M 2006 Effects of STATSGO and SSURGO as inputs on SWAT models snowmelt simulation *J. Am. Water Resour. Assoc.* **42** 1217–36

Wang X, Shang S, Yang W, Clary C R and Yang D 2010 Simulation of land use-soil interactive effects on water and sediment yields at watershed scale *Ecol. Eng.* **36** 328–44

Wang X, Yang X, Liu T, Li F, Gao R, Duan L and Luo Y 2013 Trend and extreme occurrence of precipitation in a midlatitude Eurasian steppe watershed at various time scales *Hydrol. Processes* **28** 5547–60

Yang X 2014 Hydro-climate temporal variations and their influences on net primary production in a Eurasian steppe watershed *MS thesis* Old Dominion University, Norfolk, VA

Zhou H, Li S, Sun S, Xu X, Lei J, Liu S, Du W and Wang Y 2008 Effects of natural covers on soil evaporation of the shelterbelt along the Tarim Desert Highway *Chin. Sci. Bull.* **53** 137–45

Zhang Y, Munkhtsetseg E, Kadota T and Ohata T 2005 An observational study of ecohydrology of a sparse grassland at the edge of the Eurasian cryosphere in Mongolia *J. Geophys. Res.* **110** D14103

Zhao M, Han G and Mei H 2006 Grassland resource and its situation in Inner Mongolia, China *Bull. Fac. Agric. Niigata Univ* **58** 129–32

Zhou G, Wang Y and Wang S 2002 Responses of grassland ecosystems to precipitation and land use along the northeast China transect *J. Veg. Sci.* **13** 361–8

IOP Publishing

Xixi Wang

Appendix A

Derivation of equation (3.26)

See figure 3.2 and choose the ground surface as the datum.

Applying Bernoulli equation between the ground surface and water surface yields

$$h_{gs,\,t} = -y_{gs,\,t} \tag{A.1}$$

The hydraulic head at the wetting front is determined as

$$h_{wf,\,t} = z_f + |\psi_f| \tag{A.2}$$

Applying Darcy's law between the ground surface and wetting front yields

$$i = -\frac{K_s}{2}\frac{h_{gs,\,t} - h_{wf,\,t}}{z_f} = \frac{K_s}{2}\frac{z_f + |\psi_f| + y_{gs,\,t}}{z_f} \tag{A.3}$$

In terms of mass conservation, z_f is computed as

$$z_f = \frac{F}{\theta_{ns} - \theta_a} = \frac{F}{0.95\theta_{sat} - \theta_a} \tag{A.4}$$

At any time t, the penetration rate capacity, i, satisfies

$$i = \frac{dF}{dt} \tag{A.5}$$

Substituting equations (A.4) and (A.5) into equation (A.3) and rearranging yields

$$\frac{dF}{dt} = \frac{K_s}{2}\left[1 + \frac{\left(|\psi_f| + y_{gs,\,t}\right)(0.95\theta_{sat} - \theta_a)}{F}\right] \tag{A.6}$$

$$\left[1 - \frac{\left(|\psi_f| + y_{gs,\,t}\right)(0.95\theta_{sat} - \theta_a)}{F + \left(|\psi_f| + y_{gs,\,t}\right)(0.95\theta_{sat} - \theta_a)}\right]dF = \frac{K_s}{2}dt \tag{A.7}$$

doi:10.1088/978-0-7503-4020-5ch7

For the illustration in figure A.1, integrating equation (A.7) from t_0 to $t_0 + \Delta t$ and simplifying yields

$$\int_{F_0}^{F_{\Delta t}+F_0}\left[1-\frac{\left(|\psi_f|+y_{gs,t}\right)(0.95\theta_{sat}-\theta_a)}{F+\left(|\psi_f|+y_{gs,t}\right)(0.95\theta_{sat}-\theta_a)}\right]dF=\int_{t_0}^{t_0+\Delta t}\frac{K_s}{2}dt \tag{A.8}$$

$$\left[F-\left(|\psi_f|+y_{gs,t}\right)(0.95\theta_{sat}-\theta_a)\ln\left(F+\left(|\psi_f|+y_{gs,t}\right)(0.95\theta_{sat}-\theta_a)\right)\right]\Bigg|_{F_0}^{F_{\Delta t}+F_0}=\frac{K_s}{2}t\Bigg|_{t_0}^{t_0+\Delta t} \tag{A.9}$$

$$F_{\Delta t}=\frac{K_s}{2}(\Delta t)+\left(|\psi_f|+y_{gs,t}\right)(0.95\theta_{sat}-\theta_a)\ln\left[1+\frac{F_{\Delta t}}{F_0+\left(|\psi_f|+y_{gs,t}\right)(0.95\theta_{sat}-\theta_a)}\right] \tag{A.10}$$

Here, equation (A.10) is equation (3.26).

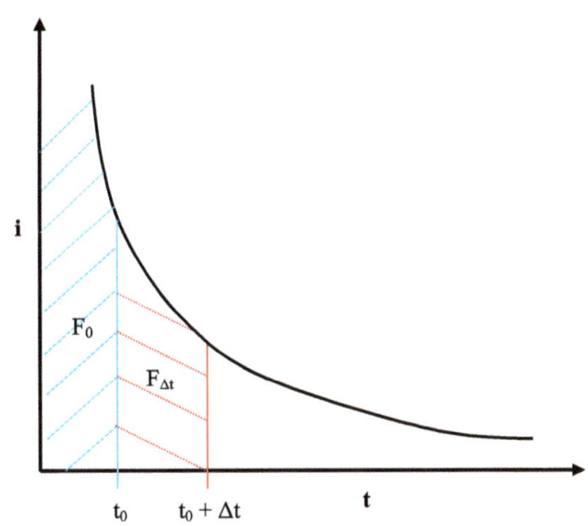

Figure A.1. Soil penetration capacity curve for the G–A method. i [L T^{-1}]: penetration rate; t [T]: time; F_0 [L]: total penetration until time t_0; $F_{\Delta t}$ [L]: penetration capacity from time t_0 to $t_0 + \Delta t$.

Xixi Wang

Appendix B

Derivation of equation (3.35)

Differentiating equation (3.29) with respect to time t yields

$$\frac{\partial\theta}{\partial t} = (\alpha mn)(\theta_r - \theta_{sat})\frac{|\alpha\psi|^{n-1}}{(1 + |\alpha\psi|^n)^{m+1}}\frac{\partial\,|\psi|}{\partial t} \tag{B.1}$$

Substituting equation (B.1) into equation (3.28) yields

$$(\alpha mn)(\theta_r - \theta_{sat})\frac{|\alpha\psi|^{n-1}}{(1 + |\alpha\psi|^n)^{m+1}}\frac{\partial\,|\psi|}{\partial t} = S' - \frac{\partial K}{\partial z}\left(1 + \frac{\partial\,|\psi|}{\partial z}\right) - K\frac{\partial^2\,|\psi|}{\partial z^2} \tag{B.2}$$

With reference to figure 3.3, the backward formats of the corresponding terms in equation (B.2) are written as

$$\frac{|\alpha\psi|^{n-1}}{(1 + |\alpha\psi|^n)^{m+1}} \approx \frac{|\alpha\psi|^{n-1}_{t-\Delta t,\, j}}{\left(1 + |\alpha\psi|^{n}_{t-\Delta t,\, j}\right)^{m+1}} \tag{B.3}$$

$$\frac{\partial\,|\psi|}{\partial t} \approx \frac{|\psi|_{t,\, j} - |\psi|_{t-\Delta t,\, j}}{\Delta t} \tag{B.4}$$

$$S' \approx S'_{t-\Delta t,\, j} \tag{B.5}$$

$$\frac{\partial K}{\partial z} \approx \frac{2\left(K_{t-\Delta t,\, j} - K_{t-\Delta t,\, j-1}\right)}{(\Delta z)_j + (\Delta z)_{j-1}} \tag{B.6}$$

$$\frac{\partial\,|\psi|}{\partial z} \approx \frac{2\left(|\psi|_{t-\Delta t,\, j} - |\psi|_{t-\Delta t,\, j-1}\right)}{(\Delta z)_j + (\Delta z)_{j-1}} \tag{B.7}$$

doi:10.1088/978-0-7503-4020-5ch8

$$K \approx K_{t-\Delta t,\, j} \tag{B.8}$$

$$\frac{\partial^2 |\psi|}{\partial z^2} \approx \frac{\dfrac{2\left(|\psi|_{t-\Delta t,\, j+1} - |\psi|_{t-\Delta t,\, j}\right)}{(\Delta z)_{j+1} + (\Delta z)_j} - \dfrac{2\left(|\psi|_{t-\Delta t,\, j} - |\psi|_{t-\Delta t,\, j-1}\right)}{(\Delta z)_j + (\Delta z)_{j-1}}}{\dfrac{(\Delta z)_{j+1}}{2} + (\Delta z)_j + \dfrac{(\Delta z)_{j-1}}{2}} \tag{B.9}$$

Substituting equations (B.3) through (B.9) into equation (B.2) yields

$$\begin{aligned}(\alpha mn)(\theta_r - \theta_{sat})\frac{|\alpha\psi|^{n-1}_{t-\Delta t,\, j}}{\left(1 + |\alpha\psi|^{n}_{t-\Delta t,\, j}\right)^{m+1}}\frac{|\psi|_{t,\, j} - |\psi|_{t-\Delta t,\, j}}{\Delta t} &= S'_{t-\Delta t,\, j} \\ - \frac{2\left(K_{t-\Delta t,\, j} - K_{t-\Delta t,\, j-1}\right)}{(\Delta z)_j + (\Delta z)_{j-1}}&\left(1 + \frac{2\left(|\psi|_{t-\Delta t,\, j} - |\psi|_{t-\Delta t,\, j-1}\right)}{(\Delta z)_j + (\Delta z)_{j-1}}\right) \\ - K_{t-\Delta t,\, j}&\frac{\dfrac{2\left(|\psi|_{t-\Delta t,\, j+1} - |\psi|_{t-\Delta t,\, j}\right)}{(\Delta z)_{j+1} + (\Delta z)_j} - \dfrac{2\left(|\psi|_{t-\Delta t,\, j+1} - |\psi|_{t-\Delta t,\, j}\right)}{(\Delta z)_{j+1} + (\Delta z)_j}}{\dfrac{(\Delta z)_{j+1}}{2} + (\Delta z)_j + \dfrac{(\Delta z)_{j-1}}{2}}\end{aligned} \tag{B.10}$$

By collecting the knowns to the right-hand side, equation (B.10) can be rewritten as

$$\begin{aligned}|\psi|_{t,\, j} = |\psi|_{t-\Delta t,\, j} &+ \frac{\Delta t}{(\alpha mn)(\theta_r - \theta_{sat})}\frac{\left(1 + |\alpha\psi|^{n}_{t-\Delta t,\, j}\right)^{m+1}}{|\alpha\psi|^{n-1}_{t-\Delta t,\, j}} \\ &\left[\begin{aligned} S'_{t-\Delta t,\, j} &- \frac{2\left(K_{t-\Delta t,\, j} - K_{t-\Delta t,\, j-1}\right)}{(\Delta z)_j + (\Delta z)_{j-1}}\left(1 + \frac{2\left(|\psi|_{t-\Delta t,\, j} - |\psi|_{t-\Delta t,\, j-1}\right)}{(\Delta z)_j + (\Delta z)_{j-1}}\right) \\ &- K_{t-\Delta t,\, j}\frac{\dfrac{2\left(|\psi|_{t-\Delta t,\, j+1} - |\psi|_{t-\Delta t,\, j}\right)}{(\Delta z)_{j+1} + (\Delta z)_j} - \dfrac{2\left(|\psi|_{t-\Delta t,\, j+1} - |\psi|_{t-\Delta t,\, j}\right)}{(\Delta z)_{j+1} + (\Delta z)_j}}{\dfrac{(\Delta z)_{j+1}}{2} + (\Delta z)_j + \dfrac{(\Delta z)_{j-1}}{2}}\end{aligned}\right]\end{aligned} \tag{B.11}$$

In equation (B.11) $K_{t-\Delta t,\, j-1}$ is computed by substituting $|\psi|_{t-\Delta t,\, j-1}$ for $|\psi|$ in equation (3.33), while $K_{t-\Delta t,\, j}$ is computed by substituting $|\psi|_{t-\Delta t,\, j}$ for $|\psi|$.

Here, equation (B.11) is equation (3.35).

IOP Publishing

Modeling Semi-arid Water–Soil–Vegetation Systems in a Changing Climate

Xixi Wang

Appendix C

Derivation of the first part of equations (4.13)

The first part of equation (4.13) is applicable for suction head $|\psi| = 152.96–336.52$ m with the hypothetical form

$$\theta = a + b\frac{|\psi_b|}{|\psi|} + c\ln\left(\frac{|\psi_b|}{|\psi|}\right) \tag{C.1}$$

Differentiating equation (C.1) with respect to $|\psi|$ yields

$$\frac{d\theta}{d|\psi|} = -b\frac{|\psi_b|}{|\psi|^2} - c\frac{1}{|\psi|} \tag{C.2}$$

$$\theta = \theta_r \rightarrow \left|\psi\right| = 336.52 \text{ m}, \frac{d\theta}{d|\psi|} = 0 \rightarrow -b\frac{|\psi_b|}{336.52^2} - c\frac{1}{336.52} = 0 \rightarrow b = -\frac{336.52}{|\psi_b|}c$$

The second part of equation (4.13) is the modified van Genuchen retention curve applicable for suction head $|\psi| = |\psi_b|$ to 152.96 m. It is expressed as

$$\theta = \theta_r + (\theta_{sat} - \theta_r)(1 + |\alpha\psi|^n)^{-m} \tag{C.3}$$

Differentiating equation (C.3) with respect to $|\psi|$ yields

$$\frac{d\theta}{d|\psi|} = (\alpha mn)(\theta_r - \theta_{sat})\frac{|\alpha\psi|^{n-1}}{(1 + |\alpha\psi|^n)^{m+1}} \tag{C.4}$$

doi:10.1088/978-0-7503-4020-5ch9

Zero-order geometric continuity at θ_ψ and $|\psi| = 152.96$ m → equation (C.1) = equation (C.3)

$$\rightarrow a + b\frac{|\psi_b|}{152.96} + c\ln\left(\frac{|\psi_b|}{152.96}\right) = \theta_r + (\theta_{sat} - \theta_r)(1 + |152.96\alpha|^n)^{-m}$$

$$\rightarrow a + \left(-\frac{336.52}{|\psi_b|}c\right)\frac{|\psi_b|}{152.96} + c\ln\left(\frac{|\psi_b|}{152.96}\right) = \theta_r + (\theta_{sat} - \theta_r)(1 + |152.96\alpha|^n)^{-m}$$

$$\rightarrow a + \left(-\frac{336.52}{|\psi_b|}c\right)\frac{|\psi_b|}{152.96} + c\ln\left(\frac{|\psi_b|}{152.96}\right) = \theta_r + (\theta_{sat} - \theta_r)(1 + |152.96\alpha|^n)^{-m}$$

$$\rightarrow a = \theta_r + (\theta_{sat} - \theta_r)(1 + |152.96\alpha|^n)^{-m} + \left(\frac{336.52}{152.96} - \ln\left(\frac{|\psi_b|}{152.96}\right)\right)c$$

First-order geometric continuity at θ_ψ and $|\psi| = 152.96$ m → equation (C.2) = equation (C.4)

$$\rightarrow \frac{336.52}{|\psi_b|}c\frac{|\psi_b|}{152.96^2} - c\frac{1}{152.96} = (\alpha mn)(\theta_r - \theta_{sat})\frac{|152.96\alpha|^{n-1}}{(1 + |152.96\alpha|^n)^{m+1}}$$

$$\rightarrow c = \frac{152.96^2}{183.56}(\alpha mn)(\theta_r - \theta_{sat})\frac{|152.96\alpha|^{n-1}}{(1 + |152.96\alpha|^n)^{m+1}}$$

In summary, the three coefficients in equation (C.1) are determined as

$$c = \frac{152.96^2}{183.56}(\alpha mn)(\theta_r - \theta_{sat})\frac{|152.96\alpha|^{n-1}}{(1 + |152.96\alpha|^n)^{m+1}} \tag{C.5}$$

$$a = \theta_r + (\theta_{sat} - \theta_r)(1 + |152.96\alpha|^n)^{-m} + \left(\frac{336.52}{152.96} - \ln\left(\frac{|\psi_b|}{152.96}\right)\right)c \tag{C.6}$$

$$b = -\frac{336.52}{|\psi_b|}c \tag{C.7}$$

α, n, and m are computed using equations (3.30)–(3.32), respectively. For convenience, these three equations are reproduced as follows:

$$\alpha = \frac{\xi}{|\psi_b|} \tag{C.8}$$

$$n = \lambda + 1 \tag{C.9}$$

$$m = 1 - \frac{1}{n} \tag{C.10}$$

www.ingramcontent.com/pod-product-compliance
Ingram Content Group UK Ltd.
Pitfield, Milton Keynes, MK11 3LW, UK
UKHW051905100926
8556IPUK00005B/776

9 780750 340182